VOLUME FOUR HUNDRED AND FIFTY-FOUR

METHODS IN ENZYMOLOGY

Computer Methods, Part A

METHODS IN ENZYMOLOGY

Editors-in-Chief

JOHN N. ABELSON AND MELVIN I. SIMON

Division of Biology
California Institute of Technology
Pasadena, California

Founding Editors

SIDNEY P. COLOWICK AND NATHAN O. KAPLAN

VOLUME FOUR HUNDRED AND FIFTY-FOUR

METHODS IN ENZYMOLOGY
Computer Methods, Part A

EDITED BY

MICHAEL L. JOHNSON
Departments of Pharmacology and Internal Medicine
University of Virginia Health System
Charlottesville, Virginia

LUDWIG BRAND
Department of Biology
Johns Hopkins University
Baltimore, MD, USA

AMSTERDAM • BOSTON • HEIDELBERG • LONDON
NEW YORK • OXFORD • PARIS • SAN DIEGO
SAN FRANCISCO • SINGAPORE • SYDNEY • TOKYO
Academic Press is an imprint of Elsevier

ELSEVIER

Academic Press is an imprint of Elsevier
525 B Street, Suite 1900, San Diego, California 92101-4495, USA
30 Corporate Drive, Suite 400, Burlington, MA 01803, USA
32 Jamestown Road, London NW17BY, UK

Copyright © 2009, Elsevier Inc. All Rights Reserved.

No part of this publication may be reproduced or transmitted in any form or by any means, electronic or mechanical, including photocopy, recording, or any information storage and retrieval system, without permission in writing from the Publisher.

The appearance of the code at the bottom of the first page of a chapter in this book indicates the Publisher's consent that copies of the chapter may be made for personal or internal use of specific clients. This consent is given on the condition, however, that the copier pay the stated per copy fee through the Copyright Clearance Center, Inc. (www.copyright.com), for copying beyond that permitted by Sections 107 or 108 of the U.S. Copyright Law. This consent does not extend to other kinds of copying, such as copying for general distribution, for advertising or promotional purposes, for creating new collective works, or for resale. Copy fees for pre-2008 chapters are as shown on the title pages. If no fee code appears on the title page, the copy fee is the same as for current chapters. 0076-6879/2008 $35.00

Permissions may be sought directly from Elsevier's Science & Technology Rights Department in Oxford, UK: phone: (+44) 1865 843830, fax: (+44) 1865 853333, E-mail: permissions@elsevier.com. You may also complete your request on-line via the Elsevier homepage (http://elsevier.com), by selecting "Support & Contact" then "Copyright and Permission" and then "Obtaining Permissions."

For information on all Elsevier Academic Press publications visit our Web site at elsevierdirect.com

ISBN-13: 978-0-12-374552-1

PRINTED IN THE UNITED STATES OF AMERICA
09 10 11 12   9 8 7 6 5 4 3 2 1

Working together to grow
libraries in developing countries

www.elsevier.com | www.bookaid.org | www.sabre.org

ELSEVIER   BOOK AID International   Sabre Foundation

Contents

Contributors	xi
Preface	xv
Volumes in Series	xvii

1. Phase Response Curves: Elucidating the Dynamics of Coupled Oscillators — 1
A. Granada, R. M. Hennig, B. Ronacher, A. Kramer, and H. Herzel

1. Introduction	2
2. Estimation of Phase Response Curves	11
3. Specific Applications	17
4. Discussion	20
Appendix I	21
Appendix II	23
Acknowledgments	24
References	24

2. Multiple Ion Binding Equilibria, Reaction Kinetics, and Thermodynamics in Dynamic Models of Biochemical Pathways — 29
Kalyan C. Vinnakota, Fan Wu, Martin J. Kushmerick, and Daniel A. Beard

1. Introduction	30
2. Biochemical Conventions and Calculations	34
3. Application to Physiological Systems	57
4. Discussion	64
Acknowledgment	66
References	66

3. Analytical Methods for the Retrieval and Interpretation of Continuous Glucose Monitoring Data in Diabetes — 69
Boris Kovatchev, Marc Breton, and William Clarke

1. Introduction	70
2. Decomposition of Sensor Errors	73
3. Measures of Average Glycemia and Deviation from Target	74
4. Risk and Variability Assessment	76
5. Measures and Plots of System Stability	80

6.	Time-Series-Based Prediction of Future BG Values	81
7.	Conclusions	84
	Acknowledgments	84
	References	84

4. Analysis of Heterogeneity in Molecular Weight and Shape by Analytical Ultracentrifugation Using Parallel Distributed Computing   87

Borries Demeler, Emre Brookes, and Luitgard Nagel-Steger

1.	Introduction	88
2.	Methodology	89
3.	Job Submission	97
4.	Results	100
5.	Conclusions	109
	Acknowledgments	111
	References	111

5. Discrete Stochastic Simulation Methods for Chemically Reacting Systems   115

Yang Cao and David C. Samuels

1.	Introduction	116
2.	The Chemical Master Equation	117
3.	The Stochastic Simulation Algorithm	119
4.	The Tau-Leaping Method	122
5.	Measurement of Simulation Error	132
6.	Software and Two Numerical Experiments	134
7.	Conclusion	137
	Acknowledgments	139
	References	139

6. Analyses for Physiological and Behavioral Rhythmicity   141

Harold B. Dowse

1.	Introduction	142
2.	Types of Biological Data and Their Acquisition	143
3.	Analysis in the Time Domain	145
4.	Analysis in the Frequency Domain	151
5.	Time/Frequency Analysis and the Wavelet Transform	161
6.	Signal Conditioning	164
7.	Strength and Regularity of a Signal	169
8.	Conclusions	171
	References	171

7. A Computational Approach for the Rational Design of Stable Proteins and Enzymes: Optimization of Surface Charge–Charge Interactions ... 175

Katrina L. Schweiker and George I. Makhatadze

1. Introduction ... 176
2. Computational Design of Surface Charge–Charge Interactions ... 183
3. Experimental Verification of Computational Predictions ... 190
4. Closing Remarks ... 202
Acknowledgments ... 204
References ... 204

8. Efficient Computation of Confidence Intervals for Bayesian Model Predictions Based on Multidimensional Parameter Space ... 213

Amber D. Smith, Alan Genz, David M. Freiberger, Gregory Belenky, and Hans P. A. Van Dongen

1. Introduction ... 214
2. Height of the Probability Density Function at the Boundary of the Smallest Multidimensional Confidence Region ... 215
3. Approximating a One-Dimensional Slice of the Probability Density Function by Means of Normal Curve Spline Pieces ... 217
4. Locating the Boundary of the Smallest Multidimensional Confidence Region ... 219
5. Finding the Minimum and Maximum of the Prediction Model over the Confidence Region ... 220
6. An Application: Bayesian Forecasting of Cognitive Performance Impairment during Sleep Deprivation ... 221
7. 95% Confidence Intervals for Bayesian Predictions of Cognitive Performance Impairment during Sleep Deprivation ... 223
8. Conclusion ... 227
Acknowledgments ... 229
Appendix ... 229
References ... 230

9. Analyzing Enzymatic pH Activity Profiles and Protein Titration Curves Using Structure-Based pK_a Calculations and Titration Curve Fitting ... 233

Jens Erik Nielsen

1. Introduction ... 234
2. Calculating the pH Dependence of Protein Characteristics ... 235
3. Setting up and Running a pK_a Calculation ... 243

4. Analyzing the Results of a pK_a Calculation	244
5. How Reliable Are Calculated pK_a Values?	246
6. Predicting pH Activity Profiles	248
7. Decomposition Analysis	249
8. Predicting Protein Stability Profiles	251
9. Fitting pH Titration Curves, pH Activity Profiles, and pH Stability Profiles	252
10. Conclusion	255
References	256

10. Least Squares in Calibration: Weights, Nonlinearity, and Other Nuisances — 259

Joel Tellinghuisen

1. Introduction	260
2. Review of Least Squares	263
3. Experiment Design Using V_{prior}—Numerical Illustrations	269
4. Conclusion	282
References	283

11. Evaluation and Comparison of Computational Models — 287

Jay I. Myung, Yun Tang, and Mark A. Pitt

1. Introduction	288
2. Conceptual Overview of Model Evaluation and Comparison	288
3. Model Comparison Methods	292
4. Model Comparison at Work: Choosing between Protein Folding Models	297
5. Conclusions	301
Acknowledgments	303
References	303

12. Desegregating Undergraduate Mathematics and Biology—Interdisciplinary Instruction with Emphasis on Ongoing Biomedical Research — 305

Raina Robeva

1. Introduction	306
2. Course Description	309
3. Discussion	317
Acknowledgments	320
References	320

13. Mathematical Algorithms for High-Resolution DNA Melting Analysis   323

Robert Palais and Carl T. Wittwer

1. Introduction   324
2. Extracting Melting Curves from Raw Fluorescence   325
3. Methods Used for Clustering and Classifying Melting Curves by Genotype   332
4. Methods Used for Modeling Melting Curves   335
References   342

14. Biomathematical Modeling of Pulsatile Hormone Secretion: A Historical Perspective   345

William S. Evans, Leon S. Farhy, and Michael L. Johnson

1. Introduction   346
2. Early Attempts to Identify and Characterize Pulsatile Hormone Release   348
3. Impact of Sampling Protocol and Pulse Detection Algorithm on Hormone Pulse Detection   349
4. Application of Deconvolution Procedures for the Identification and Characterization of Hormone Secretory Bursts   351
5. Limitations and Subsequent Improvements in Deconvolution Procedures   352
6. Evaluation of Pulsatile and Basal Hormone Secretion Using a Stochastic Differential Equations Model   353
7. Characterization of Regulation of Signal and Response Elements: Estimation of Approximate Entropy   355
8. Evaluation of Coupled Systems   356
9. Evaluation of Hormonal Networks with Feedback Interactions   360
10. Future Directions   362
Acknowledgments   362
References   363

15. *AutoDecon*: A Robust Numerical Method for the Quantification of Pulsatile Events   367

Michael L. Johnson, Lenore Pipes, Paula P. Veldhuis, Leon S. Farhy, Ralf Nass, Michael O. Thorner, and William S. Evans

1. Introduction   368
2. Methods   370
3. Results   384
4. Discussion   399
Acknowledgments   403
References   403

16. Modeling Fatigue over Sleep Deprivation, Circadian Rhythm, and Caffeine with a Minimal Performance Inhibitor Model 405

Patrick L. Benitez, Gary H. Kamimori, Thomas J. Balkin, Alexander Greene, and Michael L. Johnson

1. Introduction	406
2. Methods	407
3. Results	412
4. Discussion	414
Acknowledgments	419
References	419

Author Index   *423*
Subject Index   *437*

Contributors

Thomas J. Balkin
Department of Behavioral Biology, Walter Reed Army Institute of Research, Division of Neuroscience, Silver Spring, Maryland

Daniel A. Beard
Biotechnology and Bioengineering Center and Department of Physiology, Medical College of Wisconsin, Milwaukee, Wisconsin

Gregory Belenky
Sleep and Performance Research Center, Washington State University, Spokane, Washington

Patrick L. Benitez
Departments of Pharmacology and Medicine, University of Virginia Health System, Charlottesville, VA

Marc Breton
University of Virginia Health System, Charlottesville, Virginia

Emre Brookes
Department of Biochemistry, The University of Texas Health Science Center at San Antonio, San Antonio, Texas

Yang Cao
Department of Computer Science, Virginia Tech, Blacksburg, Virginia

William Clarke
University of Virginia Health System, Charlottesville, Virginia

Borries Demeler
Department of Biochemistry, The University of Texas Health Science Center at San Antonio, San Antonio, Texas

Harold B. Dowse
School of Biology and Ecology and Department of Mathematics and Statistics, University of Maine, Orono, Maine

William S. Evans
Endocrinology and Metabolism Department of Medicine, and Department of Obstetrics and Gynecology, University of Virginia Health System, Charlottesville, Virginia

Leon S. Farhy
Endocrinology and Metabolism Department of Medicine, University of Virginia Health System, Charlottesville, Virginia

David M. Freiberger
Sleep and Performance Research Center, Washington State University, Spokane, Washington

Alan Genz
Department of Mathematics, Washington State University, Pullman, Washington

A. Granada
Institute for Theoretical Biology, Humboldt-Universität zu Berlin, Berlin, Germany

Alexander Greene
School of Medicine, University of Florida, Gainesville, Florida

R. M. Hennig
Behavioral Physiology, Biology Department, Humboldt-Universität zu Berlin, Berlin, Germany

H. Herzel
Institute for Theoretical Biology, Humboldt-Universität zu Berlin, Berlin, Germany

Michael L. Johnson
Departments of Pharmacology and Medicine, University of Virginia Health System, Charlottesville, VA

Boris Kovatchev
University of Virginia Health System, Charlottesville, Virginia

A. Kramer
Laboratory of Chronobiology, Charité Universitätsmedizin Berlin, Berlin, Germany

Gary H. Kamimori
Department of Behavioral Biology, Walter Reed Army Institute of Research, Division of Neuroscience, Silver Spring, Maryland

Martin J. Kushmerick
Departments of Radiology, Bioengineering, Physiology and Biophysics, University of Washington, Seattle, Washington

George I. Makhatadze
Department of Biology and Center for Biotechnology and Interdisciplinary Studies, Rensselaer Polytechnic Institute, Troy, New York

Jay I. Myung
Department of Psychology, Ohio State University, Columbus, Ohio

Luitgard Nagel-Steger
Heinrich-Heine-Universität Düsseldorf, Institut für Physikalische Biologie, Düsseldorf, Germany

Ralf Nass
Endocrinology and Metabolism Department of Medicine, University of Virginia Health System, Charlottesville, Virginia

Jens Erik Nielsen
School of Biomolecular and Biomedical Science, Centre Synthesis and Chemical Biology, UCD Conway Institute, University College Dublin, Belfield, Dublin, Ireland

Robert Palais
Department of Pathology, University of Utah, Salt Lake City, Utah, and Department of Mathematics, University of Utah, Salt Lake City, Utah

Lenore Pipes
Departments of Pharmacology and Medicine, University of Virginia Health System, Charlottesville, VA

Mark A. Pitt
Department of Psychology, Ohio State University, Columbus, Ohio

Raina Robeva
Department of Mathematical Sciences, Sweet Briar College, Sweet Briar, Virginia

B. Ronacher
Behavioral Physiology, Biology Department, Humboldt-Universität zu Berlin, Berlin, Germany

David C. Samuels
Center for Human Genetics Research, Department of Molecular Physiology and Biophysics, Nashville, TN

Katrina L. Schweiker
Department of Biochemistry and Molecular Biology, Penn State University College of Medicine, Hershey, Pennsylvania, and Department of Biology and Center for Biotechnology and Interdisciplinary Studies, Rensselaer Polytechnic Institute, Troy, New York

Amber D. Smith
Sleep and Performance Research Center, Washington State University, Spokane, Washington

Yun Tang
Department of Psychology, Ohio State University, Columbus, Ohio

Joel Tellinghuisen
Department of Chemistry, Vanderbilt University, Nashville, Tennessee

Michael O. Thorner
Endocrinology and Metabolism Department of Medicine, University of Virginia Health System, Charlottesville, Virginia

Hans P. A. Van Dongen
Sleep and Performance Research Center, Washington State University, Spokane, Washington

Paula P. Veldhuis
Departments of Pharmacology and Medicine, University of Virginia Health System, Charlottesville, VA

Kalyan C. Vinnakota
Biotechnology and Bioengineering Center and Department of Physiology, Medical College of Wisconsin, Milwaukee, Wisconsin

Carl T. Wittwer
Department of Pathology, University of Utah, Salt Lake City, Utah

Fan Wu
Biotechnology and Bioengineering Center and Department of Physiology, Medical College of Wisconsin, Milwaukee, Wisconsin

Preface

The use of computers and computational methods has become ubiquitous in biological and biomedical research. This has been driven by numerous factors, a few of which follow: One primary reason is the emphasis being placed on computers and computational methods within the National Institutes of Health (NIH) roadmap; another factor is the increased level of mathematical and computational sophistication among researchers, particularly among junior scientists, students, journal reviewers, and NIH study section members; and another is the rapid advances in computer hardware and software, which make these methods far more accessible to the rank-and-file research community.

A general perception exists that the only applications of computers and computer methods in biological and biomedical research are either basic statistical analysis or the searching of DNA sequence data bases. While these are important applications, they only scratch the surface of the current and potential applications of computers and computer methods in biomedical research. The various chapters within this volume include a wide variety of applications that extend this limited perception.

The training of the majority of senior M.D.'s and Ph.D.'s in clinical or basic disciplines at academic medical centers rarely includes advanced coursework in mathematics, numerical analysis, statistics, or computer science. Generally, their hardware and software are maintained by a hospital staff that installs all hardware and software and even restricts what is available on their computers. Therefore, a critical aspect of this volume is information and methodology transfer to this target audience. This specific audience is indifferent as to whether the hardware and software are modern, object-oriented, portable, reusable, use the latest markup language, interchangeable, or easily maintained. These users are only interested in analyzing their data. The chapters within this volume have been written in order to be accessible to this target audience.

Michael L. Johnson and Ludwig Brand

METHODS IN ENZYMOLOGY

VOLUME I. Preparation and Assay of Enzymes
Edited by SIDNEY P. COLOWICK AND NATHAN O. KAPLAN

VOLUME II. Preparation and Assay of Enzymes
Edited by SIDNEY P. COLOWICK AND NATHAN O. KAPLAN

VOLUME III. Preparation and Assay of Substrates
Edited by SIDNEY P. COLOWICK AND NATHAN O. KAPLAN

VOLUME IV. Special Techniques for the Enzymologist
Edited by SIDNEY P. COLOWICK AND NATHAN O. KAPLAN

VOLUME V. Preparation and Assay of Enzymes
Edited by SIDNEY P. COLOWICK AND NATHAN O. KAPLAN

VOLUME VI. Preparation and Assay of Enzymes *(Continued)*
Preparation and Assay of Substrates
Special Techniques
Edited by SIDNEY P. COLOWICK AND NATHAN O. KAPLAN

VOLUME VII. Cumulative Subject Index
Edited by SIDNEY P. COLOWICK AND NATHAN O. KAPLAN

VOLUME VIII. Complex Carbohydrates
Edited by ELIZABETH F. NEUFELD AND VICTOR GINSBURG

VOLUME IX. Carbohydrate Metabolism
Edited by WILLIS A. WOOD

VOLUME X. Oxidation and Phosphorylation
Edited by RONALD W. ESTABROOK AND MAYNARD E. PULLMAN

VOLUME XI. Enzyme Structure
Edited by C. H. W. HIRS

VOLUME XII. Nucleic Acids (Parts A and B)
Edited by LAWRENCE GROSSMAN AND KIVIE MOLDAVE

VOLUME XIII. Citric Acid Cycle
Edited by J. M. LOWENSTEIN

VOLUME XIV. Lipids
Edited by J. M. LOWENSTEIN

VOLUME XV. Steroids and Terpenoids
Edited by RAYMOND B. CLAYTON

VOLUME XVI. Fast Reactions
Edited by KENNETH KUSTIN

VOLUME XVII. Metabolism of Amino Acids and Amines (Parts A and B)
Edited by HERBERT TABOR AND CELIA WHITE TABOR

VOLUME XVIII. Vitamins and Coenzymes (Parts A, B, and C)
Edited by DONALD B. MCCORMICK AND LEMUEL D. WRIGHT

VOLUME XIX. Proteolytic Enzymes
Edited by GERTRUDE E. PERLMANN AND LASZLO LORAND

VOLUME XX. Nucleic Acids and Protein Synthesis (Part C)
Edited by KIVIE MOLDAVE AND LAWRENCE GROSSMAN

VOLUME XXI. Nucleic Acids (Part D)
Edited by LAWRENCE GROSSMAN AND KIVIE MOLDAVE

VOLUME XXII. Enzyme Purification and Related Techniques
Edited by WILLIAM B. JAKOBY

VOLUME XXIII. Photosynthesis (Part A)
Edited by ANTHONY SAN PIETRO

VOLUME XXIV. Photosynthesis and Nitrogen Fixation (Part B)
Edited by ANTHONY SAN PIETRO

VOLUME XXV. Enzyme Structure (Part B)
Edited by C. H. W. HIRS AND SERGE N. TIMASHEFF

VOLUME XXVI. Enzyme Structure (Part C)
Edited by C. H. W. HIRS AND SERGE N. TIMASHEFF

VOLUME XXVII. Enzyme Structure (Part D)
Edited by C. H. W. HIRS AND SERGE N. TIMASHEFF

VOLUME XXVIII. Complex Carbohydrates (Part B)
Edited by VICTOR GINSBURG

VOLUME XXIX. Nucleic Acids and Protein Synthesis (Part E)
Edited by LAWRENCE GROSSMAN AND KIVIE MOLDAVE

VOLUME XXX. Nucleic Acids and Protein Synthesis (Part F)
Edited by KIVIE MOLDAVE AND LAWRENCE GROSSMAN

VOLUME XXXI. Biomembranes (Part A)
Edited by SIDNEY FLEISCHER AND LESTER PACKER

VOLUME XXXII. Biomembranes (Part B)
Edited by SIDNEY FLEISCHER AND LESTER PACKER

VOLUME XXXIII. Cumulative Subject Index Volumes I-XXX
Edited by MARTHA G. DENNIS AND EDWARD A. DENNIS

VOLUME XXXIV. Affinity Techniques (Enzyme Purification: Part B)
Edited by WILLIAM B. JAKOBY AND MEIR WILCHEK

VOLUME XXXV. Lipids (Part B)
Edited by JOHN M. LOWENSTEIN

VOLUME XXXVI. Hormone Action (Part A: Steroid Hormones)
Edited by BERT W. O'MALLEY AND JOEL G. HARDMAN

VOLUME XXXVII. Hormone Action (Part B: Peptide Hormones)
Edited by BERT W. O'MALLEY AND JOEL G. HARDMAN

VOLUME XXXVIII. Hormone Action (Part C: Cyclic Nucleotides)
Edited by JOEL G. HARDMAN AND BERT W. O'MALLEY

VOLUME XXXIX. Hormone Action (Part D: Isolated Cells, Tissues, and Organ Systems)
Edited by JOEL G. HARDMAN AND BERT W. O'MALLEY

VOLUME XL. Hormone Action (Part E: Nuclear Structure and Function)
Edited by BERT W. O'MALLEY AND JOEL G. HARDMAN

VOLUME XLI. Carbohydrate Metabolism (Part B)
Edited by W. A. WOOD

VOLUME XLII. Carbohydrate Metabolism (Part C)
Edited by W. A. WOOD

VOLUME XLIII. Antibiotics
Edited by JOHN H. HASH

VOLUME XLIV. Immobilized Enzymes
Edited by KLAUS MOSBACH

VOLUME XLV. Proteolytic Enzymes (Part B)
Edited by LASZLO LORAND

VOLUME XLVI. Affinity Labeling
Edited by WILLIAM B. JAKOBY AND MEIR WILCHEK

VOLUME XLVII. Enzyme Structure (Part E)
Edited by C. H. W. HIRS AND SERGE N. TIMASHEFF

VOLUME XLVIII. Enzyme Structure (Part F)
Edited by C. H. W. HIRS AND SERGE N. TIMASHEFF

VOLUME XLIX. Enzyme Structure (Part G)
Edited by C. H. W. HIRS AND SERGE N. TIMASHEFF

VOLUME L. Complex Carbohydrates (Part C)
Edited by VICTOR GINSBURG

VOLUME LI. Purine and Pyrimidine Nucleotide Metabolism
Edited by PATRICIA A. HOFFEE AND MARY ELLEN JONES

VOLUME LII. Biomembranes (Part C: Biological Oxidations)
Edited by SIDNEY FLEISCHER AND LESTER PACKER

VOLUME LIII. Biomembranes (Part D: Biological Oxidations)
Edited by SIDNEY FLEISCHER AND LESTER PACKER

VOLUME LIV. Biomembranes (Part E: Biological Oxidations)
Edited by SIDNEY FLEISCHER AND LESTER PACKER

VOLUME LV. Biomembranes (Part F: Bioenergetics)
Edited by SIDNEY FLEISCHER AND LESTER PACKER

VOLUME LVI. Biomembranes (Part G: Bioenergetics)
Edited by SIDNEY FLEISCHER AND LESTER PACKER

VOLUME LVII. Bioluminescence and Chemiluminescence
Edited by MARLENE A. DELUCA

VOLUME LVIII. Cell Culture
Edited by WILLIAM B. JAKOBY AND IRA PASTAN

VOLUME LIX. Nucleic Acids and Protein Synthesis (Part G)
Edited by KIVIE MOLDAVE AND LAWRENCE GROSSMAN

VOLUME LX. Nucleic Acids and Protein Synthesis (Part H)
Edited by KIVIE MOLDAVE AND LAWRENCE GROSSMAN

VOLUME 61. Enzyme Structure (Part H)
Edited by C. H. W. HIRS AND SERGE N. TIMASHEFF

VOLUME 62. Vitamins and Coenzymes (Part D)
Edited by DONALD B. MCCORMICK AND LEMUEL D. WRIGHT

VOLUME 63. Enzyme Kinetics and Mechanism (Part A: Initial Rate and Inhibitor Methods)
Edited by DANIEL L. PURICH

VOLUME 64. Enzyme Kinetics and Mechanism
(Part B: Isotopic Probes and Complex Enzyme Systems)
Edited by DANIEL L. PURICH

VOLUME 65. Nucleic Acids (Part I)
Edited by LAWRENCE GROSSMAN AND KIVIE MOLDAVE

VOLUME 66. Vitamins and Coenzymes (Part E)
Edited by DONALD B. MCCORMICK AND LEMUEL D. WRIGHT

VOLUME 67. Vitamins and Coenzymes (Part F)
Edited by DONALD B. MCCORMICK AND LEMUEL D. WRIGHT

VOLUME 68. Recombinant DNA
Edited by RAY WU

VOLUME 69. Photosynthesis and Nitrogen Fixation (Part C)
Edited by ANTHONY SAN PIETRO

VOLUME 70. Immunochemical Techniques (Part A)
Edited by HELEN VAN VUNAKIS AND JOHN J. LANGONE

VOLUME 71. Lipids (Part C)
Edited by JOHN M. LOWENSTEIN

VOLUME 72. Lipids (Part D)
Edited by JOHN M. LOWENSTEIN

VOLUME 73. Immunochemical Techniques (Part B)
Edited by JOHN J. LANGONE AND HELEN VAN VUNAKIS

VOLUME 74. Immunochemical Techniques (Part C)
Edited by JOHN J. LANGONE AND HELEN VAN VUNAKIS

VOLUME 75. Cumulative Subject Index Volumes XXXI, XXXII, XXXIV–LX
Edited by EDWARD A. DENNIS AND MARTHA G. DENNIS

VOLUME 76. Hemoglobins
Edited by ERALDO ANTONINI, LUIGI ROSSI-BERNARDI, AND EMILIA CHIANCONE

VOLUME 77. Detoxication and Drug Metabolism
Edited by WILLIAM B. JAKOBY

VOLUME 78. Interferons (Part A)
Edited by SIDNEY PESTKA

VOLUME 79. Interferons (Part B)
Edited by SIDNEY PESTKA

VOLUME 80. Proteolytic Enzymes (Part C)
Edited by LASZLO LORAND

VOLUME 81. Biomembranes (Part H: Visual Pigments and Purple Membranes, I)
Edited by LESTER PACKER

VOLUME 82. Structural and Contractile Proteins (Part A: Extracellular Matrix)
Edited by LEON W. CUNNINGHAM AND DIXIE W. FREDERIKSEN

VOLUME 83. Complex Carbohydrates (Part D)
Edited by VICTOR GINSBURG

VOLUME 84. Immunochemical Techniques (Part D: Selected Immunoassays)
Edited by JOHN J. LANGONE AND HELEN VAN VUNAKIS

VOLUME 85. Structural and Contractile Proteins (Part B: The Contractile Apparatus and the Cytoskeleton)
Edited by DIXIE W. FREDERIKSEN AND LEON W. CUNNINGHAM

VOLUME 86. Prostaglandins and Arachidonate Metabolites
Edited by WILLIAM E. M. LANDS AND WILLIAM L. SMITH

VOLUME 87. Enzyme Kinetics and Mechanism (Part C: Intermediates, Stereo-chemistry, and Rate Studies)
Edited by DANIEL L. PURICH

VOLUME 88. Biomembranes (Part I: Visual Pigments and Purple Membranes, II)
Edited by LESTER PACKER

VOLUME 89. Carbohydrate Metabolism (Part D)
Edited by WILLIS A. WOOD

VOLUME 90. Carbohydrate Metabolism (Part E)
Edited by WILLIS A. WOOD

VOLUME 91. Enzyme Structure (Part I)
Edited by C. H. W. HIRS AND SERGE N. TIMASHEFF

VOLUME 92. Immunochemical Techniques (Part E: Monoclonal Antibodies and General Immunoassay Methods)
Edited by JOHN J. LANGONE AND HELEN VAN VUNAKIS

VOLUME 93. Immunochemical Techniques (Part F: Conventional Antibodies, Fc Receptors, and Cytotoxicity)
Edited by JOHN J. LANGONE AND HELEN VAN VUNAKIS

VOLUME 94. Polyamines
Edited by HERBERT TABOR AND CELIA WHITE TABOR

VOLUME 95. Cumulative Subject Index Volumes 61–74, 76–80
Edited by EDWARD A. DENNIS AND MARTHA G. DENNIS

VOLUME 96. Biomembranes [Part J: Membrane Biogenesis: Assembly and Targeting (General Methods; Eukaryotes)]
Edited by SIDNEY FLEISCHER AND BECCA FLEISCHER

VOLUME 97. Biomembranes [Part K: Membrane Biogenesis: Assembly and Targeting (Prokaryotes, Mitochondria, and Chloroplasts)]
Edited by SIDNEY FLEISCHER AND BECCA FLEISCHER

VOLUME 98. Biomembranes (Part L: Membrane Biogenesis: Processing and Recycling)
Edited by SIDNEY FLEISCHER AND BECCA FLEISCHER

VOLUME 99. Hormone Action (Part F: Protein Kinases)
Edited by JACKIE D. CORBIN AND JOEL G. HARDMAN

VOLUME 100. Recombinant DNA (Part B)
Edited by RAY WU, LAWRENCE GROSSMAN, AND KIVIE MOLDAVE

VOLUME 101. Recombinant DNA (Part C)
Edited by RAY WU, LAWRENCE GROSSMAN, AND KIVIE MOLDAVE

VOLUME 102. Hormone Action (Part G: Calmodulin and Calcium-Binding Proteins)
Edited by ANTHONY R. MEANS AND BERT W. O'MALLEY

VOLUME 103. Hormone Action (Part H: Neuroendocrine Peptides)
Edited by P. MICHAEL CONN

VOLUME 104. Enzyme Purification and Related Techniques (Part C)
Edited by WILLIAM B. JAKOBY

VOLUME 105. Oxygen Radicals in Biological Systems
Edited by LESTER PACKER

VOLUME 106. Posttranslational Modifications (Part A)
Edited by FINN WOLD AND KIVIE MOLDAVE

VOLUME 107. Posttranslational Modifications (Part B)
Edited by FINN WOLD AND KIVIE MOLDAVE

VOLUME 108. Immunochemical Techniques (Part G: Separation and Characterization of Lymphoid Cells)
Edited by GIOVANNI DI SABATO, JOHN J. LANGONE, AND HELEN VAN VUNAKIS

VOLUME 109. Hormone Action (Part I: Peptide Hormones)
Edited by LUTZ BIRNBAUMER AND BERT W. O'MALLEY

VOLUME 110. Steroids and Isoprenoids (Part A)
Edited by JOHN H. LAW AND HANS C. RILLING

VOLUME 111. Steroids and Isoprenoids (Part B)
Edited by JOHN H. LAW AND HANS C. RILLING

VOLUME 112. Drug and Enzyme Targeting (Part A)
Edited by KENNETH J. WIDDER AND RALPH GREEN

VOLUME 113. Glutamate, Glutamine, Glutathione, and Related Compounds
Edited by ALTON MEISTER

VOLUME 114. Diffraction Methods for Biological Macromolecules (Part A)
Edited by HAROLD W. WYCKOFF, C. H. W. HIRS, AND SERGE N. TIMASHEFF

VOLUME 115. Diffraction Methods for Biological Macromolecules (Part B)
Edited by HAROLD W. WYCKOFF, C. H. W. HIRS, AND SERGE N. TIMASHEFF

VOLUME 116. Immunochemical Techniques
(Part H: Effectors and Mediators of Lymphoid Cell Functions)
Edited by GIOVANNI DI SABATO, JOHN J. LANGONE, AND HELEN VAN VUNAKIS

VOLUME 117. Enzyme Structure (Part J)
Edited by C. H. W. HIRS AND SERGE N. TIMASHEFF

VOLUME 118. Plant Molecular Biology
Edited by ARTHUR WEISSBACH AND HERBERT WEISSBACH

VOLUME 119. Interferons (Part C)
Edited by SIDNEY PESTKA

VOLUME 120. Cumulative Subject Index Volumes 81–94, 96–101

VOLUME 121. Immunochemical Techniques (Part I: Hybridoma Technology and Monoclonal Antibodies)
Edited by JOHN J. LANGONE AND HELEN VAN VUNAKIS

VOLUME 122. Vitamins and Coenzymes (Part G)
Edited by FRANK CHYTIL AND DONALD B. MCCORMICK

VOLUME 123. Vitamins and Coenzymes (Part H)
Edited by FRANK CHYTIL AND DONALD B. MCCORMICK

VOLUME 124. Hormone Action (Part J: Neuroendocrine Peptides)
Edited by P. MICHAEL CONN

VOLUME 125. Biomembranes (Part M: Transport in Bacteria, Mitochondria, and Chloroplasts: General Approaches and Transport Systems)
Edited by SIDNEY FLEISCHER AND BECCA FLEISCHER

VOLUME 126. Biomembranes (Part N: Transport in Bacteria, Mitochondria, and Chloroplasts: Protonmotive Force)
Edited by SIDNEY FLEISCHER AND BECCA FLEISCHER

VOLUME 127. Biomembranes (Part O: Protons and Water: Structure and Translocation)
Edited by LESTER PACKER

VOLUME 128. Plasma Lipoproteins (Part A: Preparation, Structure, and Molecular Biology)
Edited by JERE P. SEGREST AND JOHN J. ALBERS

VOLUME 129. Plasma Lipoproteins (Part B: Characterization, Cell Biology, and Metabolism)
Edited by JOHN J. ALBERS AND JERE P. SEGREST

VOLUME 130. Enzyme Structure (Part K)
Edited by C. H. W. HIRS AND SERGE N. TIMASHEFF

VOLUME 131. Enzyme Structure (Part L)
Edited by C. H. W. HIRS AND SERGE N. TIMASHEFF

VOLUME 132. Immunochemical Techniques (Part J: Phagocytosis and Cell-Mediated Cytotoxicity)
Edited by GIOVANNI DI SABATO AND JOHANNES EVERSE

VOLUME 133. Bioluminescence and Chemiluminescence (Part B)
Edited by MARLENE DELUCA AND WILLIAM D. MCELROY

VOLUME 134. Structural and Contractile Proteins (Part C: The Contractile Apparatus and the Cytoskeleton)
Edited by RICHARD B. VALLEE

VOLUME 135. Immobilized Enzymes and Cells (Part B)
Edited by KLAUS MOSBACH

VOLUME 136. Immobilized Enzymes and Cells (Part C)
Edited by KLAUS MOSBACH

VOLUME 137. Immobilized Enzymes and Cells (Part D)
Edited by KLAUS MOSBACH

VOLUME 138. Complex Carbohydrates (Part E)
Edited by VICTOR GINSBURG

VOLUME 139. Cellular Regulators (Part A: Calcium- and Calmodulin-Binding Proteins)
Edited by ANTHONY R. MEANS AND P. MICHAEL CONN

VOLUME 140. Cumulative Subject Index Volumes 102–119, 121–134

VOLUME 141. Cellular Regulators (Part B: Calcium and Lipids)
Edited by P. MICHAEL CONN AND ANTHONY R. MEANS

VOLUME 142. Metabolism of Aromatic Amino Acids and Amines
Edited by SEYMOUR KAUFMAN

VOLUME 143. Sulfur and Sulfur Amino Acids
Edited by WILLIAM B. JAKOBY AND OWEN GRIFFITH

VOLUME 144. Structural and Contractile Proteins (Part D: Extracellular Matrix)
Edited by LEON W. CUNNINGHAM

VOLUME 145. Structural and Contractile Proteins (Part E: Extracellular Matrix)
Edited by LEON W. CUNNINGHAM

VOLUME 146. Peptide Growth Factors (Part A)
Edited by DAVID BARNES AND DAVID A. SIRBASKU

VOLUME 147. Peptide Growth Factors (Part B)
Edited by DAVID BARNES AND DAVID A. SIRBASKU

VOLUME 148. Plant Cell Membranes
Edited by LESTER PACKER AND ROLAND DOUCE

VOLUME 149. Drug and Enzyme Targeting (Part B)
Edited by RALPH GREEN AND KENNETH J. WIDDER

VOLUME 150. Immunochemical Techniques (Part K: *In Vitro* Models of B and T Cell Functions and Lymphoid Cell Receptors)
Edited by GIOVANNI DI SABATO

VOLUME 151. Molecular Genetics of Mammalian Cells
Edited by MICHAEL M. GOTTESMAN

VOLUME 152. Guide to Molecular Cloning Techniques
Edited by SHELBY L. BERGER AND ALAN R. KIMMEL

VOLUME 153. Recombinant DNA (Part D)
Edited by RAY WU AND LAWRENCE GROSSMAN

VOLUME 154. Recombinant DNA (Part E)
Edited by RAY WU AND LAWRENCE GROSSMAN

VOLUME 155. Recombinant DNA (Part F)
Edited by RAY WU

VOLUME 156. Biomembranes (Part P: ATP-Driven Pumps and Related Transport: The Na, K-Pump)
Edited by SIDNEY FLEISCHER AND BECCA FLEISCHER

VOLUME 157. Biomembranes (Part Q: ATP-Driven Pumps and Related Transport: Calcium, Proton, and Potassium Pumps)
Edited by SIDNEY FLEISCHER AND BECCA FLEISCHER

VOLUME 158. Metalloproteins (Part A)
Edited by JAMES F. RIORDAN AND BERT L. VALLEE

VOLUME 159. Initiation and Termination of Cyclic Nucleotide Action
Edited by JACKIE D. CORBIN AND ROGER A. JOHNSON

VOLUME 160. Biomass (Part A: Cellulose and Hemicellulose)
Edited by WILLIS A. WOOD AND SCOTT T. KELLOGG

VOLUME 161. Biomass (Part B: Lignin, Pectin, and Chitin)
Edited by WILLIS A. WOOD AND SCOTT T. KELLOGG

VOLUME 162. Immunochemical Techniques (Part L: Chemotaxis and Inflammation)
Edited by GIOVANNI DI SABATO

VOLUME 163. Immunochemical Techniques (Part M: Chemotaxis and Inflammation)
Edited by GIOVANNI DI SABATO

VOLUME 164. Ribosomes
Edited by HARRY F. NOLLER, JR., AND KIVIE MOLDAVE

VOLUME 165. Microbial Toxins: Tools for Enzymology
Edited by SIDNEY HARSHMAN

VOLUME 166. Branched-Chain Amino Acids
Edited by ROBERT HARRIS AND JOHN R. SOKATCH

VOLUME 167. Cyanobacteria
Edited by LESTER PACKER AND ALEXANDER N. GLAZER

VOLUME 168. Hormone Action (Part K: Neuroendocrine Peptides)
Edited by P. MICHAEL CONN

VOLUME 169. Platelets: Receptors, Adhesion, Secretion (Part A)
Edited by JACEK HAWIGER

VOLUME 170. Nucleosomes
Edited by PAUL M. WASSARMAN AND ROGER D. KORNBERG

VOLUME 171. Biomembranes (Part R: Transport Theory: Cells and Model Membranes)
Edited by SIDNEY FLEISCHER AND BECCA FLEISCHER

VOLUME 172. Biomembranes (Part S: Transport: Membrane Isolation and Characterization)
Edited by SIDNEY FLEISCHER AND BECCA FLEISCHER

VOLUME 173. Biomembranes [Part T: Cellular and Subcellular Transport: Eukaryotic (Nonepithelial) Cells]
Edited by SIDNEY FLEISCHER AND BECCA FLEISCHER

VOLUME 174. Biomembranes [Part U: Cellular and Subcellular Transport: Eukaryotic (Nonepithelial) Cells]
Edited by SIDNEY FLEISCHER AND BECCA FLEISCHER

VOLUME 175. Cumulative Subject Index Volumes 135–139, 141–167

VOLUME 176. Nuclear Magnetic Resonance (Part A: Spectral Techniques and Dynamics)
Edited by NORMAN J. OPPENHEIMER AND THOMAS L. JAMES

VOLUME 177. Nuclear Magnetic Resonance (Part B: Structure and Mechanism)
Edited by NORMAN J. OPPENHEIMER AND THOMAS L. JAMES

VOLUME 178. Antibodies, Antigens, and Molecular Mimicry
Edited by JOHN J. LANGONE

VOLUME 179. Complex Carbohydrates (Part F)
Edited by VICTOR GINSBURG

VOLUME 180. RNA Processing (Part A: General Methods)
Edited by JAMES E. DAHLBERG AND JOHN N. ABELSON

VOLUME 181. RNA Processing (Part B: Specific Methods)
Edited by JAMES E. DAHLBERG AND JOHN N. ABELSON

VOLUME 182. Guide to Protein Purification
Edited by MURRAY P. DEUTSCHER

VOLUME 183. Molecular Evolution: Computer Analysis of Protein and Nucleic Acid Sequences
Edited by RUSSELL F. DOOLITTLE

VOLUME 184. Avidin-Biotin Technology
Edited by MEIR WILCHEK AND EDWARD A. BAYER

VOLUME 185. Gene Expression Technology
Edited by DAVID V. GOEDDEL

VOLUME 186. Oxygen Radicals in Biological Systems (Part B: Oxygen Radicals and Antioxidants)
Edited by LESTER PACKER AND ALEXANDER N. GLAZER

VOLUME 187. Arachidonate Related Lipid Mediators
Edited by ROBERT C. MURPHY AND FRANK A. FITZPATRICK

VOLUME 188. Hydrocarbons and Methylotrophy
Edited by MARY E. LIDSTROM

VOLUME 189. Retinoids (Part A: Molecular and Metabolic Aspects)
Edited by LESTER PACKER

VOLUME 190. Retinoids (Part B: Cell Differentiation and Clinical Applications)
Edited by LESTER PACKER

VOLUME 191. Biomembranes (Part V: Cellular and Subcellular Transport: Epithelial Cells)
Edited by SIDNEY FLEISCHER AND BECCA FLEISCHER

VOLUME 192. Biomembranes (Part W: Cellular and Subcellular Transport: Epithelial Cells)
Edited by SIDNEY FLEISCHER AND BECCA FLEISCHER

VOLUME 193. Mass Spectrometry
Edited by JAMES A. MCCLOSKEY

VOLUME 194. Guide to Yeast Genetics and Molecular Biology
Edited by CHRISTINE GUTHRIE AND GERALD R. FINK

VOLUME 195. Adenylyl Cyclase, G Proteins, and Guanylyl Cyclase
Edited by ROGER A. JOHNSON AND JACKIE D. CORBIN

VOLUME 196. Molecular Motors and the Cytoskeleton
Edited by RICHARD B. VALLEE

VOLUME 197. Phospholipases
Edited by EDWARD A. DENNIS

VOLUME 198. Peptide Growth Factors (Part C)
Edited by DAVID BARNES, J. P. MATHER, AND GORDON H. SATO

VOLUME 199. Cumulative Subject Index Volumes 168–174, 176–194

VOLUME 200. Protein Phosphorylation (Part A: Protein Kinases: Assays, Purification, Antibodies, Functional Analysis, Cloning, and Expression)
Edited by TONY HUNTER AND BARTHOLOMEW M. SEFTON

VOLUME 201. Protein Phosphorylation (Part B: Analysis of Protein Phosphorylation, Protein Kinase Inhibitors, and Protein Phosphatases)
Edited by TONY HUNTER AND BARTHOLOMEW M. SEFTON

VOLUME 202. Molecular Design and Modeling: Concepts and Applications (Part A: Proteins, Peptides, and Enzymes)
Edited by JOHN J. LANGONE

VOLUME 203. Molecular Design and Modeling: Concepts and Applications (Part B: Antibodies and Antigens, Nucleic Acids, Polysaccharides, and Drugs)
Edited by JOHN J. LANGONE

VOLUME 204. Bacterial Genetic Systems
Edited by JEFFREY H. MILLER

VOLUME 205. Metallobiochemistry (Part B: Metallothionein and Related Molecules)
Edited by JAMES F. RIORDAN AND BERT L. VALLEE

VOLUME 206. Cytochrome P450
Edited by MICHAEL R. WATERMAN AND ERIC F. JOHNSON

VOLUME 207. Ion Channels
Edited by BERNARDO RUDY AND LINDA E. IVERSON

VOLUME 208. Protein–DNA Interactions
Edited by ROBERT T. SAUER

VOLUME 209. Phospholipid Biosynthesis
Edited by EDWARD A. DENNIS AND DENNIS E. VANCE

VOLUME 210. Numerical Computer Methods
Edited by LUDWIG BRAND AND MICHAEL L. JOHNSON

VOLUME 211. DNA Structures (Part A: Synthesis and Physical Analysis of DNA)
Edited by DAVID M. J. LILLEY AND JAMES E. DAHLBERG

VOLUME 212. DNA Structures (Part B: Chemical and Electrophoretic Analysis of DNA)
Edited by DAVID M. J. LILLEY AND JAMES E. DAHLBERG

VOLUME 213. Carotenoids (Part A: Chemistry, Separation, Quantitation, and Antioxidation)
Edited by LESTER PACKER

VOLUME 214. Carotenoids (Part B: Metabolism, Genetics, and Biosynthesis)
Edited by LESTER PACKER

VOLUME 215. Platelets: Receptors, Adhesion, Secretion (Part B)
Edited by JACEK J. HAWIGER

VOLUME 216. Recombinant DNA (Part G)
Edited by RAY WU

VOLUME 217. Recombinant DNA (Part H)
Edited by RAY WU

VOLUME 218. Recombinant DNA (Part I)
Edited by RAY WU

VOLUME 219. Reconstitution of Intracellular Transport
Edited by JAMES E. ROTHMAN

VOLUME 220. Membrane Fusion Techniques (Part A)
Edited by NEJAT DÜZGÜNEŞ

VOLUME 221. Membrane Fusion Techniques (Part B)
Edited by NEJAT DÜZGÜNEŞ

VOLUME 222. Proteolytic Enzymes in Coagulation, Fibrinolysis, and Complement Activation (Part A: Mammalian Blood Coagulation Factors and Inhibitors)
Edited by LASZLO LORAND AND KENNETH G. MANN

VOLUME 223. Proteolytic Enzymes in Coagulation, Fibrinolysis, and Complement Activation (Part B: Complement Activation, Fibrinolysis, and Nonmammalian Blood Coagulation Factors)
Edited by LASZLO LORAND AND KENNETH G. MANN

VOLUME 224. Molecular Evolution: Producing the Biochemical Data
Edited by ELIZABETH ANNE ZIMMER, THOMAS J. WHITE, REBECCA L. CANN, AND ALLAN C. WILSON

VOLUME 225. Guide to Techniques in Mouse Development
Edited by PAUL M. WASSARMAN AND MELVIN L. DEPAMPHILIS

VOLUME 226. Metallobiochemistry (Part C: Spectroscopic and Physical Methods for Probing Metal Ion Environments in Metalloenzymes and Metalloproteins)
Edited by JAMES F. RIORDAN AND BERT L. VALLEE

VOLUME 227. Metallobiochemistry (Part D: Physical and Spectroscopic Methods for Probing Metal Ion Environments in Metalloproteins)
Edited by JAMES F. RIORDAN AND BERT L. VALLEE

VOLUME 228. Aqueous Two-Phase Systems
Edited by HARRY WALTER AND GÖTE JOHANSSON

VOLUME 229. Cumulative Subject Index Volumes 195–198, 200–227

VOLUME 230. Guide to Techniques in Glycobiology
Edited by WILLIAM J. LENNARZ AND GERALD W. HART

VOLUME 231. Hemoglobins (Part B: Biochemical and Analytical Methods)
Edited by JOHANNES EVERSE, KIM D. VANDEGRIFF, AND ROBERT M. WINSLOW

VOLUME 232. Hemoglobins (Part C: Biophysical Methods)
Edited by JOHANNES EVERSE, KIM D. VANDEGRIFF, AND ROBERT M. WINSLOW

VOLUME 233. Oxygen Radicals in Biological Systems (Part C)
Edited by LESTER PACKER

VOLUME 234. Oxygen Radicals in Biological Systems (Part D)
Edited by LESTER PACKER

VOLUME 235. Bacterial Pathogenesis (Part A: Identification and Regulation of Virulence Factors)
Edited by VIRGINIA L. CLARK AND PATRIK M. BAVOIL

VOLUME 236. Bacterial Pathogenesis (Part B: Integration of Pathogenic Bacteria with Host Cells)
Edited by VIRGINIA L. CLARK AND PATRIK M. BAVOIL

VOLUME 237. Heterotrimeric G Proteins
Edited by RAVI IYENGAR

VOLUME 238. Heterotrimeric G-Protein Effectors
Edited by RAVI IYENGAR

VOLUME 239. Nuclear Magnetic Resonance (Part C)
Edited by THOMAS L. JAMES AND NORMAN J. OPPENHEIMER

VOLUME 240. Numerical Computer Methods (Part B)
Edited by MICHAEL L. JOHNSON AND LUDWIG BRAND

VOLUME 241. Retroviral Proteases
Edited by LAWRENCE C. KUO AND JULES A. SHAFER

VOLUME 242. Neoglycoconjugates (Part A)
Edited by Y. C. LEE AND REIKO T. LEE

VOLUME 243. Inorganic Microbial Sulfur Metabolism
Edited by HARRY D. PECK, JR., AND JEAN LEGALL

VOLUME 244. Proteolytic Enzymes: Serine and Cysteine Peptidases
Edited by ALAN J. BARRETT

VOLUME 245. Extracellular Matrix Components
Edited by E. RUOSLAHTI AND E. ENGVALL

VOLUME 246. Biochemical Spectroscopy
Edited by KENNETH SAUER

VOLUME 247. Neoglycoconjugates (Part B: Biomedical Applications)
Edited by Y. C. LEE AND REIKO T. LEE

VOLUME 248. Proteolytic Enzymes: Aspartic and Metallo Peptidases
Edited by ALAN J. BARRETT

VOLUME 249. Enzyme Kinetics and Mechanism (Part D: Developments in Enzyme Dynamics)
Edited by DANIEL L. PURICH

VOLUME 250. Lipid Modifications of Proteins
Edited by PATRICK J. CASEY AND JANICE E. BUSS

VOLUME 251. Biothiols (Part A: Monothiols and Dithiols, Protein Thiols, and Thiyl Radicals)
Edited by LESTER PACKER

VOLUME 252. Biothiols (Part B: Glutathione and Thioredoxin; Thiols in Signal Transduction and Gene Regulation)
Edited by LESTER PACKER

VOLUME 253. Adhesion of Microbial Pathogens
Edited by RON J. DOYLE AND ITZHAK OFEK

VOLUME 254. Oncogene Techniques
Edited by PETER K. VOGT AND INDER M. VERMA

VOLUME 255. Small GTPases and Their Regulators (Part A: Ras Family)
Edited by W. E. BALCH, CHANNING J. DER, AND ALAN HALL

VOLUME 256. Small GTPases and Their Regulators (Part B: Rho Family)
Edited by W. E. BALCH, CHANNING J. DER, AND ALAN HALL

VOLUME 257. Small GTPases and Their Regulators (Part C: Proteins Involved in Transport)
Edited by W. E. BALCH, CHANNING J. DER, AND ALAN HALL

VOLUME 258. Redox-Active Amino Acids in Biology
Edited by JUDITH P. KLINMAN

VOLUME 259. Energetics of Biological Macromolecules
Edited by MICHAEL L. JOHNSON AND GARY K. ACKERS

VOLUME 260. Mitochondrial Biogenesis and Genetics (Part A)
Edited by GIUSEPPE M. ATTARDI AND ANNE CHOMYN

VOLUME 261. Nuclear Magnetic Resonance and Nucleic Acids
Edited by THOMAS L. JAMES

VOLUME 262. DNA Replication
Edited by JUDITH L. CAMPBELL

VOLUME 263. Plasma Lipoproteins (Part C: Quantitation)
Edited by WILLIAM A. BRADLEY, SANDRA H. GIANTURCO, AND JERE P. SEGREST

VOLUME 264. Mitochondrial Biogenesis and Genetics (Part B)
Edited by GIUSEPPE M. ATTARDI AND ANNE CHOMYN

VOLUME 265. Cumulative Subject Index Volumes 228, 230–262

VOLUME 266. Computer Methods for Macromolecular Sequence Analysis
Edited by RUSSELL F. DOOLITTLE

VOLUME 267. Combinatorial Chemistry
Edited by JOHN N. ABELSON

VOLUME 268. Nitric Oxide (Part A: Sources and Detection of NO; NO Synthase)
Edited by LESTER PACKER

VOLUME 269. Nitric Oxide (Part B: Physiological and Pathological Processes)
Edited by LESTER PACKER

VOLUME 270. High Resolution Separation and Analysis of Biological Macromolecules (Part A: Fundamentals)
Edited by BARRY L. KARGER AND WILLIAM S. HANCOCK

VOLUME 271. High Resolution Separation and Analysis of Biological Macromolecules (Part B: Applications)
Edited by BARRY L. KARGER AND WILLIAM S. HANCOCK

VOLUME 272. Cytochrome P450 (Part B)
Edited by ERIC F. JOHNSON AND MICHAEL R. WATERMAN

VOLUME 273. RNA Polymerase and Associated Factors (Part A)
Edited by SANKAR ADHYA

VOLUME 274. RNA Polymerase and Associated Factors (Part B)
Edited by SANKAR ADHYA

VOLUME 275. Viral Polymerases and Related Proteins
Edited by LAWRENCE C. KUO, DAVID B. OLSEN, AND STEVEN S. CARROLL

VOLUME 276. Macromolecular Crystallography (Part A)
Edited by CHARLES W. CARTER, JR., AND ROBERT M. SWEET

VOLUME 277. Macromolecular Crystallography (Part B)
Edited by CHARLES W. CARTER, JR., AND ROBERT M. SWEET

VOLUME 278. Fluorescence Spectroscopy
Edited by LUDWIG BRAND AND MICHAEL L. JOHNSON

VOLUME 279. Vitamins and Coenzymes (Part I)
Edited by DONALD B. MCCORMICK, JOHN W. SUTTIE, AND CONRAD WAGNER

VOLUME 280. Vitamins and Coenzymes (Part J)
Edited by DONALD B. MCCORMICK, JOHN W. SUTTIE, AND CONRAD WAGNER

VOLUME 281. Vitamins and Coenzymes (Part K)
Edited by DONALD B. MCCORMICK, JOHN W. SUTTIE, AND CONRAD WAGNER

VOLUME 282. Vitamins and Coenzymes (Part L)
Edited by DONALD B. MCCORMICK, JOHN W. SUTTIE, AND CONRAD WAGNER

VOLUME 283. Cell Cycle Control
Edited by WILLIAM G. DUNPHY

VOLUME 284. Lipases (Part A: Biotechnology)
Edited by BYRON RUBIN AND EDWARD A. DENNIS

VOLUME 285. Cumulative Subject Index Volumes 263, 264, 266–284, 286–289

VOLUME 286. Lipases (Part B: Enzyme Characterization and Utilization)
Edited by BYRON RUBIN AND EDWARD A. DENNIS

VOLUME 287. Chemokines
Edited by RICHARD HORUK

VOLUME 288. Chemokine Receptors
Edited by RICHARD HORUK

VOLUME 289. Solid Phase Peptide Synthesis
Edited by GREGG B. FIELDS

VOLUME 290. Molecular Chaperones
Edited by GEORGE H. LORIMER AND THOMAS BALDWIN

VOLUME 291. Caged Compounds
Edited by GERARD MARRIOTT

VOLUME 292. ABC Transporters: Biochemical, Cellular, and Molecular Aspects
Edited by SURESH V. AMBUDKAR AND MICHAEL M. GOTTESMAN

VOLUME 293. Ion Channels (Part B)
Edited by P. MICHAEL CONN

VOLUME 294. Ion Channels (Part C)
Edited by P. MICHAEL CONN

VOLUME 295. Energetics of Biological Macromolecules (Part B)
Edited by GARY K. ACKERS AND MICHAEL L. JOHNSON

VOLUME 296. Neurotransmitter Transporters
Edited by SUSAN G. AMARA

VOLUME 297. Photosynthesis: Molecular Biology of Energy Capture
Edited by LEE MCINTOSH

VOLUME 298. Molecular Motors and the Cytoskeleton (Part B)
Edited by RICHARD B. VALLEE

VOLUME 299. Oxidants and Antioxidants (Part A)
Edited by LESTER PACKER

VOLUME 300. Oxidants and Antioxidants (Part B)
Edited by LESTER PACKER

VOLUME 301. Nitric Oxide: Biological and Antioxidant Activities (Part C)
Edited by LESTER PACKER

VOLUME 302. Green Fluorescent Protein
Edited by P. MICHAEL CONN

VOLUME 303. cDNA Preparation and Display
Edited by SHERMAN M. WEISSMAN

VOLUME 304. Chromatin
Edited by PAUL M. WASSARMAN AND ALAN P. WOLFFE

VOLUME 305. Bioluminescence and Chemiluminescence (Part C)
Edited by THOMAS O. BALDWIN AND MIRIAM M. ZIEGLER

VOLUME 306. Expression of Recombinant Genes in Eukaryotic Systems
Edited by JOSEPH C. GLORIOSO AND MARTIN C. SCHMIDT

VOLUME 307. Confocal Microscopy
Edited by P. MICHAEL CONN

VOLUME 308. Enzyme Kinetics and Mechanism (Part E: Energetics of Enzyme Catalysis)
Edited by DANIEL L. PURICH AND VERN L. SCHRAMM

VOLUME 309. Amyloid, Prions, and Other Protein Aggregates
Edited by RONALD WETZEL

VOLUME 310. Biofilms
Edited by RON J. DOYLE

VOLUME 311. Sphingolipid Metabolism and Cell Signaling (Part A)
Edited by ALFRED H. MERRILL, JR., AND YUSUF A. HANNUN

VOLUME 312. Sphingolipid Metabolism and Cell Signaling (Part B)
Edited by ALFRED H. MERRILL, JR., AND YUSUF A. HANNUN

VOLUME 313. Antisense Technology (Part A: General Methods, Methods of Delivery, and RNA Studies)
Edited by M. IAN PHILLIPS

VOLUME 314. Antisense Technology (Part B: Applications)
Edited by M. IAN PHILLIPS

VOLUME 315. Vertebrate Phototransduction and the Visual Cycle (Part A)
Edited by KRZYSZTOF PALCZEWSKI

VOLUME 316. Vertebrate Phototransduction and the Visual Cycle (Part B)
Edited by KRZYSZTOF PALCZEWSKI

VOLUME 317. RNA–Ligand Interactions (Part A: Structural Biology Methods)
Edited by DANIEL W. CELANDER AND JOHN N. ABELSON

VOLUME 318. RNA–Ligand Interactions (Part B: Molecular Biology Methods)
Edited by DANIEL W. CELANDER AND JOHN N. ABELSON

VOLUME 319. Singlet Oxygen, UV-A, and Ozone
Edited by LESTER PACKER AND HELMUT SIES

VOLUME 320. Cumulative Subject Index Volumes 290–319

VOLUME 321. Numerical Computer Methods (Part C)
Edited by MICHAEL L. JOHNSON AND LUDWIG BRAND

VOLUME 322. Apoptosis
Edited by JOHN C. REED

VOLUME 323. Energetics of Biological Macromolecules (Part C)
Edited by MICHAEL L. JOHNSON AND GARY K. ACKERS

VOLUME 324. Branched-Chain Amino Acids (Part B)
Edited by ROBERT A. HARRIS AND JOHN R. SOKATCH

VOLUME 325. Regulators and Effectors of Small GTPases (Part D: Rho Family)
Edited by W. E. BALCH, CHANNING J. DER, AND ALAN HALL

VOLUME 326. Applications of Chimeric Genes and Hybrid Proteins (Part A: Gene Expression and Protein Purification)
Edited by JEREMY THORNER, SCOTT D. EMR, AND JOHN N. ABELSON

VOLUME 327. Applications of Chimeric Genes and Hybrid Proteins (Part B: Cell Biology and Physiology)
Edited by JEREMY THORNER, SCOTT D. EMR, AND JOHN N. ABELSON

VOLUME 328. Applications of Chimeric Genes and Hybrid Proteins (Part C: Protein–Protein Interactions and Genomics)
Edited by JEREMY THORNER, SCOTT D. EMR, AND JOHN N. ABELSON

VOLUME 329. Regulators and Effectors of Small GTPases (Part E: GTPases Involved in Vesicular Traffic)
Edited by W. E. BALCH, CHANNING J. DER, AND ALAN HALL

VOLUME 330. Hyperthermophilic Enzymes (Part A)
Edited by MICHAEL W. W. ADAMS AND ROBERT M. KELLY

VOLUME 331. Hyperthermophilic Enzymes (Part B)
Edited by MICHAEL W. W. ADAMS AND ROBERT M. KELLY

VOLUME 332. Regulators and Effectors of Small GTPases (Part F: Ras Family I)
Edited by W. E. BALCH, CHANNING J. DER, AND ALAN HALL

VOLUME 333. Regulators and Effectors of Small GTPases (Part G: Ras Family II)
Edited by W. E. BALCH, CHANNING J. DER, AND ALAN HALL

VOLUME 334. Hyperthermophilic Enzymes (Part C)
Edited by MICHAEL W. W. ADAMS AND ROBERT M. KELLY

VOLUME 335. Flavonoids and Other Polyphenols
Edited by LESTER PACKER

VOLUME 336. Microbial Growth in Biofilms (Part A: Developmental and Molecular Biological Aspects)
Edited by RON J. DOYLE

VOLUME 337. Microbial Growth in Biofilms (Part B: Special Environments and Physicochemical Aspects)
Edited by RON J. DOYLE

VOLUME 338. Nuclear Magnetic Resonance of Biological Macromolecules (Part A)
Edited by THOMAS L. JAMES, VOLKER DÖTSCH, AND ULI SCHMITZ

VOLUME 339. Nuclear Magnetic Resonance of Biological Macromolecules (Part B)
Edited by THOMAS L. JAMES, VOLKER DÖTSCH, AND ULI SCHMITZ

VOLUME 340. Drug–Nucleic Acid Interactions
Edited by JONATHAN B. CHAIRES AND MICHAEL J. WARING

VOLUME 341. Ribonucleases (Part A)
Edited by ALLEN W. NICHOLSON

VOLUME 342. Ribonucleases (Part B)
Edited by ALLEN W. NICHOLSON

VOLUME 343. G Protein Pathways (Part A: Receptors)
Edited by RAVI IYENGAR AND JOHN D. HILDEBRANDT

VOLUME 344. G Protein Pathways (Part B: G Proteins and Their Regulators)
Edited by RAVI IYENGAR AND JOHN D. HILDEBRANDT

VOLUME 345. G Protein Pathways (Part C: Effector Mechanisms)
Edited by RAVI IYENGAR AND JOHN D. HILDEBRANDT

VOLUME 346. Gene Therapy Methods
Edited by M. IAN PHILLIPS

VOLUME 347. Protein Sensors and Reactive Oxygen Species (Part A: Selenoproteins and Thioredoxin)
Edited by HELMUT SIES AND LESTER PACKER

VOLUME 348. Protein Sensors and Reactive Oxygen Species (Part B: Thiol Enzymes and Proteins)
Edited by HELMUT SIES AND LESTER PACKER

VOLUME 349. Superoxide Dismutase
Edited by LESTER PACKER

VOLUME 350. Guide to Yeast Genetics and Molecular and Cell Biology (Part B)
Edited by CHRISTINE GUTHRIE AND GERALD R. FINK

VOLUME 351. Guide to Yeast Genetics and Molecular and Cell Biology (Part C)
Edited by CHRISTINE GUTHRIE AND GERALD R. FINK

VOLUME 352. Redox Cell Biology and Genetics (Part A)
Edited by CHANDAN K. SEN AND LESTER PACKER

VOLUME 353. Redox Cell Biology and Genetics (Part B)
Edited by CHANDAN K. SEN AND LESTER PACKER

VOLUME 354. Enzyme Kinetics and Mechanisms (Part F: Detection and Characterization of Enzyme Reaction Intermediates)
Edited by DANIEL L. PURICH

VOLUME 355. Cumulative Subject Index Volumes 321–354

VOLUME 356. Laser Capture Microscopy and Microdissection
Edited by P. MICHAEL CONN

VOLUME 357. Cytochrome P450, Part C
Edited by ERIC F. JOHNSON AND MICHAEL R. WATERMAN

VOLUME 358. Bacterial Pathogenesis (Part C: Identification, Regulation, and Function of Virulence Factors)
Edited by VIRGINIA L. CLARK AND PATRIK M. BAVOIL

VOLUME 359. Nitric Oxide (Part D)
Edited by ENRIQUE CADENAS AND LESTER PACKER

VOLUME 360. Biophotonics (Part A)
Edited by GERARD MARRIOTT AND IAN PARKER

VOLUME 361. Biophotonics (Part B)
Edited by GERARD MARRIOTT AND IAN PARKER

VOLUME 362. Recognition of Carbohydrates in Biological Systems (Part A)
Edited by YUAN C. LEE AND REIKO T. LEE

VOLUME 363. Recognition of Carbohydrates in Biological Systems (Part B)
Edited by YUAN C. LEE AND REIKO T. LEE

VOLUME 364. Nuclear Receptors
Edited by DAVID W. RUSSELL AND DAVID J. MANGELSDORF

VOLUME 365. Differentiation of Embryonic Stem Cells
Edited by PAUL M. WASSAUMAN AND GORDON M. KELLER

VOLUME 366. Protein Phosphatases
Edited by SUSANNE KLUMPP AND JOSEF KRIEGLSTEIN

VOLUME 367. Liposomes (Part A)
Edited by NEJAT DÜZGÜNEŞ

VOLUME 368. Macromolecular Crystallography (Part C)
Edited by CHARLES W. CARTER, JR., AND ROBERT M. SWEET

VOLUME 369. Combinational Chemistry (Part B)
Edited by GUILLERMO A. MORALES AND BARRY A. BUNIN

VOLUME 370. RNA Polymerases and Associated Factors (Part C)
Edited by SANKAR L. ADHYA AND SUSAN GARGES

VOLUME 371. RNA Polymerases and Associated Factors (Part D)
Edited by SANKAR L. ADHYA AND SUSAN GARGES

VOLUME 372. Liposomes (Part B)
Edited by NEJAT DÜZGÜNEŞ

VOLUME 373. Liposomes (Part C)
Edited by NEJAT DÜZGÜNEŞ

VOLUME 374. Macromolecular Crystallography (Part D)
Edited by CHARLES W. CARTER, JR., AND ROBERT W. SWEET

VOLUME 375. Chromatin and Chromatin Remodeling Enzymes (Part A)
Edited by C. DAVID ALLIS AND CARL WU

VOLUME 376. Chromatin and Chromatin Remodeling Enzymes (Part B)
Edited by C. DAVID ALLIS AND CARL WU

VOLUME 377. Chromatin and Chromatin Remodeling Enzymes (Part C)
Edited by C. DAVID ALLIS AND CARL WU

VOLUME 378. Quinones and Quinone Enzymes (Part A)
Edited by HELMUT SIES AND LESTER PACKER

VOLUME 379. Energetics of Biological Macromolecules (Part D)
Edited by JO M. HOLT, MICHAEL L. JOHNSON, AND GARY K. ACKERS

VOLUME 380. Energetics of Biological Macromolecules (Part E)
Edited by JO M. HOLT, MICHAEL L. JOHNSON, AND GARY K. ACKERS

VOLUME 381. Oxygen Sensing
Edited by CHANDAN K. SEN AND GREGG L. SEMENZA

VOLUME 382. Quinones and Quinone Enzymes (Part B)
Edited by HELMUT SIES AND LESTER PACKER

VOLUME 383. Numerical Computer Methods (Part D)
Edited by LUDWIG BRAND AND MICHAEL L. JOHNSON

VOLUME 384. Numerical Computer Methods (Part E)
Edited by LUDWIG BRAND AND MICHAEL L. JOHNSON

VOLUME 385. Imaging in Biological Research (Part A)
Edited by P. MICHAEL CONN

VOLUME 386. Imaging in Biological Research (Part B)
Edited by P. MICHAEL CONN

VOLUME 387. Liposomes (Part D)
Edited by NEJAT DÜZGÜNEŞ

VOLUME 388. Protein Engineering
Edited by DAN E. ROBERTSON AND JOSEPH P. NOEL

VOLUME 389. Regulators of G-Protein Signaling (Part A)
Edited by DAVID P. SIDEROVSKI

VOLUME 390. Regulators of G-Protein Signaling (Part B)
Edited by DAVID P. SIDEROVSKI

VOLUME 391. Liposomes (Part E)
Edited by NEJAT DÜZGÜNEŞ

VOLUME 392. RNA Interference
Edited by ENGELKE ROSSI

VOLUME 393. Circadian Rhythms
Edited by MICHAEL W. YOUNG

VOLUME 394. Nuclear Magnetic Resonance of Biological Macromolecules (Part C)
Edited by THOMAS L. JAMES

VOLUME 395. Producing the Biochemical Data (Part B)
Edited by ELIZABETH A. ZIMMER AND ERIC H. ROALSON

VOLUME 396. Nitric Oxide (Part E)
Edited by LESTER PACKER AND ENRIQUE CADENAS

VOLUME 397. Environmental Microbiology
Edited by JARED R. LEADBETTER

VOLUME 398. Ubiquitin and Protein Degradation (Part A)
Edited by RAYMOND J. DESHAIES

VOLUME 399. Ubiquitin and Protein Degradation (Part B)
Edited by RAYMOND J. DESHAIES

VOLUME 400. Phase II Conjugation Enzymes and Transport Systems
Edited by HELMUT SIES AND LESTER PACKER

VOLUME 401. Glutathione Transferases and Gamma Glutamyl Transpeptidases
Edited by HELMUT SIES AND LESTER PACKER

VOLUME 402. Biological Mass Spectrometry
Edited by A. L. BURLINGAME

VOLUME 403. GTPases Regulating Membrane Targeting and Fusion
Edited by WILLIAM E. BALCH, CHANNING J. DER, AND ALAN HALL

VOLUME 404. GTPases Regulating Membrane Dynamics
Edited by WILLIAM E. BALCH, CHANNING J. DER, AND ALAN HALL

VOLUME 405. Mass Spectrometry: Modified Proteins and Glycoconjugates
Edited by A. L. BURLINGAME

VOLUME 406. Regulators and Effectors of Small GTPases: Rho Family
Edited by WILLIAM E. BALCH, CHANNING J. DER, AND ALAN HALL

VOLUME 407. Regulators and Effectors of Small GTPases: Ras Family
Edited by WILLIAM E. BALCH, CHANNING J. DER, AND ALAN HALL

VOLUME 408. DNA Repair (Part A)
Edited by JUDITH L. CAMPBELL AND PAUL MODRICH

VOLUME 409. DNA Repair (Part B)
Edited by JUDITH L. CAMPBELL AND PAUL MODRICH

VOLUME 410. DNA Microarrays (Part A: Array Platforms and Web-Bench Protocols)
Edited by ALAN KIMMEL AND BRIAN OLIVER

VOLUME 411. DNA Microarrays (Part B: Databases and Statistics)
Edited by ALAN KIMMEL AND BRIAN OLIVER

VOLUME 412. Amyloid, Prions, and Other Protein Aggregates (Part B)
Edited by INDU KHETERPAL AND RONALD WETZEL

VOLUME 413. Amyloid, Prions, and Other Protein Aggregates (Part C)
Edited by INDU KHETERPAL AND RONALD WETZEL

VOLUME 414. Measuring Biological Responses with Automated Microscopy
Edited by JAMES INGLESE

VOLUME 415. Glycobiology
Edited by MINORU FUKUDA

VOLUME 416. Glycomics
Edited by MINORU FUKUDA

VOLUME 417. Functional Glycomics
Edited by MINORU FUKUDA

VOLUME 418. Embryonic Stem Cells
Edited by IRINA KLIMANSKAYA AND ROBERT LANZA

VOLUME 419. Adult Stem Cells
Edited by IRINA KLIMANSKAYA AND ROBERT LANZA

VOLUME 420. Stem Cell Tools and Other Experimental Protocols
Edited by IRINA KLIMANSKAYA AND ROBERT LANZA

VOLUME 421. Advanced Bacterial Genetics: Use of Transposons and Phage for Genomic Engineering
Edited by KELLY T. HUGHES

VOLUME 422. Two-Component Signaling Systems, Part A
Edited by MELVIN I. SIMON, BRIAN R. CRANE, AND ALEXANDRINE CRANE

VOLUME 423. Two-Component Signaling Systems, Part B
Edited by MELVIN I. SIMON, BRIAN R. CRANE, AND ALEXANDRINE CRANE

VOLUME 424. RNA Editing
Edited by JONATHA M. GOTT

VOLUME 425. RNA Modification
Edited by JONATHA M. GOTT

VOLUME 426. Integrins
Edited by DAVID CHERESH

VOLUME 427. MicroRNA Methods
Edited by JOHN J. ROSSI

VOLUME 428. Osmosensing and Osmosignaling
Edited by HELMUT SIES AND DIETER HAUSSINGER

VOLUME 429. Translation Initiation: Extract Systems and Molecular Genetics
Edited by JON LORSCH

VOLUME 430. Translation Initiation: Reconstituted Systems and Biophysical Methods
Edited by JON LORSCH

VOLUME 431. Translation Initiation: Cell Biology, High-Throughput and Chemical-Based Approaches
Edited by JON LORSCH

VOLUME 432. Lipidomics and Bioactive Lipids: Mass-Spectrometry–Based Lipid Analysis
Edited by H. ALEX BROWN

VOLUME 433. Lipidomics and Bioactive Lipids: Specialized Analytical Methods and Lipids in Disease
Edited by H. ALEX BROWN

VOLUME 434. Lipidomics and Bioactive Lipids: Lipids and Cell Signaling
Edited by H. ALEX BROWN

VOLUME 435. Oxygen Biology and Hypoxia
Edited by HELMUT SIES AND BERNHARD BRÜNE

VOLUME 436. Globins and Other Nitric Oxide-Reactive Protiens (Part A)
Edited by ROBERT K. POOLE

VOLUME 437. Globins and Other Nitric Oxide-Reactive Protiens (Part B)
Edited by ROBERT K. POOLE

VOLUME 438. Small GTPases in Disease (Part A)
Edited by WILLIAM E. BALCH, CHANNING J. DER, AND ALAN HALL

VOLUME 439. Small GTPases in Disease (Part B)
Edited by WILLIAM E. BALCH, CHANNING J. DER, AND ALAN HALL

VOLUME 440. Nitric Oxide, Part F Oxidative and Nitrosative Stress in Redox Regulation of Cell Signaling
Edited by ENRIQUE CADENAS AND LESTER PACKER

VOLUME 441. Nitric Oxide, Part G Oxidative and Nitrosative Stress in Redox Regulation of Cell Signaling
Edited by ENRIQUE CADENAS AND LESTER PACKER

VOLUME 442. Programmed Cell Death, General Principles for Studying Cell Death (Part A)
Edited by ROYA KHOSRAVI-FAR, ZAHRA ZAKERI, RICHARD A. LOCKSHIN, AND MAURO PIACENTINI

VOLUME 443. Angiogenesis: *In Vitro* Systems
Edited by DAVID A. CHERESH

VOLUME 444. Angiogenesis: *In Vivo* Systems (Part A)
Edited by DAVID A. CHERESH

VOLUME 445. Angiogenesis: *In Vivo* Systems (Part B)
Edited by DAVID A. CHERESH

VOLUME 446. Programmed Cell Death, The Biology and Therapeutic Implications of Cell Death (Part B)
Edited by ROYA KHOSRAVI-FAR, ZAHRA ZAKERI, RICHARD A. LOCKSHIN, AND MAURO PIACENTINI

VOLUME 447. RNA Turnover in Prokaryotes, Archae and Organelles
Edited by LYNNE E. MAQUAT AND CECILIA M. ARRAIANO

VOLUME 448. RNA Turnover in Eukaryotes: Nucleases, Pathways and Anaylsis of mRNA Decay
Edited by LYNNE E. MAQUAT AND MEGERDITCH KILEDJIAN

VOLUME 449. RNA Turnover in Eukaryotes: Analysis of Specialized and Quality Control RNA Decay Pathways
Edited by LYNNE E. MAQUAT AND MEGERDITCH KILEDJIAN

VOLUME 450. Fluorescence Spectroscopy
Edited by LUDWIG BRAND AND MICHAEL L. JOHNSON

VOLUME 451. Autophagy: Lower Eukaryotes and Non-mammalian Systems (Part A)
Edited by DANIEL J. KLIONSKY

VOLUME 452. Autophagy: Mammalian and Clinical (Part B)
Edited by DANIEL J. KLIONSKY

VOLUME 453. Autophagy: Disease and Clinical Applications (Part C)
Edited by DANIEL J. KLIONSKY

VOLUME 454. Computer Methods (Part A)
Edited by MICHAEL L. JOHNSON AND LUDWIG BRAND

CHAPTER ONE

Phase Response Curves: Elucidating the Dynamics of Coupled Oscillators

A. Granada,* R. M. Hennig,† B. Ronacher,† A. Kramer,‡ and H. Herzel*

Contents

1. Introduction 2
   1.1. An example—phase response curve of cellular circadian rhythms 2
   1.2. Self-sustained biological rhythms 4
   1.3. Coupled and entrained oscillators 5
   1.4. Phase response curves 7
   1.5. Entrainment zones and circle map 8
   1.6. Limit cycle in phase space and phase response curves 10
2. Estimation of Phase Response Curves 11
   2.1. Definitions 11
   2.2. A skeleton protocol—a minimal phase response curve recipe 12
3. Specific Applications 17
   3.1. Nonlinear dynamics of the heart 17
   3.2. Classifications of neurons 18
   3.3. Classification of central pattern generators 19
   3.4. Consequences of entrainment for insect communication 19
4. Discussion 20
   Appendix I 21
   Appendix II 23
   Acknowledgments 24
   References 24

Abstract

Phase response curves (PRCs) are widely used in circadian clocks, neuroscience, and heart physiology. They quantify the response of an oscillator to pulse-like perturbations. Phase response curves provide valuable information on the properties of oscillators and their synchronization. This chapter discusses

* Institute for Theoretical Biology, Humboldt-Universität zu Berlin, Berlin, Germany
† Behavioral Physiology, Biology Department, Humboldt-Universität zu Berlin, Berlin, Germany
‡ Laboratory of Chronobiology, Charité Universitätsmedizin Berlin, Berlin, Germany

Methods in Enzymology, Volume 454 © 2009 Elsevier Inc.
ISSN 0076-6879, DOI: 10.1016/S0076-6879(08)03801-9 All rights reserved.

biological self-sustained oscillators (circadian clock, physiological rhythms, etc.) in the context of nonlinear dynamics theory. Coupled oscillators can synchronize with different frequency ratios, can generate toroidal dynamics (superposition of independent frequencies), and may lead to deterministic chaos. These nonlinear phenomena can be analyzed with the aid of a phase transition curve, which is intimately related to the phase response curve. For illustration purposes, this chapter discusses a model of circadian oscillations based on a delayed negative feedback. In a second part, the chapter provides a step-by-step recipe to measure phase response curves. It discusses specifications of this recipe for circadian rhythms, heart rhythms, neuronal spikes, central pattern generators, and insect communication. Finally, it stresses the predictive power of measured phase response curves. PRCs can be used to quantify the coupling strength of oscillations, to classify oscillator types, and to predict the complex dynamics of periodically driven oscillations.

1. Introduction

1.1. An example—phase response curve of cellular circadian rhythms

The concept of phase response curves (PRCs) has been introduced in a variety of research fields, ranging from cardiac rhythms (Winfree, 1980) and neurophysiology (Reyes and Fetz, 1993a) to animal communication (Buck, 1988; Sismondo, 1990). A PRC describes the magnitude of phase changes after perturbing an oscillatory system. The aim of this chapter is to provide a recipe of how to estimate PRCs in different biological systems. Furthermore, we will show that PRCs are helpful in understanding the complexity of coupled oscillators.

Prominent applications of PRCs are in the field of circadian clocks (Pittendrigh and Daan, 1976). This section describes a recent example from this field to introduce some terminology. Circadian clocks are endogenous biological oscillators that generate rhythms with a period of about 24 h (from the Latin circa diem, "about a day"). In mammals, these oscillations are cell autonomous and essentially based on a negative transcriptional–translational feedback loop. Hence circadian rhythms can be detected in single cells such as neurons (Welsh *et al.*, 1995) or fibroblasts (Yagita *et al.*, 2001). This allows the quantitative analysis of clock properties, for example, using luciferase reporters in cultured cells (Brown *et al.*, 2005). Here we present data from a recent analysis of dermal fibroblasts from skin biopsies of human subjects who where either early ("larks") or late chronotypes ("owls") in their behavior (Brown *et al.*, 2008). The aim of this study was to investigate whether these different types of behavior have a correlation in the dynamical properties of the circadian clocks present in dermal fibroblasts. During the course of the study, phase-shifting experiments were performed to analyze potential differences between cells derived from "larks" or "owls." Figure 1.1A shows bioluminescence oscillations of fibroblasts synchronized with dexamethasone [for experimental details, see Brown *et al.* (2008)]. The different lines represent perturbations with a phase-shifting chemical (forskolin) applied at different time points to identical plates of fibroblast after dexamethasome synchronization. In Figure 1.1B, the resulting phase shift with respect to unperturbed oscillations is plotted as a function of perturbation time. This phase response curve indicates that the given stimuli can lead to phase advances and phase delays of up to 5 h. Similar phase shifts can be induced by temperature pulses. Figure 1.1C illustrates the so-called entrainment of fibroblasts to external temperature rhythms. Even though the endogenous period of 24.5 h (see days 5–10 in constant 37°) deviates from the 24-h temperature cycle, coupling induces an entrainment (days 1–5) with a fixed phase relation and frequency locking (1:1 entrainment). In the study of Brown *et al.* (2008), PRCs and entrainment of human fibroblasts via temperature rhythms characterize the chronotype of human beings. Whereas most early chronotypes have a relatively short endogenous period, large phase-shifting properties (Fig. 1.1B) can also be associated with this behavior.

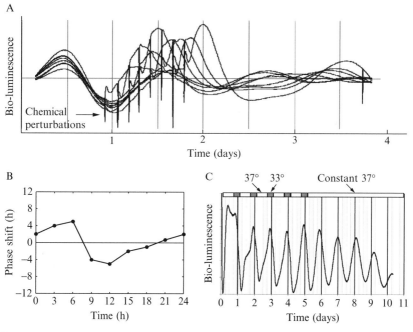

Figure 1.1 Time series, phase response curve, and entrainment of human dermal fibroblasts [modified from Brown *et al.* (2008)]. (A) Bioluminescence oscillations of fibroblasts synchronized with dexamethasone and later phase shifted by perturbations. (B) PRC: resulting phase shift relative to unperturbed oscillations as a function of perturbation time. (C) Entrainment of fibroblasts to external temperature rhythms.

The following theoretical sections discuss how phase response curves provide valuable information on the synchronization and entrainment of self-sustained oscillations. In particular, we will point out that phase response curves lead to iterated maps termed phase transition curves (PTCs). These one-dimensional models can be used to predict the complex dynamics of coupled oscillators, including 1:1 synchronization, alternations, and deterministic chaos (Glass and Mackey, 1988).

1.2. Self-sustained biological rhythms

Endogenous rhythms are widespread in biological systems, with periods ranging from milliseconds (neuronal spikes, vocal folds oscillations) to years [hibernation cycles (Mrosovsky, 1977), insect populations (Alexander and Moore, 1962)]. Self-sustained rhythms can be generated on the physiological level (heart beat, respiration, hormones), in intracellular biochemical networks (calcium oscillations, glycolytic oscillations), and via transcriptional feedback loops (somitic clock, NF-κB oscillations, circadian clock). Characteristic properties of endogenous oscillations are an autonomously determined period and amplitude and relaxation of amplitude perturbations. Such self-sustained oscillations are termed stable limit cycles in nonlinear dynamics theory.

To be specific, we introduce an example of limit cycle oscillations as a consequence of negative feedback. A popular example of such oscillator models has been introduced by Goodwin (1965) to describe genetic feedback regulation. Here we discuss a modified Goodwin model, termed the Gonze model, that was used to simulate circadian rhythms in mammalian cells (Gonze et al., 2005). In this context the dynamic variables X, Y, and Z represent the mRNA levels of clock genes, their cytoplasmatic protein concentrations, and nuclear inhibitor concentrations, respectively. The corresponding equations contain production and degradation terms:

$$\frac{dX}{dt} = v_1 \frac{K_1^4}{K_1^4 + Z^4} - v_2 \frac{X}{K_2 + X} \tag{1.1}$$

$$\frac{dY}{dt} = v_3 X - v_4 \frac{Y}{K_4 + Y} \tag{1.2}$$

$$\frac{dZ}{dt} = v_5 Y - v_6 \frac{Z}{K_6 + Z} \tag{1.3}$$

Michaelis–Menten kinetics is used for the decay, since the degradation processes are controlled enzymatically. Another nonlinear term refers to inhibition of the transcription due to the Z variable in Eq. (1.1). Solving the aforementioned equations for appropriate parameter values, any initial condition will approach after some time a unique, strictly periodic solution—a stable limit cycle (see Fig. 1.2). Note that the inhibitor Z is delayed by more than 6 h with respect to the mRNA oscillations represented by X.

1.3. Coupled and entrained oscillators

Biological rhythms typically interact with other oscillators. Examples are coupled rhythms of heart, respiration and movement (Glass and Mackey, 1988), acoustic communication of insects (Hartbauer et al., 2005), synchronous blinking of fireflies (Greenfield et al., 1997), and the synchronization of neuronal activities (Hopfield and Herz, 1995). Circadian rhythms, as described by the model given earlier, can be observed on the single cell level in neurons of the suprachiasmatic nucleus (SCN)(Honma et al., 1998). Synaptic connections, gap junctions, and neurotransmitters are believed to synchronize SCN neurons in a robust manner (Yamaguchi et al., 2003). Moreover, peripheral organs such as heart and liver are coupled to the master clock in the SCN (Hastings et al., 2003). Synchronization via bidirectional coupling can lead to oscillations with the same period but different phases. As an illustration, Fig. 1.2B shows simulations of two coupled Gonze oscillators [see Eqs. (1.1)–(1.3)] representing cells with different autonomous periods. The coupling synchronizes both cells and leads to a 1:1 frequency locking with a constant phase shift between the cells. External periodic stimulation of biological oscillators can lead to entrainment for sufficiently strong coupling. In such a case the external driver determines the period. A p:q (p and q being small integers) frequency ratio implies that after p cycles of oscillator 1 and q cycles of oscillator 2 the initial state is reached again. For example, the frequency locking between respiration and heart beats might be 4:1 (Schäfer et al., 1998; Seidel and Herzel, 1998), meaning that there are four heart beats (oscillator 1) during one respiration cycle (oscillator 2). Prominent examples are the entrainment of autonomous circadian rhythms by the light–dark cycle (compare with Fig. 1.2C) and the periodic stimulation of heart cells (Guevara et al., 1981). The latter example illustrates the complex dynamics of coupled oscillators: Apart from 1:1 frequency locking between external stimuli and internal beats, other entrainment ratios, such as 1:2, 2:3, and 3:2, can be detected. Such a diversity of frequency ratios has also been measured in motor patterns (von Holst, 1939) and in voice disorders due to the asymmetry of left and right vocal folds (Mergell et al., 2000). An entrainment with a frequency ratio of p:q is still a limit cycle characterized by long periods and subharmonics in the frequency spectrum (Berge et al., 1984). In addition to p:q synchronization, coupled oscillators can also oscillate with independent frequencies. Such a dynamics has been termed torus. For example, slow modulations are known as "beating" or "relative coordination" (von Holst, 1939). Furthermore, coupling of oscillators can lead to "deterministic chaos," that is, internally generated irregular behavior without any random input.

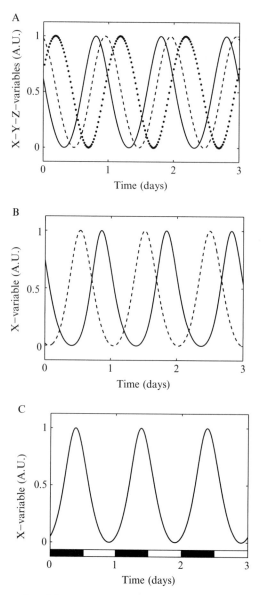

Figure 1.2 Time series of the Gonze model for circadian rhythms. (A) Time series of the Gonze model [see Eqs. (1.1)–(1.3)]. The variable "X" (solid line) represents mRNA levels of clock genes, "Y" (dashed) their cytoplasmatic protein concentrations, and "Z" (dots) its nuclear inhibitor. Amplitudes are normalized to the unit interval. (B) Two coupled Gonze oscillators with different autonomous periods (T_1 = 23.5-h solid line and T_2 = 24.5-h dashed line). When coupled, both oscillators synchronize with a constant phase difference. (C) Entrainment of an autonomous circadian Gonze oscillator by a 12:12-h light–dark cycle. Arbitrary units (A.U.) are used. Computational details are given in Appendix II.

Chaos has been associated with cardiac arrhythmias (Glass and Mackey, 1988) and a variety of voice disorders (Herzel et al., 1994). The remarkable complexity of coupled oscillators is described in detail in nonlinear dynamics textbooks (Pikovsky et al., 2001; Solari et al., 1996). Here we show how the plethora of responses can be understood by unifying principles related to the measurement of phase response curves.

1.4. Phase response curves

Fortunately, the overwhelming complexity of coupled oscillators can be reduced in many cases to a few basic principles and simple mathematical models. Instead of studying the continuous interaction of rhythmic processes, this section considers single pulse-like perturbations of limit cycle oscillators. Understanding single pulse effects will guide us later to a discussion of periodic stimuli that may induce entrainment. Note that PRCs have some analogy to the impulse-response concept applied successfully in linear systems. For both linear and nonlinear oscillators, valuable information on the dynamical system can be extracted by studying the response to single pulses. A perturbation of an oscillator by a pulse will lead to amplitude and phase changes. For a stable limit cycle, amplitude perturbations will decay but phase changes persist. Thus we can monitor the phase change $\Delta\phi$ due to a pulse given at a phase ϕ. The effect of the perturbations on the phase usually depends strongly on the phase within the cycle. For example, current injection in neurons and heart cells will have only minor effects during action potentials. In circadian clocks, a light pulse in the morning advances the circadian phase in humans, whereas a light pulse at midnight delays the phase. The corresponding graphical representation of phase shift versus pulse phase is called phase response curve.

Figure 1.3A shows an example calculated for the Gonze oscillator [see Eqs. (1.1)–(1.3)]. In the middle of the cycle (phases 0.3 to 0.8), pulses advance the phase, whereas for small and large phases a delay is observed. Such a phase response curve leads directly to the PTC displayed in Figure 1.3B. Here the old phase refers to the phase immediately before the perturbation and the new phase denotes the phase after the pulse. A phase transition curve can be interpreted as a one-dimensional model describing the mapping of an old phase to a new phase. Because this model can be applied again and again it has been termed the "iterated map"

$$\phi_{n+1} = f(\phi_n). \quad (1.4)$$

If we start, for example, with an initial phase $\phi_0 = 0.6$ we obtain $\phi_1 \approx 0.75$ (see Fig. 1.3B). Applying the PTC with the new $\phi_1 = 0.75$ we obtain $\phi_2 = f(\phi_1) \approx 0.86$. Iterating further we stay at the same point $\phi^* \approx 0.86$, meaning that we reached the stable solution. Generally, at the intersections of PTCs

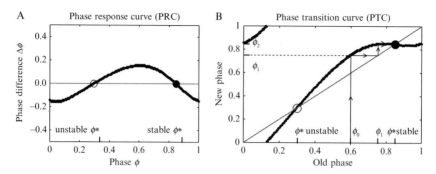

Figure 1.3 Phase transition curve and phase response curve from the Gonze model. (A) Phase response curve and (B) phase transition curve described by Eqs. (1.1)–(1.3). Intersections with $\Delta\phi = 0$ (for PRC) and the diagonal correspond to unstable (○) and stable (●) fixed points. Simulations details are given in Appendix II.

with the diagonal we find $\phi^* = f(\phi^*)$, which implies that the new phase equals the old phase. These are so-called fixed points of the iterated map. Linear stability analysis reveals that these fixed points are stable for small slopes, that is, for $|\frac{df}{d\phi}(\phi^*)| < 1$ and unstable for $|\frac{df}{d\phi}(\phi^*)| > 1$ (Kaplan and Glass, 1995).

Phase response curves describe the effects of single pulses. If subsequent perturbations can be regarded as approximately independent, the associated phase response curve can be applied iteratively. Iterated maps allow us to generalize effects of single pulses to series of perturbations. For subsequent pulses we can apply the iterated map repeatedly and obtain in this way a series of phases $\{\phi_0, \phi_1, \phi_2, \ldots\}$. Along these lines, 1:1 entrainment can be related to the stable fixed point of the iterated map.

1.5. Entrainment zones and circle map

As discussed earlier, two coupled oscillators can exhibit a variety of dynamics, ranging from synchronization with rational p:q frequency ratios to toroidal oscillations and deterministic chaos. This section discusses how these dynamical regimes depend on the parameters of the system, such as the coupling strength and the frequency ratio. Evidently, oscillations remain independent for zero coupling strength, but weak coupling can lead to 1:1 synchronization if the frequencies are very close. This has been discovered by Huygens (1673) while observing pendulum clocks coupled via vibrations of the wall. Similarly, p:q frequency locking can be found if the autonomous frequencies are sufficiently close to a p:q ratio. For increasing coupling strength, synchronization is observed more easily, that is, larger deviations from perfect p:q ratios can still lead to frequency locking.

Figure 1.4A shows p:q synchronization for a periodically driven Gonze oscillator introduced earlier [see Eqs. (1.1)–(1.3)]. The horizontal axis denotes the frequency ratio of the external forcing and the autonomous oscillation, and the vertical axis refers to the forcing strength k relative to the total oscillator amplitude. The dark regions mark frequency locking with ratios 1:2, 1:1, and 3:2 (often termed "Arnold tongues," referring to the Russian mathematician V.I. Arnold). For increasing values of k the width of most entrainment regions increases, as expected. Between the entrainment zones toroidal oscillations dominate. For large k values, period doubling and deterministic chaos also occur [compare Gonze et al. (2005)]. In the previous section we argued that phase response curves and the associated phase transition curves are useful tools in understanding the complex behavior of coupled oscillators. An intensively studied example of a PTC is the sine map, an example of an iterated map, described by the following equation:

$$\phi_{n+1} = \phi_n + \Omega + k \sin(2\pi\phi_n). \tag{1.5}$$

Here ϕ_n is a phase variable $\phi \in [0, 1]$, the parameter Ω can be related to the frequency ratio, and k represents the coupling strength. Plots of the graphs resemble phase transition curves as shown in Fig. 1.3B. For iterated maps such as the sine map, entrainment zones can be calculated easily. Figure 1.4B shows the frequency locking regions of this discrete one-dimensional model. The qualitative features of the differential equation system analyzed in Fig. 1.4A are very similar. In both cases the 1:1 synchronization zone is most prominent and Arnold tongues at 1:2 and 3:2

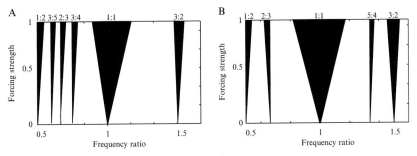

Figure 1.4 Comparison of entrainment regions. Entrainment regions for the Gonze model of circadian rhythms [see Eq. (1.1)] and the sine map [see Eq. (1.5)]. The horizontal axis denotes the frequency ratio of the external forcing and the autonomous oscillation, and the vertical axis refers to the forcing strength k relative to the total oscillator amplitude. Dark regions mark frequency locking ratios. (A) Entrainment regions for the Gonze model where 1:2, 3:5, 2:3, 3:4, 1:1, and 3:2 frequency locking is shown. (B) Entrainment regions for the sine map where 1:2, 2:3, 1:1, 5:4, and 3:2 frequency locking is marked. Computational details are given in Appendix II.

frequency ratios are clearly visible. Figure 1.4 illustrates that coupled oscillators exhibit universal features independent of the specific details of the system. Frequency ratios and coupling strength are the most essential system parameters determining synchronization behavior. Consequently, phase response curves (or phase transition curves as the sine map) are central elements in understanding coupled biological oscillators.

1.6. Limit cycle in phase space and phase response curves

Figure 1.2 represents self-sustained oscillations as periodic time series. In nonlinear dynamics, the so-called phase space plays a central role. The dynamical variables of the system of interest, for example, X, Y, and Z in the example given earlier, serve as coordinates of the phase space. A stable limit cycle corresponds to a closed curve in phase space attracting nearby orbits. In other words, small perturbations will relax back to the limit cycle. This section shows that phase space representations (or "phase portraits") elucidate the universal role of phase response curves. For simplicity we assume here a circular limit cycle and fast radial relaxation of pulse-like perturbations. The corresponding mathematical model is described in Appendix II. Figure 1.5 shows a stable limit cycle with counterclockwise rotation. Relatively small perturbations (horizontal arrows) relax quickly to the limit cycle (dashed arrows). Perturbations in the upper half of the cycle, that is, for $\phi \in (0, 0.5)$, lead to delay of the phase, whereas perturbations in the lower part advance the phase. The corresponding PRC in Fig. 1.5C has similarities to the PRC in Fig. 1.3A.

The lower part of Fig. 1.5 shows the effect of large perturbations. For small ϕ we get again a pronounced phase delay. The situation changes drastically around $\phi = 0.5$ at the left side of the limit cycle.

Perturbations at phases slightly above $\phi = 0.5$ induce a strong phase advance and thus the PRC exhibits a large discontinuity at $\phi = 0.5$ (see Fig. 1.5F). These phase portraits reveal that we can expect two different types of PRCs if we vary the perturbation strength. Small continuous PRCs, such as in Fig. 1.5C, are termed type 1 PRCs, whereas discontinuous PRCs as in Fig. 1.5F represent type 0 PRCs (Winfree, 1980). Both types of PRCs are observed in the field of circadian clocks (Pittendrigh and Daan, 1976), in electrically stimulated heart cells (Guevara *et al.*, 1986), and in insect communication (Hartbauer *et al.*, 2005; Sismondo, 1990). The phase portraits in Figure 1.5 demonstrate that both types of PRCs can be expected generically if the strength of the perturbation is varied. The phase space analysis also predicts another interesting feature: Specific perturbations pointing to the midpoint of our radial-symmetric limit cycle might lead to a vanishing amplitude and an undefined phase. Such "phase singularities" have been predicted by Winfree (1980) and were discussed in cardiology (Jalife and Antzelevitch, 1979) and circadian rhythms (Ukai *et al.*, 2007).

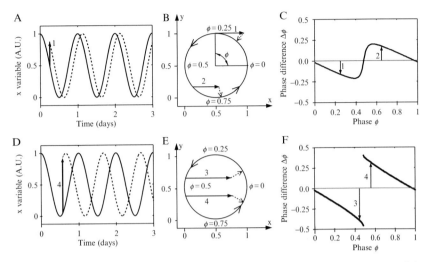

Figure 1.5 Strong and small perturbations in the time domain, in phase space, and the resulting PRC types. (A) Time series of an unperturbed oscillation (solid line) and a perturbed one (dashed line) delayed due to a small perturbation (arrow 1). (B) Two small perturbations (arrows 1 and 2) and the radial relaxation back to the limit cycle (dashed arrows) and (C) the corresponding type 1 PRC with the phase delay (arrow 1) and advance (arrow 2) resulting from small perturbations of the limit cycle. (D) Unperturbed (solid) and perturbed time series (dashed) with a strong advance (arrow 4). (E) Two strong perturbations (arrows 3 and 4) with large delay (3) and advance (4), and (F) the associated type 0 PRC with a discontinuity at $\phi = 0.5$ from delay (3) to advance (4). The corresponding mathematical model is described in Appendix II.

2. ESTIMATION OF PHASE RESPONSE CURVES

2.1. Definitions

A phase response curve is obtained by systematically applying the same perturbation at different phases ϕ and measuring the resulting phase shifts $\Delta\phi$. The PRC is the plot of all those phase shifts against the phase ϕ at which each perturbation was applied. Before going further into the experimental protocols of how to obtain a PRC, it is necessary to clarify some associated concepts. The oscillatory system under study should have at least one variable from which one can obtain a time series (see Fig. 1.2A) with its amplitude A_0, phase ϕ, and free running period T_0. The amplitude is considered here as the distance between the maximum and the minimum values during a complete unperturbed cycle $A_0 = A_{max} - A_{min}$. The phase of an oscillator can be defined in many equivalent ways and here we follow the definition given by Winfree (1980): the elapsed time measured from a reference divided by the intrinsic period. Typically the phase ϕ is normalized in a periodic way between 0 and 1 as follows:

$$\phi = \frac{\text{time } t \text{ since marker event}}{\text{intrinsic period}} \text{ modulo } 1. \qquad (1.6)$$

In practice, this can be done by considering two consecutive unperturbed peak times t_1 and t_2 with t_1 corresponding to $\phi = 0$ and t_2 to $\phi = 1$. So actually the phase ϕ is a fraction of the period $T = t_2 - t_1$. The pulse perturbation is characterized by its strength P_s and duration P_d. In the circadian example of Fig. 1.2A, the X variable has $T_0 = 1$ day and $A_0 = 1$.

2.2. A skeleton protocol—a minimal phase response curve recipe

We present here a general protocol to obtain a phase response curve pointing to similarities between different approaches. We describe a basic four-step PRC recipe and illustrate it with examples from different fields, including circadian rhythms, heart dynamics, neuronal spikes, central pattern generators, and insect communication.

1. Characterize the oscillator: Select an oscillatory output of interest from which the unperturbed amplitude A_0 and period T_0 can be measured. The unperturbed situation can be achieved by uncoupling the system from its external inputs. Note that the resulting PRC will be associated with the selected output variable and not necessarily with the oscillator system as a whole.
2. Specify the perturbation: The common approach is a short (relative to the period T_0) pulse-like perturbation. Choose a perturbation of interest and define a certain pulse strength P_s and duration P_d. Generally, systems can be perturbed in different ways, and most of these perturbations may result in a phase change of the selected output variable. The perturbation to use should mainly depend on the information we want to extract from the PRC. The resulting PRC will be associated with this particular perturbation. Some discussions of the selection of appropriate perturbations can be found in Oprisan et al. (2003).
3. Measure phase changes: The same perturbation must be applied at different phases ϕ leading to phase differences $\Delta\phi$ between the perturbed and the unperturbed systems. In Fig. 1.5 we assumed an almost immediate relaxation of the perturbation leading to certain phase shifts. In many systems, however, the relaxation of perturbations may last several cycles. In dynamical systems theory relaxations to attractors such as limit cycles are termed transients. Depending on transients in the system under study, the phase of the perturbed system may be determined by the immediate peak after the perturbation or several cycles later.

4. Characterize the phase response curve: Once we have all the phase changes $\Delta\phi$ generated from the applied perturbations at different phases ϕ we are ready to plot the PRC with $\Delta\phi$ on the y axis, to quantify phase advances and delays, and to classify the PRC (e.g., type 1 or type 0).

As illustrative examples we describe here these four steps applied in different fields. Five applications are discussed: (a) circadian clocks, (b) heart rhythms, (c) spike dynamics, (d) central pattern generators (CPGs), and (e) insect communication.

1. Characterize the oscillatory system:
   (a) Circadian clocks: Circadian rhythms can be observed in biochemical, physiological, or behavioral processes. The period T_0 is roughly 24 h. Often rodents are used as model organisms and their locomotor activity in a running wheel is measured as an output variable. In order to characterize the unperturbed oscillator the animals are kept in constant conditions, such as constant darkness, and are housed in isolated compartments with a monitorized running wheel. The revolutions of the running wheel are counted, and phases ϕ and periods T of locomotor activity are derived. For a review of PRCs in the field of circadian clocks, see Johnson (1999).
   (b) Heart rhythms: Normal heart rhythms are triggered by pacemaker cells of the sinus node. The period T_0 is typically in the order of 1 s and can be regulated by autonomous neuronal activity. Human electrocardiograms measure heart activity directly. In model systems such as embryonic chicken cell aggregates (Guevara et al., 1981), intracellular electrodes record transmembrane potentials with an amplitude A_0 of about 100 mV.
   (c) Spike dynamics: The membrane potential of individual neurons can be used as an output variable with a period T_0 ranging from a few to hundreds of milliseconds. Spikes are measured from brain slices (Netoff et al., 2005; Reyes and Fetz, 1993b; Tateno and Robinson, 2007) or via chronical electrode implantation for in vivo recordings (Velazquez et al., 2007). Current clamp and dynamic clamp are often used as measuring method. Current perturbations are applied in order to investigate the properties of these oscillatory circuits. Frequently, neurons do not exhibit spontaneously sustained and stable spiking frequency and thus the spiking activity may be controlled by an externally injected constant current. Once the quality of the resulting periodic spiking is tested, the perturbation session can start. Typical values of periods are $T_0 = 10$ to 50 ms with amplitudes of about $A_0 = 100$ mV (e.g., in Tateno and Robinson, 2007).
   (d) Central pattern generators: CPGs are neuronal circuits that can generate rhythmic motor patterns in vertebrates and invertebrates such as locomotion, feeding, respiration, and scratching [for reviews,

see Selverston and Moulins (1985) and Marder et al. (2005)]. Because CPGs in the animal kingdom are as numerous as diverse, only a single motor system that is rather well investigated at the cellular level was selected for the present study in order to exemplify measurements of a PRC: locust flight. For this system, detailed studies of the neuronal circuitry underlying pattern generation were performed *in vitro* and *in vivo* and the output variable of the system measured consisted of peripheral wing depressor activity. A characteristic period is $T_0 = 100$ ms (Ausborn et al., 2007).

(e) Insect communication: Many insects use periodic sound signals for long-range communication, with the ultimate goal of attracting sexual partners. To characterize the oscillatory system the signals ("chirps") produced by an undisturbed animal are used as the output variable. The animal (typically a male) is placed in a sound-proof chamber equipped with a microphone and a computer-controlled automatic recording device. It is important to hold the temperature constant, as the chirp period depends strongly on temperature. The periods are then measured for several individuals in order to obtain the interindividual range of intrinsic chirp periods as well as their intraindividual variation. In a particularly well-investigated example, the bush cricket *Mecopoda elongata*, the intrinsic chirp period of solo singers is around $T_0 = 2$ s (Hartbauer et al., 2005; Sismondo, 1990).

2. Select the perturbation:
   (a) Circadian clocks: A variety of perturbations can be used in plants as well as in animals. Temperature, feeding, and light pulses are generally used in systemic approaches. When focusing on the central pacemaker of mammals, the suprachiasmatic nucleus, neurotransmitters, and temperature pulses are mainly applied. In running-wheel experiments, light pulses are usually selected as perturbation stimuli. After releasing the animals in constant conditions the perturbation session starts after around 10 circadian cycles. Light pulses are applied at different phases and approximately 10 cycles are left between each perturbation (Spoelstra et al., 2004). Pulse duration ranges from $P_d = 0.01\ T_0$ (15 min) to $P_d = 0.25\ T_0$ (6 h) and pulse strength is between $P_s = 1$ lux and $P_s = 100$ lux (Comas et al., 2006).
   (b) Heart rhythms: Typically, short current pulses are delivered to heart cells. For example, chicken heart cells have been stimulated with a pulse strength of $P_s = 27$ mA and a pulse duration of $P_d = 0.02\ T_0$ (20 ms)(Glass and Mackey, 1988). Such perturbations might mimic ectopic pacemakers or bursts of vagal activity.
   (c) Spike dynamics: Square current pulses are typical stimuli with a pulse duration in the order of $P_d = 0.2\ T_0$ (0.5 ms) and a pulse strength in the range of $P_s = 0.1$ pA to $P_s = 100$ pA. The strength of the

perturbation is classified into "strong" and "weak" depending on its resulting PRC type.

(d) Central pattern generators: The pulse duration used to investigate this system consisted either of intracellularly applied depolarizing current pulses to interneurons (Robertson and Pearson, 1985) or trains of short electrical stimulation pulses applied to sensory nerve fibers from the tegula at the wing base (Ausborn et al., 2007). The typical duration of a current pulse or a stimulation train (at 220 Hz) was about 30% of the wing beat cycle duration (ca. 100 ms) in a deafferented locust with pulse duration $P_d = $ 25–30 ms and pulse strength $P_s = $ 10 nA.

(e) Insect communication: Normally the most critical maskers—and signals that cause entrainment—are the sounds of other conspecifics. Hence, the species-specific chirp is typically used as the perturbation signal. In the example introduced previously, the chirp as a perturbation signal occupies only about 15% of the chirp period ($P_d = $ 0.3 s). In cases where the duration of the sound chirp is large compared to the chirp period, however, it may not be possible to use the species-specific signal. After a male begins to sing, the disturbance pulses are played back via a loudspeaker at various phases of its chirp rhythm. Depending on the stability of the undisturbed rhythm, a pulse is given every 7th to 12th cycle. The intensity of the perturbation pulse is preferably adjusted as to mimic the sound pressures observed at the typical intermale distances observed in the field. In the example of *M. elongata*, 50, 60, and 70 dB SPL were used (Hartbauer et al., 2005).

3. Measure phase changes:
  (a) Circadian clocks: After each perturbation, a stable new phase typically evolves during several cycles due to the transients observed in circadian systems. Phase changes can be estimated via forward and backward extrapolation using around 10 periods before and after the perturbation (Spoelstra et al., 2004). It should be noted that transients are a dynamical property of many oscillators and can be also found in any other oscillatory system outside circadian clocks.
  (b) Heart rhythms: Phase shifts are measured typically within the same heartbeat in which the pulse is delivered, as the coordinated contraction of the heart resets all membrane properties. Consequently, transients after single pulses are short-lived.
  (c) Spike dynamics: The perturbed phase ϕ is typically taken from the immediate phase after the stimulus and the free running period T_0 is taken from the preceding unperturbed cycle. Trials that show free period standard deviations larger than around 15% are typically discarded from data.

(d) Central pattern generators: Perturbation pulses were applied at different phases during a cycle and the phase shift was calculated for the cycle in which the stimulus was given. Usually the mean period of 5–10 cycles before the perturbation was measured as a reference for the introduced phase shift (Ausborn et al., 2007; Robertson and Pearson, 1985).

(e) Insect communication: Because the sound signals are discrete and highly stereotypic events, the automatic recording device allows one to determine any induced phase shifts with high precision. In *M. elongata*, which exhibits a very low variation of chirp periods, five cycles before the perturbation pulse were used to determine the intrinsic chirp period T_0, and phase shifts were determined for the perturbed cycle and the following one.

4. Characterize the PRC:
   (a) Circadian clocks: Phases are cyclic variables, that is, a phase of $\phi = 0.6$ can be regarded as $\phi = -0.4$ as well. In circadian clocks, phase changes $\Delta\phi$ are often plotted in the phase interval $[-0.5, 0.5]$ (compare Fig. 1.3). This representation can induce artificial discontinuities in the plots (i.e., a continuous phase increase above $\phi = 0.5$ will jump from to $\phi = -0.5$) and have to be distinguished from the real phase jumps of type 0 PRCs. In such cases, monotonically plotted PRCs are preferred where all phases are considered as delays or advances (Johnson, 1999). Because pulses have a certain duration, the assignment of the phase where the pulse is delivered is somewhat arbitrary (Comas et al., 2006).
   (b) Heart rhythms: Current pulses applied directly have strong effects on the heart cell period. For example, the stimulation of spontaneously oscillating Purkinje fiber induced period lengthening of up to 0.3 of the period and period shortening of up to -0.4. The increase of the pulse strength from $P_s = 2.6\ \mu A$ to $P_s = 5.0\ \mu A$ changed the PRC from type 1 to type 0. A comparable transition from type 1 to type 0 PRC was reported during vagal stimulation of the sinus node in dogs (Yang and Levy, 1984). A discontinuous PTC from electrically stimulated chicken cells was used to compare p:q frequency locking of model simulations with experimentally observed rhythms (Glass et al., 1987).
   (c) Spike dynamics: Both monophasic (advance only) or biphasic (advance and delay) PRCs are observed. Most PRCs are of type 1, but strong stimulations can change a type 1 PRC into a type 0 PRC with a distinct discontinuity. To finally plot the PRC, fitting techniques are usually used and different protocols have been suggested (Galán et al., 2005; Netoff et al., 2005; Oprisan et al., 2003).

(d) Central pattern generators: For most interneurons in the circuitry, as well as for the feedback loop from sensory nerves, phase delays and advances can be induced. For some interneurons, only phase delays were observed (Robertson and Pearson, 1985). For interneurons type 1 PRCs and a magnitude of the phase shifts below $\Delta\phi = 0.3$ were reported, whereas for sensory nerve stimulation from the tegula corresponding to a stronger stimulation of the whole network as compared to the current injection into single interneurons, a type 0 PRC was shown and phase shifts up to $\Delta\phi = 0.6$ were reported. In this system, all interneurons capable of phase resetting are known to receive excitatory or inhibitory input originating from the tegula afferents (Wolf and Pearson, 1987). The negative sensory feedback loop of the tegula serves to control the oscillation frequency of the central pattern generator in a flying locust, as frequency is increased only at very low levels of feedback but reduced with increasing activity of the sensory organ (Ausborn et al., 2007).

(e) Insect communication: In *M. elongata*, a sound pulse occurring at a phase between 0.2 and 0.6 results in a phase delay of the disturbed period, whereas in response to a stimulus at a phase between 0.7 and 0.9 results in a phase advance. Remarkably, the following cycle is not at all affected by these disturbing signals. Sound pulses delivered at phases 0.95 to 0.2 do not affect the rhythm markedly. The PRC of this insect exhibits an exponentially increasing left branch (with phase shifts up to $\Delta\phi = 0.4$) and a nearly linear right branch (up to $\Delta\phi = -0.2$) and has been classified as type 0 PRC (Sismondo, 1990). However, at low intensities the responses of some individuals were classified rather as type 1 PRC. The PRC has been successfully used to predict behavioral interactions between two or more males that call at different distances y (see later).

It is important to note that the measured PRCs correspond to the selected output variable and the specific perturbation. PRC information can be generalized for other variables and perturbations but this depends strongly on the system under study.

3. Specific Applications

3.1. Nonlinear dynamics of the heart

What is the use of PRCs in heart physiology? Glass and Mackey (1988) provide an excellent introductory review on physiological rhythms and mathematic modeling. This monograph contains numerous examples of PRCs and their applications. It was shown that iterated maps provide insight to cardiac arrhythmias. For example, the interaction of sinus and ectopic pacemakers can be modeled using phase transition curves (Moe et al., 1977). The resulting complex entrainment ratios 2:1, 3:1, and so on are termed bigeminy or trigeminy in the clinical context. Another application of iterated maps is arrhythmias due to a pathological prolongation of atrioventricular conduction delay (Glass and Hunter, 1991). For a normal conduction delay the sinus and the atrioventricular node are entrained in a 1:1 rhythm. Long delays lead to a higher order p:q synchronization clinically known as Wenckeback rhythms. The baroreceptor reflex couples heart rhythms, respiration, and blood pressure regulation. Seidel et al. (1997) studied cardiac responses of brief (350 ms) neck suction pulses in healthy human subjects. Phase response curves revealed that respiratory neurons modulate the reflex independent of the cardiac phase.

Our phase space analysis indicated that specific perturbations might lead to a vanishing amplitude and an undefined phase. It was suggested by Winfree (1980) that this "singularity" may be associated with sudden cardiac death. Indeed, it was shown that brief depolarization pulses can annihilate spontaneous activity of the sinus node (Jalife and Antzelevitch, 1979).

These examples illustrate that the phase response curve can help in understanding complex interactions of physiological rhythms such as heartbeat and respiration.

3.2. Classifications of neurons

Synchronized activity in neural networks is known to be the base for coding and storing information in the brain (Rieke et al., 1997). Models in which just the phase of the oscillator is considered (Hoppensteadt and Izhikevich, 1997; Kuramoto, 2003) proved to be suitable for predicting many of the observed phase-locking regimes and more complex dynamics (Pikovsky et al., 2001). It should be noted that these so-called phase models can be used safely when the considered perturbations have a negligible effect on the neural oscillator amplitude and period (weak perturbations); in that case the PRC is known as an infinitesimal phase response curve, which is a subclass of type 1 PRCs.

Hodgkin (1948) classified periodically spiking neurons into two groups depending on their frequency response to injected currents (excitability class I and class II). Later these two classes have been associated with dynamical properties of oscillators (Rinzel and Ermentrout, 1989), and under certain conditions these association can be extended to the specific PRC shape (Hansel et al., 1995). Monotonic PRCs (only phase advance or only phase delay) can be associated with class I neurons and biphasic PRCs (phase advance and delay) can be associated with class II neurons. One interesting application of this theory has been published by Velazquez et al. (2007) where the authors measured PRCs and used them to predict how mitral cells of the mice olfactory bulb achieve synchronization. Furthermore, Velazquez et al. (2007) classified neurons into class I and class II according to their specific PRC shape. Class I and II neurons should not be confused with type 0 and 1 PRCs.

3.3. Classification of central pattern generators

Typically, CPGs integrate sensory information to produce a meaningful motor output, but rhythm generation by a CPG circuit is possible without feedback from the periphery and can therefore, in most cases, be studied *in vitro*. Known CPGs can roughly be classified according to the following schemes: (1) pattern generating networks that are driven by continuous excitation or started by trigger neurons (vertebrate walking, lamprey swimming, locust flight, leech swimming) and reset or modify their rhythmic output upon sensory stimulation; (2) pattern generating networks that are driven by a pacemaker cell intrinsic to the CPG (stomatogastric rhythm in crabs, heartbeat leech, insects); and (3) coupled pattern generating networks that act in different segments (leech), body parts (legs in insects), or serve different functions (bats: call, wing beat and respiration, vertebrates: running and respiration, crickets: singing and respiration). In *sensu strictu* only the latter class of CPGs represents examples of coupled oscillators, but the source of coupling varies greatly among these systems (sensory feedback in stick insect walking, central coupling in crayfish swimmeret beating, central and mechanical coupling in leech swimming). PRCs were determined most commonly for the first and second classes in order to elucidate the principles of rhythm generation in these neuronal circuits [for further references, see Selverston and Moulins (1985) and Marder et al. (2005)].

3.4. Consequences of entrainment for insect communication

In acoustically communicating insects, the message is often conveyed by the specific time pattern of the signal, which is produced by a central pattern generator. In dense aggregations, however, signals from different individuals may mask the specific time pattern, which is crucial for signal recognition by a potential mate. Several insect species relieved this masking problem by synchronizing or alternating their signals. Synchrony or alternating requires a (reciprocal) entraining of the CPGs of different individuals, and the features of this entraining have been investigated in quite some detail in some grasshopper and bush cricket species. In particular, PRC responses have been used to predict entrainment by periodical stimuli in *M. elongata* (Hartbauer et al., 2005; Sismondo, 1990). In this species, the chirp periods are very stable within individuals, and standard deviation is around 40 ms (i.e., only 2%). In contrast, within a population, different individuals exhibit a large range of intrinsic periods, between 1.5 and 3 s. Depending on the ratio between the chirp periods of the individual oscillators, synchronous signaling, bistable alternation, or toroidal oscillation with permanently changing chirp periods have been obtained in model calculations, and these types of coupling have also been observed in real interactions among males (Hartbauer *et al.*, 2005; Sismondo, 1990). Remarkably, the intensity dependence of the slope of the PRC can be used to predict the outcome of entrainment interactions: at low sound intensities, the slope of the PRC becomes less steep and the transition point between phase delay and advance moves to a higher phase value (Hartbauer *et al.*, 2005). As a consequence, the model predicts that males interacting at greater distances are more likely to alternate their calls (phase around 0.5). Exactly this has been observed in field studies (Sismondo, 1990). Usually, singing males compete for attracting receptive females. In most investigated cases the females show a preference for the leading signaler. Thus we expect strong male–male competition for the leader role (Greenfield, 1994; Greenfield *et al.*, 1997). Indeed, frequent changes of leader and follower roles have been reported in interactions between males of another bush cricket species, *Neoconocephalus spiza* (Greenfield and Roizen, 1993). Interestingly, a different implementation of the entrainment seems to be realized than in *N. spiza*, for which an inhibitory resetting model has been proposed (Greenfield and Roizen, 1993; Hartbauer *et al.*, 2005). Usually it is assumed that the onset of a disturbing sound pulse resets the CPG oscillator to its basal level. In *N. spiza*, evidence was found that the sound pulse exerts a tonic inhibition over its whole duration. This mechanism, however, cannot be realized in *M. elongata*, as here we find partial overlaps between disturbing stimulus and chirps produced in response (Hartbauer *et al.*, 2005).

4. Discussion

Oscillatory dynamics is an essential element of living systems. Examples are physiological rhythms, neuronal oscillations, and the circadian clock. From a dynamical systems point of view, self-sustained oscillations are limit cycles due to intrinsic nonlinearities and delays of regulatory systems. Interacting oscillators give rise to highly complex dynamics, including p:q frequency locking, toroidal oscillations, and deterministic chaos. It was argued in this chapter that phase response curves are helpful in characterizing oscillators and in predicting the dynamics of coupled oscillators. We have shown that a four-step recipe to estimate PRCs applies to seemingly totally different oscillators: circadian clock, heart beats, neuronal spikes, central pattern generators, and insect communication. Thus PRCs provide a unifying concept to study these biological systems from a common perspective. The comparison revealed many similarities of the approaches: In most cases, perturbation durations P_d were a few percent of the period T_0 and often an increase of the perturbation strength P_s transformed type 1 into type 0 phase response curves. The associated phase transition curves can be regarded as iterated maps, such as the well-studied sine map. From such a model, effects of periodic stimuli such as stable entrainment or alternation might be predicted.

The predictive power of PTCs is limited if subsequent pulses cannot be considered as independent due to transients. In such cases one might extend PTCs to two-dimensional models describing amplitude and phase changes (Kunysz et al., 1995; Schuster, 1988). Alternatively, time-continuous models as our Eqs. (1.1)–(1.3) can describe the complexity of coupled oscillators. However, Figs. 1.2–1.5 demonstrate that many phenomena found in systems of differential equations can be traced back to simple models based on phase response curves.

A striking feature of phase response curves is their universality. Despite enormous differences of timescales (milliseconds to days) and a variety of oscillation-generating mechanisms (pacemaker cells, inhibitory neurons, negative transcriptional feedback loops), many properties of PRCs and coupled oscillators appear quite similar. For example, many type 1 PRCs resemble each other (see Figs. 1.1, 1.3, and 1.5), and Arnold tongues as shown in Fig. 1.4 have been found in many biological systems (Glass and Mackey, 1988). This universality implies that an observed PRC gives no direct information on mechanistic details, as different mechanisms result in comparable PRCs. Thus PRCs alone cannot reveal the underlying genetic, neuronal, or physiological regulations. Fortunately, PRCs provide predictions on the dynamical properties of oscillators. For example, measuring a type 0 PRC suggests that a reduction of the pulse strength should lead to a type 1 PRC, and specific perturbations might lead to a phase singularity (Winfree, 1980). An even more powerful prediction concerns bifurcations of coupled oscillators. As illustrated nicely in Guevara et al. (1981) and Hartbauer et al. (2005), the measurement of PRCs leads to quantitative predictions of p:q frequency locking and alternations. In this sense, phase response curves elucidate the dynamics of coupled oscillators.

APPENDIX I

Since the dynamics of oscillators has been developed in several fields more or less independently, the terminology appears sometimes confusing. Thus this appendix provides a collection of widely used terms embedded into the framework of nonlinear dynamics theory.

Dynamical system: System in which the time dependencies are crucial. This includes models (differential equations, iterated maps) and time-series data.

Phase space: Coordinate system spanned by the essential time-dependent variables of the dynamical system. Examples are chemical concentrations [compare Eqs. (1.1)–(1.3)] or voltage, conductivities, and current in case of membrane (de)polarization.

Attractor: Attractive asymptotic state of a dynamical system assuming stationary external conditions. Known attractor types are stable steady states ("homeostasis"), limit cycles (self-sustained oscillations), tori (superposition of independent oscillations), and deterministic chaos (intrinsic irregular dynamics).

Transient: Relaxation dynamics to the attractor. Often linear stability analysis can quantify the relaxation rate.

Bifurcations: Qualitative changes of the dynamics due to variation of external parameters. An example is the onset of oscillations (Hopf bifurcation) due to external constant currents.

Synchronization: Coordinated temporal dynamics of interacting oscillators. Related terms are frequency locking, phase locking, entrainment, absolute coordination, and Arnold tongues.

Phase response curve types: Type 1—Continuous dependence of phase shift $\Delta\phi$ on phase ϕ at which pulse has been applied. Type 0—Discontinuous PRC. "0" and "1" refer to the average slope of the associated phase transition curve.

Phase response curve classes: Class I and class II neurons were introduced by Hodgkin (1948) to characterize neurons response to external injected currents. Often the corresponding PRCs are monophasic (only advance) or biphasic (advance and delay) upon current stimulation. This classification should not be confused with type 0 and type 1 PRCs.

Isochrones: As illustrated in Fig. 1.5, perturbations relax after some transients to the limit cycle exhibiting a shifted phase. This resulting new phase can be assigned to the end point of the perturbation. Lines in phase space leading to the same asymptotic phase are termed isochrones (Winfree, 1980). Depending on the oscillator under study, the calculation of isochrones requires some effort but from these PRCs can be deduced easily for various perturbations.

Geometric phase and temporal phase: The phase defined in Section II refers to the relative elapsed time within a cycle and is known as temporal phase. The geometric phase can be defined as the angle of rotation around the center of the limit cycle. Note that both phases differ considerably for systems with fast and slow dynamics ("relaxation oscillators").

Dead zones: It has been observed frequently that pulses have almost no effect during certain phases. An example is the reduced sensitivity of circadian oscillators to light during the day. Such phase intervals have been termed dead zones. Insensitivities during dead zones may be related to adaptation, saturation, or gating.

Negative feedback oscillators: Weak negative feedbacks tend to stabilize steady states. Strong and delayed feedbacks, however, may induce self-sustained oscillation. This design principle applies to transcriptional oscillators and to neuronal networks with inhibitory connections.

APPENDIX II

Numerical Methods

The numerical simulations used to generate Figs. 1.2, 1.3, and 1.4 make use of two mathematical models of oscillators: a modified version of the Goodwin oscillator (Goodwin, 1965), termed the Gonze model (Gonze et al., 2005), and a variation of the Poincaré oscillator (Winfree, 1980) described by Eq. (1.7). The Goodwin model was created to describe a genetic feedback regulation system (Goodwin, 1965) and Gonze et al. (2005) modified it to simulate circadian rhythms in mammalian cells [see Eqs. (1.1)–(1.3)]. Equation (1.8) represents the Gonze model with an additional coupling term proportional to the coupling strength C_s and an external periodic force proportional to E_s.

$$\frac{dr}{dt} = -\gamma(r - 1) \tag{1.7a}$$

$$\frac{d\phi}{dt} = \frac{2\pi}{24} \tag{1.7b}$$

The variable r in Eq. (1.7) represents the radial dynamics, ϕ is the phase dynamics, and γ is the radial relaxation rate. The Gonze model was used with the same parameters values used by Gonze et al. (2005) with amplitude normalized between 0 and 1. All models and programs were implemented in both programming languages MatLab and Bad Ermentrout's XPPAUT (Ermentrout, 2003) with the use of Rob Clewley's XPP-MatLab interface.

$$\frac{dX_i}{dt} = v_1 \frac{K_1^4}{K_1^4 + Z_i^4} - v_2 \frac{X_i}{K_2 + X_i} + C_s(X_j - X_i) + E_s\left(\sin\left(\frac{2\pi}{T_e}t\right)\right) \tag{1.8a}$$

$$\frac{dY_i}{dt} = v_3 X_i - v_4 \frac{Y_i}{K_4 + Y_i} \tag{1.8b}$$

$$\frac{dZ_i}{dt} = v_5 Y_i - v_6 \frac{Z_i}{K_6 + Z_i} \tag{1.8c}$$

Computational Details

Figure 1.2A: Time series of the Gonze model in the uncloupled case ($C_s = 0$) and without entrainment ($E_s = 0$)[see Eq. (1.8)]. Figure 1.2B: Two coupled Gonze oscillators rescaled to have different autonomous periods (23.5 and 24.5 h) coupled through diffusive coupling in the X coordinate with a coupling strength $C_s = 0.5$ and without entrainment ($E_s = 0$). Figure 1.2C: Obtained by entraining the Gonze oscillator to a 24-h rhythm with an additive sinusoidal forcing term in the X variable with strength $E_s = 0.125$ and without coupling ($C_s = 0$). Figure 1.3: A phase response curve (PRC) and phase transition curve (PTC) of the Gonze oscillator obtained by applying square pulses with duration $P_d = 1$ h and pulse strength $P_s = 0.35$ without coupling ($C_s = 0$) and entrainment ($E_s = 0$). The new phase was measured 10 cycles after each perturbation and the pulse onset was considered the perturbation phase. Figure 1.4A: Entrainment regions for the Gonze model of circadian rhythms. The Gonze oscillator was entrained within a period range T_e from 12 to 40 h (frequency ratio from 0.5 to 1.7) and entrainment strength E_s from 0 to 1. Frequencies were considered to be locked if, for six random initial conditions, their phase difference was smaller than 10 min during at least five cycles. Figure 1.4B: Entrainment regions for the sine map [see Eq. (1.5)] within a frequency ratio of 0.5 to 1.7 and entrainment strength k from 0 to 1. Figure 1.5: Strong and small perturbations to the Poincaré oscillator [see Eqs. (1.7)] and its associated limit cycle representation and phase response curves. For all simulations in Fig. 1.5, 1 h pulses were used and a relaxation rate $\gamma = 10$. The phase response curves were calculated following the same procedure as for Fig. 1.3.

ACKNOWLEDGMENTS

We thank Steven Brown for providing data and Jan Benda, Marian Comas, Manfred Hartbauer, and Leon Glass for stimulating discussions. This work was supported by the Deutsche Forschungsgemeinschaft (SFB 618). Research in Achim Kramer's laboratory is supported by the 6th EU framework program EUCLOCK.

REFERENCES

Alexander, R., and Moore, T. (1962). The evolutionary relationships of 17-year and 13-year cicadas, and three new species (Homoptera, Cicadidae, Magicicada). *Miscell. Pub. Museum Zool. Michigan* **121**, 1–59.

Ausborn, J., Stein, W., and Wolf, H. (2007). Frequency control of motor patterning by negative sensory feedback. *J. Neurosci.* **27**, 9319–9328.

Berge, P., Pomeau, Y., and Vidal, C. (1984). "Order within Chaos: Towards a Deterministic Approach to Turbulence." Wiley, New York.

Brown, S. A., Fleury-Olela, F., Nagoshi, E., Hauser, C., Juge, C., Meier, C. A., Chicheportiche, R., Dayer, J. M., Albrecht, U., and Schibler, U. (2005). The period length of fibroblast circadian gene expression varies widely among human individuals. *PLoS Biol.* **3**, e338.

Brown, S. A., Kunz, D., Dumas, A., Westermark, P. O., Vanselow, K., Tilmann-Wahnschaffe, A., Herzel, H., and Kramer, A. (2008). Molecular insights into human daily behavior. *Proc. Natl. Acad. Sci. USA* **105**, 1602–1607.

Buck, J. (1988). Synchronous rhythmic flashing of fireflies. II. *Q. Rev. Biol.* **63**, 265–289.

Comas, M., Beersma, D. G. M., Spoelstra, K., and Daan, S. (2006). Phase and period responses of the circadian system of mice (*Mus musculus*) to light stimuli of different duration. *J. Biol. Rhythms* **21**, 362–372.

Ermentrout, B. (2003). Simulating, analyzing, and animating dynamical systems: A guide to XPPAUT for researchers and students. *Appl. Mech. Rev.* **56**, B53.

Galán, R. F., Ermentrout, G. B., and Urban, N. N. (2005). Efficient estimation of phase-resetting curves in real neurons and its significance for neural-network modeling. *Phys. Rev. Lett.* **94**, 158101.

Glass, L., Guevara, M. R., and Shrier, A. (1987). Universal bifurcations and the classification of cardiac arrhythmias. *Ann. N.Y. Acad. Sci.* **504**, 168–178.

Glass, L., and Hunter, P. (1991). *In* "Theory of Heart: Biomechanics, Biophysics, and Nonlinear Dynamics of Cardiac Function" (A. McCulloch, ed.). Springer-Verlag, New York.

Glass, L., and Mackey, M. M. (1988). "From Clocks to Chaos: The Rhythms of Life." Princeton University Press, Princeton, NJ.

Gonze, D., Bernard, S., Waltermann, C., Kramer, A., and Herzel, H. (2005). Spontaneous synchronization of coupled circadian oscillators. *Biophys. J.* **89**, 120–129.

Goodwin, B. C. (1965). Oscillatory behavior in enzymatic control processes. *Adv. Enzyme Regul.* **3**, 425–438.

Greenfield, M. (1994). Cooperation and conflict in the evolution of signal interactions. *Annu. Rev. Ecol. Syst.* **25**, 97–126.

Greenfield, M., and Roizen, I. (1993). Katydid synchronbous chorusing is an evolutionary stable outcome of female choice. *Nature* **364**, 618–620.

Greenfield, M., Tourtellot, M., and Snedden, W. (1997). Precedence effects and the evolution of chorusing. *Proc. Roy. Soc. Lond. B* **264**, 1355–1361.

Guevara, M. R., Glass, L., and Shrier, A. (1981). Phase locking, period-doubling bifurcations, and irregular dynamics in periodically stimulated cardiac cells. *Science* **214**, 1350–1353.

Guevara, M. R., Shrier, A., and Glass, L. (1986). Phase resetting of spontaneously beating embryonic ventricular heart cell aggregates. *Am. J. Physiol.* **251**, H1298–H1305.

Hansel, D., Mato, G., and Meunier, C. (1995). Synchrony in excitatory neural networks. *Neural. Comput.* **7**, 307–337.

Hartbauer, M., Kratzer, S., Steiner, K., and Römer, H. (2005). Mechanisms for synchrony and alternation in song interactions of the bushcricket *Mecopoda elongata* (Tettigoniidae: Orthoptera). *J. Comp. Physiol. A Neuroethol. Sens. Neural Behav. Physiol.* **191**, 175–188.

Hastings, M. H., Reddy, A. B., and Maywood, E. S. (2003). A clockwork web: Circadian timing in brain and periphery, in health and disease. *Nat. Rev. Neurosci.* **4**, 649–661.

Herzel, H., Berry, D., Titze, I. R., and Saleh, M. (1994). Analysis of vocal disorders with methods from nonlinear dynamics. *J. Speech Hear. Res.* **37**, 1008–1019.

Hodgkin, A. (1948). The local electric charges associated with repetitive action in a non-medullated axon. *J. Physiol. (London)* **107**, 165–181.

Honma, S., Shirakawa, T., Katsuno, Y., Namihira, M., and Honma, K. (1998). Circadian periods of single suprachiasmatic neurons in rats. *Neurosci. Lett.* **250**, 157–160.

Hopfield, J. J., and Herz, A. V. (1995). Rapid local synchronization of action potentials: Toward computation with coupled integrate-and-fire neurons. *Proc. Natl. Acad. Sci. USA* **92**, 6655–6662.

Hoppensteadt, F., and Izhikevich, E. (1997). "Weakly Connected Neural Networks." Springer-Verlag, New York.
Huygens, C. (1673). "Horologium Oscillatorium." Apud F. Muguet, Paris, France.
Jalife, J., and Antzelevitch, C. (1979). Phase resetting and annihilation of pacemaker activity in cardiac tissue. *Science* **206**, 695–697.
Johnson, C. H. (1999). Forty years of PRCs: What have we learned? *Chronobiol. Int.* **16**, 711–743.
Kaplan, D., and Glass, L. (1995). "Understanding Nonlinear Dynamics." Springer-Verlag, New York.
Kunysz, A., Glass, L., and Shrier, A. (1995). Overdrive suppression of spontaneously beating chick heart cell aggregates: Experiment and theory. *Am. J. Physiol.* **269**, H1153–H1164.
Kuramoto, Y. (2003). "Chemical Oscillations, Waves, and Turbulence." Dover, New York.
Marder, E., Bucher, D., Schulz, D. J., and Taylor, A. L. (2005). Invertebrate central pattern generation moves along. *Curr. Biol.* **15**, R685–R699.
Mergell, P., Herzel, H., and Titze, I. R. (2000). Irregular vocal-fold vibration–high-speed observation and modeling. *J. Acoust. Soc. Am.* **108**, 2996–3002.
Moe, G. K., Jalife, J., Mueller, W. J., and Moe, B. (1977). A mathematical model of parasystole and its application to clinical arrhythmias. *Circulation* **56**, 968–979.
Mrosovsky, N. (1977). Strategies in cold: Natural torpidity and thermogenesis. *In* "Circannual Cycles in Hibernators," Vol. 1, pp. 21–65. Academic Press, New York.
Netoff, T. I., Banks, M. I., Dorval, A. D., Acker, C. D., Haas, J. S., Kopell, N., and White, J. A. (2005). Synchronization in hybrid neuronal networks of the hippocampal formation. *J. Neurophysiol.* **93**, 1197–1208.
Oprisan, S. A., Thirumalai, V., and Canavier, C. C. (2003). Dynamics from a time series: Can we extract the phase resetting curve from a time series? *Biophys. J.* **84**, 2919–2928.
Pikovsky, A., Rosenblum, M., and Kurths, J. (2001). Synchronization: A Universal Concept in Nonlinear Sciences Cambridge University Press, Cambridge.
Pittendrigh, C., and Daan, S. (1976). The entrainment of circadian pacemakers in nocturnal rodents. IV. Entrainment: Pacemaker as clock. *J. Comp. Physiol. A.* **106**, 291–331.
Reyes, A. D., and Fetz, E. E. (1993a). Effects of transient depolarizing potentials on the firing rate of cat neocortical neurons. *J. Neurophysiol.* **69**, 1673–1683.
Reyes, A. D., and Fetz, E. E. (1993b). Two modes of interspike interval shortening by brief transient depolarizations in cat neocortical neurons. *J. Neurophysiol.* **69**, 1661–1672.
Rieke, F., Warland, D., de Ruyter van Steveninck, R., and Bialek, W. (1997). "Spikes-Exploring the Neural Code." Computational Neurosciences series. MIT Press, Cambridge, MA.
Rinzel, J., and Ermentrout, G. (1989). Analysis of neural excitability and oscillations. *In* "Methods in Neuronal Modeling: From Synapses to Networks" (C. Koch and I. Segev, eds.), pp. 135–169. MIT Press, Cambridge, MA.
Robertson, R. M., and Pearson, K. G. (1985). Neural circuits in the flight system of the locust. *J. Neurophysiol.* **53**, 110–128.
Schuster, H. G. (1988). "Deterministic Chaos." Physik Verlag, Weinheim.
Schäfer, C., Rosenblum, M. G., Kurths, J., and Abel, H. H. (1998). Heartbeat synchronized with ventilation. *Nature* **392**, 239–240.
Seidel, H., and Herzel, H. (1998). Analyzing entrainment of heartbeat and respiration with surrogates. *IEEE Eng. Med. Biol. Mag.* **17**, 54–57.
Seidel, H., Herzel, H., and Eckberg, D. L. (1997). Phase dependencies of the human baroreceptor reflex. *Am. J. Physiol.* **272**, H2040–H2053.
Selverston, A. I., and Moulins, M. (1985). Oscillatory neural networks. *Annu. Rev. Physiol.* **47**, 29–48.
Sismondo, E. (1990). Synchronous, alternating, and phase-locked stridulation by a tropical katydid. *Science* **249**, 55–58.

Solari, H. G., Natiello, M. A., and Mindlin, G. B. (1996). "Nonlinear Dynamics: A Two-Way Trip from Physics to Math." CRC Press, Boca Raton, FL.

Spoelstra, K., Albrecht, U., van der Horst, G. T. J., Brauer, V., and Daan, S. (2004). Phase responses to light pulses in mice lacking functional per or cry genes. *J. Biol. Rhythms* **19**, 518–529.

Tateno, T., and Robinson, H. P. C. (2007). Phase resetting curves and oscillatory stability in interneurons of rat somatosensory cortex. *Biophys. J.* **92**, 683–695.

Ukai, H., Kobayashi, T. J., Nagano, M., Hei Masumoto, K., Sujino, M., Kondo, T., Yagita, K., Shigeyoshi, Y., and Ueda, H. R. (2007). Melanopsin-dependent photoperturbation reveals desynchronization underlying the singularity of mammalian circadian clocks. *Nat. Cell Biol.* **9**, 1327–1334.

Velazquez, J. L. P., Galán, R. F., Dominguez, L. G., Leshchenko, Y., Lo, S., Belkas, J., and Erra, R. G. (2007). Phase response curves in the characterization of epileptiform activity. *Phys. Rev. E Stat. Nonlin. Soft Matter Phys.* **76**, 061912.

von Holst, E. (1939). Die relative Koordination als Phänomen und als Methode zentralnervöser Funktionsanalyse. *Ergebnisse Physiol.* **42**, 228–306.

Welsh, D. K., Logothetis, D. E., Meister, M., and Reppert, S. M. (1995). Individual neurons dissociated from rat suprachiasmatic nucleus express independently phased circadian firing rhythms. *Neuron* **14**, 697–706.

Winfree, A. (1980). "The Geometry of Biological Time." Springer-Verlag, New York.

Wolf, H., and Pearson, K. G. (1987). Intracellular recordings from interneurons and motoneurons in intact flying locusts. *J. Neurosci. Methods* **21**, 345–354.

Yagita, K., Tamanini, F., van Der Horst, G. T., and Okamura, H. (2001). Molecular mechanisms of the biological clock in cultured fibroblasts. *Science* **292**, 278–281.

Yamaguchi, S., Isejima, H., Matsuo, T., Okura, R., Yagita, K., Kobayashi, M., and Okamura, H. (2003). Synchronization of cellular clocks in the suprachiasmatic nucleus. *Science* **302**, 1408–1412.

Yang, T., and Levy, M. N. (1984). The phase-dependency of the cardiac chronotropic responses to vagal stimulation as a factor in sympathetic-vagal interactions. *Circ. Res.* **54**, 703–710.

CHAPTER TWO

MULTIPLE ION BINDING EQUILIBRIA, REACTION KINETICS, AND THERMODYNAMICS IN DYNAMIC MODELS OF BIOCHEMICAL PATHWAYS

Kalyan C. Vinnakota,* Fan Wu,* Martin J. Kushmerick,[†] *and* Daniel A. Beard*

Contents

1. Introduction	30
1.1. The physicochemical basis of biology	30
1.2. Basic principles of chemical thermodynamics	31
1.3. Simulating biochemical systems	32
2. Biochemical Conventions and Calculations	34
2.1. Thermodynamics of biochemical reactions	34
2.2. Enzyme kinetics and the Haldane constraint	35
2.3. Multiple cation equilibria in solutions: Apparent equilibrium constant, buffering of cations, and computing time courses of cations as a consequence of biochemical reaction networks	36
2.4. Multiple cation equilibria in solutions: Proton, magnesium, and potassium ion binding in the creatine kinase reaction	48
2.5. Example of detailed kinetics of a biochemical reaction: Citrate synthase	53
3. Application to Physiological Systems	57
3.1. pH dynamics and glycogenolysis in cell free reconstituted systems and in skeletal muscle	57
3.2. Cardiac energetics	58
3.3. Physiological significance of multiple ion binding equilibria, reaction kinetics, and thermodynamics in cardiac energetics	63

* Biotechnology and Bioengineering Center and Department of Physiology, Medical College of Wisconsin, Milwaukee, Wisconsin
[†] Departments of Radiology, Bioengineering, Physiology and Biophysics, University of Washington, Seattle, Washington

4. Discussion	64
4.1. Integrating thermodynamic information in biochemical model building	64
4.2. Databases of biochemical information	65
Acknowledgment	66
References	66

Abstract

The operation of biochemical systems *in vivo* and *in vitro* is strongly influenced by complex interactions between biochemical reactants and ions such as H^+, Mg^{2+}, K^+, and Ca^{2+}. These are important second messengers in metabolic and signaling pathways that directly influence the kinetics and thermodynamics of biochemical systems. Herein we describe the biophysical theory and computational methods to account for multiple ion binding to biochemical reactants and demonstrate the crucial effects of ion binding on biochemical reaction kinetics and thermodynamics. In simulations of realistic systems, the concentrations of these ions change with time due to dynamic buffering and competitive binding. In turn, the effective thermodynamic properties vary as functions of cation concentrations and important environmental variables such as temperature and overall ionic strength. Physically realistic simulations of biochemical systems require incorporating all of these phenomena into a coherent mathematical description. Several applications to physiological systems are demonstrated based on this coherent simulation framework.

1. INTRODUCTION

1.1. The physicochemical basis of biology

It is essential that modern research in biology and biomedical science, which is increasingly focused on gaining quantitative understanding of the behavior of biochemical systems, be grounded in the physical chemical principles that govern the organization and behavior of matter. In particular, it is crucial to recognize that living systems, which continuously transport and transform material and transduce free energy among chemical, electrical, and mechanical forms, operate in nonequilibrium thermodynamic states. Thus effective characterization of the operation of living systems, through quantitative analysis and computational simulation, must account for the chemical thermodynamics of such systems. From this perspective, a useful characterization of a biological network includes an investigation and appreciation of its thermodynamic properties. Similarly, a realistic simulation of a biological system must be constrained to operate in a thermodynamically feasible manner.

1.2. Basic principles of chemical thermodynamics

The basic laws of thermodynamics are familiar to most readers. The two laws most relevant to systems biology may be stated as follow. (1) Energy is neither created nor destroyed; a net change in energy of a system must be due to a net transport of energy in or out. (2) In systems away from equilibrium, entropy is produced and never consumed; entropy may decrease only if the rate of export exceeds the rate of import.

By the second law, isolated systems move naturally in the direction of increasing entropy, always heading toward a thermodynamic equilibrium where the entropy is maximized. Yet biological systems are not isolated and, in the context of biology, equilibrium can be equated with death. Thus increasing entropy does not provide a gauge to predict or constrain the operation of biological systems that exchange material and energy with their environment. For such systems the concept of *free energy* provides a useful metric for describing thermodynamic status.

Various forms of free energy are defined for various sorts of systems. The most useful form of free energy for biological systems is Gibbs free energy, which is defined

$$G = E + PV - TS, \qquad (2.1)$$

where E, P, V, T, and S, are the internal energy, pressure, volume, temperature, and entropy of the system. Gibbs free energy is a useful metric for biological systems because it can be shown that systems without import or export of chemical substances, held at constant pressure and temperature, spontaneously move down the gradient in G (Beard and Qian, 2008b).

For a chemical reaction in dilute solution the change in Gibbs free energy per number of times the reaction turns over is given by the well-known formula (Alberty, 2003; Beard and Qian, 2008b)

$$\Delta_r G = \Delta_r G^o + RT \sum_{i=1}^{N_s} v_i \ln(C_i/C_o), \qquad (2.2)$$

where $\Delta_r G^o$ is the equilibrium free energy for the reaction, which does not depend on the concentrations of the chemical components, R is the gas constant, T is the absolute temperature, v_i is the stoichiometric number of species i for the reaction, C_i is the concentration of species I, and C_o is the reference concentration, taken to be $1\ M$. The summation in Eq. (2.2) is over all species in the system; N_s is the number of species.

The equilibrium free energy for a reaction may be computed based on the values of free energy of formation, $\Delta_f G_i^o$, for the species involved in a reaction:

$$\Delta_r G^o = \sum_{i=1}^{N_s} \Delta_f G_i^o. \tag{2.3}$$

The value of $\Delta_f G_i^o$ for a given species depends on the environmental conditions, most notably temperature, pressure, and the ionic solution strength, as discussed later. Chemical equilibrium is achieved when the driving force for the reaction $\Delta_r G$ goes to zero. Equilibrium yields

$$-\frac{\Delta_r G^o}{RT} = \sum_{i=1}^{N_s} v_i \ln\left(C_i/C_o\right) = \ln \prod_{i=1}^{N_s} \left(C_i/C_o\right)^{v_i}$$
$$\exp\left(-\frac{\Delta_r G^o}{RT}\right) = K_{eq} = \prod_{i=1}^{N_s} \left(C_i/C_o\right)^{v_i} \tag{2.4}$$

where K_{eq} is the equilibrium constant for the reaction.

1.3. Simulating biochemical systems

In living systems operating away from equilibrium, each chemical process (including reactions and transport processes) obeys the following relationship between flux and free energy:

$$J^+/J^- = e^{-\Delta G/RT}, \tag{2.5}$$

where $J = J^+ - J^-$ is the net flux for the process and J^+ and J^- are the forward and reverse fluxes. For example, for an enzyme-catalyzed reaction, J^+ and J^- are the forward and reverse rates of turnover for the enzyme's catalytic cycle.

Equation (2.5) reveals that nonzero net flux occurs only when the free energy change for a given process is nonzero. Also, it is apparent that the quantity

$$-\Delta G \cdot J = RT \ln\left(J^+/J^-\right) \cdot \left(J^+ - J^-\right) \tag{2.6}$$

is always positive. In fact, $-\Delta G \cdot J$ is the process's rate of free energy dissipation. Thus it follows that systems maintained away from equilibrium dissipate free energy. Dissipation of free energy is a hallmark of nonequilibrium systems and thus a hallmark of life.

At the center of our biochemical systems modeling approach is an explicit accounting of the fact that biochemical reactants (e.g., ATP) exist in solution as a number of rapidly interconverting species (e.g., ATP^{4-}, $HATP^{3-}$, $MgATP^{2-}$, and $KATP^{3-}$)(Beard and Qian, 2008a,c). Accounting for all of these species for systems composed of many (tens to hundreds) of reactants is theoretically straightforward, but practically messy. Notably, Wu and colleagues (2007b) have detailed how a model of nontrivial complexity is constructed. However, before going into computational details, the following list is a number of benefits of accounting for this level of chemical detail in biochemical systems simulation.

1. Nonambiguous meaning. By tracking the species-level distribution of biochemical reactants, model variables have a nonambiguous physical meaning. Because species participate in enzyme-catalyzed reactions, a species-level model allows us to represent chemical reactions with greater physical realism than is possible when the species distributions of reactants are not calculated. By expressing enzyme mechanisms in terms of species, changes in apparent kinetic properties brought on by changes in pH and ion concentration are handled explicitly. More details on this point are provided in Section 2.5 and in Beard *et al.* (2008).
2. Nonambiguous integration. In biosystems modeling it is common to track certain species but not others. For example, some models of cellular energetics account for Mg^{2+}-bound and -unbound ATP (Korzeniewski, 2001; Vendelin *et al.*, 2000, 2004). However, in these cases the unbound state is a mixture of many states. Other modeling applications may be based on treating all reactants, including ATP, as the total sum of species (Zhou *et al.*, 2005). As a result, there is often not one unique way to integrate models that simulate different components of a larger system. By imposing a formalism where model variables have nonambiguous meaning this problem of nonuniqueness is resolved.
3. Resulting models are thermodynamically feasible. By computing species distribution of reactants, we can track how changes in pH and cationic concentrations affect overall thermodynamic driving forces and ensure that these changes are handled by our models. Thus simulation at this level of detail ensures thermodynamically validity, an important requirement for model reliability.
4. Model uncertainty is reduced. One might imagine that imposing the level of detail outlined previously may introduce unnecessary complexity and increase the number of unknown parameters into biochemical systems models. However, the opposite is true when ionic dissociation constants and basic thermodynamic data are available for the reactions and reactants in a given model. The thermodynamic and dissociation data constrain the enzyme mechanisms and reduce the number of adjustable parameters necessary to describe their behavior.

In addition to these practical advantages related to model building and identification, the major advantage is scientific.

5. Model reliability is improved. We have greater confidence in the behavior predicted by physically grounded models than in models that invoke phenomenological descriptions of components or in models that do not account for chemical species distribution and/or biochemical thermodynamics. This is particularly important when using models to predict behavior outside of the operational regime of the data set(s) used for parameterization and validation.

2. Biochemical Conventions and Calculations

Equation (2.2) is the formula for Gibbs free energy for a chemical reaction. However, biochemical reactions are not typically expressed as mass- and charge-balanced chemical reactions Alberty (2003).

2.1. Thermodynamics of biochemical reactions

Metabolites in intracellular milieu exist as protonated, metal ion bound and free unbound forms described by ionic equilibria in solution with multiple cations. Each metabolite concentration is therefore described as a sum of its individual constituent species concentrations computed by ionic equilibria principles. A biochemical reaction involves sums of species on both the reactant and the product sides, which results in each reaction having effective proton and metal ion stoichiometries due to binding changes across the biochemical reaction. The proton generation stoichiometry of a biochemical reaction is the total difference of average proton binding between the reactants and the products, plus the proton generation stoichiometry of the reaction defined in terms of its most unbound species in the pH range of interest, also known as the reference reaction. The apparent equilibrium constant of a biochemical reaction defined in terms of sums of species for each metabolite therefore depends on proton and metal binding of the reactants and products.

Alberty developed these concepts in terms of transformed variables by using Legendre transforms to gain a global view of biochemical systems by treating pH and free magnesium ion concentrations as independent variables. Fundamental principles underlying this work are reviewed in Alberty (2004). More detailed treatment of this subject matter and basic biochemical data and calculations are given in Alberty (2003). Vinnakota et al. (2006) extended the application of concepts developed by Alberty to a system with finite buffer capacity and variable pH. The pH variation itself is computed from proton binding changes in each of the biochemical reactions. Using mass balance constraint for protons and Mg^{2+} ions, we applied these concepts to compute a pH time course while accounting for pH, Mg^{2+}, and K^+ effects on kinetics and thermodynamics of the reactions. The methodology in Vinnakota *et al.* (2006) was extended and improved by Wu *et al.* (2007b) and Beard and Qian (2008a) to account for more metal ions and was applied to coupled electrophysiological and metabolic systems.

During the past decade, tables of fundamental biochemical data have been generated by Alberty (2003), which enabled the computation of standard free energies of many biochemical reactions in terms of their reference species and therefore their apparent equilibrium constants at specified pH and metal ion concentrations. The methods presented here demonstrate that the thermodynamic analysis and the biochemical network representation are unified through proton and other metal ion mass balance constraints.

2.2. Enzyme kinetics and the Haldane constraint

Thermodynamic analyses predict equilibria for pathways at specified pH, temperature, ionic strength, ion concentrations, and cofactor concentrations. However, this gives only part of the picture, because the kinetics govern the dynamics of approach of biochemical systems to these equilibria in a closed system and in an open system the combined effect of kinetics constrained by thermodynamics govern the fluxes and the metabolite transients. Likewise, kinetic models with inappropriate thermodynamic constraints will not yield a realistic prediction of fluxes and concentrations, as the apparent equilibrium constant can change by orders of magnitude per unit pH depending on the proton stoichiometry of the biochemical reaction. These two approaches can be combined by incorporating the apparent equilibrium constant computed by biochemical thermodynamic principles into the Haldane relationship, which relates the forward and reverse maximal velocities of the enzyme kinetic flux. This unified approach combining both kinetics and thermodynamics is applied in Vinnakota *et al.* (2006), Wu *et al.* (2007b, 2008), and Beard and Qian (2008a). Alberty (2006) has demonstrated that the overall apparent equilibrium constant of a biochemical reaction is independent of the internal enzyme kinetic mechanisms for simple enzyme kinetic mechanisms.

While the study of the catalytic kinetics of enzymes represents one of the most established and well-documented fields in biochemical research, the impact of the biochemical state (pH, ionic strength, temperature, and certain cation concentrations) is typically not formally accounted for in kinetic studies (Alberty, 1993, 2006). *In vitro* experiments using purified proteins and controlled substrate concentrations to characterize enzyme kinetics are conducted under conditions that do not necessarily match the physiological environment, but are determined based on a number of factors, including the requirements of the assays used to measure the kinetics. Therefore, it is difficult to compare results obtained from different studies and to use available kinetic data to predict *in vivo* function without ambiguity.

Outlining these and other issues in somewhat greater detail, the following specific challenges associated with interpreting *in vitro* kinetic data must be overcome to make optimal use of them.

1. While a great deal of high-quality data may be available for a particular enzyme, much of these data were obtained in the 1960s and 1970s when tools for proper analysis of data were not available. As a result, the reported kinetic parameter values [typically obtained from double reciprocal plots of inverse flux versus inverse substrate (Lineweaver and Burk, 1934)] may not optimally match reported data.
2. Data on biochemical kinetics are typically obtained under nonphysiological pH and ionic conditions. Therefore the reported kinetic constants must be corrected to apply to simulations of physiological systems.
3. A third problem related to the second is that kinetic constants are associated with apparent mechanisms that operate on biochemical reactants, which are sums of biochemical species (Alberty, 1993). The result is that the reported mechanisms and associated parameter values are dependent on the biochemical state and not easily translated to apply to different biochemical states or to simulations in which the biochemical state changes.
4. The reported kinetic mechanisms and parameters are often not constrained to match thermodynamic data for a given reaction. Since the basic thermodynamics of a given reaction is typically characterized with greater precision than the kinetics of an enzyme catalyzing the reaction, putative kinetic mechanisms should be constrained to match the biochemical reaction thermodynamics.

A study on citrate synthase addressed and corrected these problems by posing reaction mechanisms in terms of species and ensuring that mechanisms properly account for thermodynamics. This basic approach was first introduced by Frieden and Alberty (1955), yet has received little attention. By reanalyzing legacy data from a variety of sources of kinetic data on citrate synthase, for example, we are able to show that data used to support the consensus model of the mechanism for this enzyme (random bi–bi mechanism) are all consistent with the compulsory-order ternary-complex mechanism and not consistent with the random bi–bi model (Beard *et al.*, 2008).

2.3. Multiple cation equilibria in solutions: Apparent equilibrium constant, buffering of cations, and computing time courses of cations as a consequence of biochemical reaction networks

Most metabolites in physiological milieu are anionic and bind to H^+, Mg^{2+}, K^+, and Ca^{2+} ions in rapid equilibrium. To compute the distribution of a metabolite into ion bound and unbound species, we choose as a reference form the species with no dissociable protons or metal ions in the pH range of 5–9 for physiological models. Based on this definition, the following steps and calculations are needed to account for H$^+$ and metal cation binding to metabolites, buffer capacity and pH change, and the effects of these parameters on reaction equilibria.

1. Based on binding equilibria, the total concentration of a biochemical reactant is defined in terms of its reference species concentration and the free concentrations of ions and their binding affinities.
2. On the basis of the definition of reference and biochemical reactions, the reference and apparent equilibrium constants of each biochemical reaction are computed. The apparent equilibrium constant is a function of the reference equilibrium constant and the binding polynomials of the reactants and the products. The apparent equilibrium constant constrains the enzyme kinetic flux through the Haldane relationship.
3. The flux through each biochemical reaction is derived on the basis of a detailed catalytic scheme. The reference reaction is balanced with respect to mass and charge and has a reference proton stoichiometry.
4. The coupled differential equations for ion and biochemical reactant concentrations are integrated to simulate the system dynamics.

2.3.1. Calculation of proton binding fraction for each metabolite using multiple cation equilibria

A biochemical reactant in physiological milieu exists in different ion bound species in rapid equilibrium. These ions include H$^+$, Mg^{2+}, K$^+$, and other cations. Multiple cation equilibria principles are used to describe the distribution of each biochemical reactant in various ion bound species. For example, the reactant phosphocreatine (PCr) will exist in the following forms:

$$[PCr_{total}] = [HPCr^{2-}] + [H_2PCr^-] + [MgHPCr] + [KHPCr^-]. \tag{2.7}$$

Each ion bound species is in turn computed in terms of an equilibrium relationship between the most unbound form of the reactant and the metal ion that it binds to. Detailed examples are provided in Sections 2.4 and 2.5. Table 2.1 defines the conventions and nomenclature used in our biochemical models.

In general, let [L] represent the concentration of unbound anionic form of biochemical reactant L and [L$_{total}$] be its total concentration in all forms. The total concentration is expressed as

Table 2.1 Terminology

Symbol	Definition	Unit
I	Ionic strength	M
$K_{a,l}$	Dissociation constant for the lth protonation reaction	M
pKa_l	$-\log_{10}(K_{a,l}/c_0)$	Dimensionless
$K_{M,m}$	Dissociation constant for metabolite binding to metal M	M
$pKa_{M,m}$	$-\log_{10}(K_{M,m}/c_0)$	Dimensionless
c_0	Reference concentration for all species (1 M)	M
P_i	Binding polynomial for ith metabolite	Dimensionless
$\bar{N}_H^i$	Average proton binding of ith metabolite at specified T, P, pH, pMg, and I	Dimensionless, noninteger
n_k	Proton generation stoichiometry of the kth reference reaction	Dimensionless
$\Delta_r N_H^k$	Proton generation stoichiometry of the kth biochemical reaction at specified T, P, pH, pMg, and I	Dimensionless, noninteger
$N_H(j)$	Number of H atoms in species j	Dimensionless, integer
z_j	Charge on species j	Dimensionless, integer
$\Delta_f G_j^0 (I=0)$	Gibbs energy of formation at zero ionic strength of species j	kJ/mol
v_j^k	Stoichiometric coefficient of species j in the kth reference reaction	
K_{ref}^k	Equilibrium constant of kth reference reaction	Dimensionless
$\Delta_r G_k^0$	Standard free energy of kth reference reaction	kJ/mol
K_{app}^k	Apparent equilibrium constant of kth biochemical reaction	Dimensionless
$\Delta_r G'^0_k$	Standard free energy of kth biochemical reaction	kJ/mol
$\Delta_r G_k$	Free energy span of kth biochemical reaction	kJ/mol
$\Delta_r H_j^0$	Standard Enthalpy of jth proton dissociation reaction at a given ionic strength	kJ/mol
J_r^k	Flux through reaction kth biochemical reaction	M/min
$[L_i]$	Concentration of biochemical reactant i	M

$$[L_{total}] = [L] + \sum_{p=1}^{N_p}[LH_p] + \sum_{m=1}^{N_m}[LM^m], \qquad (2.8)$$

where $[LH_p]$ is L bound to p $[H^+]$ ions and N_p is the highest number of protons that can bind to L; $[LM^m]$ is L bound to the mth metal ion. The second term on the right refers to the sum of the proton bound forms, and the third term refers to the sum of the metal bound forms. Any number of cations can be included in this scheme if their dissociation constants are known for a given reactant. For simplicity, we assume that at most only one metal ion of each type binds to L. Note that the net charge on each of the species is not stated here for simplicity of notation, but this quantity is equal to the sum of the charge of the reference species and the total charge of the metal ions bound for each species. The concentration $[LH_p]$ is given by the equilibrium relation:

$$[LH_p] = \frac{[L][H^+]^p}{\prod_{l=1}^{p} K_{a,l}}, p \geq 1, \qquad (2.9)$$

where $K_{a,l}$ is the dissociation constant for the reaction, $[LH_l] \rightleftharpoons [LH_{l-1}] + [H^+]$ and $K_{a,l} = \left(\frac{[H^+][LH_{l-1}]}{[LH_l]}\right)_{eq}$ at a given temperature and ionic strength.

For example, for phosphocreatine, $[H_2PCr^-] = \frac{[HPCr^{2-}][H^+]}{K_{a,1}}$.

The dissociation constants [usually expressed as $pKa \stackrel{.}{=} -\log_{10}(K_a)$] are obtained from sources such as the NIST Database 46. Typically, these pK values must be recalculated for specific temperature and ionic strength to be simulated. The following equation can be applied for the temperature correction (Alberty, 2003):

$$pKa_{T2} = pKa_{T1} + \left(\frac{1}{T_2} - \frac{1}{T_1}\right) \cdot \frac{\Delta H^0}{2.303 \cdot R}, \qquad (2.10)$$

where T_1 is the temperature at which the pK was reported in the database, T_2 is the temperature at which pKa_{T2} is calculated, and $R = 8.314$ JK^{-1} mol^{-1} is the universal gas constant.

The enthalpy of dissociation reactions is a function of ionic strength. The following empirical equation (Alberty, 2003) gives ionic strength dependence of enthalpy at 298.15 K:

$$\Delta H^0 = \Delta H^0(I=0) + \frac{1.4775 I^{1/2} \sum v_i z_i^2}{1 + 1.6 I^{1/2}}. \qquad (2.11)$$

The effects of ionic strength on pK can be approximated using the following empirical equation (Alberty, 2003):

$$pKa(I) = pKa(I_1) + \frac{1.17582}{2.303}\left(\frac{I_1^{1/2}}{1+1.6I_1^{1/2}} - \frac{I^{1/2}}{1+1.6I^{1/2}}\right)\sum v_i z_i^2. \quad (2.12)$$

The numerical constants in Eqs. (2.11) and (2.12) are derived from Alberty's fits to experimental data from Clarke and Glew (1980) as described in Alberty (2003). In summary, the pK at a given temperature and ionic strength is corrected for the desired ionic strength using Eq. (2.12). A temperature-corrected value is then obtained from Eq. (2.10) in which the dissociation enthalpy is first corrected for ionic strength using Eq. (2.11).

The concentration of each of the mth metal bound species $[LM^m]$ is given by

$$[LM^m] = \frac{[L][M^m]}{K_{M,m}}, \quad (2.13)$$

where $K_{M,m}$ is the dissociation constant for metal ion $[M^m]$ binding to $[L]$. We assume in Eq. (2.13) that each species binds at most one metal ion per molecule.

The concentration of bound protons due to each bound form $[LH_p]$ with p dissociable protons is given by $p[LH_p]$. The sum total of bound proton concentration is given by $\sum_{p=1}^{N_p} p[LH_p]$ for the biochemical reactant L.

For example, for phosphocreatine, $[MgHPCr] = \frac{[HPCr^{2-}][Mg^{2+}]}{K_{Mg}}$.

The average proton binding for biochemical reactant L is the ratio of bound proton concentration to total concentration of L, which is given by

$$\bar{N}_H^L = \frac{\sum_{p=1}^{N_p} p[LH_p]}{[L_{total}]} = \frac{\sum_{p=1}^{N_p} p[LH_p]}{[L] + \sum_{p=1}^{N_p}[LH_p] + \sum_{m=1}^{N_m}[LM^m]}$$

$$= \frac{\sum_{p=1}^{N_p} \frac{p[L][H]^p}{\prod_{l=1}^{p} K_{a,l}}}{[L] + \sum_{p=1}^{N_p} \frac{[L][H^+]^p}{\prod_{l=1}^{p} K_{a,l}} + \sum_{m=1}^{N_m} \frac{[L][M^m]}{K_{M,m}}} = \frac{\sum_{p=1}^{N_p} \frac{p[H^+]^p}{\prod_{l=1}^{p} K_{a,l}}}{P_L} \quad (2.14)$$

where P_L is defined as the binding polynomial of L.

$$P_L = 1 + \sum_{p=1}^{N_p} \frac{[H^+]^p}{\prod_{l=1}^{p} K_{a,l}} + \sum_{m=1}^{N_m} \frac{[M^m]}{K_{M,m}} \qquad (2.15)$$

For example, the binding polynomial for phosphocreatine would be written as

$$P_L = 1 + \frac{[HPCr^{2-}][H^+]}{K_{a,1}} + \frac{[HPCr^{2-}][Mg^{2+}]}{K_{Mg}} + \frac{[HPCr^{2-}][K^+]}{K_K}. \qquad (2.16)$$

A similar expression can be written for average metal ion binding, where the numerator contains metal ion binding equilibria instead of just proton binding equilibria. The number p in Eqs. (2.8) and (2.14) denotes the number of dissociable protons bound in each proton bound form of the metabolite L. Note that the product of [L] and P_L gives [L_{total}] and that each term the binding polynomial represents the ratio of the unbound, proton bound, and metal ion bound forms of L to the unbound form of L.

2.3.2. Reference and biochemical reactions and the flux through an enzyme catalyzed reaction

We define the reference species of the biochemical reactant as the most deprotonated species in the pH range 5.5–8.5, which are used to calculate the average proton binding $\bar{N}_H^L$ for each biochemical reactant L. As an example the creatine kinase reaction is shown here in terms of its reference species to illustrate this point:

$$HPCr^{2-} + ADP^{3-} + H^+ \rightleftarrows HCr^0 + ATP^{4-}. \qquad (2.17)$$

The biochemical reaction associated with this reference reaction is defined in terms of sums of species constituting each of the reactants and products:

$$PCr_{total} + ADP_{total} \rightleftarrows Cr_{total} + ATP_{total}. \qquad (2.18)$$

The average proton generation stoichiometry, $\Delta_r N_H$ of the biochemical reaction, is the difference between the average proton binding of the reactants and the products plus the proton generation stoichiometry, n, of the reference reaction:

$$\Delta_r N_H = \sum_{\text{reactants}} \bar{N}_H^{\text{reactant}} - \sum_{\text{products}} \bar{N}_H^{\text{product}} + n. \qquad (2.19)$$

Defined in this way, n is an integer but $\Delta_r N_H$ is not because the first two terms on the right-hand side of Eq. (2.19), which are sums of average proton binding of the reactants and the products computed from Eq. (2.14), are nonintegers. Note that the first two terms on the right-hand side of Eq. (2.19) go to zero at highly alkaline pH values. For a system of reactions we define $\Delta_r N_H^k$ and n_k as the proton generation stoichiometries of the biochemical and reference reactions, respectively, for the kth reaction.

The proton generation flux through a reaction is given by the product of the proton stoichiometry for that reaction and the flux through the reaction. For a set of reactions in a system, the total proton generation flux is given by a summation of the individual proton fluxes for all biochemical reactions:

$$\text{Proton flux} = \sum_{k=1}^{N_r} \Delta_r N_H^k J_k, \qquad (2.20)$$

where J_k is the flux through the kth biochemical reaction.

2.3.3. Apparent equilibrium constant and the Haldane constraint

Changes in the apparent equilibrium constant due to pH and other cations will impact the reverse V_{max} as defined in the model by the Haldane relationship. The standard free energy of the kth reference reaction is the difference between the free energies of formation of the products and the reactants in that reaction:

$$\Delta_r G_k^0 = \sum_j v_j^k \Delta_f G_j^0. \qquad (2.21)$$

The free energy of formation of each reference species is a function of ionic strength:

$$\Delta_f G_j^0 = \Delta_f G_j^0(I=0) - \frac{2.91482 z_j^2 I^{1/2}}{1 + 1.6 I^{1/2}}. \qquad (2.22)$$

Since the reference reaction is balanced with respect to protons, the free energy of the reference reaction, $\Delta_r G_k^0$, is independent of pH and therefore a constant at a given ionic strength and temperature.

The equilibrium constant of the reference reaction is given by

$$K_{ref}^k = e^{-\Delta_r G_k^0/RT}. \tag{2.23}$$

The apparent equilibrium constant of a biochemical reaction is defined in terms of the metabolite concentrations at equilibrium [indicated by the subscript "eq" in Eq. (24)], which are sums of their constituent species:

$$K_{app}^k = \frac{\prod [L_{total}^{product}/c_0]_{eq}^{v_{product}^k}}{\prod [L_{total}^{reactant}/c_0]_{eq}^{v_{reactant}^k}} = K_{ref}^k [H^+]^{-n_k} \frac{\prod P_{products}}{\prod P_{reactants}}, \tag{2.24}$$

where n_k is the proton stoichiometry of the reference reaction and P is the binding polynomial for a biochemical reactant as defined in Eq. (2.15).

The standard free energy for a biochemical reaction is given by

$$\Delta_r G_k'^0 = -RT \ln\left(K_{app}^k\right). \tag{2.25}$$

2.3.4. Ion mass balance and the differential equations for ion concentrations

The differential equation for a particular ion concentration is derived from a statement of mass conservation for that ion. For metal ions, the total concentration (bound plus free) concentration is

$$[M_{j,total}^{z_j+}] = [M_j^{z_j+}] + [M_{j,bound}^{z_j+}], \tag{2.26}$$

where $[M_j^{z_j+}]$ is the jth free metal ion concentration and $[M_{j,total}^{z_j+}]$ is the metabolite bound dissociable metal ion concentration.

The conservation equation for protons includes an additional term for covalently bound protons in reference species:

$$[H_{total}^+] = [H^+] + [H_{bound}^+] + [H_{reference}^+], \tag{2.27}$$

where $[H_{reference}^+]$ denotes the protons in reference species.

In a closed system, the total proton pool is constant ($\frac{d[H_{total}^+]}{dt} = 0$). Differentiating Eq. (2.27) for proton conservation with respect to time, we get

$$0 = \frac{d[H^+]}{dt} + \frac{d[H_{bound}^+]}{dt} + \frac{d[H_{reference}^+]}{dt}. \tag{2.28}$$

The second term in Eq. (2.28) can be expanded by using the chain rule:

$$\frac{d[H^+_{bound}]}{dt} = \frac{\partial[H^+_{bound}]}{\partial[H^+]}\frac{d[H^+]}{dt} + \sum_j \frac{\partial[H^+_{bound}]}{\partial[M^{z_j+}_j]}\frac{d[M^{z_j+}_j]}{dt} + \sum_i \frac{\partial[H^+_{bound}]}{\partial[L_i]}\frac{d[L_i]}{dt}. \tag{2.29}$$

The third term is obtained from summation of the product of proton stoichiometry of each of the reference reactions with the flux through the reaction (generation of free protons from the reference pool diminishes the reference pool):

$$\frac{d[H^+_{reference}]}{dt} = -\sum_{k=1}^{N_r} n_k J_k. \tag{2.30}$$

Using Eqs. (2.29) and (2.30) in Eq. (2.28) and transposing terms, we obtain the following equation for $d[H^+]/dt$:

$$\frac{d[H^+]}{dt} = \frac{-\sum_j \frac{\partial[H^+_{bound}]}{\partial[M^{z_j+}_j]}\frac{d[M^{z_j+}_j]}{dt} - \sum_i \frac{\partial[H^+_{bound}]}{\partial[L_i]}\frac{d[L_i]}{dt} + \sum_{k=1}^{N_r} n_k J_k}{1 + \frac{\partial[H^+_{bound}]}{\partial[H^+]}}. \tag{2.31}$$

When hydrogen ion is transported into and out of the system, this equation becomes

$$\frac{d[H^+]}{dt} = \frac{-\sum_j \frac{\partial[H^+_{bound}]}{\partial[M^{z_j+}_j]}\frac{d[M^{z_j+}_j]}{dt} - \sum_i \frac{\partial[H^+_{bound}]}{\partial[L_i]}\frac{d[L_i]}{dt} + \sum_{k=1}^{N_r} n_k J_k + J^H_t}{1 + \frac{\partial[H^+_{bound}]}{\partial[H^+]}}, \tag{2.32}$$

where J^H_t is the flux of $[H^+]$ transport into the system.

Similarly, the rate of change of each metal ion $[M^{z_j+}_j]$ is written as

$$\frac{d[M^{z_j+}_j]}{dt} = \frac{-\frac{\partial[M^{z_j+}_{j,bound}]}{\partial[H^+]}\frac{d[H^+]}{dt} - \sum_l \frac{\partial[M^{z_j+}_{j,bound}]}{\partial[M^{z_l+}_l]}\frac{d[M^{z_l+}_l]}{dt} - \sum_i \frac{\partial[M^{z_j+}_{j,bound}]}{\partial[L_i]}\frac{d[L_i]}{dt} + J^{M_j}_t}{1 + \frac{\partial[M^{z_j+}_{j,bound}]}{\partial[M^{z_j+}_j]}}.$$

$$\tag{2.33}$$

For a total of N_j metal ions, Eq. (2.33) plus either Eq. (2.31) or Eq. (2.32) results in $N_j + 1$ coupled differential equations for the ion concentrations. Solving for the time derivatives, we can obtain $N_j + 1$ differential equations for the ion concentration time derivatives that may be numerically integrated simultaneously with the N_L differential equations for the biochemical reactants $[L_{total, i}]$.

The set of differential equations for biochemical reactant L_i is given by

$$\frac{d[L_{total, i}]}{dt} = \sum_k v_k^i J_k, \qquad (2.34)$$

where v_k^i is the stoichiometry of the ith metabolite in the kth reaction and J_k is the flux through the kth reaction.

As an example, we apply these general equations to treat muscle cell cytoplasm. The cation species in cardiac and skeletal muscle cell cytoplasm that significantly influence the kinetics and apparent equilibrium constants of biochemical reactions are H^+, Mg^{2+}, and K^+. The Ca^{2+} ion is also an important second messenger for cellular processes whose concentration changes by at least two orders of magnitude during rest and mechanical contraction (Guyton and Hall, 2006). However, the contribution of Ca^{2+} to the binding polynomial of most metabolites is much smaller when compared to that of H^+, Mg^{2+}, and K^+. For each reactant in the cytoplasm, the binding polynomial is

$$P_i([H^+], [Mg^{2+}], [K^+]) = 1 + [H^+]/K_i^H + [Mg^{2+}]/K_i^{Mg} + [K^+]/K_i^K, \qquad (2.35)$$

where K_i^H, K_i^{Mg}, and K_i^K are the binding constants for H^+, Mg^{2+}, and K^+. Here we consider only one binding reaction per ion per reactant. Following Eqs. (2.32) and (2.33), we can write the differential equations for the H^+, Mg^{2+}, and K^+ concentrations:

$$\frac{d[H^+]}{dt} = \frac{-\frac{\partial[H_{bound}]}{\partial[Mg^{2+}]}\frac{d[Mg^{2+}]}{dt} - \frac{\partial[H_{bound}]}{\partial[K^+]}\frac{d[K^+]}{dt} - \frac{\partial[H_{bound}]}{\partial[L_i]}\frac{d[L_i]}{dt} + \sum_{k=1}^{N_f} n_k J_k + J_t^H}{1 + \frac{\partial[H_{bound}]}{\partial[H^+]}}$$

$$(2.36)$$

$$\frac{d[Mg^{2+}]}{dt} = \frac{-\frac{\partial[Mg_{bound}]}{\partial[H^+]}\frac{d[H^+]}{dt} - \frac{\partial[Mg_{bound}]}{\partial[K^+]}\frac{d[K^+]}{dt} - \frac{\partial[Mg_{bound}]}{\partial[L_i]}\frac{d[L_i]}{dt} + J_t^{Mg}}{1 + \frac{\partial[Mg_{bound}]}{\partial[Mg^{2+}]}}$$

$$(2.37)$$

$$\frac{d[K^+]}{dt} = \frac{-\frac{\partial[K_{bound}]}{\partial[H^+]}\frac{d[H^+]}{dt} - \frac{\partial[K_{bound}]}{\partial[Mg^{2+}]}\frac{d[Mg^{2+}]}{dt} - \frac{\partial[K_{bound}]}{\partial[L_i]}\frac{d[L_i]}{dt} + J_t^K}{1 + \frac{\partial[K_{bound}]}{\partial[K^+]}}. \quad (2.38)$$

These equations can be solved as a system of linear equations for the derivatives of $[H^+]$, $[Mg^{2+}]$, and $[K^+]$. We now define the expressions for partial derivatives and buffering terms based on the binding polynomial defined in Eq. (2.35):

$$\frac{\partial[H_{bound}]}{\partial[Mg^{2+}]} = -\sum_{i=1}^{N_r} \frac{[L_i][H^+]/K_i^H}{K_i^{Mg}\left(P_i([H^+],[Mg^{2+}],[K^+])\right)^2}, \quad (2.39)$$

$$\frac{\partial[H_{bound}]}{\partial[K^+]} = -\sum_{i=1}^{N_r} \frac{[L_i][H^+]/K_i^H}{K_i^K\left(P_i([H^+],[Mg^{2+}],[K^+])\right)^2}, \quad (2.40)$$

$$\frac{\partial[H_{bound}]}{\partial[H^+]} = \sum_{i=1}^{N_r} \frac{[L_i](1+[Mg^{2+}]/K_i^{Mg}+[K^+]/K_i^K)}{K_i^H\left(P_i([H^+],[Mg^{2+}],[K^+])\right)^2}, \quad (2.41)$$

$$\frac{\partial[Mg_{bound}]}{\partial[H^+]} = -\sum_{i=1}^{N_r} \frac{[L_i][Mg^{2+}]/K_i^{Mg}}{K_i^H\left(P_i([H^+],[Mg^{2+}],[K^+])\right)^2}, \quad (2.42)$$

$$\frac{\partial[Mg_{bound}]}{\partial[K^+]} = -\sum_{i=1}^{N_r} \frac{[L_i][Mg^{2+}]/K_i^{Mg}}{K_i^K\left(P_i([H^+],[Mg^{2+}],[K^+])\right)^2}, \quad (2.43)$$

$$\frac{\partial[Mg_{bound}]}{\partial[Mg^{2+}]} = \sum_{i=1}^{N_r} \frac{[L_i](1+[H^+]/K_i^H+[K^+]/K_i^K)}{K_i^{Mg}\left(P_i([H^+],[Mg^{2+}],[K^+])\right)^2}, \quad (2.44)$$

$$\frac{\partial[K_{bound}]}{\partial[H^+]} = -\sum_{i=1}^{N_r} \frac{[L_i][K^+]/K_i^K}{K_i^H\left(P_i([H^+],[Mg^{2+}],[K^+])\right)^2}, \quad (2.45)$$

$$\frac{\partial[K_{bound}]}{\partial[Mg^{2+}]} = -\sum_{i=1}^{N_r} \frac{[L_i][K^+]/K_i^K}{K_i^{Mg}\left(P_i([H^+],[Mg^{2+}],[K^+])\right)^2}, \quad (2.46)$$

$$\frac{\partial[K_{bound}]}{\partial[K^+]} = \sum_{i=1}^{N_r} \frac{[L_i](1+[H^+]/K_i^H+[Mg^{2+}]/K_i^{Mg})}{K_i^K\left(P_i([H^+],[Mg^{2+}],[K^+])\right)^2}. \quad (2.47)$$

The flux terms for H^+, Mg^{2+}, and K^+ are

$$\Phi^H = -\sum_{i=1}^{N_r} \frac{\partial[H_{bound}]}{\partial[L_i]} \frac{d[L_i]}{dt} + \sum_{k=1}^{N_f} n_k J_k + J_t^H, \qquad (2.48)$$

$$\Phi^{Mg} = -\sum_{i=1}^{N_r} \frac{\partial[Mg_{bound}]}{\partial[L_i]} \frac{d[L_i]}{dt} + J_t^{Mg}, \qquad (2.49)$$

$$\Phi^K = -\sum_{i=1}^{N_r} \frac{\partial[K_{bound}]}{\partial[L_i]} \frac{d[L_i]}{dt} + J_t^K, \qquad (2.50)$$

where N_r is the number of reactants, N_f is the number of reactions, n_k is the stoichiometric coefficient of the kth reaction, J_k is the flux of the kth reaction, and J_t^H (J_t^{Mg}, J_t^K) is the transport flux of [H^+]([Mg^{2+}], [K^+]) into the system.

The buffering terms are

$$\alpha_H = 1 + \frac{\partial[H_{bound}]}{\partial[H^+]}, \qquad (2.51)$$

$$\alpha_{Mg} = 1 + \frac{\partial[Mg_{bound}]}{\partial[Mg^{2+}]}, \qquad (2.52)$$

$$\alpha_K = 1 + \frac{\partial[K_{bound}]}{\partial[K^+]}, \qquad (2.53)$$

The time derivatives of [H^+], [Mg^{2+}], and [K^+] obtained from solving Eqs. (2.36), (2.37), and (2.38) are given in terms of the buffering terms and the flux terms defined in the preceding equations:

$$\frac{d[H^+]}{dt} = \left[\left(\frac{\partial[K_{bound}]}{\partial[Mg^{2+}]} \cdot \frac{\partial[Mg_{bound}]}{\partial[K^+]} - \alpha_{Mg}\alpha_K \right) \Phi^H \right.$$

$$+ \left(\alpha_K \frac{\partial[H_{bound}]}{\partial[Mg^{2+}]} - \frac{\partial[H_{bound}]}{\partial[K^+]} \cdot \frac{\partial[K_{bound}]}{\partial[Mg^{2+}]} \right) \Phi^{Mg}$$

$$\left. + \left(\alpha_{Mg} \frac{\partial[H_{bound}]}{\partial[K^+]} - \frac{\partial[H_{bound}]}{\partial[Mg^{2+}]} \cdot \frac{\partial[Mg_{bound}]}{\partial[K^+]} \right) \Phi^K \right] / D,$$

$$(2.54)$$

$$\frac{d[\text{Mg}^{2+}]}{dt} = \left[\left(\alpha_K \frac{\partial [\text{Mg}_{bound}]}{\partial [\text{H}^+]} - \frac{\partial [\text{K}_{bound}]}{\partial [\text{H}^+]} \cdot \frac{\partial [\text{Mg}_{bound}]}{\partial [\text{K}^+]}\right) \Phi^H \right.$$
$$+ \left(\frac{\partial [\text{K}_{bound}]}{\partial [\text{H}^+]} \cdot \frac{\partial [\text{H}_{bound}]}{\partial [\text{K}^+]} - \alpha_H \alpha_K\right) \Phi^{Mg}$$
$$\left. + \left(\alpha_H \frac{\partial [\text{Mg}_{bound}]}{\partial [\text{K}^+]} - \frac{\partial [\text{H}_{bound}]}{\partial [\text{K}^+]} \cdot \frac{\partial [\text{Mg}_{bound}]}{\partial [\text{H}^+]}\right) \Phi^K \right] \Big/ D,$$

(2.55)

$$\frac{d[\text{K}^+]}{dt} = \left[\left(\alpha_{Mg} \frac{\partial [\text{K}_{bound}]}{\partial [\text{H}^+]} - \frac{\partial [\text{K}_{bound}]}{\partial [\text{Mg}^{2+}]} \cdot \frac{\partial [\text{Mg}_{bound}]}{\partial [\text{H}^+]}\right) \Phi^H \right.$$
$$+ \left(\alpha_H \frac{\partial [\text{K}_{bound}]}{\partial [\text{Mg}^{2+}]} - \frac{\partial [\text{K}_{bound}]}{\partial [\text{H}^+]} \cdot \frac{\partial [\text{H}_{bound}]}{\partial [\text{Mg}^{2+}]}\right) \Phi^{Mg}$$
$$\left. + \left(\frac{\partial [\text{Mg}_{bound}]}{\partial [\text{H}^+]} \cdot \frac{\partial [\text{H}_{bound}]}{\partial [\text{Mg}^{2+}]} - \alpha_H \alpha_{Mg}\right) \Phi^K \right] \Big/ D,$$

(2.56)

where

$$D = \alpha_H \frac{\partial [\text{K}_{bound}]}{\partial [\text{Mg}^{2+}]} \cdot \frac{\partial [\text{Mg}_{bound}]}{\partial [\text{K}^+]} + \alpha_K \frac{\partial [\text{H}_{bound}]}{\partial [\text{Mg}^{2+}]} \cdot \frac{\partial [\text{Mg}_{bound}]}{\partial [\text{H}^+]}$$
$$+ \alpha_{Mg} \frac{\partial [\text{H}_{bound}]}{\partial [\text{K}^+]} \cdot \frac{\partial [\text{K}_{bound}]}{\partial [\text{H}^+]} - \alpha_{Mg} \alpha_K \alpha_H$$
$$- \frac{\partial [\text{H}_{bound}]}{\partial [\text{K}^+]} \cdot \frac{\partial [\text{K}_{bound}]}{\partial [\text{Mg}^{2+}]} \cdot \frac{\partial [\text{Mg}_{bound}]}{\partial [\text{H}^+]}$$
$$- \frac{\partial [\text{H}_{bound}]}{\partial [\text{Mg}^{2+}]} \cdot \frac{\partial [\text{Mg}_{bound}]}{\partial [\text{K}^+]} \cdot \frac{\partial [\text{K}_{bound}]}{\partial [\text{H}^+]}.$$

(2.57)

2.4. Multiple cation equilibria in solutions: Proton, magnesium, and potassium ion binding in the creatine kinase reaction

The creatine kinase reaction catalyzes the phosphorylation of ADP by PCr to produce ATP and creatine (Cr). At given pH and free Mg^{2+} and K^+ concentrations, ATP, ADP, PCr, and Cr are distributed as a mixture of different ion bound forms. In a solution with given total amounts of ions and metabolites, the free ion concentrations are determined by the simultaneous binding equilibria of the metabolites with the cations. Experimentally, total concentrations of metabolites and ions are measured in solution and the free ion concentrations are calculated from ion mass balance equations derived from multiple cation equilibria. Lawson and Veech (1979) determined experimentally the effect of Mg^{2+} and pH on the creatine kinase equilibrium constant at 38 °C by starting with known quantities of either ATP and Cr or PCr and ADP and adding the creatine kinase enzyme to let the reaction proceed to equilibrium. The free magnesium ion concentration was calculated by subtracting total bound magnesium from total magnesium ion concentration. Ionic strength was maintained at 0.25 M in these solutions by the addition of almost 250 mM of KCl. From the observed equilibrium constants and theoretical calculations, they calculated the creatine kinase apparent equilibrium constant at pH 7, 1 mM Mg^{2+}, and 0.25 M ionic strength to be 166. This is the number that is often used as the *in vivo* creatine kinase equilibrium constant. However, they did not take potassium binding into account, which was later shown to be comparable to magnesium binding to ATP at physiological concentrations of the potassium ion by Kushmerick (1997). The consequence of not accounting for potassium binding is the underestimation of free magnesium concentration. The ionic strength of I = 0.25 M used in these experiments is also very high when compared to values computed from the ionic composition of myoplasm. Godt and Maughan (1988) estimated I = 0.18 M for frog myoplasm. Teague and Dobson (1992) studied the effect of temperature on creatine kinase equilibrium experiment under similar conditions as in the Lawson and Veech (1979) study, but varied the temperature. From binding polynomials of metabolites, the reference equilibrium constant was computed at different temperatures. The slope of the plot of log of apparent equilibrium constant versus $1/T$ was used to estimate the reaction enthalpy from the van't Hoff relationship. Subsequent computations of the creatine kinase equilibrium as a function of temperature, Mg^{2+}, and ionic strength (Golding *et al.*, 1995; Teague *et al.*, 1996) were based on Teague and Dobson (1992) data. Vinnakota (2006) reinterpreted data of Teague and Dobson (1992) with potassium binding in addition to magnesium and proton binding and obtained the values of equilibrium constant and the enthalpy change of the reference reaction.

The reference reaction for the creatine kinase reaction is written in terms of its most deprotonated species in the physiological pH range:

$$HPCr^{2-} + ADP^{3-} + H^+ \rightleftharpoons HCr^0 + ATP^{4-}. \tag{2.58}$$

The biochemical reaction is written in terms of sums of species as they exist in solution governed by their individual equilibria with protons, magnesium, and potassium ions:

$$\text{PCr} + \text{ADP} \rightleftarrows \text{ATP} + \text{Cr}. \qquad (2.59)$$

Table 2.2 lists the binding reactions, pK values, and dissociation enthalpies for each metabolite in the experimental system of Teague and Dobson (1992).

The concentrations of free magnesium and potassium ions are calculated by solving ion mass conservation relation equations at a given pH:

$$[\text{Mg}^{2+}_{total}] = [\text{Mg}^{2+}] + [\text{MgATP}^{2-}] + 2[\text{Mg}_2\text{ATP}] + [\text{MgHATP}^-] \\ + [\text{MgADP}^-] + [\text{Mg}_2\text{ADP}^+] + [\text{MgHPCr}] + [\text{MgHPi}] \\ + [\text{MgH}_2\text{Pi}^+] \qquad (2.60)$$

$$[\text{K}^+_{total}] = [\text{K}^+] + [\text{KATP}^{3-}] + [\text{KADP}^{2-}] + [\text{KHPCr}^-] + [\text{KHPi}^-]. \qquad (2.61)$$

Table 2.2 Binding reactions and equilibrium constants

Metabolite	Binding reaction	pK ($T = 298.15K$, $I = 0.1$)	ΔH ($I = 0$) kJ/mol
ATP	$\text{HATP}^{3-} \rightleftarrows \text{ATP}^{4-} + \text{H}^+$	6.48	−5
	$\text{MgATP}^{2-} \rightleftarrows \text{ATP}^{4-} + \text{Mg}^{2+}$	4.19	−18
	$\text{MgHATP}^{2-} \rightleftarrows \text{HATP}^{3-} + \text{Mg}^{2+}$	2.32	—
	$\text{Mg}_2\text{ATP} \rightleftarrows \text{MgATP}^{2-} + \text{Mg}^{2+}$	1.7	—
	$\text{KATP}^{3-} \rightleftarrows \text{ATP}^{4-} + \text{K}^+$	1.17	−1
ADP	$\text{HADP}^{2-} \rightleftarrows \text{ADP}^{3-} + \text{H}^+$	6.38	−3
	$\text{MgADP}^- \rightleftarrows \text{ADP}^{3-} + \text{Mg}^{2+}$	3.25	−15
	$\text{Mg}_2\text{ADP}^+ \rightleftarrows \text{MgADP}^- + \text{Mg}^{2+}$	1	—
	$\text{KADP}^{2-} \rightleftarrows \text{ADP}^{3-} + \text{K}^+$	1	—
PCr	$\text{H}_2\text{PCr}^- \rightleftarrows \text{HPCr}^{2-} + \text{H}^+$	4.5	2.66
	$\text{MgHPCr} \rightleftarrows \text{HPCr}^{2-} + \text{Mg}^{2+}$	1.6	8.19
	$\text{KHPCr}^- \rightleftarrows \text{HPCr}^{2-} + \text{K}^+$	0.31	—
Cr	$\text{H}_2\text{Cr}^+ \rightleftarrows \text{HCr} + \text{H}^+$	2.3	—
Pi	$\text{H}_2\text{Pi}^- \rightleftarrows \text{HPi}^{2-} + \text{H}^+$	6.75	3
	$\text{MgHPi} \rightleftarrows \text{HPi}^{2-} + \text{Mg}^{2+}$	1.65	−12
	$\text{MgH}_2\text{Pi}^+ \rightleftarrows \text{H}_2\text{Pi}^- + \text{Mg}^{2+}$	1.19	—
	$\text{KHPi}^- \rightleftarrows \text{HPi}^{2-} + \text{K}^+$	0.5	—

At each temperature, Teague and Dobson's (1992) data give the final equilibrium composition in terms of total Mg^{2+} and K^+ concentrations, total ATP, ATP, PCr and Cr concentrations, and the final pH. We treat the measured final pH as the negative logarithm of $[H^+]$ activity ($-\log_{10}(\gamma[H^+])$) and compute the $[H^+]$ concentration and pH as $-\log_{10}([H^+])$ using an approximate activity coefficient γ calculated from the extended Debye-Huckel equation (Alberty, 2003). Phosphate was added to the medium as one of the buffers. Therefore, phosphate binding to ions is accounted in the overall ion mass balance.

The apparent equilibrium constant of the creatine kinase reaction is given by

$$K_{app}^{CK} = \frac{[ATP]_{eq} \cdot [Cr]_{eq}}{[PCr]_{eq} \cdot [ADP]_{eq}}$$

$$= \frac{([ATP^{4-}]_{eq} + [HATP^{3-}]_{eq} + [MgATP^{2-}]_{eq} + [MgHATP^-]_{eq} + [Mg_2ATP]_{eq} + [KATP^{3-}]_{eq})([HCr]_{eq} + [H_2Cr^+]_{eq})}{([ADP^{3-}]_{eq} + [HADP^{2-}]_{eq} + [MgADP^-]_{eq} + [Mg_2ADP^+]_{eq} + [KADP^{2-}]_{eq})([HPCr^{2-}]_{eq} + [H_2PCr^-]_{eq} + [MgHPCr]_{eq} + [KHPCr^-]_{eq})}$$

(2.62)

Each species of a metabolite can be expressed in terms of the most unbound species through ionic equilibrium relationships defined by binding reactions in Table 2.2. For example, ATP^{4-} is given by

$$[ATP^{4-}]_{eq} = \frac{[ATP]_{eq}}{1 + \frac{[H^+]}{K_{HATP}} + \frac{[Mg^{2+}]}{K_{MgATP}} + \frac{[H^+][Mg^{2+}]}{K_{HATP} K_{MgHATP}} + \frac{[Mg^{2+}]^2}{K_{MgATP} K_{Mg2ATP}} + \frac{[K^+]}{K_{KATP}}}.$$

(2.63)

The denominator of the RHS of Eq. (2.63) is known as the binding polynomial (P_{ATP}) for ATP:

$$P_{ATP} = 1 + \frac{[H^+]}{K_{HATP}} + \frac{[Mg^{2+}]}{K_{MgATP}} + \frac{[H^+][Mg^{2+}]}{K_{HATP} K_{MgHATP}} \qquad (2.64)$$
$$+ \frac{[Mg^{2+}]^2}{K_{MgATP} K_{Mg2ATP}} + \frac{[K^+]}{K_{KATP}}.$$

Therefore, $[ATP]_{eq} = [ATP^{4-}]_{eq} P_{ATP}$. Similarly, other metabolite concentrations can be expressed in terms of their most unbound forms and binding polynomials. From Eqs. (2.62), (2.63), and (2.64), the apparent equilibrium constant of creatine kinase reaction can be expressed as

$$K_{app}^{CK} = \frac{[ATP^{4-}]_{eq}[HCr]_{eq}}{[HPCr^{2-}]_{eq}[ADP^{3-}]_{eq}[H^+]}[H^+]\frac{P_{ATP}P_{Cr}}{P_{PCr}P_{ADP}} = K_{ref}^{CK}[H^+]\frac{P_{ATP}P_{Cr}}{P_{PCr}P_{ADP}}.$$

(2.65)

Binding equilibria were corrected to experimental ionic strength and temperature using the methods described in Section 2.3.1. These constants are also corrected to experimental temperatures using the van't Hoff relationship. We analyzed the equilibrium compositions of solutions in Teague and Dobson's study to compute the reference equilibrium constant of creatine kinase plotted as a function of $1/T$ in Fig. 2.1. From our analysis, we estimate the equilibrium constant of the reference reaction, K_{ref}^{CK}, to be 9.82e8 at 298.15 K and 0.25 M ionic strength, whereas the value of this constant was estimated to be 3.77e8 by Teague and Dobson (1992). From the slope of the plot in Fig. 2.1, we estimate the enthalpy of the reference reaction to be -17.02 ± 3.32 kJ/mol as opposed to the estimate of -16.73 kJ/mol by Teague and Dobson (1992).

After estimating the K_{ref}^{CK}, we then compute the apparent equilibrium constant of the creatine kinase reaction as a function of pH at physiological conditions found in muscle, that is, 310.15 K, 0.18 M ionic strength, 1 mM free magnesium (Konishi, 1998), and 140 mM free potassium (Kushmerick, 1997)(Fig. 2.2). At pH 7 the apparent equilibrium constant under physiological conditions is 166.2. This is very close to the previous estimates of 166 by Lawson and Veech (1979) and Teague and Dobson (1992) at $-\log_{10}(\gamma[H^+]) = 7$ under unphysiological experimental conditions at a high potassium

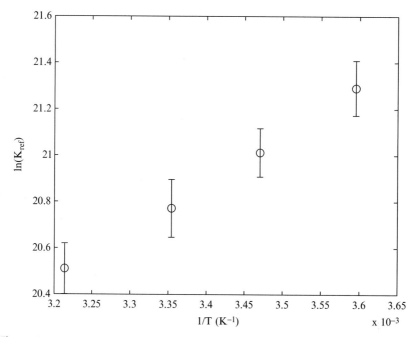

Figure 2.1 $\ln K_{ref}$ of creatine kinase reaction vs $1/T$. $-R$ times the slope of this plot gives the enthalpy of the reference reaction.

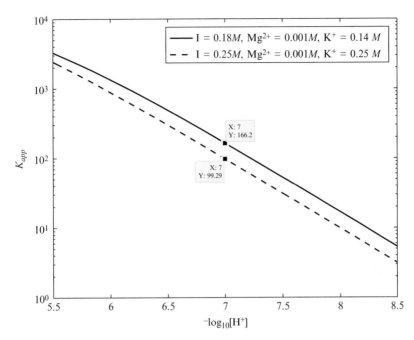

Figure 2.2 Apparent equilibrium constant of creatine kinase reaction as a function of pH at 310 K, 1 mM free magnesium, and two different ionic strength and potassium concentrations: I = 0.18 M, K$^+$ = 140 mM; I = 0.25 M, K$^+$ = 250 mM.

concentration and a high ionic strength. Errors in the analyses by Lawson and Veech (1979) and Teague and Dobson (1992)(ignoring potassium binding and underestimating free magnesium ion concentration and the effect of high ionic strength) tended to cancel out, resulting in an estimate of K_{app}^{CK} at $-\log_{10}(\gamma[H^+]) = 7$ similar to ours at pH 7. Under the experimental conditions of the previous studies (Lawson and Veech, 1979; Teague and Dobson, 1992), that is, 0.25 M ionic strength and 250 mM K$^+$ concentration, we compute the apparent equilibrium constant at pH 7 to be 99.3, which is very low compared to the physiological value.

2.5. Example of detailed kinetics of a biochemical reaction: Citrate synthase

Here we present details of modeling a single-reaction system—citrate synthase—applying the principles detailed in the preceding sections. The biochemical reaction is

$$\text{oxaloacetate} + \text{acetyl-CoA} \rightleftarrows \text{coenzyme A} + \text{citrate}. \quad (2.66)$$

If we denote the flux of this reaction as a function of reactant concentrations by the variable J, then the concentration kinetics of the reactants is governed by the differential equation:

$$\begin{aligned} d[\text{oxaloacetate}]/dt &= -J \\ d[\text{acetyl-CoA}]/dt &= -J \\ d[\text{coenzyme A}]/dt &= +J \\ d[\text{citrate}]/dt &= +J. \end{aligned} \quad (2.67)$$

What remains is to impose an appropriate mathematical expression for J that effectively models the kinetics of the reaction in Eq. (2.66). Such expressions are the domain of enzyme kinetics, in which mechanisms are proposed and data are fit to obtain estimates of associated parameters. In searching the literature, we are likely to discover that the existing kinetic models for the enzyme are parameterized from *in vitro* experiments conducted at nonphysiological pH, ionic strength, and [Mg^{2+}]. Furthermore, the given mechanism will either violate the Haldane constraint or, if it does not, obey the constraint with a value of the apparent equilibrium constant that matches the nonphysiological conditions of the *in vitro* assay. Accounting and correcting for these facts require treating the detailed biochemical thermodynamics of the reaction system.

To do this we, define reactions in terms of the reference stoichiometry, which in contrast to the biochemical reaction of Eq. (2.66) conserves mass (in terms of elements) and charge:

$$OAA^{2-} + ACCOA^{0} + H_2O \rightleftharpoons COAS^{-} + CIT^{3-} + 2H^{+}. \quad (2.68)$$

Reference species associated with reactants in Eq. (2.66) are listed in Table 2.3. A thermodynamically balanced model for the kinetics of this reaction relies on a calculation of the thermodynamic properties of the reaction. Using data in Table 2.3, the thermodynamic properties are computed:

$$\begin{aligned} \Delta_r G^\circ &= \Delta_f G^\circ_{COASH} + \Delta_f G^\circ_{CIT} - \Delta_f G^\circ_{OAA} - \Delta_f G^\circ_{ACCOA} - \Delta_f G^\circ_{H_2O} \\ &= +42.03 \text{kJ/mol}, \end{aligned} \quad (2.69)$$

$$K^{CS}_{app} = e^{-\Delta_r G'^\circ/RT} = \frac{P_{COA} \cdot P_{CIT}}{P_{OAA} \cdot P_{ACCOA} \cdot (10^{-pH})^2} e^{-\Delta_r G^\circ/RT}, \quad (2.70)$$

where K^{CS}_{app} is the apparent equilibrium constant for the reaction of Eq. (2.66). In other words, in chemical equilibrium ([CoA][citrate])/([oxaloacetate][acetyl-CoA]) $= K^{CS}_{app}$, where these concentrations correspond to sums over

Table 2.3 Reactants, reference species, Gibbs free energy of formation, and binding constants for Eqs. (2.66) and (2.76) (298.15 K, 1 M reactants, I = 0.17 M, P = 1 atm)

Reactant	Reference species	$\Delta_f G°$ (kJ/mol)	Ion bound species	pK^a
Water	H_2O	−235.74	—	—
Oxaloacete	OAA^{2-}	−794.41	$MgOAA^0$	0.8629[b]
Acetyl-coenzyme A	$ACCOA^0$	−178.19	—	—
Citrate	CIT^{3-}	−1165.59	$HCIT^{2-}$	5.63
			$MgCIT^-$	3.37[b]
			$KCIT^{2-}$	0.339[b]
Coenzyme A	$COAS^-$	−0.72	$COASH^0$	8.13
Adenosine triphosphate (ATP)	ATP^{4-}	−2771.00	$HATP^{3-}$	6.59
			$MgATP^{2-}$	3.82
			$KATP^{3-}$	1.01
Adenosine diphosphate (ADP)	ADP^{3-}	−1903.96	$HADP^{2-}$	6.42
			$MgADP^-$	2.79
			$KADP^{2-}$	0.882
Inorganic phosphate (Pi)	HPO_4^{2-}	−1098.27	$H_2PO_4^-$	6.71
			$MgHPO_4^0$	1.69
			$KHPO_4^-$	−0.0074

[a] All values from Alberty (2003) unless otherwise noted.
[b] NIST Database 46: Critical Stability Constants.

all species associated with each reactant. The binding polynomials in Eq. (2.71) depend on pH and binding ion concentration through the following relationships:

$$P_{OAA} = 1 + [Mg^{2+}]/K_{Mg, OAA},$$
$$P_{CIT} = 1 + 10^{-pH}/K_{H, CIT} + [Mg^{2+}]/K_{Mg, CIT} + [K^+]/K_{K, CIT}$$
$$P_{ACCOA} = 1, P_{COASH} = 1 + 10^{-pH}/K_{H, COASH}$$
(2.71)

Setting pH = 7, $[K^+]$ = 150 mM, and $[Mg^{2+}]$ = 1 mM, we obtain the apparent thermodynamic properties:

$$\begin{aligned} \Delta_r G'^o &= -47.41 \text{kJ} \cdot \text{mol}^{-1} \\ K_{app}^{CS} &= 2.02 \times 10^8. \end{aligned}$$
(2.72)

The relationships between the reference species concentrations and the reactant concentrations are

$$\begin{aligned}
[OAA^{2-}] &= [OAA]/P_{OAA} \\
[ACCOA^{0}] &= [ACCOA]/P_{ACCOA} \\
[CIT^{3-}] &= [CIT]/P_{CIT} \\
[COAS^{-}] &= [COASH]/P_{COASH}.
\end{aligned} \quad (2.73)$$

The computed species concentrations and thermodynamic values are used in the kinetic model of this enzyme.

The enzyme kinetic flux, which is based on a compulsory-order ternary-complex mechanism,

$$\begin{aligned}
E + OAA^{2-} &\underset{k_{21}}{\overset{k_{12}}{\rightleftharpoons}} E \cdot OAA^{2-} \\
E \cdot OAA^{2-} + ACCOA^{0} &\underset{k_{32}}{\overset{k_{23}}{\rightleftharpoons}} E \cdot OAA^{2-} \cdot ACCOA^{0}. \\
E \cdot OAA^{2-} \cdot ACCOA^{0} + H_2O &\underset{k_{43}}{\overset{k_{34}}{\rightleftharpoons}} E \cdot CIT^{3-} + COAS^{-} + 2\,H^{+} \\
E \cdot CIT^{3-} &\underset{k_{14}}{\overset{k_{41}}{\rightleftharpoons}} E + CIT^{3-}
\end{aligned} \quad (2.74)$$

where E denotes the enzyme and is tested and parameterized against experimental data in Beard *et al.* (2008). In addition to treating the enzyme kinetic mechanism explicitly in terms of biochemical species, the Beard *et al.* (2008) model also accounts for pH dependence of enzyme activity by treating the enzyme as a monobasic acid with the unbound form of the enzyme as the active form. The following expression for V_{max} as a function of [H$^+$] concentration accounts for the pH dependence of enzyme activity:

$$V_{max} = \frac{V_{max}^{opt}}{1 + \frac{[H^+]}{K_{iH}}}, \quad (2.75)$$

where V_{max}^{opt} is the maximum enzyme velocity attained when all of the enzyme is unbound and K_{iH} is the proton dissociation constant.

3. Application to Physiological Systems

3.1. pH dynamics and glycogenolysis in cell free reconstituted systems and in skeletal muscle

Intracellular pH is an important physiological variable that is regulated within a narrow range. The preceding sections showed that pH significantly affects enzyme activities and apparent equilibrium constants of biochemical reactions. These in turn may influence the energetic state of muscle and the metabolic fluxes. Computational models of biochemical pathways with the capability to predict the metabolic proton load and the consequent changes in muscle pH are useful tools for investigating pH regulation during physiological situations such as exercise and pathological conditions such as ischemia/reperfusion. The models of glycogenolysis developed by Vinnakota (2006) and Vinnakota *et al.* (2006) compute the metabolic proton load due to glycogenolysis in isolated cell free reconstituted systems and in intact mouse skeletal muscle during anoxia/reperfusion. Vinnakota *et al.* (2006) considered H^+, Mg^{2+}, and K^+ ion binding to the metabolites in the pathways modeled. However, the K^+ concentration was kept constant because of its low relative change during the transients modeled.

Scopes (1974) simulated postmortem glycogenolysis in a reconstituted system with rabbit and porcine enzymes from the glycogenolysis pathway, breaking down glycogen into lactate. ATPase and AMPdeaminase were added to this system with an initial concentration of PCr at 24 mM and an initial pH of 7.25. Figures 2.3 and 2.4 show pH time course data and model simulation and the computed proton generation fluxes due to biochemical reactions from the study of Vinnakota *et al.* (2006). The pH time course shows a slight initial rise in pH and thereafter a nearly linear decline until a final pH is reached near 5.5 at 190 min that continues until 300 min. The proton fluxes due to CK, ATPase, glycogenolysis, and AMP deaminase are plotted along with the sum total of all fluxes in Fig. 2.4. These fluxes identify the role of each biochemical reaction in shaping the pH transient. The creatine kinase flux initially consumes protons, which explains the initial rise in pH. The sum of proton fluxes due to the ATPase reaction and the glycogenolysis pathway has a flat time course, which explains the nearly linear decline of pH from 20 to 190 min. However, the important features that emerge from these computations are that (1) the ATPase proton flux is acidifying in the physiological pH range 7.2—6.8 and (2) the contribution of proton fluxes from the glycogenolysis pathway is relatively small during this pH range. The sharp decline in glycogenolysis at the end is due to a reduction in glycogen and a degradation of the adenine nucleotide pool. In summary, near-normal resting pH ATP hydrolysis is the source of metabolic acid load during anaerobic glycogenolysis. However, it

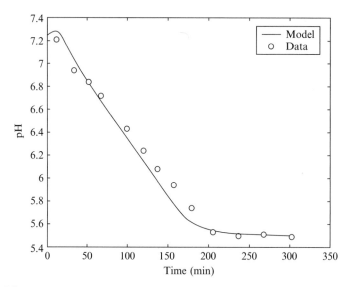

Figure 2.3 pH transient during simulated postmortem glycogenolysis and model-derived proton fluxes. Reproduced from Vinnakota et al. (2006) with permission.

is important to note that proton stoichiometries of biochemical reactions are variable with pH, which means that ATP hydrolysis is not always acidifying and that the combined proton stoichiometry of all biochemical reactions in the pathway breaking down glycogen into lactate is not always slightly alkalinizing at all pH values. These results are important for computing metabolic acid load during ischemia and anoxia in cardiac and skeletal muscle, where oxidative ATP generation is impaired, and the creatine kinase and the ATP generation from glycogenolysis work toward maintaining energy balance. Calculations following the principles outlined here provide mechanistic and accurate answers to continued discussions in the physiological literature on the source of protons in muscle glycogenolysis and glycolysis.

3.2. Cardiac energetics

The heart in healthy mammals is a highly oxidative organ in which normal function is tightly coupled with energetic state, which is maintained by a continuous supply of substrates to fuel energy metabolism (Katz, 2006). When under stresses or disease, such as ischemia or chronic heart failure, impaired cardiac function is associated with the altered energetic state of the heart (Ingwall and Weiss, 2004; Stanley et al., 2005; Zhang et al., 1993, 2001).

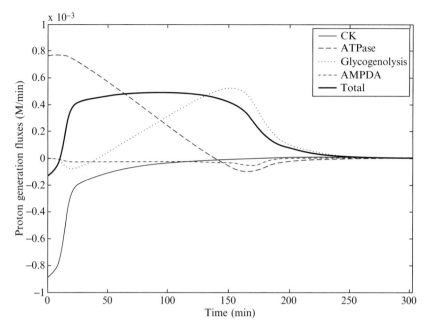

Figure 2.4 Proton generation fluxes during simulated postmortem glycogenolysis in a cell free reconstituted system. Reproduced from Vinnakota *et al.* (2006) with permission.

3.2.1. The gibbs free energy of ATP hydrolysis in the heart

The Gibbs free energy of ATP hydrolysis reaction, $\Delta_r G_{ATPase}$, represents an effective marker of the cardiac energetic state. Based on the chemical reference reaction of the ATP hydrolysis,

$$ATP^{4-} + H_2O = ADP^{3-} + PI^{2-} + H^+, \quad (2.76)$$

the standard Gibbs free energy of ATP hydrolysis, $\Delta_r G^o_{ATPase}$, can be calculated from Eq. (2.3) as

$$\Delta_r G^o_{ATPase} = \Delta_f G^o_{ADP} + \Delta_f G^o_{PI} + \Delta_f G^o_{H} - \Delta_f G^o_{ATP} - \Delta_f G^o_{H2O}, \quad (2.77)$$

with the corresponding equilibrium constant

$$K^{ATPase}_{ref} = \exp\left(-\frac{\Delta_r G^o_{ATPase}}{RT}\right) = \left(\frac{[ADP^{3-}]_c [PI^{2-}][H^+]}{[ATP]^{4-}}\right)_{eq}. \quad (2.78)$$

For the biochemical reaction of ATP hydrolysis,

$$\text{ATP} = \text{ADP} + \text{PI}, \tag{2.79}$$

the transformed equilibrium constant and standard Gibbs free energy can be calculated as

$$K_{app}^{ATPase} = \left(\frac{[\text{ADP}][\text{PI}]}{[\text{ATP}]}\right)_{eq} = \frac{K_{ref}^{ATPase}}{[\text{H}^+]} \frac{P_{ADP} P_{PI}}{P_{ATP}}, \tag{2.80}$$

and

$$\Delta_r G_{ATPase}^{\prime o} = -RT \ln K_{app}^{ATPase}, \tag{2.81}$$

where P_{ATP}, P_{ADP}, and P_{PI} are the binding polynomials as functions of ionic concentrations (e.g., [H$^+$], [Mg^{2+}], and [K$^+$]) and $\Delta_r G_{ATPase}$ is obtained from the definition stated in Eq. (2.2),

$$\Delta_r G_{ATPase} = \Delta_r G_{ATPase}^{\prime o} + RT \ln \left(\frac{[\text{ADP}][\text{PI}]}{[\text{ATP}]}\right). \tag{2.82}$$

3.2.2. Calculations of $\Delta_r G_{ATPase}^{\prime o}$ and ΔG_{ATPase} in the heart *in vivo*

Cardiomyocytes contain multiple subcellular organelles (Katz , 2006), including myofibrils, mitochondria, sarcoplasmic reticulum, and nuclei, among which myofibrils and mitochondria occupy the largest fractions of intracellular volume (Vinnakota and Bassingthwaighte, 2004). Most metabolites and ions require specific protein carriers to traverse the inner mitochondrial membrane (LaNoue and Schoolwerth, 1979; Nelson and Cox, 2005), which constitutes a compartment. To account for this compartmentation, we divide the cellular volume of cardiomyocytes into cytoplasm and mitochondria, where the local concentrations of metabolites drive local biochemical reactions. The free energies for a biochemical reaction are generally different in different intracellular compartments (Gibbs, 1978). Since most of the most energy-consuming processes coupled to ATP hydrolysis, such as mechanical contraction due to myosin ATPase, calcium handling, and maintenance of ion gradients across the sarcolemma, are located in the cytoplasm, the cytoplasmic $\Delta_r G_{ATPase}$ provides the driving force for normal cardiac function (Gibbs, 1985).

Under the physiological conditions with assumed temperature $T = 310.15$ K, ionic strength $I = 0.17$ M, pH$_c$ = 7.1, [Mg^{2+}]$_c$ = 1 mM, and

$[K^+]_c = 130$ mM, we are able to compute $\Delta G'^o_{eq,ATPase} = -34.7$ kJ mol^{-1} in the cytoplasm of the cardiomyocytes based on thermodynamic data listed in Table 2.2. Here the subscript "c" denotes cytoplasm. A more detailed process for estimating $\Delta G'^o_{eq,ATPase}$ is given later.

3.2.3. Example 1: Calculating cytoplasmic $\Delta_r G'^o_{ATPase}$ in the heart under *in vivo* physiological conditions

Step 1: Based on basic thermodynamic data and computational algorithm presented elsewhere (Alberty, 2003; Beard and Qian, 2007; Golding et al., 1995; Teague et al., 1996), K_{ref}^{ATPase} under the physiological conditions ($T = 310.15$ K and $I = 0.17$ M) is calculated to be 1.16×10^{-1}.

Step 2: Using the binding constants presented in Table 2.2, we can calculate the binding polynomials under the environmental conditions and cytoplasmic ion concentrations to be $P_{ADP} = 2.82$, $P_{PI} = 1.58$, and $P_{ATP} = 9.28$, respectively.

Step 3: From Eq. (2.80),

$$K_{app}^{ATPase} = \left(\frac{[ADP][PI]}{[ATP]}\right)_{eq} = \frac{K_{ref}^{ATPase}}{[H^+]_c} \frac{P_{ADP} P_{PI}}{P_{ATP}} = 7.03 \times 10^5. \tag{2.83}$$

Step 4: From Eq. (2.81),

$$\Delta_r G'^o_{ATPase} = -RT \ln K_{app}^{ATPase} = -34.72 \text{ kJ mol}^{-1}. \tag{2.84}$$

3.2.4. Example 2: Calculating $\Delta_r G_{ATPase}$ at basal work rate in the heart *in vivo*

The values of basal $[ATP]_c$, $[ADP]_c$, and $[PI]_c$ are 9.67×10^{-3}, 4.20×10^{-5}, and 2.90×10^{-4} M, respectively, as reported previously (Wu et al., 2008). Thus,

$$\Delta_r G_{ATPase} = \Delta_r G'^o_{ATPase} + RT \ln \left(\frac{[ADP][PI]}{[ATP]}\right)$$
$$= -34.7 + 2.58 \times (-13.6) = -69.8 \text{ kJ mol}^{-1}. \tag{2.85}$$

[Note that in Wu et al.(2008) this value was reported as -70.0 kJ mol^{-1} because $\Delta_r G'^o_{ATPase}$ was computed based on slightly different dissociation constants.] A previous computational analysis (Wu et al., 2008) showed that

$\Delta_r G_{ATPase}$ changes from -70 kJ mol^{-1} at the basal work rate to -64 kJ mol^{-1} at the maximal work rate in the normal dog heart *in vivo*, which means that the threefold of work increase results in an increase of around 2.5 RT in $\Delta_r G_{ATPase}$. The values of $\Delta_r G_{ATPase}$ over the threefold increase of workload are calculated by assuming pH$_c$, [Mg^{2+}]$_c$, and [K$^+$]$_c$ fixed at the basal values.

3.2.5. Effects of changes in ion concentrations on $\Delta_r G'^o$ of biochemical reactions in the mitochondrial matrix

The preceding sections showed that the standard free energy of a biochemical reaction is a function of the binding polynomials of the biochemical reactants and is therefore dependent on free ion concentrations. In the mitochondrial matrix, the temporal changes in free ion concentrations are determined by ion transport fluxes across the mitochondrial inner membrane and the ion fluxes are due to biochemical reactions. The differential equations for the time derivatives of ion concentrations derived based on the principles and methods described in the preceding sections are presented in detail elsewhere (Beard and Qian, 2007; Vinnakota, 2006). Figure 2.5 shows the changes in ion concentrations in the mitochondrial matrix (A) and the resulting changes in standard free energies of selected biochemical reactions (B). In Fig. 2.5A, steady-state values of [H$^+$]$_{matrix}$, [Mg^{2+}]$_{matrix}$, and [K$^+$]$_{matrix}$ are plotted against an increased *in vivo* cardiac work rate, that is, oxygen consumption rate (MVO$_2$). As MVO$_2$ increases from its baseline workload value [3.5 μmol min^{-1} (g tissue)$^{-1}$] to its maximal workload value [10.7 μmol min^{-1} (g tissue)$^{-1}$], [H$^+$]$_{matrix}$ increases from 5.77×10^{-7} to 6.12×10^{-8} M, [Mg^{2+}]$_{matrix}$ decreases from 1.18 to 0.80×10^{-3} M, and [K$^+$]$_{matrix}$ increases from 9.45×10^{-2} to 1.00×10^{-1} M, respectively. Since the K$^+$/H$^+$ antiporter on the inner mitochondrial membrane is operated near equilibrium in the computational model (Wu *et al.*, 2007b), the increase of [K$^+$]$_{matrix}$ is almost linearly proportional to that of [H$^+$]$_{matrix}$. Figure 2.5B shows that changes in [H$^+$]$_{matrix}$, [Mg^{2+}]$_{matrix}$, and [K$^+$]$_{matrix}$ impact the values of $\Delta_r G'^o$ of mitochondrial ATP hydrolysis (ATPase), α-ketoglutarate dehydrogenase (AKGD), and citrate synthase (CITS) differently over the range of the *in vivo* work rate. When the workload is elevated from the baseline value to the maximal value, $-\Delta_r G'^o_{ATPase}/RT$ decreases from 20.70 to 19.77, $-\Delta_r G'^o_{AKGD}/RT$ decreases from 15.47 to 15.42, and $-\Delta_r G'^o_{CITS}/RT$ decreases from 20.46 to 20.15. $\Delta_r G'^o_{ATPase}$ changes the most by around 1 RT unit (i.e., 2.58 kJ mol^{-1}), which means that the apparent equilibrium constant of the mitochondrial ATP hydrolysis varies around threefold. For a biochemical reaction operating near equilibrium, the ratio of forward and reverse reaction fluxes is closely coupled to changes in the apparent equilibrium constants via the mass action law (Beard and Qian, 2007). Equation (81) shows that a small change in $\Delta_r G'^o_{ATPase}$ results in a large change in K^{ATPase}_{app} due to exponentiation. For example, the reaction catalyzed by aconitase in the citric acid cycle, which converts citrate into isocitrate, is maintained near equilibrium under physiological conditions

(Wu et al., 2007b) with $\Delta_r G'^o_{ACON} = 9.42 \text{kJ mol}^{-1}$ at the baseline work rate and $\Delta_r G'^o_{ACON} = 8.99 \text{kJ mol}^{-1}$ at the maximal work rate. The decrease of $\Delta_r G'^o_{ACON}$ by 0.43 kJ mol^{-1} leads to the increase of corresponding K^{ACON}_{app} by about 18 %.

3.3. Physiological significance of multiple ion binding equilibria, reaction kinetics, and thermodynamics in cardiac energetics

The advantages of accounting for multiple ion binding equilibria, reaction kinetics, and thermodynamics are demonstrated in recent cardiac energetics studies by Wu et al. (2007b, 2008). As shown in Wu et al. (2007b), the enzyme mechanisms are presented in terms of chemical species, $\Delta_r G'^o$ of the biochemical reactions are computed by explicitly accounting for effects of ion concentrations, and the corresponding apparent equilibrium constants are used to establish the Haldane constraints for modeling the enzyme kinetics. Accurate estimation of the free energy of ATP hydrolysis is important for determining the energetic state of the heart under various conditions.

The application of the methodologies explained in this chapter in a detailed computational model of cardiac energetics (Wu et al., 2008) has resulted in the first estimates of free energy of ATP hydrolysis in the heart *in vivo* at rest and during increased workloads that account for intracellular compartmentation. The capability of the heart to maintain its "energy homeostasis" in a wide range of cardiac workloads has been a puzzle since the mid-1980s (Balaban et al., 1986). The ^{31}P magnetic resonance spectroscopy (^{31}P-MRS) technique has been applied widely and successfully as a noninvasive technique in studying *in vivo* cardiac energetics, albeit with difficulties in detecting subcellular concentrations of chemical species and low concentrations of phosphate metabolites, such as ADP and resting Pi. Because of these limitations, the ^{31}P-MRS experimental observations may not provide quantitative measurement on concentration changes of certain metabolites (e.g., cytosolic [Pi]) and may lead to an inaccurate estimation of energy states in the heart *in vivo*. Thus, it is necessary to use a computational tool to unambiguously determine the concentrations and energetic states based on the experimental observations. Also, modeling reveals details of the underlying mechanisms that are not amenable to direct experimental observation and quantification.

The model of Wu et al. (2008) predicts significant changes in cytosolic [Pi] from the basal work rate to the maximal work rate and leads to two novel conclusions that will be contentious and inspire more thinking and experimenting. The first conclusion is that oxidative phosphorylation is controlled by Pi feedback in the heart *in vivo*; the second conclusion is that decreased ATP hydrolysis potential—caused by increased cytoplasmic Pi and ADP concentrations—limits ATP utilization rates at a high workload and during ischemia. These two hypotheses are supported by agreement

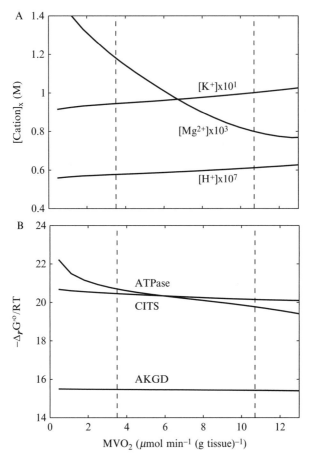

Figure 2.5 Model-predicted steady-state cation concentrations in the mitochondrial matrix and corresponding $\Delta G'^{o}_{eq}$ of ATP hydrolysis, AKGD, and CITS at various work rates in the heart *in vivo*. (A) Changes in $[H^+]_x$, $[Mg^{2+}]_x$, and $[K^+]_x$, where subscript "x" denotes the mitochondrial matrix. (B) Changes in $\Delta G'^{o}_{ATPase}$ of ATP hydrolysis (ATPase), AKGD, and CITS.

between the model predictions and experimental observations on independent experimental data on transient changes and steady states in ischemic heart *in vivo* (Wu *et al.*, 2008).

4. Discussion

4.1. Integrating thermodynamic information in biochemical model building

We have described the methods for accounting for detailed thermodynamics in kinetic simulations of biochemical systems. These methods are crucial for a number of applications, including identifying and parameterizing catalytic mechanisms for enzymes based on *in vitro* data and translating from *in vitro* experiments to predicting *in vivo* function of enzymes. In addition, in analyzing *ex vivo* data (such as from purified mitochondria) or simulating extreme *in vivo* states (such as ischemia/hypoxia), even normally irreversible reactions approach or go to equilibrium. Thus it is always important to accurately account for thermodynamics.

Studies of specific enzymes and transporters (Beard *et al.*, 2008; Dash and Beard, 2008; Qi *et al.*, 2008) have shown that raw data in the literature are very useful because they are consistent and may be explained with consensus mechanistic models. However, the kinetic parameters originally reported based on those raw data are not always consistent. By reanalyzing raw data from several kinetic studies for enzymes and transporters, we have been able to show that data introduced to support competing hypotheses (different mechanisms) are actually consistent with one consensus mechanism for each enzyme or transporter, and obtain associated kinetic parameter estimates. In general, it is necessary to reanalyze reported data and determine a mechanism (or mechanisms) and associated parameter estimates that account for biochemical state and the kinetic data for each enzyme and transporter.

4.2. Databases of biochemical information

Simulation of large-scale systems will require the development of databases of kinetic and thermochemical data from which models can be assembled. Recently developed models of glycolysis (Vinnakota, 2006; Vinnakota *et al.*, 2006) and mitochondrial energetics (Wu *et al.*, 2007a,b, 2008) have been built in part based on thermodynamic data complied by Alberty (2003). Analysis and simulation of larger scale systems will require extending Alberty's thermodynamic database for species and reactions not currently included and developing and parameterizing mechanistic models for large numbers of enzymes and metabolite transporters.

In addition to data on thermodynamic constants, databases of kinetic mechanisms and parameters are needed. For many enzymes, detailed mechanistic models have been previously developed based on kinetic measurements made using purified enzymes *in vitro*. However, in nearly all cases much work remains to adapt and extend these mechanistic models to account for the biochemical state (pH, ionic strength, etc.) so that each enzyme model may be incorporated into an integrated system model in a rigorous way. Therefore, it will be necessary to collect kinetic mechanisms and associated parameters.

A number of databases on enzyme kinetics currently exist, with BRENDA (www.brenda.uni-koeln.de) being perhaps the most widely recognized. BRENDA is a clearinghouse of a vast amount of information on enzymes, including estimates of functional parameters (i.e., apparent kinetic constants) and pointers to the primary literature. A recent effort called System for the Analysis of Biochemical Pathways (SABIO-RK, sabio.villa-bosch.de) aims to collect kinetic models and data from the literature. While these tools are useful, much more focused and specialized tools are needed as well. Along these lines we have begun construction of a Biological Components Databank (www.biocoda.org) to provide an online clearinghouse of mechanistic models of functional biological components to use as building blocks to construct computational models of biological systems. Here, rather than compiling the reported estimates of kinetic constants obtained from the original sources, we are reanalyzing reported data and determining mechanisms and associated parameter estimates for a number of enzymes and transporters. As pilot studies, we have gone through this process for several components.

Our goal is to build a data bank where each entry meets detailed validation and mechanistic standards. However, it is important to realize that it is unavoidable that each entry will pose a unique challenge, requiring a critical assessment of the literature and an expert analysis of data. Therefore, building this data bank will be a huge effort, requiring widespread collaboration within the community. Information on how to contribute to the data bank can be found on the Web site.

ACKNOWLEDGMENT

This work supported by NIH grant HL072011.

REFERENCES

Alberty, R. A. (1993). Levels of thermodynamic treatment of biochemical reaction systems. *Biophys. J.* **65,** 1243–1254.
Alberty, R. A. (2003). "Thermodynamics of Biochemical Reactions." Wiley-Interscience, Hoboken, NJ.
Alberty, R. A. (2004). A short history of the thermodynamics of enzyme-catalyzed reactions. *J. Biol. Chem.* **279,** 27831–27836.
Alberty, R. A. (2006). Relations between biochemical thermodynamics and biochemical kinetics. *Biophys. Chem.* **124,** 11–17.
Balaban, R. S., Kantor, H. L., Katz, L. A., and Briggs, R. W. (1986). Relation between work and phosphate metabolite in the *in vivo* paced mammalian heart. *Science* **232,** 1121–1123.
Beard, D. A., and Qian, H. (2007). Relationship between thermodynamic driving force and one-way fluxes in reversible processes. *PLoS ONE* **2,** e144.
Beard, D. A., and Qian, H. (2008a). Biochemical reaction networks. In "Chemical Biophysics: Quantitative Analysis of Cellular Processes," pp. 128–161. Cambridge Univ. Press, Cambridge, UK.
Beard, D. A., and Qian, H. (2008b). Concepts from physical chemistry. In "Chemical Biophysics: Quantitative Analysis of Cellular Systems," pp. 7–23. Cambridge Univ. Press, Cambridge, UK.

Beard, D. A., and Qian, H. (2008c). Conventions and calculations for biochemical systems. *In* "Chemical Biophysics: Quantitative Analysis of Cellular Systems," pp. 24–40. Cambridge Univ. Press, Cambridge, UK.

Beard, D. A., Vinnakota, K. C., and Wu, F. (2008). Detailed enzyme kinetics in terms of biochemical species: Study of citrate synthase. *PLoS ONE* **3,** e1825.

Clarke, E. C. W., and Glew, D. N. (1980). Evaluation of Debye-Huckel limiting slopes for water between 0 and 50 C. *J. Chem. Soc. Faraday Trans.* 1 **76,** 1911–1916.

Dash, R. K., and Beard, D. A. (2008). Analysis of cardiac mitochondrial Na^+/Ca^{2+} exchanger kinetics with a biophysical model of mitochondrial Ca^{2+} handing suggests a 3:1 stoichiometry. *J. Physiol.* **586,** 3267–3285.

Frieden, C., and Alberty, R. A. (1955). The effect of pH on fumarase activity in acetate buffer. *J. Biol. Chem.* **212,** 859–868.

Gibbs, C. (1985). The cytoplasmic phosphorylation potential: Its possible role in the control of myocardial respiration and cardiac contractility. *J. Mol. Cell Cardiol.* **17,** 727–731.

Gibbs, C. L. (1978). Cardiac energetics. *Physiol. Rev.* **58,** 174–254.

Godt, R. E., and Maughan, D. W. (1988). On the composition of the cytosol of relaxed skeletal muscle of the frog. *Am. J. Physiol.* **254,** C591–C604.

Golding, E. M., Teague, W. E. Jr., and Dobson, G. P. (1995). Adjustment of K′ to varying pH and pMg for the creatine kinase, adenylate kinase and ATP hydrolysis equilibria permitting quantitative bioenergetic assessment. *J. Exp. Biol.* **198,** 1775–1782.

Guyton, A. C., and Hall, J. E. (2006). "Textbook of Medical Physiology." Elsevier Saunders, Philadelphia.

Ingwall, J. S., and Weiss, R. G. (2004). Is the failing heart energy starved? On using chemical energy to support cardiac function. *Circ. Res.* **95,** 135–145.

Katz, A. M. (2006). "Physiology of the Heart." Lippincott Williams & Wilkins, Philadelphia.

Konishi, M. (1998). Cytoplasmic free concentrations of Ca^{2+} and Mg^{2+} in skeletal muscle fibers at rest and during contraction. *Japan. J. Physiol.* **48,** 421–438.

Korzeniewski, B. (2001). Theoretical studies on the regulation of oxidative phosphorylation in intact tissues. *Biochim. Biophys. Acta* **1504,** 31–45.

Kushmerick, M. J. (1997). Multiple equilibria of cations with metabolites in muscle bioenergetics. *Am. J. Physiol.* **272,** C1739–C1747.

LaNoue, K. F., and Schoolwerth, A. C. (1979). Metabolite transport in mitochondria. *Annu. Rev. Biochem.* **48,** 871–922.

Lawson, J. W., and Veech, R. L. (1979). Effects of pH and free Mg^{2+} on the Keq of the creatine kinase reaction and other phosphate hydrolyses and phosphate transfer reactions. *J. Biol. Chem.* **254,** 6528–6537.

Lineweaver, H., and Burk, D. (1934). The determination of enzyme dissociation constants. *J. Am. Chem. Soc.* **56,** 658–666.

Nelson, D. L., and Cox, M. M. (2005). "Lehninger Principles of Biochemistry." W.H. Freeman, New York.

Qi, F., Chen, X., and Beard, D. A. (2008). Detailed kinetics and regulation of mammalian NAD-linked isocitrate dehydrogenase. *Biochim. Biophys. Acta* **1784,** 1641–1651.

Scopes, R. K. (1974). Studies with a reconstituted muscle glycolytic system: The rate and extent of glycolysis in simulated post-mortem conditions. *Biochem. J.* **142,** 79–86.

Stanley, W. C., Recchia, F. A., and Lopaschuk, G. D. (2005). Myocardial substrate metabolism in the normal and failing heart. *Physiol. Rev.* **85,** 1093–1129.

Teague, W. E. Jr., and Dobson, G. P. (1992). Effect of temperature on the creatine kinase equilibrium. *J. Biol. Chem.* **267,** 14084–14093.

Teague, W. E. Jr., Golding, E. M., and Dobson, G. P. (1996). Adjustment of K′ for the creatine kinase, adenylate kinase and ATP hydrolysis equilibria to varying temperature and ionic strength. *J. Exp. Biol.* **199,** 509–512.

Vendelin, M., Kongas, O., and Saks, V. (2000). Regulation of mitochondrial respiration in heart cells analyzed by reaction-diffusion model of energy transfer. *Am. J. Physiol. Cell Physiol.* **278,** C747–C764.

Vendelin, M., Lemba, M., and Saks, V. A. (2004). Analysis of functional coupling: Mitochondrial creatine kinase and adenine nucleotide translocase. *Biophys. J.* **87,** 696–713.

Vinnakota, K., Kemp, M. L., and Kushmerick, M. J. (2006). Dynamics of muscle glycogenolysis modeled with pH time course computation and pH-dependent reaction equilibria and enzyme kinetics. *Biophys. J.* **91,** 1264–1287.

Vinnakota, K. C. (2006). "pH Dynamics, Glycogenolysis and Phosphoenergetics in Isolated Cell Free Reconstituted Systems and in Mouse Skeletal Muscle." University of Washington, Seattle, WA.

Vinnakota, K. C., and Bassingthwaighte, J. B. (2004). Myocardial density and composition: A basis for calculating intracellular metabolite concentrations. *Am. J. Physiol. Heart Circ. Physiol.* **286,** H1742–H1749.

Wu, F., Jeneson, J. A., and Beard, D. A. (2007a). Oxidative ATP synthesis in skeletal muscle is controlled by substrate feedback. *Am. J. Physiol. Cell Physiol.* **292,** C115–C1247.

Wu, F., Yang, F., Vinnakota, K. C., and Beard, D. A. (2007b). Computer modeling of mitochondrial tricarboxylic acid cycle, oxidative phosphorylation, metabolite transport, and electrophysiology. *J. Biol. Chem.* **282,** 24525–24537.

Wu, F., Zhang, E. Y., Zhang, J., Bache, R. J., and Beard, D. A. (2008). Phosphate metabolite concentrations and ATP hydrolysis potential in normal and ischemic hearts. *J. Physiol.* **586,** 4193–4208.

Zhang, J., Merkle, H., Hendrich, K., Garwood, M., From, A. H., Ugurbil, K., and Bache, R. J. (1993). Bioenergetic abnormalities associated with severe left ventricular hypertrophy. *J. Clin. Invest.* **92,** 993–1003.

Zhang, J., Ugurbil, K., From, A. H., and Bache, R. J. (2001). Myocardial oxygenation and high-energy phosphate levels during graded coronary hypoperfusion. *Am. J. Physiol. Heart Circ. Physiol.* **280,** H318–H326.

Zhou, L., Stanley, W. C., Saidel, G. M., Yu, X., and Cabrera, M. E. (2005). Regulation of lactate production at the onset of ischaemia is independent of mitochondrial NADH/NAD+: Insights from in silico studies. *J. Physiol.* **569,** 925–937.

CHAPTER THREE

Analytical Methods for the Retrieval and Interpretation of Continuous Glucose Monitoring Data in Diabetes

Boris Kovatchev, Marc Breton, *and* William Clarke

Contents

1. Introduction	70
2. Decomposition of Sensor Errors	73
3. Measures of Average Glycemia and Deviation from Target	74
4. Risk and Variability Assessment	76
5. Measures and Plots of System Stability	80
6. Time-Series-Based Prediction of Future BG Values	81
7. Conclusions	84
Acknowledgments	84
References	84

Abstract

Scientific and industrial effort is now increasingly focused on the development of closed-loop control systems (artificial pancreas) to control glucose metabolism of people with diabetes, particularly type 1 diabetes mellitus. The primary prerequisite to a successful artificial pancreas, and to optimal diabetes control in general, is the continuous glucose monitor (CGM), which measures glucose levels frequently (e.g., every 5 min). Thus, a CGM collects detailed glucose time series, which carry significant information about the dynamics of glucose fluctuations. However, a CGM assesses blood glucose indirectly via subcutaneous determinations. As a result, two types of analytical problems arise for the retrieval and interpretation of CGM data: (1) the order and the timing of CGM readings and (2) sensor errors, time lag, and deviations from BG need to be accounted for. In order to improve the quality of information extracted from CGM data, we suggest several analytical and data visualization methods. These analyses evaluate CGM errors, assess risks associated with glucose variability, quantify glucose system stability, and predict glucose fluctuation. All analyses are illustrated with data collected using MiniMed CGMS (Medtronic, Northridge, CA)

University of Virginia Health System, Charlottesville, Virginia

Methods in Enzymology, Volume 454 © 2009 Elsevier Inc.
ISSN 0076-6879, DOI: 10.1016/S0076-6879(08)03803-2 All rights reserved.

and Freestyle Navigator (Abbott Diabetes Care, Alameda, CA). It is important to remember that traditional statistics do not work well with CGM data because consecutive CGM readings are highly interdependent. In conclusion, advanced analysis and visualization of CGM data allow for evaluation of dynamical characteristics of diabetes and reveal clinical information that is inaccessible via standard statistics, which do not take into account the temporal structure of data. The use of such methods has the potential to enable optimal glycemic control in diabetes and, in the future, artificial pancreas systems.

1. INTRODUCTION

In health, the metabolic network responds to ambient glucose concentration and glucose variability (Hirsch and Brownlee, 2005). The goals of the network are to reduce basal and postprandial glucose elevations and to avoid overdelivery of insulin and hypoglycemia. In both type 1 and type 2 diabetes mellitus (T1DM, T2DM), this internal self-regulation is disrupted, leading to higher average glucose levels and dramatic increases in glucose variability. Recent national data show that nearly 21 million Americans have diabetes, and one in three American children born today will develop the disease. In individuals with T1DM, the immune system destroys the pancreatic β cells. As a result, 50,000 insulin shots are needed over a lifetime with T1DM, accompanied by testing of blood glucose (BG) levels several times a day. In T2DM, increased insulin resistance is amplified by the failure of the β cell to compensate with adequate insulin delivery. Both T1DM and T2DM are lifelong conditions that affect people of every race and nationality and are leading causes of kidney failure, blindness, and amputations not related to injury. The only treatment of diabetes proven to reduce the risks of serious complications is tight control of BG levels (DCCT 1993; UKPDS, 1998). It is estimated that diabetes and its comorbidities account for more than $132 billion of our nation's annual health care costs and one out of every three Medicare dollars (U.S. Senate hearing, 2006).

Monitoring of BG levels and accurate interpretation of these data are critical to the achievement of tight glycemic control. Since measuring mean BG directly is not practical, the assessment of glycemic control is with a single simple test, namely, hemoglobin A1c (HbA1c)(Santiago, 1993). However, the development of continuous glucose monitoring (CGM) technology has changed this conclusion (Klonoff, 2005). It is now feasible to observe temporal glucose fluctuations in real time and to use these data for feedback control of BG levels. Such a control could be patient initiated or actuated by a control algorithm via variable insulin infusion. Increasing industrial and research effort is concentrated on the development of CGM that sample and record frequent (e.g., every 1–5 min) glucose level estimates. Several devices are currently on the U.S. and European markets (Klonoff, 2005, 2007). While the accuracy of these devices in terms of approximating any particular glucose level is still inferior to self-monitoring of blood glucose (SMBG)(Clarke and Kovatchev, 2007), CGM yields a wealth of information not only about current glucose levels but also about the BG rate and direction of change (Kovatchev et al., 2005). This is why a recent comprehensive review of this technology's clinical implications, accuracy, and current problems rightfully placed CGM on the roadmap for 21st-century diabetes therapy (Klonoff, 2005, 2007).

While CGM is new and the artificial pancreas is still experimental Hovorka (2004, 2005), Steil (2006), Weinzimer (2006), the current gold-standard clinical practice of assessment is SMBG. Contemporary SMBG memory meters store up to several hundred self-monitoring BG readings, calculate various statistics, and visualize some testing results. However, SMBG data are generally one dimensional, registering only the amplitude of BG fluctuations at intermittent points in time. Thus the corresponding analytical methods are also one dimensional, emphasizing the concept of risk related to BG amplitude, but are incapable of capturing the process of BG fluctuations over time (Kovatchev et al., 2002, 2003). In contrast, CGM can capture the *temporal dimension* of BG fluctuations, enabling detailed tracking of this process. As a result, the statistical methods traditionally applied to SMBG data become unsuitable for the analysis and interpretation of continuous monitoring time series (Kollman et al., 2005; Kovatchev et al., 2005). New analytical tools are needed and are being introduced, ranging from variability analysis (McDonnell et al., 2005) and risk tracking (Kovatchev et al., 2005) to time series and Fourier approaches (Miller and Strange, 2007). Before proceeding with the description of the analytical methods of this chapter, we will first formulate the principal requirements and challenges posed by the specifics of CGM data.

1. CGM assesses blood glucose fluctuations *indirectly* by measuring the concentration of interstitial glucose, but is calibrated via self-monitoring to approximate BG (King et al., 2007). Because CGM operates in the interstitial compartment, which is presumably related to blood via diffusion across the capillary wall (Boyne et al., 2003; Steil et al., 2005), they face a number of significant challenges in terms of sensitivity, stability, calibration, and physiological time lag between blood and interstitial glucose concentration (Cheyne et al., 2002; Kulcu et al., 2003; Stout et al., 2004). Thus, analytical methods are needed to assess and evaluate different types of sensor errors due to calibration, interstitial delay, or random noise.

2. The CGM data stream has some inherent characteristics that allow for advanced data analysis approaches, but also call for caution if standard statistical methods are used. Most importantly, CGM data represent *time series*, that is, sequential readings that are ordered in time. This leads to two fundamental requirements to their analysis: First, consecutive sensor readings taken from the same subject within a relatively short time are highly interdependent. Second, the order of the CGM data points is essential for clinical decision making. For example, the sequences 90→ 82→ 72 mg/dl and 72→ 82→ 90 mg/dl are clinically very different. In other words, while a random reshuffling of CGM data in time will not change traditional statistics, such as mean and variance, it will have a profound impact on the temporal interpretation of CGM data. It is therefore imperative to extract CGM information across several dimensions, including risk associated with BG amplitude as well as time.
3. A most important critical feature of contemporary CGM studies is their limited duration. Because the sensors of the CGM devices are generally short-lived (5–7 days), the initial clinical trials of CGM are bound to be relatively short term (days), and therefore their results cannot be assessed by slow measures, such as HbA1c, which takes 2–3 months to react to changes in average glycemia (Santiago, 1993). Thus, it is important to establish an array of clinical and numerical metrics that would allow testing of the effectiveness of CGM over the relatively short-term, few-day life span of the first CGM sensors. Before defining such an array, we would reiterate that the primary goal of CGM and the artificial pancreas is β (and possibly α)-cell replacement. Thus, the effectiveness of CGM needs to be judged via assessment of their ability to approximate nondiabetic BG concentration and fluctuation.

These principal requirements are reflected by the analytical methods presented in this chapter, which include (i) decomposition of sensor errors into errors due to calibration and blood-to-interstitial time lag, (ii) analysis of average glycemia and deviations from normoglycemia, (iii) risk and variability analysis of CGM traces that uses a nonlinear data transformation of the BG scale to transfer data into a risk space, (iv) measures of system stability, and (v) prediction of glucose trends and events using time-series-based forecast methods. The proposed analyses are accompanied by graphs, including glucose and risk traces and system dynamics plots. The presented methods are illustrated by CGM data collected during clinical trials using MiniMed CGMS (Medtronic, Northridge, CA) and Freestyle Navigator (Abbott Diabetes Care, Alameda, CA).

Before proceeding further, it is important to note that the basic unit for most analyses is the glucose trace of an individual, that is, a time-stamped series of CGM or blood glucose data recorded for one person. Summary characteristics and group-level analyses are derived after the individual traces are processed to produce meaningful individual markers of average glycemia, risk, or glucose variation. The analytical methodology is driven by the understanding that BG fluctuations are a continuous process in time, BG (t). Each point of this process is characterized by its value (BG level) and by its rate/direction of BG change. CGM presents the process BG(t) as a discrete time series $\{BG(t_n), n = 1, 2, \ldots\}$ that approximates BG(t) in steps determined by the resolution of the particular device (e.g., a new value displayed every 5 min).

2. DECOMPOSITION OF SENSOR ERRORS

Figure 3.1 presents the components of the error of MiniMed CGMS assessed during a hyperinsulinemic hypoglycemic clamp involving 39 subjects with T1DM. In this study reference, BG was sampled every 5 min and then reference data were synchronized with data from the CGMS. The calibration error was estimated as the difference between CGMS readings and computer-simulated recalibration of the raw CGMS current using *all* reference BG points to yield an approximation of the dynamics of interstitial glucose (IG) adjusted for the BG-to-IG gradient (King *et al.*, 2007). The physiologic BG-to-IG time lag was estimated as the difference between reference BG and the "perfectly" recalibrated CGMS signal. The mean absolute deviation (MAD) of sensor data was 20.9 mg/dl during euglycemia and 24.5 mg/dl during descent into and recovery from hypoglycemia. Computer-simulated recalibration reduced MAD to 10.6

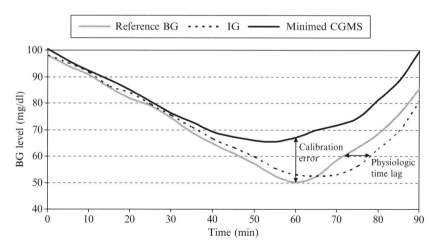

Figure 3.1 CGMS error during hypoglycemic clamp decomposed into error of calibration and physiologic deviation caused by blood-to-interstitial time lag.

and 14.6 mg/dl, respectively. Thus, during this experiment, approximately half of the sensor deviation from reference BG was attributed to calibration error; the rest was attributed to BG-to-IG gradient and random sensor deviations.

A diffusion model was fitted for each individual subject's data, as well as globally across all subjects. While the details of this model have been reported previously (King *et al.*, 2007), of particular importance is the finding that "global," across-subjects parameters describe the observed blood-to-interstitial delays reasonably well (King *et al.*, 2007). The availability of global parameters allows glucose concentration in the interstitium to be numerically estimated directly from reference BG data. This in turn allows for (i) setting the accuracy benchmark by simulating a sensor that does not have calibration errors, (ii) tracking and correction of errors due to calibration, and (iii) numerical compensation for BG-to-IG differential through an inverted diffusion model estimating BG from the sensor's IG approximation (in essence, the reference BG line in Fig. 3.1 could be derived numerically from the IG line, thereby reducing the influence of interstitial time lag).

3. Measures of Average Glycemia and Deviation from Target

Certain traditional data characteristics are clinically useful for the representation of CGM data. The computation of mean glucose values from CGM data and/or BG data points is straightforward and is generally suggested as a descriptor of overall control. Computing of pre- and postmeal averages and their difference can serve as an indication of the overall effectiveness of meal control. Computing percentage of time spent within, below, or above preset target limits has been proposed as well. The suggested cutoff limits are 70 and 180 mg/dl, which create three commonly accepted glucose ranges: hypoglycemia (BG $\leq$ 70 mg/dl)(ADA, 2005); normoglycemia (70 mg/dl < BG $\leq$ 180 mg/dl), and hyperglycemia (BG > 180 mg/dl). Percentage of time within additional bands can be computed as well to emphasize the frequency of extreme glucose excursions. For example, when it is important to distinguish between postprandial and postabsorptive (fasting) conditions, a fasting target range of 70–145 mg/dl is suggested. Further, %time <50 mg/dl would quantify the frequency of severe hypoglycemia, whereas %time >300 mg/dl would quantify the frequency of severe hyperglycemia. Table 3.1A includes the numerical measures of average glycemia and measures of deviation from target.

Plotting glucose traces observed during a set period of time represents the general pattern of a person's BG fluctuation. To illustrate the effect

Table 3.1 Summary measures representing CGM traces

A: Average glycemia and deviations from target	
Mean BG	Computed from CGM or blood glucose data for the entire test
% time spent within target range of 70–180 mg/dl; below 70 and above 180 mg/dl	For CGM, this generally equals to % readings within each of these ranges. For BG measurements that are not equally spaced in time, we suggest calculating the % time within each range via linear interpolation between consecutive glucose readings
% time ≤ 50 mg/dl	Reflects the occurrence of extreme hypoglycemia
% time >300 mg/dl	Reflects the occurrence of extreme hyperglycemia
B: Variability and risk assessment	
BG risk index	= LBGI + HBGI – measure of overall variability in "risk space"
Low BG index (LBGI)	Measure of the frequency and extent of low BG readings
High BG index (HBGI)	Measure of the frequency and extent of high BG readings
SD of BG rate of change	A measure of the stability of closed-loop control over time

of treatment observed via CGM we use previously published 72-h glucose data collected pre- and 4 weeks postislet transplantation (Kovatchev et al., 2005). Figure 3.2 presents glucose traces [of the process BG(t)] pre- and post-transplantation with superimposed aggregated glucose traces. The aggregated traces represent the time spent below/within/above target range.

The premise behind aggregation is as follows: Frequently one is not particularly interested in the exact BG value because close values such as 150 and 156 mg/dl are clinically indistinguishable. It is, however, important whether and when BG crosses certain thresholds, for example, 70 and 180 mg/dl as specified in the previous section. Thus, the entire process BG(t) can be aggregated into a process described only by the crossings of the thresholds of hypoglycemia and hyperglycemia. In Figs. 3.2A and B the aggregated process is depicted by squares that are black for hypoglycemia, white for normoglycemia (euglycemia), and gray for hyperglycemia. Each square represents the average of 1 h of CGM data. It is evident that

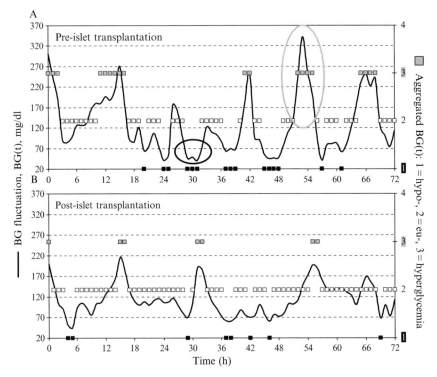

Figure 3.2 Glucose traces pre- and postislet transplantation with superimposed aggregated glucose traces. Aggregated traces represent the time spent below/within/above target range.

the aggregated process presents a clearer visual interpretation of changes resulting from islet transplantation: post-treatment of most of the BG fluctuations are within target, leading to a higher density of green squares. Possible versions of this plot include adding thresholds, such as 50 and 300 mg/dl, which would increase the levels of the aggregated process to five, and a higher resolution of the plot in the hypoglycemic range where one square of the aggregated process would be the average of 30 min of data. Table 3.2A includes a summary of the suggested graphs.

4. Risk and Variability Assessment

Computing standard deviation (SD) as a measure of glucose variability is not recommended because the BG measurement scale is highly asymmetric, the hypoglycemic range is numerically narrower than the hyperglycemic range, and the distribution of the glucose values of an individual is typically

Table 3.2 Graphs visualizing CGM traces and the effectiveness of treatment[a]

A: Average glycemia and deviations from target	
Glucose trace (Fig. 3.2)	Traditional plot of frequently sampled glucose data
Aggregated glucose trace (Fig. 3.2)	Corresponds to time spent below/within/above a preset target range. Visualizes the crossing of glycemic thresholds
B: Variability and risk assessment	
Risk trace (Fig. 3.3)	Corresponds to LBGI, HBGI, and BGRI. Designed to equalize the size of glucose deviations toward hypo- and hyperglycemia, emphasize large glucose excursions, and suppress fluctuation within target range, thereby highlighting essential variance
Histogram of BG rate of change (Fig. 3.4)	Represents the spread and range of glucose transitions. Related to system stability. Corresponds to SD of BG rate of change
Poincaré plot (Fig. 3.5)	Represents the spread of the system attractors and can be used for detection of cyclic glucose fluctuations

[a] Each graph corresponds to a numerical measure from Table 1.

quite skewed (Kovatchev et al., 1997). As a result from this asymmetry, SD would be predominantly influenced by hyperglycemic excursions and would not be sensitive to hypoglycemia. It is also possible for confidence intervals based on SD to assume unrealistic negative values. Thus, instead of reporting traditional measures of glucose variability, we suggest using risk indices based on a symmetrization transformation of the BG scale into a risk space (Kovatchev et al., 1997, 2001). The symmetrization formulas, published a decade ago, are data independent and have been used successfully in numerous studies. In brief, for any BG reading, we first compute:

$$f(BG) = 1.509 \times \left[\left(\ln(BG)\right)^{1.084} - 5.381\right]$$

if BG is measured in mg/dl or

$$f(BG) = 1.509 \times \left[\left(\ln(18 \times BG)\right)^{1.084} - 5.381\right]$$

if BG is measured in mmol/liter. Then we compute the BG risk function using

$$r(BG) = 10 \times f(BG)^2$$

and separate its left and right branches as follows:

$$rl(BG) = r(BG) \text{ if } f(BG) < 0 \text{ and } 0 \text{ otherwise}$$

$$rh(BG) = r(BG) \text{ if } f(BG) > 0 \text{ and } 0 \text{ otherwise.}$$

Given a series of CGM readings $BG_1, BG_2, \ldots BG_n$, we compute the low and high BG indices (LBGI, HBGI) as the average of $rl(BG)$ and $rh(BG)$, respectively (Kovatchev et al., 2001, 2005):

$$LBGI = \frac{1}{n} \sum_{i=1}^{n} rl(BG_i)$$

and

$$HBGI = \frac{1}{n} \sum_{i=1}^{n} rh(BG_i).$$

The BG risk index is then defined as $BGRI = LBGI + HBGI$.

In essence, the LBGI and the HBGI split the overall glucose variation into two independent sections related to excursions into hypo- and hyperglycemia and, at the same time, equalize the amplitude of these excursions with respect to the risk they carry. For example, in BG space, a transition from 180 to 250 mg/dl would appear threefold larger than a transition from 70 to 50 mg/dl, while in risk space these fluctuations would appear equal. Using the LBGI, HBGI, and their sum BGRI complements, the use of thresholds described earlier by adding information about the *extent of BG fluctuations*. A simple example would clarify this point. Assume two sets of BG readings: (110,65) and (110,40) mg/dl. In both cases we have 50% of readings below the threshold of 70 mg/dl; thus the percentage readings below target are 50% in both cases. However, the two scenarios are hardly equivalent in terms of risk for hypoglycemia, which is clearly depicted by the difference in their respective LBGI values: 5.1 and 18.2. Table 3.1B includes the suggested measures of glucose variability and associated risks.

Figure 3.3A and B present 72-h traces of BG dynamics in risk space corresponding to the glucose traces of Fig. 3.2A and B at baseline and postislet transplantation. Each figure includes fluctuations of the LBGI (lower half) and HBGI (upper half), with both indices computed from

Analysis of Continuous Glucose Monitoring Data 79

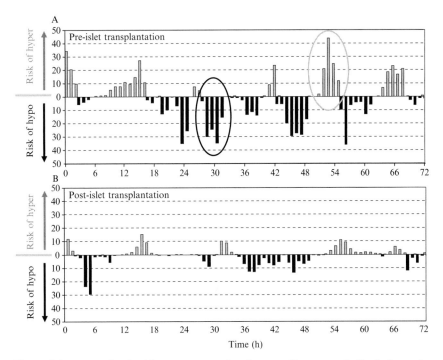

Figure 3.3 CGM data in risk space: converting data equalizes numerically the hypoglycemic and hyperglycemic ranges and suppresses the variance in the safe euglycemic range.

1-h time blocks (Kovatchev et al., 2005). In particular, Figs 3.2A and 3.3A demonstrate the effect of transforming BG fluctuations from glucose to risk values. A hypoglycemic event at hour 30 of study in Fig. 3.2A is visually expanded and emphasized in Fig. 3.3A (black circles). In contrast, the magnitude a hyperglycemic event at hour 54 in Fig. 3.2A is reduced in Fig. 3.3A to reflect the risk associated with that event (gray circles). Further, comparing Fig. 3.3A to Fig. 3.3B, it becomes evident that the magnitude of risk associated with glucose fluctuations decreases as a result of treatment. The average LBGI was 6.72 at baseline and 2.90 post-transplantation. Similarly, the HBGI was reduced from 5.53 at baseline to 1.73 after islet transplantation.

Thus, the advantages of a risk plot include the following: (i) the variance carried by hypo- and hyperglycemic readings is equalized; (ii) excursions into extreme hypo- and hyperglycemia get progressively increasing risk values; and (iii) the variance within the safe euglycemic range is attenuated, which reduces noise during data analysis. In essence, Figs. 3.3A and B link better glycemic control to a narrower pattern of risk fluctuations. Because the LBGI, HBGI, and the combined BGRI can theoretically range from 0 to 100, their values can be interpreted as percentages of maximum possible risk.

5. Measures and Plots of System Stability

Analysis of BG rate of change (measured in mg/dl/min) is suggested as a way to evaluate the dynamics of BG fluctuations on the timescale of minutes. In mathematical terms, this is an evaluation of the "local" properties of the system as opposed to "global" properties discussed earlier. Being the focus of differential calculus, local functional properties are assessed at a neighborhood of any point in time t_0 by the value $BG(t_0)$ and the derivatives of $BG(t)$ at t_0. The *BG rate of change* at t_i - is computed as the ratio $[BG(t_i) - BG(t_{i-1})]/(t_i - t_{i-1})$, where $BG(t_i)$ and $BG(t_{i-1})$ are CGM or reference BG readings taken at times t_i and t_{i-1} that are close in time. Recent investigations of the frequency of glucose fluctuations show that optimal evaluation of the BG rate of change would be achieved over time periods of 15 min (Miller and Strange, 2007; Shields and Breton, 2007), for example, $\Delta t = t_i - t_{i-1} = 15$. For data points equally spaced in time, this computation provides a sliding approximation of the first derivative (slope) of $BG(t)$. A larger variation of the BG rate of change indicates rapid and more pronounced BG fluctuations and therefore a less stable system. Thus, we use the standard deviation of the BG rate of change as a measure of stability of closed-loop control. Two points are worth noting: (i) as opposed to the distribution of BG levels, distribution of the BG rate of change is *symmetric* and, therefore, using SD is statistically accurate (Fig. 3.4) and (ii) the SD of BG rate of change has been introduced as a measure of stability computed from CGM data and is known as CONGA of order 1 (McDonnel *et al.*, 2005).

Figure 3.4A and B present histograms of the distribution of the BG rate of change over 15 min, computed from MiniMed CGMS data of our

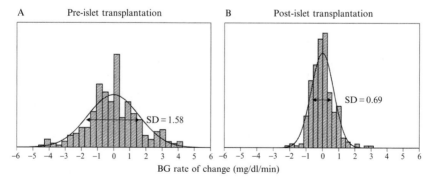

Figure 3.4 Histograms of the distribution of the BG rate of change over 15 min, computed from MiniMed CGMS data pre- and postislet transplantation.

transplantation case. It is apparent that the baseline distribution is more widespread than the distribution post-transplantation. Numerically, this effect is reflected by 19.3% of BG rates outside of the [−2, 2] mg/dl/min range in Fig. 3.4A versus only 0.6% BG rates outside that range in Fig. 3.4B. Thus, pretransplantation the patient experienced rapid BG fluctuations, whereas post-transplantation the rate of fluctuations was reduced dramatically. This effect is also captured by the SD of the BG rate of change, which is reduced from 1.58 to 0.69 mg/dl/min as a result of treatment.

Another look at system stability is provided by the Poincaré plot (lag plot) used in nonlinear dynamics to visualize the attractor of the investigated system (Brennan et al., 2001): a smaller, more concentrated attractor indicates system stability, whereas a more scattered Poincaré plot indicates system irregularity, reflecting poorer glucose control. Each point of the plot has coordinates $BG(t_{i-1})$ on the X axis and $BG(t_i)$ on the Y axis. Thus, the difference (Y-X) coordinates of each data point represents the BG rate of change occurring between times t_{i-1} and t_{i-1}. Figure 3.5A and B present Poincaré plots of CGM data at baseline and postislet transplantation. It is evident that the spread of the system attractor is substantially larger before treatment compared with post-treatment. Thus, the principal axes of the Poincaré plot can be used as numerical metrics of system stability.

Another use of the Poincaré plot is to scan data for *patterns of oscillation*. Because the plot of an oscillator is an ellipse (see Fig. 3.5C), elliptical configuration of data points would indicate cyclic glucose fluctuations. Table 3.2B includes a summary of the suggested graphs.

6. TIME-SERIES-BASED PREDICTION OF FUTURE BG VALUES

Most contemporary CGM systems include glucose prediction capabilities, in particular hypoglycemia and hyperglycemia alarms. Practically all predictions are currently based on a linear extrapolation of glucose values, for example, *projection* ahead of the current glucose trend. Because glucose fluctuations are generally nonlinear, such projections frequently result in errors and typically have a high false-alarm rate. In contrast, a time-series model-based sliding algorithm designed to continually predict glucose levels 30–45 min ahead had substantially higher accuracy that typical linear projections (Zanderigo et al., 2007). The sliding algorithm works as follows. For each time series, a linear model is fitted, continually at any sampling time, against past glucose data by weighted least squares. Then, the model is used to predict the glucose level at a preset prediction horizon. In model fitting, data points are "weighted" using a forgetting factor of 0.8 (which was determined to be optimal in numerical experiments), that is, the weight

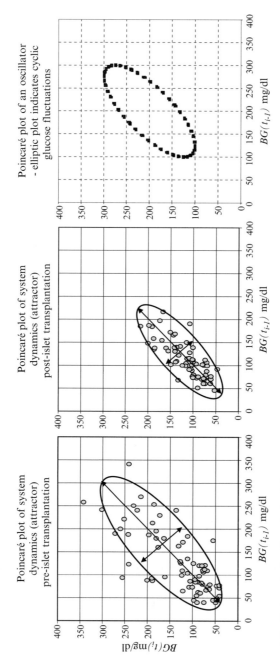

Figure 3.5 Poincaré plot (lag plot) of glucose fluctuations: a smaller, more concentrated spread of data indicates system stability, whereas a more scattered Poincaré plot indicates system irregularity, resulting from poorer glucose control.

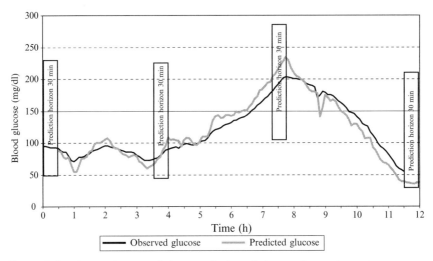

Figure 3.6 Thirty-minute real-time prediction of glucose fluctuation using an autoregression algorithm.

of the kth point before the actual sampling time is $(0.8)^k$. Figure 3.6 presents the action of the prediction algorithm at a prediction horizon of 30 min. While the algorithm tends to exaggerate transient peak and nadir glucose values, its overall predictive capability is very good. Judged by continuous glucose error-grid analysis (CG-EGA; Kovatchev et al., 2004), >80% of predicted vs actual values fall in CG-EGA zone A and >85% fall in zones A+B. As would be expected, the accuracy of the prediction is highest during euglycemia, with 97% of predicted vs actual values falling in CG-EGA zones A+B.

Finally, a statistical disadvantage of the CGM data stream is the high interdependence between data points taken from the same subject within a relatively short time. As a result, standard statistical analyses, such as t tests, while appropriate for independent data points, will produce inaccurate results if applied directly to CGS data. The reason is a severe violation of the statistical assumptions behind the calculation of degrees of freedom, which are essential to compute the p value of any statistical test. In order to clarify the dependence of consecutive CGM data points, we have computed their autocorrelation and have shown that it remains significant for approximately 1 h, after which its significance drops below the level of 0.05 (Kovatchev and Clarke, 2008). Thus, CGM readings separated by more than 1 h in time could be considered *linearly* independent, which is sufficient for some statistical tests. A note of caution is that linear independence does not imply stochastic independence, which might be essential in some cases. Another conclusion from this autocorrelation analysis is that CGM data aggregated in 1-h blocks would be reasonably approximated by a Markov chain, which opens a number of possibilities for the analysis of aggregated data. Finally, the linear dependence between consecutive data points practically disappears at a time lag of ≈30 min. Therefore, a projection of BG levels more than 30 min ahead, which is using linear methods, would be inaccurate. This last point has significant clinical impact on the settings of hypo- or hyperglycemia alarms, many of which are based on linear projections of past CGM data.

7. Conclusions

The intent of this chapter was to introduce a set of mathematically rigorous methods that provide statistical and visual interpretation of frequently sampled BG data, which might serve as a basis for evaluation of the effectiveness of closed-loop control algorithms. Because a major purpose of closed-loop control is to reduce BG fluctuations, these methods augment the traditional approaches with understanding and analysis of variability-associated risks and the temporal structure of glucose data. Table 3.1 presents a summary of the metrics suggested for the analysis of CGM data, whereas Table 3.2 presents corresponding graphs. It is envisioned that this system of methods would be employed both in *in silico* computer simulation trials (Kovatchev et al., 2008) and in clinical trials involving patients.

ACKNOWLEDGMENTS

This chapter was prepared with support of the JDRF Artificial Pancreas Consortium. The theoretical development of some of the presented metrics was supported by Grant RO1 DK 51562 from the National Institutes of Health. Data and material support were provided by Abbott Diabetes Care (Alameda, CA).

REFERENCES

American Diabetes Association (ADA) Workgroup on Hypoglycemia (2005). Defining and reporting hypoglycemia in diabetes: A report from the American Diabetes Association workgroup on hypoglycemia. *Diabetes Care* **28,** 1245–1249.
Boyne, M., Silver, D., Kaplan, J., and Saudek, C. (2003). Timing of changes in interstitial and venous blood glucose measured with a continuous subcutaneous glucose sensor. *Diabetes* **52,** 2790–2794.
Brennan, M., Palaniswami, M., and Kamen, P (2001). Do existing measures of Poincare plot geometry reflect nonlinear features of heart rate variability? *IEEE Trans. Biomed. Eng.* **48,** 1342–1347.
Cheyne, E. H., Cavan, D. A., and Kerr, D (2002). Performance of continuous glucose monitoring system during controlled hypoglycemia in healthy volunteers. *Diabetes Technol. Ther.* **4,** 607–613.

Clarke, W. L., and Kovatchev, B. P. (2007). Continuous glucose sensors continuing questions about clinical accuracy. *J. Diabetes Sci. Technol.* **1,** 164–170.

Diabetes Control and Complications Trial (DCCT) Research Group (1993). The effect of intensive treatment of diabetes on the development and progression of long-term complications of insulin-dependent diabetes mellitus. *N. Engl. J. Med.* **329,** 978–986.

Hirsch, I. B., and Brownlee, M. (2005). Should minimal blood glucose variability become the gold standard of glycemic control? *J. Diabetes Complications* **19,** 178–181.

Hovorka, R., Chassin, L. J., Wilinska, M. E., Canonico, V., Akwi, J. A., Federici, M. O., Massi-Benedetti, M., Hutzli, I., Zaugg, C., Kaufmann, H., Both, M., Vering, T., et al. (2004). Closing the loop: The Adicol experience. *Diabetes Technol. Ther.* **6,** 307–318.

Hovorka, R. (2005). Continuous glucose monitoring and closed-loop systems. *Diabet. Med.* **23,** 1–12.

King, C. R., Anderson, S. M., Breton, M. D., Clarke, W. L., and Kovatchev, B. P. (2007). Modeling of calibration effectiveness and blood-to-interstitial glucose dynamics as potential confounders of the accuracy of continuous glucose sensors during hyperinsulinemic clamp. *J. Diabetes Sci. Technol.* **1,** 317–322.

Klonoff, D. C. (2005). Continuous glucose monitoring: Roadmap for 21st century diabetes therapy. *Diabetes Care* **28,** 1231–1239.

Klonoff, D. C. (2007). The artificial pancreas: How sweet engineering will solve bitter problems. *J. Diabetes Sci. Technol.* **1,** 72–81.

Kollman, C., Wilson, D. M., Wysocki, T., Tamborlane, W. V., and Beck, R. W. (2005). Diabetes Research in Children Network: Limitation of statistical measures of error in assessing the accuracy of continuous glucose sensors. *Diabetes Technol. Ther.* **7,** 665–672.

Kovatchev, B. P., and Clarke, W. L. (2008). Peculiarities of the continuous glucose monitoring data stream and their impact on developing closed-loop control technology. *J. Diabetes Sci. Technol.* **2,** 158–163.

Kovatchev, B. P., Clarke, W. L., Breton, M., Brayman, K., and McCall, A. (2005). Quantifying temporal glucose variability in diabetes via continuous glucose monitoring: Mathematical methods and clinical applications. *Diabetes Technol. Ther.* **7,** 849–862.

Kovatchev, B. P., Cox, D. J., Gonder-Frederick, L. A., and Clarke, W. L. (1997). Symmetrization of the blood glucose measurement scale and its applications. *Diabetes Care* **20,** 1655–1658.

Kovatchev, B. P., Cox, D. J., Gonder-Frederick, L. A., and Clarke, W. L. (2002). Methods for quantifying self-monitoring blood glucose profiles exemplified by an examination of blood glucose patterns in patients with type 1 and type 2 diabetes. *Diabetes Technol. Ther.* **4,** 295–303.

Kovatchev, B. P., Cox, D. J., Kumar, A., Gonder-Frederick, L. A., and Clarke, W. L. (2003). Algorithmic evaluation of metabolic control and risk of severe hypoglycemia in type 1 and type 2 diabetes using self-monitoring blood glucose (SMBG) data. *Diabetes Technol. Ther.* **5,** 817–828.

Kovatchev, B. P., Gonder-Frederick, L. A., Cox, D. J., and Clarke, W. L. (2004). Evaluating the accuracy of continuous glucose-monitoring sensors: Continuous glucose-error grid analysis illustrated by TheraSense Freestyle Navigator data. *Diabetes Care* **27,** 1922–1928.

Kovatchev, B. P., Straume, M., Cox, D. J., and Farhi, L. S. (2001). Risk analysis of blood glucose data: A quantitative approach to optimizing the control of insulin dependent diabetes. *J. Theor. Med.* **3,** 1–10.

Kulcu, E., Tamada, J. A., Reach, G., Potts, R. O., and Lesho, M. J. (2003). Physiological differences between interstitial glucose and blood glucose measured in human subjects. *Diabetes Care* **26,** 2405–2409.

McDonnell, C. M., Donath, S. M., Vidmar, S. I., Werther, G. A., and Cameron, F. J. (2005). A novel approach to continuous glucose analysis utilizing glycemic variation. *Diabetes Technol. Ther.* **7,** 253–263.

Miller, M., and Strange, P. (2007). Use of Fourier models for analysis and interpretation of continuous glucose monitoring glucose profiles. *J. Diabetes Sci. Technol.* **1,** 630–638.

Santiago, J. V. (1993). Lessons from the Diabetes Control and Complications Trial. *Diabetes* **42,** 1549–1554.

Shields, D., and Breton, M. D. (2007). "Blood vs. Interstitial Glucose Dynamic Fluctuations: The Nyquist Frequency of Continuous Glucose Monitors." Proc. 7th Diabetes Technol Mtg, p. A87. San Francisco, CA.

Steil, G. M., Rebrin, K., Darwin, C., Hariri, F., and Saad, M. F. (2006). Feasibility of automating insulin delivery for the treatment of type 1 diabetes. *Diabetes* **55,** 3344–3350.

Steil, G. M., Rebrin, K., Hariri, F., Jinagonda, S., Tadros, S., Darwin, C., and Saad, M. F. (2005). Interstitial fluid glucose dynamics during insulin-induced hypoglycaemia. *Diabetologia* **48,** 1833–1840.

Stout, P. J., Racchini, J. R., and Hilgers, M. E. (2004). A novel approach to mitigating the physiological lag between blood and interstitial fluid glucose measurements. *Diabetes Technol. Ther.* **6,** 635–644.

UK Prospective Diabetes Study Group (1998). Intensive blood-glucose control with sulphonylureas or insulin compared with conventional treatment and risk of complications in patients with type 2 diabetes. *Lancet* **352,** 837–853.

U.S. Senate hearing (September 27, 2006). "The Potential of an Artificial Pancreas: Improving Care for People with Diabetes."

Weinzimer, S. (2006). "Closed-Loop Artificial Pancreas: Feasibility Studies in Pediatric Patients With Type 1 Diabetes." Proc. 6th Diabetes Technology Meeting, p. S55. Atlanta, GA.

Zanderigo, F., Sparacino, G., Kovatchev, B., and Cobelli, C. (2007). Glucose prediction algorithms from continuous monitoring data: Assessment of accuracy via continuous glucose-error grid analysis. *J. Diabetes Sci. Technol.* **1,** 645–651.

CHAPTER FOUR

Analysis of Heterogeneity in Molecular Weight and Shape by Analytical Ultracentrifugation Using Parallel Distributed Computing

Borries Demeler,* Emre Brookes,* *and* Luitgard Nagel-Steger[†]

Contents

1. Introduction	88
2. Methodology	89
3. Job Submission	97
4. Results	100
4.1. Example 1: Simulated five-component system	100
4.2. Example 2: Amyloid aggregation	104
5. Conclusions	109
Acknowledgments	111
References	111

Abstract

A computational approach for fitting sedimentation velocity experiments from an analytical ultracentrifuge in a model-independent fashion is presented. This chapter offers a recipe for obtaining high-resolution information for both the shape and the molecular weight distributions of complex mixtures that are heterogeneous in shape and molecular weight and provides suggestions for experimental design to optimize information content. A combination of three methods is used to find the solution most parsimonious in parameters and to verify the statistical confidence intervals of the determined parameters. A supercomputer implementation with a MySQL database back end is integrated into the UltraScan analysis software. The UltraScan LIMS Web portal is used to perform the calculations through a Web interface. The performance and limitations of the method when employed for the analysis of complex

---

* Department of Biochemistry, The University of Texas Health Science Center at San Antonio, San Antonio, Texas
[†] Heinrich-Heine-Universität Düsseldorf, Institut für Physikalische Biologie, Düsseldorf, Germany

Methods in Enzymology, Volume 454   © 2009 Elsevier Inc.
ISSN 0076-6879, DOI: 10.1016/S0076-6879(08)03804-4   All rights reserved.

mixtures are demonstrated using both simulated data and experimental data characterizing amyloid aggregation.

1. INTRODUCTION

Many of today's biomedical research projects studying the molecular basis for cancer and other diseases focus on the understanding of dynamic interactions among molecules implicated in the disease process. Analytical ultracentrifugation (AUC) offers an array of powerful tools to study such interactions. AUC experiments make it possible to observe macromolecules and macromolecular assemblies in *solution*, that is, in a physiological environment unconstrained by crystal packing forces or an electron microscope grid. Systems can be studied under a wide range of concentrations and buffer conditions, and the methods are applicable to a very large range of molecular weights, extending from just a few hundred daltons to systems as large as whole virus particles. The range of applications is further extended by several detection systems, which include absorbance optics for ultraviolet (UV) and visible wavelengths, Rayleigh interference optics, fluorescence intensity optics, turbidity, and schlieren optics. In addition, new detectors are currently being developed, such as multiwavelength absorbance, small angle light scattering, and Raman spectroscopy detectors. AUC experiments, in conjunction with sophisticated numerical analysis, can yield a wealth of information about a wide range of hydrodynamic and thermodynamic properties of the macromolecules under investigation, including molecular weight, association (K_{eq}) and rate (k_{off}) constants, sedimentation coefficients (s), diffusion coefficients (D), and shape factors (f/f_0), as well as partial concentrations of individual solutes. These parameters provide insight into macromolecular organization and function, oligomerization characteristics, conformation, binding stoichiometry, and sample composition. This chapter describes new computational tools and algorithms whose goal it is to identify hydrodynamic parameters with the highest possible resolution.

During an AUC experiment, a sample of interest is dissolved in a buffer, and the solution is placed into a sector-shaped cell and sedimented in a centrifugal force field. The centrifugal force is adjusted according to the size of the molecules under study and can be as high as 260,000 g, permitting resolution of components with a broad molecular weight range. During the experiment, differently sized and shaped components in the sample will sediment away from the rotor center at rates proportional to their molecular weight divided by their shape, creating a moving boundary with a distinct concentration profile that is dependent on s and D of each solute. Depending on optical system, data are collected every few seconds or minutes by monitoring the change in total concentration over the entire radial domain. An example of a multicomponent data set obtained with UV absorption

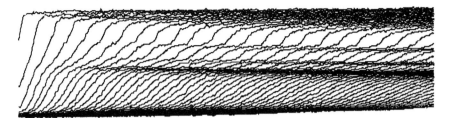

Figure 4.1 Experimental AUC data of a multicomponent mixture. The x axis reflects the radius, and the y axis represents the relative concentration. The direction of sedimentation is from left to right; each trace represents a single time point in the experiment. In this example, multiple components can be distinguished by mere visual inspection of the profile.

optics is shown in Fig. 4.1. Each component in the mixture contributes a partial concentration to the observed concentration profile. In order to analyze these data, one has to identify the contributions of individual components to the overall concentration profile. This process involves modeling the partial concentration, the sedimentation, and the diffusion transport of each solute over time and requires proper accounting of noise contributions. In the case of reacting systems, such as reversibly self- or hetero associating systems, equilibrium and rate constants need to be considered as well. The sedimentation and diffusion transport of each solute is described by a partial differential equation (PDE), the Lamm equation (Lamm, 1929), which can be solved at high resolution using an adaptive space-time finite element approach for either the noninteracting (Cao and Demeler, 2005) or the interacting (Cao and Demeler, 2008) case. Linear combinations of Lamm equation solutions are then used to approximate the entire concentration profile. Finding the correct parameter combinations of partial concentration, s and D, for each solute is accomplished by solving the inverse problem of fitting simulated finite element solutions to experimental data. To this end, we have developed new algorithms that allow us to model such experiments at the highest possible resolution. These algorithms are based on first-principle biophysical descriptions and employ advanced numerical techniques and parallel computing to accomplish this goal. Our developments have been integrated into an open-source software package called UltraScan (Demeler, 2005, 2008), which contains a comprehensive range of tools to help interpret experimental data and derive aforementioned parameters from AUC experiments.

2. Methodology

The computational task of analyzing experimental data by building a model that best represents the experimental information can be separated into four phases: (1) initialization—determining the appropriate parameter range to be searched; (2) two-dimensional spectrum analysis (2DSA)—calculating a linear combination of basis functions that covers the parameter space with subsequent linear least-squares fitting, with simultaneous elimination of systematic noise; (3) refinement—parsimonious regularization by genetic algorithm analysis (GA) or optionally nonlinear fitting with a discrete model; and (4) Monte Carlo analysis (MC)—statistical evaluation of the results and attenuation of stochastic noise contributions. In addition, a careful design of the experiment is an important consideration for the success of the experiment. A large range of different approaches exist to evaluate experimental sedimentation data, each with its own advantages and limitations on information content. Of those, model-independent approaches such as the van Holde–Weischet analysis (Demeler and van Holde, 2004) and the dC/dt method (Stafford, 1992) are preferable for the initialization steps; subsequently, direct boundary modeling by finite element solutions of the Lamm equation is desirable because unlike model-independent methods that provide only s-value distributions and, in some cases, partial concentrations, finite element solutions permit simultaneous determination of the sedimentation and diffusion coefficients of each species, as well as determination of partial concentrations. When s and D are available, additional information can be derived, including molecular weight and frictional parameters by applying the Svedberg equation, which relates s and D to the molecular weight of the particle [see Eq. (4.1), where R is the gas constant, T is the temperature, ρ is the density of the solvent, and s, D, M, and $\bar{v}$ are the sedimentation and diffusion coefficients, the molecular weight, and the partial specific volume of the solute, respectively]. Once s and D have been determined, a frictional ratio, f/f_0, can be calculated as well according to Eq. (4.2). This ratio provides a convenient parameterization for the shape of the molecule. The lower limit of 1.0 can be interpreted as a spherical molecule, whereas values greater than 1 indicate increasing nonglobularity. Values up to 1.3 are common for mostly globular proteins, whereas values between 1.8 and 2.5 are consistent with elongated, denatured, or intrinsically disordered proteins. Values larger than 2.5 can be found for very long molecules such as linear DNA fragments or long fibrils. Solving the inverse problem of fitting a model to the experimental data can be accomplished with a nonlinear least-squares fitting routine to arrive at the best-fit parameter set for the nonlinear parameters to be searched. However, this approach is suitable only for simple cases, where at most one or two solutes are present. The reason for the failure of this approach with a larger number of species is related to the complexity of the error surface, which increases when an increasing number of solutes and parameters is modeled and causes the optimization algorithm often to stall in a local minimum, preventing convergence at the global minimum. This chapter describes three alternative approaches addressing this issue, and each approach provides a complementary description of the solution space.

These approaches can be linked to provide an optimal description of data. Before detailing our approach, we should mention the importance of high-quality data. No amount of sophisticated analysis can compensate for poor quality of primary data and all efforts should be taken to eliminate unnecessary noise from data. The precision of parameter estimation is inversely correlated with the experimental noise present in primary data. It is therefore important that systematic noise contributions resulting from instrument flaws are accounted for and that stochastic noise contributions are attenuated. It has been shown previously that systematic noise contributions such as time- and radially invariant noise can be eliminated effectively using algebraic means (Schuck and Demeler, 1999) and that the effect of stochastic noise contributions can be reduced using MC methods (Demeler and Brookes, 2008). Experimental design considerations can further improve noise characteristics, for example, by using intensity measurements instead of absorbance measurements, stochastic noise is reduced by a factor of $\sqrt{2}$ by not subtracting the reference signal. This subtraction leads to the convolution of two stochastic noise vectors and an increase in the stochastic noise:

$$\frac{s}{D} = \frac{M(1 - \bar{v}\rho)}{RT} \tag{4.1}$$

$$\frac{f}{f_0} = \frac{RT}{3D\eta(6N^2\pi^2 M\bar{v})^{\frac{1}{3}}} \tag{4.2}$$

Step 1. Selecting the appropriate parameter space. The complexity of the evaluation can be reduced if a subspace of reasonable parameter values can be obtained through model-independent approaches. The enhanced van Holde–Weischet method (Demeler and van Holde, 2004) is ideally suited for this purpose, as it provides diffusion-corrected s-value distributions from sedimentation velocity experiments. The diffusion coefficient, which is required for the solution of the Lamm equation, can be initially estimated by parameterizing the shape function using the frictional ratio, f/f_0, which is a measure of the globularity of a particle [Eq. (4.3), where N is Avogadro's number, k is the frictional ratio f/f_0, η is the viscosity of the solvent, and all other symbols are the same as in Eq. (4.1)]. A reasonable assumption can be made that the shape of the particle ranges somewhere between spherical ($f/f_0 = 1.0$) and rod shaped ($f/f_0 \leq 4.0$) for most solutes. Given the limits from the s-value range determined with either the van Holde–Weischet or the dC/dt method, and the assumption on particle shape, it is now possible to define the limits of a two-dimensional parameter space over s and f/f_0:

$$D = RT\left[N18\pi(k\eta)^{3/2}\left(\frac{s\bar{v}}{2(1-\bar{v}\rho)}\right)^{1/2}\right]^{-1} \tag{4.3}$$

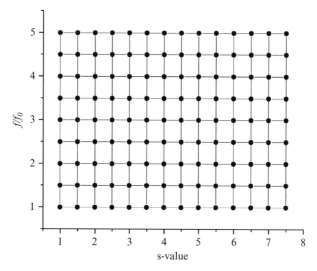

Figure 4.2 Two-dimensional grid over s and f/f_0. Each node point represents a term in the linear model, whose amplitude is determined by the 2DSA analysis through least-squares fit using NNLS.

Step 2. Evaluating the basis functions over all parameters and eliminating systematic noise. A two-dimensional grid with discrete s and f/f_0 parameter combinations covering the s-value range and the f/f_0 range is constructed. Such a grid is shown in Fig. 4.2. Here, each grid point represents a complete finite element simulation over both space and time, as well as a solute in the delimited parameter space. The solutes at each grid point are simulated at unity concentration and their contributions are summed to obtain the final concentration profile. The amplitude of each term in the sum represents the relative contribution of each solute. The fitting problem is thus reduced from a nonlinear fitting problem to a linear least-squares approach [Eq. (4.4)] that only requires determination of the amplitudes of each grid point. In this approach, $\boldsymbol{M}$ is the model [Equation (4.5)], which is compared to experimental data $\boldsymbol{b}$ over all time and radius points r and t. $c_{l,m}$ represents the amplitude of each grid point in Fig. 4.2, and L is the solution of the Lamm equation for a single, nonreacting solute (Brookes and Demeler, 2006; Cao and Demeler, 2005). The fit is accomplished with the nonnegatively constrained least-squares fitting algorithm called "NNLS" (Lawson and Hanson, 1974). This approach will result in positive amplitudes only, or zero, if a component is not present in experimental data. There are several considerations to be made relating to computer memory requirements. The finite element solution at each grid point in Fig. 4.2 requires approximately 800–1000 radial points and 50–500 time points, depending on the experiment. It is clear that fitting a high-resolution grid will require excessive amounts of memory. To circumvent this significant problem, we employed a divide-and-conquer approach, termed the multistage two-dimensional spectrum analysis, which repeatedly moves the initial low-resolution grid by small increments Δs and $\Delta f/f_0$ until the entire parameter space is covered. This approach accomplishes a higher resolution analysis without overwhelming available memory and has been shown to provide a serial speedup over standard NNLS and parallelizes with no excess computation and negligible communications overhead (Brookes et al., 2006). Scalability testing of 2DSA has shown a linear speedup from 4 to 512 processors. We formulated an equation for processor utilization (Brookes et al., 2006), which predicts the optimal number of processors to guarantee a minimum processor utilization for a specific problem size. We currently use this method of computing the number of processors required for a grid job submission by targeting 80% processor utilization for problems without time-invariant noise and 70% processor utilization for better speedup on the more computationally intensive time-invariant noise calculation. As a consequence, our implementation achieves the best quality of results without wasting computational resources on a large cluster. Since the solution is sparse, only a few parameters are returned from a coarse grid. Typically, we apply 100–300 grid movings of a 10 × 10 grid to obtain a resolution that is commensurate with the resolution of the analytical ultracentrifuge. Fewer grid movings can be used if the van Holde–Weischet analysis reports a narrow range. Solutes with positive amplitudes from different grids are then unioned with each other to form new grids with a maximum number of solutes equivalent to that of a single initial grid (generally less than 100 solutes). Each unioned grid represents a single stage in the multistage process and is refitted until all grids have been successively unioned into a single grid. An iterative variation of this algorithm unions the final grid with all initial grids and repeats the 2DSA analysis until the solution is converged. A stable solution can be reached for cases where time- and radially invariant noise is not considered, which is equivalent to performing the solution containing all grids in a single iteration. For cases where invariant noise components are calculated, three to four iterations generally converge to a mostly stable solution. Our approach is parallelized with MPI on the level of each grid calculation, and the relevant UltraScan modules have been deployed on the NSF TeraGrid and on TIGRE [the Texas Internet Grid for Research and Education project is a computational grid that integrates computing systems, storage systems and databases, visualization laboratories, displays, instruments, and sensors across Texas (http://www.hipcat.net/Projects/tigre)] sites (Vadapalli et al., 2007). Simultaneously to the parameter determination, we use the 2DSA to algebraically extract the systematic noise contributions that result from imperfections in the optical system or instrument (Schuck and Demeler, 1999). After convergence, the vector of time-invariant noise contributions is subtracted from experimental data, yielding a data set only perturbed by Gaussian random noise. At that point, a subset of the initial two-dimensional parameter space is obtained, and relative concentrations for all solutes have been determined. Although the solution provides a good qualitative representation of experimental data, the solution cannot be regarded as unique at this point because it is overdetermined and subject to degeneracy. In this solution, low-frequency false positives are common—they result from the random noise present in experimental data and because the true solutes are not necessarily aligned with the fitted grid. Additional processing is required to further refine the 2DSA solution:

$$\text{Min} \sum_{i=1}^{r} \sum_{j=1}^{t} \left[M_{ij} - b_{ij} \right]^2 \quad (4.4)$$

$$\mathbf{M} = \sum_{l=s_{min}}^{s_{max}} \sum_{m=f_{min}}^{f_{max}} c_{l,m} L(s_l, D(s_l, k_m)). \quad (4.5)$$

Step 3. Parameter refinement and regularization. We found that parameter refinement of values obtained with the 2DSA is best achieved with a genetic algorithm implementation (Brookes and Demeler, 2006). We can adapt Occam's razor for our problem, which can be stated as follows: the most parsimonious solution capable of producing nearly the same residual mean square deviation (RMSD) is the preferred solution. Because the solution obtained in Step 2 is overdetermined and not unique we need to determine the most parsimonious solution. Implementation of the genetic algorithm analysis is described in detail in Brookes and Demeler (2006). Briefly, an evolutionary process is used to optimize the solution from a pool of multiple, individual solutions. The individual solutions are derived from the 2DSA analysis. Here, each new solution contains exactly one solute for each solute determined previously in the 2DSA analysis. Each solute is initialized randomly with an s and f/f_0 pair drawn from a symmetrical, rectangular region defined by a small, user-selected Δs and $\Delta f/f_0$ range surrounding each solute determined in the 2DSA analysis. If two regions overlap, one or two new ranges are defined in which a new, additional solute is placed (see Fig. 4.3A). All new solutions are then allowed to evolve to the best-fitting parameter combination according to the rules defined in the GA. Random number generators perform mutation, crossover, deletion, and insertion operations on the parameter combinations in order to generate new individuals for the next generation. The overriding selection criterion is the RMSD of a particular solution. Several rules favor survival of the fittest solution. To guard against loss of parameter diversity, we coevolve multiple demes, each consisting typically of 100 individual solutions, and permit only limited parameter migration between demes. Regularization is

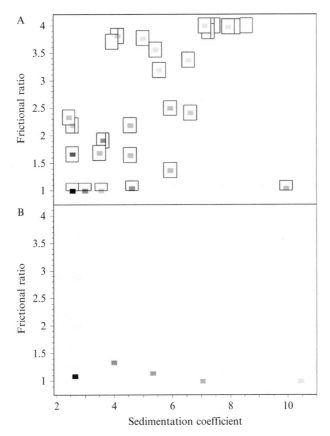

Figure 4.3 GA initialization (A) and GA analysis result (B). (A) Initializing regions formed around each solute determined in the 2DSA analysis. Each solute is shown by a rectangle drawn in a gray level corresponding to the partial concentration of the solute. Regions are clipped at a frictional ratio equal to unity. Overlapping regions are subdivided and a new GA solute is placed into the subdivided region. (B) GA analysis results from the initialization shown in A, indicating not only a successful parsimonious regularization, but also a more consistent tendency of frictional ratios.

achieved by penalizing the fitness of a particular solution in direct proportion to the number of solutes represented in this solution (Brookes and Demeler, 2007). In contrast to Tikhonov and maximum entropy regularization, we term this approach "parsimonious regularization" because of its ability to find the smallest number of necessary, discrete solutes to describe a given experimental system (see Fig. 4.3B). MPI is used to parallelize on the level of the deme calculation, with each deme occupying a different processor. This approach is embarrassingly parallel and scales linearly on all platforms. Individual generations progress asynchronously to prevent processor idling. Furthermore, we can choose to terminate evolution early when no further improvement in solutions is occurring. Importantly, we note that the GA analysis without 2DSA initialization will achieve the same result, although at much higher computational cost. Our strategy of linking 2DSA with GA has reduced the overall computational task by over 200-fold for the inverse problem of AUC experimental data fitting by drastically reducing the parameter space to interesting regions only and thus simplifying the parameter search.

Step 4. Statistical evaluation of the results. After Step 3, a discrete, parsimonious solution is obtained that represents the best fit to experimental data without any sacrifice in goodness of fit. Any uncertainty associated with the determined parameters is solely a result of Gaussian noise in the experimental observations. In order to determine the confidence interval on each parameter, a Monte Carlo analysis is performed. In this approach, synthetic noise of the same quality as observed in original experimental data is added to the best-fit solution from Step 3, and new data are fitted again with the procedure outlined in Step 3. This is typically repeated 50–100 times or until a statistically relevant description of the solution space is achieved. Each time, a slightly different result is generated, resulting in a distribution of parameter values for each parameter. These distributions can now be used to derive standard deviations for all parameters determined in Step 3, and a statistically reliable confidence interval can be assigned to the s and D values, molecular weights, frictional ratios, and partial concentrations from each solute identified in the previous steps. Monte Carlo analysis also achieves a second goal: When a stochastic signal is added to a solution, the amplification of the noise signal proceeds with a factor of $\sqrt{2}$ while the intrinsic signal of the sample contained in the analyzed system is amplified linearly. As a result, the Monte Carlo analysis attenuates the effect of stochastic noise on the solution, which can be quite apparent when the 2DSA analysis is performed at high resolution, where the analysis often finds false positives at the high end of the frictional ratio spectrum. An additional improvement in results can be obtained if signals from a high-speed and low-speed measurement are combined and the same sample is analyzed in a global fashion by combining multiple speeds in the GA analysis. In the high-speed experiment, a maximum signal from the sedimentation coefficient is obtained due to the large centrifugal force. However, sedimentation is rapid and the time allowed for diffusional boundary spreading is minimal. As a consequence, sedimentation coefficients in a high-speed experiment can be determined with high precision, whereas diffusion coefficients are often unreliable. Because of the relationships among s, D, and molecular weight shown in Eqs. (4.1) and (4.2), any lack of precision in the diffusion coefficient translates into a lack of precision of molecular weight and shape. By measuring the same sample also under low-speed conditions, sedimentation is much slower, and the sample has sufficient time to diffuse before being pelleted at the bottom of the cell. This provides a much improved signal on the diffusion transport and, as a consequence, on the accuracy with which the shape of the solute can be determined. Constraining s to the 95% confidence region of the sedimentation coefficient from each solute determined in the high-speed experiment, it is now possible to converge only on the diffusion coefficients using low-speed data.

3. Job Submission

Our methods have been implemented on a parallel computing platform in the UltraScan software package (Demeler, 2008). We have developed modules that allow submission to compute resources. These resources are local or remote clusters and grid-based supercomputers. High-performance computers are generally dedicated to specific jobs and operate in a batch mode maintained by a queue mechanism. We use the Globus-based TIGRE software stack (Vadapalli *et al.*, 2007), a grid middleware environment developed by HIPCAT (Consortium for High Performance Computing across Texas, http://www.hipcat.net), to communicate with the various compute resources. The 2DSA and GA processes are submitted by the user through a Web interface to the grid. Monte Carlo analysis can be added to each method. Analysis is performed on a target cluster and when the job completes, the results are e-mailed back to the researcher. Our submission scheme is shown in Fig. 4.4. Submission is performed from a Web page shown in Fig. 4.5. When the user submits a job, the Web server sends the user's request to us_gridpipe, which is a *named pipe*. This is a special type of file that simply holds written data until they are read. The PERL (Wall *et al.*, 2000) script us_gridpipe.pl *daemon*, a program that is always running in the background, reads from us_gridpipe, manages a global list of jobs, and controls startup of TIGRE jobs. Upon receipt of the researcher's job request, the us_gridpipe.pl daemon will first execute us_gridcontrol, a C++ program, to collect experimental data from the LIMS database in preparation for job execution. When us_gridcontrol completes, it informs the us_gridpipe.pl daemon via the named pipe that all of the experimental data have been extracted from the database and placed into a file on the disk. The us_gridpipe.pl daemon inserts the job request into its list of TIGRE jobs and begins execution of the PERL script us_tigre_job.pl, which controls the job execution. TIGRE resources are shared and it is important to select the number of processors carefully. The authors have developed a formula to compute the optimal number of processors to achieve a specific processor utilization (Brookes *et al.*, 2006) and this computation is performed for TIGRE jobs. Once the number of processors is known, us_tigre_job.pl sends experimental data and user's Web request parameters to the user-selected cluster and submits the job to

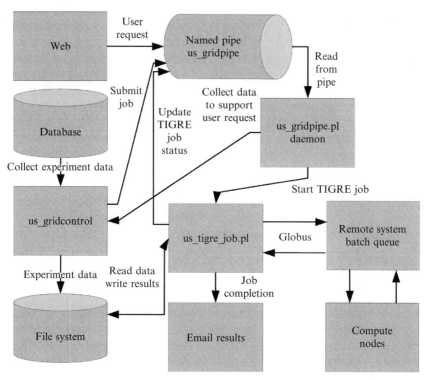

Figure 4.4 A flow diagram of the mechanism for submission of parallel analysis jobs from the USLIMS Web portal. The mechanism links Web server, database, local and remote file systems, e-mail, and remote batch queues operating on multiple supercomputers. The function of each component is explained in the text.

the cluster's queue. The execution scheduling is then controlled by the target cluster utilizing cluster-dependent queue control software. The cluster's queue control mechanism may be PBS (Portable Batch System), LSF (Load Sharing Facility), or some other queue control software. The Globus component of the TIGRE middleware hides the differences between the cluster's queue control specifics by providing a uniform application interface. The TIGRE job is monitored until completion. Upon completion, us_tigre_job.pl retrieves result data from the target cluster, e-mails the results to the researcher, and informs the us_gridpipe.pl daemon that the TIGRE job is finished. The us_tigre_job.pl script also collects all run-time statistics and stores them in a database. At this point, us_gridpipe.pl deletes the completed job from its job list and us_tigre_job.pl exits. The us_gridpipe.pl daemon also accepts requests to obtain information about the job list, which is available for viewing directly from the Web interface. The user imports the e-mailed results into the UltraScan software where the model can be visualized in 2D and 3D by a C++ GUI module of UltraScan,

Analysis of Mass and Shape Heterogeneity

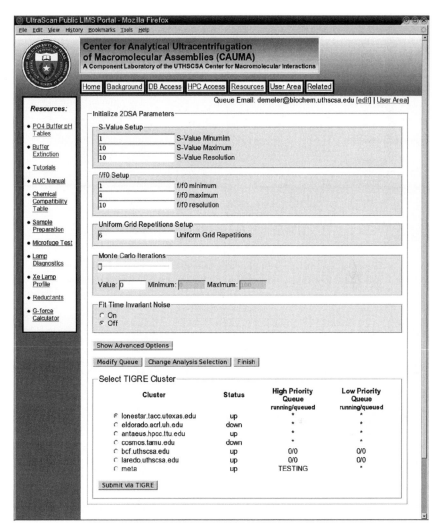

Figure 4.5 USLIMS job submission Web page for 2DSA analyses. The user selects analysis parameters, such as s-value range, f/f_0 range, number of grid movings, and Monte Carlo iterations, and selects systematic noise correction options. Under advanced options, regularization, radially invariant noise, iterative fitting, and meniscus fitting can be selected. At the bottom, a list of clusters is offered for submission of a job, and relevant system status information is provided.

and residuals and systematic noise contributions can be displayed and processed (see Fig. 4.6). All required parallelization modules are available for the Linux operating system and can be downloaded for free from the UltraScan Web site (http://www.ultrascan.uthscsa.edu).

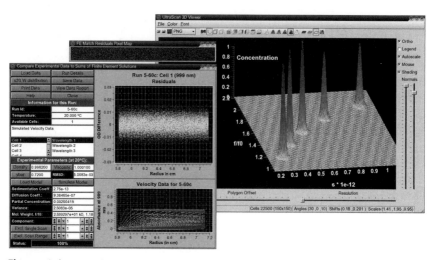

Figure 4.6 Visualization tools from the UltraScan GUI used to display models obtained in the supercomputer calculation, showing residual bitmap, residuals, experimental data and model overlay, and 3D solute distribution (f/f_0 vs s). The model shown here represents the global GA analysis of Example 1. (See Color Insert.)

4. Results

Using simulated and experimental example systems, we demonstrate here the capability of our method to resolve heterogeneity in mass and shape for complex systems and then explain additional insights gained from these improved methods. The first example shows a simulated data set with noise equivalent to noise produced in a well-maintained Beckman Optima XL-A instrument. Using simulated data allows us to determine the reliability of the method by comparing the fitting results with known target values used for the simulation. The second example explores the ability of our methodology to characterize the heterogeneity observed in the aggregation of amyloid-forming proteins and use it to detect changes in shape and mass induced by a ligand thought to interrupt the formation of larger amyloid aggregates.

4.1. Example 1: Simulated five-component system

This example presents the result of a four-step analysis on a simulated aggregating five-component system of a 25-kDa monomeric protein that is oligomerizing irreversibly up to a hexadecameric association state. During oligomerization, the protein changes frictional ratio. The target solute properties are listed in Table 4.1, the molecular masses of the simulated

Table 4.1 Monte Carlo results from a global genetic algorithm optimization using multispeed data[a]

Solute	Molecular mass (kDa)	Partial concentration	Frictional ratio, f/f_0
1	25.08 (24.75, 25.21) [25]	0.0994 (0.0980, 0.101) [0.1]	1.19 (1.187, 1.206) [1.2]
2	49.54 (49.08, 50.58) [50]	0.100 (0.987, 0.102) [0.1]	1.39 (1.386, 1.408) [1.4]
3	100.7 (99.28, 102.3) [100]	0.102 (0.0983, 0.102) [0.1]	1.61 (1.59, 1.62) [1.6]
4	204.5 (196.8, 207.4) [200]	0.0992 (0.0981, 0.100) [0.1]	1.83 (1.78, 1.84) [1.8]
5	399.1 (387.7, 409.3) [400]	0.100 (0.0998, 0.101) [0.1]	2.00 (1.96, 2.03) [2.0]

[a] Results demonstrate remarkable agreement with the original target model. Parentheses: 95% confidence intervals; square brackets: target value. All values are rounded off to three or four significant digits. In all cases, target values fall within the 95% confidence intervals of the predicted values.

solutes are 25, 50, 100, 200, and 400 kDa, and the corresponding frictional ratios are 1.2, 1.4, 1.6, 1.8, and 2.0, simulating an end-to-end aggregation event. Initially, the system is simulated at 60 krpm with 60 scans equally spaced over 4 h and is simulated over a 1.4-cm column length. The same system is then simulated at 20 krpm for 30 scans equally spaced over 23.3 h. Resulting data are fitted to ASTFEM solutions of the Lamm equation (Cao and Demeler, 2005) using first the 2DSA (see Fig. 4.7), which was initialized with the enhanced van Holde–Weischet analysis (Demeler and van Holde, 2004)(suggesting a fitting range of 2–12 s, data not shown), and frictional ratios were selected to range between 1 and 3. The 2DSA analysis for both experiments resulted in a solute distribution where the concentration signals roughly mapped out the region of the target values, indicating a system with 24 (60 krpm) and 21 (20 krpm) discrete solutes at different concentrations ranging between 2.5 and 11.5 s and frictional ratios ranging between 1 and 3. The four- to fivefold excess of the number of solutes observed in the 2DSA analysis is a consequence of low-concentration stochastic noise contributions and from the possible lack of alignment of the target values with the discrete grid. In addition, the fitting system is overdetermined, and hence a unique solution is unlikely. Results for high- and low-speed data differed as follows. First, for 60-krpm data, results indicated a large spread of frictional values, whereas 20-krpm data showed a much more narrow frictional range. This can be attributed to the lack of diffusion signal in the high-speed experiment, which is needed to get

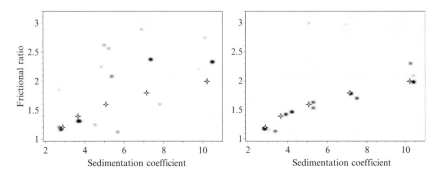

Figure 4.7 2DSA analysis of 60-krpm data (left) and 20-krpm data (right). High-speed data show more frictional ratio variation than low-speed data due to lack of a diffusion signal. Because of stochastic noise and because solutes do not necessarily align with target values, false positives are prevalent. Crosses indicate known target values, and gray levels indicate relative concentration (black: most concentrated).

accurate shape information. However, for high-speed data, a better separation resulted in more accurate determination of partial concentrations, as individual solutes were better separated and diffusing boundaries did not overlap as much. In the next step, output from the 2DSA analysis is used to initialize a GA analysis in order to obtain a parsimonious regularization (Brookes and Demeler, 2007). Results of the GA analysis are shown in Fig. 4.8. Most notably, both low- and high-speed experiments are able to correctly identify the number of target components by employing the parsimonious regularization [reducing the number of solutes from 24 (60 krpm) or 21 (20 krpm) solutes in the 2DSA to 5 in the GA]. Furthermore, the same observations regarding frictional ratio range and partial concentration made in the 2DSA analysis again apply in the GA analysis. Because the GA analysis is not restricted to a fixed grid, which may not necessarily align with the true target values, the GA analysis also comes much closer to the target values without any increase in RMSD, despite fewer parameters. Any deviation of results from the target values at this point is caused by stochastic noise in data (assuming all systematic noise sources have been eliminated). Hence, a Monte Carlo analysis will be able to map out the range of possible parameter values. To improve the results further, the signal from more precise s values and partial concentration derived from the high-speed experiment can be combined with the improved diffusion signal obtained in the low-speed experiment by globally fitting both speeds to a single model using the GA analysis. In such a global fit, sedimentation coefficients and partial concentrations will be constrained by the high-speed experiment, whereas the diffusion signal from the low-speed experiment constrains the frictional ratio range, generating optimal results for a system heterogeneous in shape and molecular weight. Results

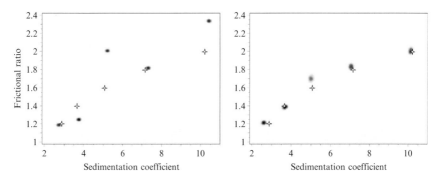

Figure 4.8 GA analysis of 60-krpm data (left) and 20-krpm data (right). As in the 2DSA, high-speed data show more frictional ratio variation than low-speed data due to lack of a diffusion signal, but reproduce the partial concentration better due to better separation. Because of stochastic noise, solutes do not necessarily align with target values, but for both speeds, parsimonous regularization achieved the correct number of total solutes. Crosses indicate known target values, and gray levels indicate relative concentration (black: most concentrated).

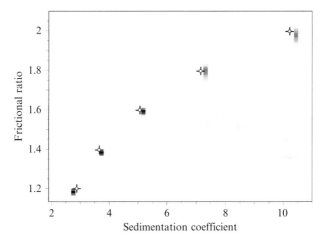

Figure 4.9 Global GA–Monte Carlo analysis for 60- and 20-krpm data. Crosses indicate the target solute position. As can be seen from this plot, target values are reproduced very closely and even partial concentrations are well matched. Gray levels indicate partial concentration (black: highest concentration).

for a global GA–Monte Carlo analysis of 60- and 20-krpm data are shown in Fig. 4.9. As can be seen here, a very faithful reproduction of the original target values is achieved in the global fit, and the combination of low- and high-speed information leads to a better definition of the shape function, as well as the partial concentration profile. For each solute, the final partial concentrations, molecular weights, and frictional ratios, as well as each parameter's 95% confidence interval, are summarized in Table 4.1. For the global multispeed Monte Carlo analysis, we find that all parameters are determined correctly within the 95% confidence interval.

4.2. Example 2: Amyloid aggregation

This example presents sedimentation studies on the Alzheimer's disease-related amyloid-β peptide. Amyloid formation (Iqbal *et al.*, 2001) is under intense scrutiny because of the role it appears to play in common neurodegenerative diseases such as Alzheimer's and Parkinson's disease (Selkoe, 2003). Amyloidogenic protein aggregates are characterized by a cross β-sheet structure and the formation of fibrils. These fibrils (Hardy and Higgins, 1992) and, to an increasing extent, also oligomeric species (Barghorn *et al.*, 2005; Klein *et al.*, 2004) are discussed as potential causes for disease formation. Current effort focuses on the characterization of the mechanism causing pathologic protein aggregation. In the case of Alzheimer's disease, the aggregating entity leading to formation of the characteristic amyloid plaques detectable postmortem in the brain tissue is a proteolytic fragment of the amyloid precursor protein. Aggregation of this 39 to 42 amino acid long fragment leads to the formation of fibrils with a typical width of about 10 nm (Holm Nielsen *et al.*, 1999) and up to several hundred nanometers in length. In addition to fibrillar structures, typically obtained as an end point in incubation studies of disease-associated amyloidogenic proteins, globular or amorphous oligomeric aggregates, ranging from dimers to several hundred-mers, are being considered as entities involved in the disease process (Rochet and Lansbury, 2000). Oligomeric species that differ in size and shape and exhibit neurotoxic properties have been characterized (Klein *et al.*, 2004). Several therapeutic strategies, as well as diagnostic methods, try to target the protein aggregates directly (Spencer *et al.*, 2007). In this context, the need for methods suitable to monitor the aggregate size and shape distributions of protein solutions is evident, and our methods promise to provide insight into the aggregation mechanism by following mass and shape changes.

4.2.1. Sample preparation

The amyloid-β peptide (1–42) is from Bachem (Bubendorf, Switzerland) and is dissolved in 2 mM NaOH (Fezoui *et al.*, 2000). After freeze-drying, aliquots are stored at $-70°$ until use. The amyloid-β peptide (1–42) labeled on the N terminus with fluorescent dye Oregon Green (Aβ (1–42)-OG) is synthesized by P. Henklein (Charite, Berlin, Germany). It is dissolved in anhydrous dimethyl sulfoxide (DMSO) and stored in aliquots at $-70°$. The inhibitor ligand (Kirsten and Schrader, 1997; Rzepecki, *et al.*, 2004) is synthesized by the group of T. Schrader (Organic Chemistry Department, University Duisbug-Essen, Germany). A 5 mM stock solution is stored at $4°$

in 100% DMSO. In order to reduce aggregate formation, protein concentrations are kept as low as possible, and low salt conditions are chosen. Aggregation mixtures containing 17.5 μM Aβ (1–42) and 3.5 μM Aβ (1–42)-OG are dissolved in 10 mM sodium phosphate, pH 7.4, and 4% anhydrous DMSO. For inhibitor studies, 200 μM inhibitor is added to the sample. Prior to sedimentation velocity centrifugation the solutions are incubated and slowly agitated for 5 days at room temperature. Sample treatment as described here minimizes loss due to the formation of large insoluble aggregates during sedimentation experiments.

4.2.2. Analytical ultracentrifugation

Sedimentation velocity experiments are performed in a Beckman Optima XL-A using the four-hole AN-60 Ti rotor. The 300- to 400-μl sample is filled into standard double sector aluminum center pieces using both sectors as sample sectors. Radial scans are taken in intensity mode at a resolution of 0.002 cm. All samples are measured at 493 nm to observe end-labeled Oregon Green and to avoid background absorbance from the aggregation inhibitor, which absorbs strongly in the ultraviolet region. The partial specific volume of the Aβ (1–42) peptide ($\bar{v} = 0.7377 \text{cm}^3/\text{g}$) is calculated on the basis of its amino acid content by a routine implemented in Ultra-Scan. Experimental intensity data are time-invariant noise corrected using the 2DSA analysis. The van Holde–Weischet analysis is used to initialize the s-value range in the 2DSA from 1 to 150 S. The frictional ratio range is initialized between 1 and 10. 2DSA analyses are performed with 24 grid movings with a 10-point resolution in both dimensions, resulting in a final s-value resolution of 0.625 S and 0.042 f/f_0 units. The 2DSA results are used to initialize the GA analysis, and parsimonously regularized GA distributions are used to initialize the GA Monte Carlo analysis.

4.2.3. Transmission electron microscopy (TEM)

Transmission electron microscopy experiments were performed by W. Meyer-Zaika in the inorganic chemistry department of the University Duisburg-Essen, Germany, with a Phillips CM 200 FEG instrument. After absorbing to the holey carbon film-coated copper grids (Plano, Wetzlar, Germany), the samples are stained negatively with a 2% (w/v) ammonium molybdate solution.

4.2.4. Experimental results

Based on the loading concentration measured before the experiments, and the first scan's plateau concentration, we concluded that approximately 10% of the dye-labeled Aβ (1–42) peptide was lost during the acceleration phase. This suggests that about 90% end-labeled Aβ (1–42) remained soluble and that our sedimentation velocity analysis will provide a representative picture for most, but not all, of the sample. The Aβ(1–42) peptide incubated at the

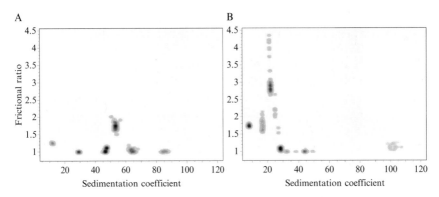

Figure 4.10 GA–Monte Carlo analysis of 21 μM Aβ (1–42)/Aβ (1–42)-OG in 10 mM sodium phosphate buffer, pH 7.4, 4% DMSO. The sample was incubated slowly and agitated for 5 days at room temperature prior to centrifugation. (A) Without inhibitor. (B) 200 μM inhibitor added. A noticeable decrease in the concentration of larger, globular structures is observed when an inhibitor is added, favoring smaller species with larger frictional coefficients.

described conditions reproducibly showed sedimentation boundaries, which indicated at least three different classes of molecular species with the majority sedimenting with an s value around 50 S, less than 20% sedimented with s values below 30 S, and a minor amount of material did not sediment and simply contributed to the baseline at the applied forces and is presumably monomer. A rotor speed of 20 krpm seemed to be optimal for obtaining a sedimentation signal for most species present in the mixture. The 2DSA resulted in an s-value range between 1 and 150 S with an RMSD of 0.003 and a total concentration of 0.15 OD for the control experiment, and 0.16 OD for the experiment where inhibitor was added. Parsimonious regularization with the GA analysis reduced the total number of detected species by a factor of 5. The GA results were used to initialize a GA/Monte Carlo analysis, and the GA/Monte Carlo results are shown in Fig. 4.10A for Aβ (1–42) without inhibitor (control) and in Fig. 4.10B for Aβ (1–42) with inhibitor. In the control sample all species appear to be mostly globular, consistent with the globular and slightly nonglobular structures observed in TEM images shown in Figs. 4.11 and 4.12. This sample also contained a species sedimenting with 53.0 S (+1.47 S/ −1.3 S) and a frictional ratio of 1.76 (+0.15/−0.18), representing about 35% of the material. This species appears to be more elongated, consistent with some shorter fibrils seen in Fig. 4.12, similar to fibrils reported in the literature (Antzutkin et al., 2002). When simulated as a long rod, this species is consistent with an axial ratio of 16.2, a diameter of 7.4 nm, a length of 120 nm, and a molecular mass of about 4.4×10^6 Da. Networks of large 10-nm amyloid fibrils also seen in Figs. 4.11 and 4.12 and, to a much lesser

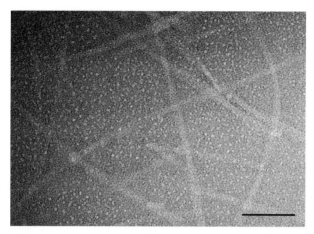

Figure 4.11 Electron micrograph of Aβ (1–42) fibrils seen together with small globular oligomers after 5 days of incubation agitated slowly at room temperature. Bar: 50 nm.

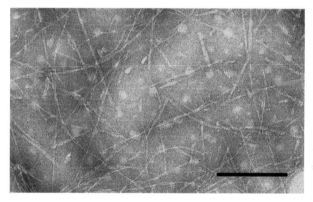

Figure 4.12 Electron micrograph of Aβ (1–42) fibrils seen together with larger globular and elongated oligomers after 5 days of incubation agitated slowly at room temperature. Bar: 200 nm.

degree, also in Fig. 4.13 are most likely insoluble and part of the fraction lost during the acceleration phase. The sedimentation experiment of the sample containing the inhibitor showed a significant decrease of the large globular structures, which is consistent with the absence of such particles in the corresponding TEM image shown in Fig. 4.13. The inhibitor-containing sample also displayed a much reduced presence of large fibril networks, suggesting that the inhibitor may successfully degrade both globular aggregates and also fibril networks. Most notably, the GA Monte Carlo analysis suggested that the majority of the material sedimented much slower when inhibitor was added to the sample, resulting in a decrease of the

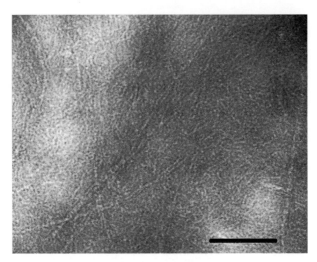

Figure 4.13 Electron micrograph of 21 μM Aβ (1–42)/Aβ (1–42)-OG mixed with 200 μM inhibitor after 5 days of incubation agitated slowly at room temperature. Bar: 100 nm.

weight-average sedimentation coefficient from 52.0 to 25.5 S and a concomitant increase in the weight average frictional coefficient from 1.29 to 1.93 when compared to the control experiment. The major species (33% sedimenting at 21.8 S) detected in the sample containing the inhibitor displayed a significantly increased frictional ratio with a large standard deviation (2.79, +1.11/−0.63), which is strongly indicative of a large range of fibril structures with varying lengths. Such structures were not measured in the control experiment (Fig. 4.10A). Again, this observation is closely matched by the corresponding TEM experiment, which shows a high concentration of varying length fibrils that are quite short in comparison to the large fibril networks observed in Figs. 4.11 and 4.12. Instead, Fig. 4.13 revealed the presence of high numbers of thin filaments (width below 5 nm). Because of their high abundance and their dimensions close to the resolution limit of TEM it was impossible to give reasonable estimates about their length. Nevertheless, a process leading to thinner filaments, for example, by preventing lateral association, should result in larger frictional ratios and increased axial ratios. For the control experiment, we also modeled the globular solutes seen in Figs. 4.11 and 4.12. Small spherical particles seen in the TEM experiment (Figs. 4.11 and 4.12) suggest diameters ranging from 4 to 20 nm. Such particles are consistent with the globular solutes found in the velocity experiment shown in Fig. 4.10A. A summary of the globular particles is shown in Table 4.2. The literature reports micellar amyloid β (Sabatè and Estelrich, 2005) as well as other globular aggregates (Hepler *et al.*, 2006; Lambert *et al.*, 1998) of the peptide,

Table 4.2 Globular species detected in sedimentation velocity experiment and their predicted diameter[a]

s value	f/f₀	Relative concentration (%)	Calculated diameter (nm)	Approximate molecular mass (Da)
11.5 S	1.25	4.20	4.3	2.7×10^5
28.7 S	1.00	7.90	12.0	7.5×10^5
46.4 S	1.00	27.80	15.5	1.5×10^6
65.2 S	1.00	17.90	18.2	2.6×10^6
84.0 S	1.00	7.39	20.7	3.8×10^6

[a] Predicted sizes correlate well with the measured particles in TEM images.

which also supports the detection of these particles by sedimentation velocity centrifugation. The very small globular particles visible in Fig. 4.11 have an estimated diameter of 3 to 4 nm.

5. CONCLUSIONS

We have developed new procedures that allow investigators to analyze AUC experiments with unsurpassed precision. We have shown that by applying computationally advanced tools we can improve AUC data analysis by a significant measure and provide information content and reliability of the results that exceed, by far, the information gleaned from traditional methods (Brown and Schuck, 2006; Demeler and Saber, 1998; Schuck, 2000; Stafford, 1992; van Holde and Weischet, 1978). The investigator can now reliably characterize mixtures of solutions that are simultaneously heterogeneous in molecular weight and also in shape. In addition, our methods make it possible to measure the statistical relevance of the results. We have developed a grid-based implementation of all necessary tools and have built the necessary environment to conveniently use these tools through user-friendly Web interfaces and to distribute compute jobs to remote supercomputers using the Globus grid technology (Foster, 2005). Example 1 demonstrated the ability of our method to extract very detailed molecular weight and shape information from sedimentation velocity experiments containing stochastic noise equivalent to that observed in a Beckman Optima XL-A by accurately resolving up to five different species ranging between 25 and 400 kDa and covering frictional ratios ranging between 1.2 and 2.0. Because these results are obtained when optimal conditions are present, we would like to stress the importance of high-quality data. We note that these results can only be obtained when data are free of systematic error sources, which need to be carefully controlled or eliminated. While our software allows for the correction of time- and radially invariant noise contributions, other systematic error sources may still be present that could significantly impact the accuracy of the results obtained. Potential distortions of data could result from concentration-dependent nonideality terms that are not considered in the model, from nonlinearities in the optics, refractive artifacts, including Wiener skewing, temperature, wavelength, and rotor speed variations, as well as buffer gradients that may be changing density and viscosity as a function of radius, and incorrect speed selections for the execution of the experiment. Other factors affecting accuracy can be controlled by the experimentalist using special care during the design of the experiment to assure that optimal conditions exist during the experiment, such as (1) cleaning the lamp before the experiment to assure maximum intensity; (2) using only carefully aligned cells (we use the Spin CAT cell alignment tool, available from Spin Analytical, P.O. Box 865, Durham, NH 03824-0865); (3) avoiding the use of worn out or deformed cell housings or scratched/damaged centerpieces and windows; (4) making sure that the optical system is aligned properly; (5) measuring at 230 nm, where light intensity is maximal, if buffer conditions permit the use of low UV wavelengths; and (6) making sure that optical density levels do not exceed the linearity of the optics, which is a function of light intensity that varies with wavelength of the light used. We also recommend using intensity measurements instead of absorbance measurements, which avoids the subtraction of two stochastic noise vectors, and hence avoids an increase of stochastic noise by a factor of $\sqrt{2}$; we also recommend avoiding buffer components that either contribute to the background absorbance or change the absorbance or refractive index during a run (certain reductants are prone to exhibit this problem when they change oxidation state midrun). Results obtained from the methods described here, when applied to the experimental amyloid aggregation system, provided new information on several levels and the presented AUC analysis methods proved to be a valuable tool in characterizing the aggregation process of amyloidogenic proteins. Particularly with respect to the evaluation of aggregation modulating compounds the method will be of great importance. Due to the improved data evaluation method, differently shaped and sized particles could be detected in one experiment, which could be validated by electron microscopic images performed with aliquots of the identical samples. A frictional ratio of 2.8 associated with the dominant species with 21.7 S in samples where the $A\beta$ (1–42) peptide was incubated for 5 days together with the inhibitor suggested an increased axial ratio of the underlying particles. This corresponds well to the TEM experiments of the same sample, which show a large number of thin filaments, which are not present in controls of $A\beta$ (1–42) or in the ligand alone. Also, based on the sedimentation velocity results we could determine the relative amount of globular aggregates present in the inhibitor-free control sample and estimate the size and molecular weight of these species. Such species were also detected in the TEM images. Since electron microscope images show only a minute section of a sample and presume that all species absorb equally well to the grid, analytical ultracentrifugation and our analysis methods can nicely complement TEM and add important information to the study of amyloids by more comprehensively representing the soluble portion of the sample and by describing the relative ratio in the mass and shape of the soluble particles. We suggest that further velocity experiments performed at lower speeds could potentially derive additional information about the presence of fibril networks that appeared to have sedimented during acceleration and were therefore not detected in our experiments and plan to investigate this further.

ACKNOWLEDGMENTS

We thank Jeremy Mann and the UTHSCSA Bioinformatics Core Facility and the Texas Advanced Computing Center (UT Austin) for computational support. We gratefully acknowledge support from the National Science Foundation through Teragrid Grant TGMCB060019T, as well as the National Institutes of Health through Grant RR022200 (both to BD). L.N.S. was supported by a grant of the Volkswagenstiftung (I/82 649) to Dieter Willbold, Head of the Institute for Physical Biology of the Heinrich-Heine-University.

REFERENCES

Antzutkin, O. N., Leapman, R. D., Balbach, J. J., and Tycko, R. (2002). Supramolecular structural constraints on Alzheimer's β-amyloid fibrils from electron microscopy and solid-state nuclear magnetic resonance. *Biochemistry* **41**, 15436–15450.

Barghorn, S., Nimmrich, V., Striebinger, A., Krantz, C., Keller, P., Janson, B., Bahr, M., Schmidt, M., Bitner, R. S., Harlan, J., Barlow, E., Ebert, U., *et al.* (2005). Globular amyloid β-peptide oligomer: A homogenous and stable neuropathological protein in Alzheimer's disease. *J. Neurochem.* **95**, 834–847.

Brookes, E., Boppana, R. V., and Demeler, B. (2006). Computing large sparse multivariate optimization problems with an application in biophysics. Supercomputing '06. *ACM* 0-7695-2700-0/06.

Brookes, E., and Demeler, B. (2006). Genetic algorithm optimization for obtaining accurate molecular weight distributions from sedimentation velocity experiments. "Analytical Ultracentrifugation VIII" (C. Wandrey and H. Cölfen, eds.). Springer. *Progr. Colloid Polym. Sci.* **131**, 78–82.

Brookes, E., and Demeler, B. (2007). Parsimonious regularization using genetic algorithms applied to the analysis of analytical ultracentrifugation experiments. *ACM GECCO Proc.* 978-1-59593-697-4/07/0007.

Brown, P. H., and Schuck, P. (2006). Macromolecular size-and-shape distributions by sedimentation velocity analytical ultracentrifugation. *Biophys. J.* **90**, 4651–4661.

Cao, W., and Demeler, B. (2005). Modeling analytical ultracentrifugation experiments with an adaptive space-time finite element solution of the Lamm equation. *Biophys. J.* **89**, 1589–1602.

Cao, W., and Demeler, B. (2008). Modeling analytical ultracentrifugation experiments with an adaptive space-time finite element solution for multi-component reacting systems. *Biophys. J.* **5,** 54–65.

Demeler, B. (2005). UltraScan: A comprehensive data analysis software package for analytical ultracentrifugation experiments. *In* "Modern Analytical Ultracentrifugation: Techniques and Methods" (D. J. Scott, S. E. Harding, and A. J. Rowe, eds.), pp. 210–229. Royal Society of Chemistry, United Kingdom.

Demeler, B. (2008). UltraScan: A comprehensive data analysis software for analytical ultracentrifugation experiments. http://www.ultrascan.uthscsa.edu.

Demeler, B., and Brookes, E. (2008). Monte Carlo analysis of sedimentation experiments. *Colloid Polym. Sci.* **286,** 129–137.

Demeler, B., and Saber, H. (1998). Determination of molecular parameters by fitting sedimentation data to finite element solutions of the Lamm equation. *Biophys. J.* **74,** 444–454.

Demeler, B., and van Holde, K. E. (2004). Sedimentation velocity analysis of highly heterogeneous systems. *Anal. Biochem.* **335,** 279–288.

Fezoui, Y., Hartley, D. M., Harper, J. D., Khurana, R., Walsh, D. M., Condron, M. M., Selkoe, D. J., Lansbury, P. T. Jr,, Fink, A. L., and Teplow, D. B. (2000). An improved method of preparing the amyloid beta-protein for fibrillogenesis and neurotoxicity experiments. *Amyloid* **7,** 166–178.

Foster, I. (2005). Globus Toolkit Version 4: Software for Service-Oriented Systems. IFIP International Conference on Network and Parallel Computing, Springer-Verlag. LNCS **3779,** 2–13.

Hardy, J. A., and Higgins, G. A. (1992). Alzheimer's disease: The amyloid cascade hypothesis. *Science* **256,** 184–185.

Hepler, R. W., Grimm, K. M., Nahas, D. D., Breese, R., Dodson, E. C., Acton, P., Keller, P. M., Yeager, M., Wang, H., Shughrue, P., Kinney, G., and Joyce, J. G. (2006). Solution state characterization of amyloid β-derived diffusible ligands. *Biochemistry* **45,** 15157–15167.

Holm Nielsen, E., Nybo, M., and Svehag, S. E. (1999). Electron microscopy of prefibrillar structures and amyloid fibrils. *Methods Enzymol.* **309,** 491–496.

Iqbal, Z., Sisosdia, S. S., and Winblad, B. (2001). *In* "Alzheimer's Disease and Related Disorders: Advances in Etiology, Pathogenesis and Therapeutics." Wiley, London.

Kirsten, C., and Schrader, T. (1997). Intermolecular β-sheet stabilization with aminopyrazoles. *J. Am. Chem. Soc.* **119,** 12061–12068.

Klein, W. L., Stine, W. B. Jr., and Teplow, D. B. (2004). Small assemblies of unmodified amyloid β-protein are the proximate neurotoxin in Alzheimer's disease. *Neurobiol. Aging* **25,** 569–580.

Lamm, O. (1929). Die Differentialgleichung der Ultrazentrifugierung. *Ark. Mat. Astron. Fys.* **21B,** 1–4.

Lawson, C. L., and Hanson, R. J. (1974). "Solving Least Squares Problems." Prentice-Hall, Englewood Cliffs, NJ.

Rochet, J. C., and Lansbury, P. T. (2000). Amyloid fibrillogenesis: Themes and variations. *Curr. Opin. Struct. Biol.* **10,** 60–68.

Rzepecki, P., Nagel-Steger, L., Feuerstein, S., Linne, U., Molt, O., Zadmard, R., Aschermann, K., Wehner, M., Schrader, T., and Riesner, D. (2004). Prevention of Alzheimer's associated Aβ aggregation by rationally designed nonpeptidic β-sheet ligands. *J. Biol. Chem.* **279,** 47497–47505.

Schuck, P. (2000). Size-distribution analysis of macromolecules by sedimentation velocity ultracentrifugation and Lamm equation modeling. *Biophys. J.* **78,** 1606–1619.

Schuck, P., and Demeler, B. (1999). Direct sedimentation boundary analysis of interference optical data in analytical ultracentrifugation. *Biophys. J.* **76,** 2288–2296.

Selkoe, D. J. (2003). Folding proteins in fatal ways. *Nature* **426,** 900–904.
Spencer, B., Rockenstein, E., Crews, L., Marr, R., and Masliah, E. (2007). Novel strategies for Alzheimer's disease treatment. *Expert Opin. Biol. Ther.* **7,** 1853–1867.
Stafford, W. (1992). Boundary analysis in sedimentation transport experiments: A procedure for obtaining sedimentation coefficient distributions using the time derivative of the concentration profile. *Anal. Biochem.* **203,** 295–301.
Vadapalli, R. K., Sill, A., Dooley, R., Murray, M., Luo, P., Kim, T., Huang, M., Thyagaraja, K., and Chaffin, D. (2007). Demonstration of TIGRE environment for Grid enabled/suitable applications. Grid2007, Austin, TX. http://www.grid2007.org/demo-vadapalli.pdf.
van Holde, K. E., and Weischet, W. O. (1978). Boundary analysis of sedimentation velocity experiments with monodisperse and paucidisperse solutes. *Biopolymers* **17,** 1387–1403.
Wall, L., Christiansen, T., and Orwant, J. (2000). *In* "Programming PERL," 3rd Ed.

CHAPTER FIVE

Discrete Stochastic Simulation Methods for Chemically Reacting Systems

Yang Cao[*] and David C. Samuels[†]

Contents

1. Introduction	116
2. The Chemical Master Equation	117
3. The Stochastic Simulation Algorithm	119
4. The Tau-Leaping Method	122
4.1. The hybrid SSA/tau-leaping strategy	125
4.2. The tau-selection formula	127
5. Measurement of Simulation Error	132
6. Software and Two Numerical Experiments	134
6.1. StochKit: A *stoch*astic simulation tool*kit*	134
6.2. The schlögl model	136
6.3. The LacZ/LacY model	136
7. Conclusion	137
Acknowledgments	139
References	139

Abstract

Discrete stochastic chemical kinetics describe the time evolution of a chemically reacting system by taking into account the fact that, in reality, chemical species are present with integer populations and exhibit some degree of randomness in their dynamical behavior. In recent years, with the development of new techniques to study biochemistry dynamics in a single cell, there are increasing studies using this approach to chemical kinetics in cellular systems, where the small copy number of some reactant species in the cell may lead to deviations from the predictions of the deterministic differential equations of classical chemical kinetics. This chapter reviews the fundamental theory related to stochastic chemical kinetics and several simulation methods based on that theory. We focus on nonstiff biochemical systems and the two most important

---

[*] Department of Computer Science, Virginia Tech, Blacksburg, Virginia
[†] Center for Human Genetics Research, Department of Molecular Physiology and Biophysics, Nashville, TN

discrete stochastic simulation methods: Gillespie's stochastic simulation algorithm (SSA) and the tau-leaping method. Different implementation strategies of these two methods are discussed. Then we recommend a relatively simple and efficient strategy that combines the strengths of the two methods: the hybrid SSA/tau-leaping method. The implementation details of the hybrid strategy are given here and a related software package is introduced. Finally, the hybrid method is applied to simple biochemical systems as a demonstration of its application.

1. Introduction

Biochemical systems have traditionally been modeled by a set of ordinary differential equations (ODEs). The general form of the reaction rate equations (RREs) in that approach can be formulated as

$$\frac{dx_i}{dt} = f_i(x_1, \cdots, x_n), \qquad (5.1)$$

for $i = 1, \ldots n$, where the state variables x_i represent the concentrations of involved species and functions f_i are inferred from the various chemical reactions of the system. This set of ODEs is usually solved with numerical methods packages such as DASSL and DASPK (Brenan *et al.*, 1996), ODEPACK (Hindmarsh, 1983), or CVODE (Cohen and Hindmarsh, 1996), for example. An important feature of the equations in this approach is that the system is deterministic and continuous. For systems where all chemical species are present in large copy numbers, it is reasonable to model every species by its concentration and the traditional ODE approaches seem to work very well. However, if the system is small enough that the molecular populations of some of the reactant species are small, from one to thousands, discreteness and stochasticity may play important roles in the dynamics of the system. Such a case occurs often in cellular systems (Arkin *et al.*, 1998; Fedoroff and Fontana, 2002; McAdams and Arkin, 1997), which typically involve copy numbers of one or two for the number of genes of a given protein, on the order of tens to hundreds for the corresponding RNAs, and on the order of thousands for regulatory proteins and enzymes. In that case, Eq. (5.1) cannot accurately describe the system's true dynamic behavior. Thus new modeling and simulation methods are needed to reflect the discrete and stochastic features of biochemical systems on a cellular scale.

To include discreteness, the most accurate way to simulate the time evolution of a system of chemically reacting molecules is to do a molecular dynamics simulation tracking the positions and velocities of all the molecules and the occurrence of all chemical reactions when molecules physically collide with each other. However, molecular dynamics simulations are generally too expensive to be practical except in the case of a relatively small number of molecules and even then only for very short timescales. Instead we consider the case where the dynamics of biochemical systems can be approximated by assuming that the reactant molecules are "well stirred" such that their positions become randomized and need not be tracked in detail. When that is true, the state of the system can be defined simply by the instantaneous molecular populations of the various chemical species. The chemical reactions can be defined as events that change the state of the system following biochemical rules, changing the molecular populations by integer numbers.

This chapter discusses numerical methods that can be applied to simulate such systems, taking into account discreteness and stochasticity at the molecular level. The focus will be on two major simulation methods: Gillespie's stochastic simulation algorithm (SSA) and the tau-leaping method. Finally, a merger of these two methods, the hybrid SSA/tau-leaping method, will be described.

2. THE CHEMICAL MASTER EQUATION

Let us consider a system of N molecular species $\{S_1, \ldots, S_N\}$ interacting through M elemental chemical reaction channels $\{R_1, \ldots, R_M\}$. We assume that the system is confined to a constant volume and is well stirred or, in other words, is in thermal (but not chemical) equilibrium at a constant temperature. Under these assumptions, the state of the system can be represented by the populations of the species involved. We denote these populations by, $\mathbf{X}(t) \equiv (X_1(t), \ldots, X_N(t))$, where $X_i(t)$ is the number of molecules of species S_i in the system at time t. The well-stirred condition is crucial. When this condition is broken, the spatial information of each species becomes important and the population information for the species will not be enough alone to determine the system dynamics. In cases where the well-stirred condition does not hold, the required simulation techniques will be different from what we discuss in this chapter. The so-called elemental reactions only include unimolecular and bimolecular reactions. Generalizations can be made to include more complicated reaction types, such as the commonly used Michaelis–Menten reaction (Cao *et al.*, 2005b; Rao and Arkin, 2003). We note that modeling these higher order reaction types using discrete stochastic methods are still under research and are not the focus of this review.

For a well-stirred system, each reaction channel R_j can be characterized by a *propensity function* a_j and a *state change vector* $\mathbf{v}_i \equiv (v_{1i}, \ldots, v_{Ni})$. The propensity function is defined by the following statement: $aj(\mathbf{x})\, dt$ is the probability, given $\boldsymbol{X}(t) = \boldsymbol{x}$; that one Rj reaction will occur in the next infinitesimal time interval $[t; t + dt)$.

v_{ij} is the change in the molecular population S_i induced by one reaction R_j. The matrix v is known as the stoichiometric matrix. The propensity function $a_j(x)$ reflects the fundamental characteristics of the stochastic chemical kinetics. Its value depends on the populations of the reactant populations and a reaction probability rate constant c_j, which is defined so that $c_j\, dt$ is the probability that a randomly chosen combination of R_j reactant molecules will react in the next infinitesimal time dt. Then a_j is the product of c_j and the number of all possible combinations of R_j reactant molecules.

The following are three simple examples of basic reactions and their propensity functions and state change vectors:

$$\text{For } S_1 \xrightarrow{c_1} S_2, \quad a_j(x) = c_1 x_1, \quad \text{and} \quad v_j = (-1, 1, 0, \cdots, 0). \tag{5.2}$$

$$\text{For } S_1 + S_2 \xrightarrow{c_1} S_3, \quad a_j(x) = c_1 x_1 x_2, \quad \text{and} \quad v_j = (-1, -1, 1, 0, \cdots, 0). \tag{5.3}$$

$$\text{For } S_1 + S_1 \xrightarrow{c_1} S_2, \quad a_j(x) = \frac{1}{2} c_1 x_1 (x_1 - 1), \quad \text{and} \quad v_j = (-2, -1, 0, \cdots, 0). \tag{5.4}$$

It is easy to see that the form of the propensity function is similar to the mass action terms in the deterministic RREs. The value of c_j is similar to its counterpart: the reaction rate k_j in the RREs. Indeed there is a connection between c_j and k_j depending on the reaction type. For a unimolecular reaction such as in the example of Eq. (5.2), $c_1 = k_1$. For a bimolecular reaction between different species such as in Eq. (5.3), $c_1 = k_1/A\,\Omega$, where A is the Avogadro number and Ω is the constant volume. For a bimolecular reaction between the same species, the forms of the propensity function and the reaction rate function have a slight difference, but when x_1 is large the difference will be negligibly small and we will have $c_1 \approx 2k_1/A\Omega$

Once the propensity functions and stoichiometric matrix are determined, the dynamics of the system obey the chemical master equation (CME):

$$\frac{\partial P(\mathbf{x}, t \mid \mathbf{x}_0, t_0)}{\partial t} = \sum_{j=1}^{M} [a_j(\mathbf{x} - \boldsymbol{v}) P(\mathbf{x} - \boldsymbol{v}_j, t \mid \mathbf{x}_0, t_0) - a_j(\mathbf{x}) P(\mathbf{x}, t \mid \mathbf{x}_0, t_0)], \tag{5.5}$$

where $P(\boldsymbol{x}; t \mid \boldsymbol{x}_0, t_0)$ denotes the probability that $\boldsymbol{X}(t)$ will be $\boldsymbol{x}$ given that $\boldsymbol{X}(t_0) = \boldsymbol{x}_0$. In principle, the CME completely determines the dynamics of $P(\boldsymbol{x}; t \mid \boldsymbol{x}_0, t_0)$ but the CME is essentially an ODE whose dimension is given by the number of all possible combinations of states of $\boldsymbol{x}$. Consider the example of a small reaction network of five species and assume that the population of each species is in the range 1 to 100. The dimension of the corresponding CME will then be $100^5 = 10^{10}$. As the number of species increases, the dimension of the corresponding CME increases exponentially, a problem known as the "curse of dimension." It is easy to see that the CME is both theoretically and computationally intractable for all but the simplest models. There has been some interesting research (Munsky and Khammash, 2006; Zhang and Watson, 2007) trying to reduce the dimension of the CME or to provide an approximate numerical solution of the CME. Progress has been made but so far these methods still can only be practically applied to simple models.

3. THE STOCHASTIC SIMULATION ALGORITHM

Another way to study the dynamics of the reaction system is to construct realizations of $\boldsymbol{X}(t)$ through numerical simulation. In the numerical simulation, the key is not to get the probabilities $P(\boldsymbol{x}; t \mid \boldsymbol{x}_0, t_0)$ but to generate a single trajectory (a realization) that the system may undergo. The most important simulation method for this is Gillespie's SSA (Gillespie, 1976, 1977). Instead of following the time evolution of the probabilities, the SSA generates a trajectory of the system step by step. In each step the SSA starts from a current state $x(t) = x$ and asks two questions:

When will the next reaction occur? We denote this time interval by τ.
When the next reaction occurs, which reaction will it be? We denote the chosen reaction by the index j.

To answer these questions, one needs to study the joint probability density function $p(\tau, j \mid \boldsymbol{x}; t)$, which is defined by

$$p(\tau, j \mid \mathbf{x}, t)dt = \text{the probability, given } \mathbf{X}(t) = \mathbf{x}, \text{ that the next reaction will occur in the infinitesimal time interval} \quad (5.6)$$
$$[t+\tau, \; t+\tau+dt], \text{ and will be an } R_j \text{ reaction.}$$

It can be derived (Gillespie, 1976, 1977) that

$$p(\tau, j \mid \mathbf{x}, t) = a_j(\mathbf{x})\exp\left(-a_0(\mathbf{x})\tau\right) \tag{5.7}$$

where $a_0(\mathbf{x}) \equiv \sum_{j=1}^{M} a_j(\mathbf{x})$. Equation (5.7) is the theoretical foundation for the SSA. It implies that the time τ to the next occurring reaction is an exponentially distributed random variable with mean value $1/a_0(\mathbf{x})$ and that the index j of that reaction is the integer random variable with point probability $a_j(\mathbf{x})/a_0(\mathbf{x})$. To advance the system from state $\mathbf{x}$ at time t, the SSA generates two random numbers, r_1 and r_2, uniformly over the unit interval and then takes the time of the next reaction to be $t + \tau$, where

$$\tau = \frac{1}{a_0(\mathbf{x})} \ln\left(\frac{1}{r_1}\right) \tag{5.8}$$

and the index for the next reaction to be the smallest integer j satisfying

$$\sum_{j'=1}^{j} a_{j'}(\mathbf{x}) > r_2 a_0(\mathbf{x}). \tag{5.9}$$

The system state is then updated according to $\mathbf{X}(t + \tau) = \mathbf{x} + \mathbf{v}_j$, and this process is repeated until the simulation final time or some other terminating condition is reached. The algorithm is listed here:

Algorithm 1: The basic SSA method

Starting from initial condition $t = t_0$ and $\mathbf{x} = \mathbf{x}_0$,

Step 1: With $\mathbf{x}(t) = \mathbf{x}$, calculate all $a_j(\mathbf{x})$ and $a_0(\mathbf{x})$.
Step 2: If $a_0(\mathbf{x}) = 0$, terminate the simulation. Otherwise generate two uniform random numbers, r_1 and r_2. Calculate τ and j according to Eqs. (5.8) and (5.9), respectively.
Step 3: Update the system by $t = t + \tau$ and $\mathbf{x} = \mathbf{x} + \mathbf{v}_j$.
Step 4: If t reaches the end time, stop. Otherwise, go to step 1.

The SSA is exact in the sense that the sample paths it generates are precisely distributed according to the solution of the CME, which makes it one of the most fundamental simulation methods for discrete stochastic biochemical systems. Although this algorithm looks quite simple, due to its importance there are several different implementation strategies proposed in the literature for the SSA. They are the direct method (DM) (Gillespie, 1977), the first reaction method (FRM)(Gillespie, 1977), the next reaction method (NRM)(Gibson and Bruck, 2000), the optimized direct method (ODM)(Cao et al., 2004), the sorted direct method (SDM)(McCollum et al., 2006), and the logarithmic direct method (LDM)(Li and Petzold, 2006). The following is a brief review for these implementation strategies.

The DM is exactly the algorithm 1 just given. The FRM is theoretically equivalent to the DM but is quite different in the implementation details. The FRM generates a potential reaction time for each reaction and chooses the "first" reaction channel that has the earliest firing time to occur. In the FRM implementation, one generates M uniform random numbers $r_1, \ldots, r_M$ in every step and calculates a time τ_k for each reaction channel R_k by

$$\tau_k = \frac{1}{a_k(\mathbf{x})} \ln\left(\frac{1}{r_k}\right). \tag{5.10}$$

Then τ and j are given by

$$\begin{aligned} \tau &= \min_{1 \leq k \leq M}(\tau_k) \\ j &= \text{the index for the smallest } \tau_k. \end{aligned} \tag{5.11}$$

It can be proved that the τ and j generated from Eq. (5.11) follow the same distributions as in Eqs. (5.8) and (5.9). Thus the DM and the FRM are statistically equivalent. However, in every step the FRM generates M τ_k values but uses only one of them. Thus the FRM is much less efficient than the DM.

Gibson and Bruck (2000) have made remarkable progress improving the implementation efficiency of the FRM. Their method is the next reaction method (NRM). The NRM uses a dependent graph to record the influence of each reaction channel on the other reaction channels. It records the absolute time $t + \tau_k$ as the expected firing time for the R_k reaction. If the firing of one reaction channel does not change the propensity of another reaction channel, the expected firing time for the latter reaction remains the same. In this way the NRM avoids unnecessary updates of the propensity function and expected firing time. For a reaction channel R_k whose reactants have been changed by the firing reaction, the NRM uses a cleverly designed formula to reuse the uniform random number r_k generated in the previous step. As a result, in every step there is only one uniform random number generated. The NRM turns out to be much more efficient than the FRM. However, using a detailed numerical analysis, it has been shown (Cao et al., 2004) that the NRM still has a higher computational cost than the direct method except for simple systems where the reactions are almost totally independent of each other.

To decrease the computational cost, the optimized direct method (ODM)(Cao et al., 2004) adopts the dependent graph to avoid the unnecessary recalculation of propensity functions and rearranges the index of the reaction channels so that the more frequent reaction channels are always indexed before the less frequent ones. With these two improvements over the DM, the ODM becomes one of the most efficient SSA implementation strategies currently in use.

The reindex technique of the ODM requires one or a few sample runs using the SSA to collect the necessary information. This is not convenient in many applications. In order to dynamically adjust the index of the reaction channels, the sorted direct method (SDM) was proposed (McCollum et al., 2006). In the SDM, a bubble-up sorting method was applied to the index of reaction channels. In the simulation, every time when one reaction occurs, its reaction index decreases by one so that in the next step it is found more quickly. Then, after a certain initial simulation time, the index list will be sorted into a form very close to the optimal one. The SDM is a little less efficient than the ODM but its adaptive feature makes it a very good strategy, particularly in simulation of oscillation systems where a fast reaction in one time period may become slow in another time period. In that case, the dynamic indexing of this method is very useful.

The logarithmic direct method (LDM) has been proposed (Li and Petzold, 2006), which applies a binary search method to the direct method. When the number of reaction channels, M, is large, the LDM can complete the search for the index j within $O[\log(M)]$ time. Thus the LDM has advantages for large biochemical system.

We note that for all of the aforementioned implementation strategies, the differences among them (except for the case of the FRM) in computation time are usually less than 20%. This is far from the computation speed needed in many applications. As the SSA is a procedure simulating every reaction event one at a time, the computational cost is inevitably high. Thus people have to consider alternative methods to gain efficiency by sacrificing exactness, as long as the approximation accuracy is kept under control.

4. THE TAU-LEAPING METHOD

The tau-leaping method (Gillespie, 2001) was designed to speed up a stochastic simulation by leaping over many reactions in one time step. This idea is illustrated in Fig. 5.1. The tau-leaping method makes the leap by

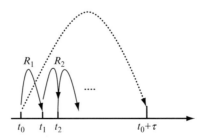

Figure 5.1 Comparison between the SSA method and the tau-leaping method. The tau-leaping method leaps over many reactions in one time step.

answering the following question: How often does each reaction channel fire in the next specified time interval τ? More precisely, let

$K_j(\tau; \mathbf{x}, t) \triangleq$ the number of times, given $\mathbf{X}(t) = \mathbf{x}$, that reaction channel R_j will fire in the time interval $[t, t + \tau], (j = 1, \cdots, M)$.
(5.12)

For arbitrary values of τ it will be about as difficult to compute $K_i(\tau;\mathbf{x},t)$ as to solve the CME. The tau-leaping method chooses a small τ value to satisfy the following leap condition: *For the current state x, require τ to be small enough that the change in the state during $[t; t + \tau)$ will be so small that no propensity function will suffer an appreciable change in its value.* Under the leap condition, a good approximation to $K_j(\tau; \mathbf{x}, t)$ will be provided by $P[a_j(\mathbf{x})\,\tau]$, the Poisson random variable with mean (and variance) $a_j(\mathbf{x})\,\tau$. So if $\mathbf{X}(t) = \mathbf{x}$ and we choose τ to satisfy the leap condition, we can update the state to time $t + \tau$ according to the approximate formula

$$\mathbf{X}(t + \tau) \doteq \mathbf{x} + \sum_{j=1}^{M} v_j P\left(a_j(\mathbf{x})\tau\right) \quad (5.13)$$

where $P[a_j(\mathbf{x})\tau]$ for each $j = 1, \ldots, M$ denotes an independent sample of the Poisson random variable with mean and variance $a_j(\mathbf{x})\tau$. This computational procedure is the *tau-leaping approximation*.

The tau-leaping method makes a natural connection between the SSA and the deterministic RREs. When τ is chosen very small such that in every time step there is at most one reaction occurring, the tau-leaping method reduces to a linear approximation of the SSA. When τ is allowed to be large such that

$$a_j(\mathbf{x})\tau \gg 1, \quad \text{for all } j = 1, \ldots, M, \quad (5.14)$$

the Poisson random number $P[a_j(\mathbf{x})\tau]$ can be approximated by the Normal random number with mean and variance $a_j(\mathbf{x})\tau$, denoted by $N[a_j(\mathbf{x})\tau, a_j(\mathbf{x})\tau]$. Then Eq. (5.13) reduces to the forward Euler method for the chemical Langevin equation (CLE) (Gillespie, 2001). Moreover, when the values $a_j(\mathbf{x})\tau$ for all $j = 1, \ldots, M$ are even larger, the standard deviation is then negligible compared to the mean value. The Poisson random number $P[a_j(\mathbf{x})\tau]$ can then be simply replaced by its mean value $a_j(\mathbf{x})\tau$. Then Eq. (5.13) becomes

$$\mathbf{X}(t + \tau) \doteq \mathbf{x} + \sum_{j=1}^{M} v_j a_j(\mathbf{x})\tau, \quad (5.15)$$

which is the forward Euler method for the corresponding RREs. Note that here the merger of the tau-leaping method into the forward Euler method is seamless. One does not need to check Eq. (5.14) for all js. The idea of using the normal random number or just the mean value to approximate the Poisson random number can be applied for any individual j. This procedure can be wrapped in a Poisson random number approximation procedure. Choose two threshold values: M_1, for which the Poisson random number $P(M_1)$ can be safely approximated by a normal random number $N(M_1, M_1)$, and M_2, for which $P(M_2)$ can be safely approximated by M_2. We then have the following algorithm.

Algorithm 2: The Poisson random number approximation procedure

Given the mean and variance value m, we follow the following procedure to generate an approximation to the Poisson random number $P(m)$:

Case 1: If $m < M_1$, return a Poisson random number $P(m)$.
Case 2: If $m \geq M_1$ and $m < M_2$, return a normal random number $N(m, m)$.
Case 3: If $m \geq M_2$, return m.

We denote the approximation function generated by algorithm 2 as $\rho(m)$. Then Eq. (5.13) becomes

$$\mathbf{X}(t+\tau) \doteq \mathbf{x} + \sum_{j=1}^{M} v_j \rho\Big(a_j(\mathbf{x})\tau\Big). \qquad (5.16)$$

There are two practical problems to be addressed before the tau-leaping method can be applied to realistic applications. First, we need a procedure to quickly determine the largest value of τ that is compatible with the leap condition. Second, we need to develop a method to avoid possible negative populations that could result from two reasons:

a. The Poisson random number (or its approximation) is unbounded. There is a small possibility that a large random number may exceed the number of some reactants and cause negative populations to occur.
b. There are multiple reaction channels consuming the same reactant. When they fire at the same time, even though neither of them separately exhausts the number of that reactant, their overall effect may do so.

The following section discusses the details on how to solve these two practical problems. As negative populations present the more serious problem, we discuss the corresponding solution first. Then we review several τ-selection formulas and present a simple and efficient one.

4.1. The hybrid SSA/tau-leaping strategy

Negative populations resulting from the original tau-leaping method have been found to happen in the simulation of certain systems in which some consumed reactant species are present in small numbers. It was believed that the reason for this error was mostly due to the fact that the Poisson random variable is unbounded such that the Poisson approximation to $K_j(\tau; x, t)$ in Eq. (5.13) might result in reaction channel R_j firing so many times that the population of one of its reactant species would be driven negative. To resolve this problem, the *binomial* tau-leaping method (Chatterjee et al., 2005; Tian and Burrage, 2004) was proposed in which bounded binomial random variables replace the unbounded Poisson random variables. However, we have developed an understanding that the negative population problem arises more often from multiple reaction channels consuming the same reactant than from the unbounded Poisson random variable. The binomial tau-leaping method becomes complicated when dealing with this case. Cao et al. (2005c) made an observation that most negative populations in both cases were related to species with a low population. This may seem to be a rather obvious observation, but it does point us toward a method of dealing with this problem. Based on this observation, an adaptive hybrid SSA/tau-leaping method was proposed, which seems to resolve the negativity problem satisfactorily.

The hybrid SSA/tau-leaping algorithm (Cao et al., 2005c) is based on the fact that negative populations typically arise from multiple firings of reactions that are only a few reaction events away from consuming all the molecules of one of their reactants. To focus on those reaction channels, the hybrid SSA/tau-leaping algorithm introduces a second control parameter n_c, a positive integer that is usually set somewhere between 2 and 20. Any reaction channel with a positive propensity function that is currently within n_c firings of exhausting one of its reactants is classified as a *critical reaction*. The hybrid algorithm chooses τ in such a way that no more than one firing of *all* the critical reactions can occur during the leap. Essentially, the algorithm simulates the critical reactions using an adapted (and thus not quite exact) version of the SSA and the remaining noncritical reactions using the Poisson tau-leaping method. Since no more than one firing of a critical reaction can occur during a leap, the probability of producing a negative population is reduced to nearly zero. On those rare occasions when a negative population does arise (from firings of some noncritical reaction), that step can simply be rejected and repeated with τ reduced by half, or else the simulation can be started over using a larger value for n_c.

It can be shown (Cao et al., 2005c) that the hybrid SSA/tau-leaping procedure becomes identical to the SSA if n_c is chosen so large that *every* reaction channel is critical and becomes identical to the tau-leaping method if $n_c = 0$ so that *none* of the reaction channels is critical. Thus, the hybrid

SSA/tau-leaping algorithm is not only more robust, but also potentially more accurate than the earlier tau-leaping algorithm.

There are some details left to discuss before giving the full description of the hybrid SSA/tau-leaping method. First, how do we decide whether a reaction is *critical* with the parameter n_c? This is done by first estimating for each reaction R_j with $a_j(x) > 0$ the maximum number of times L_j that R_j can fire before exhausting one of its reactants (Chatterjee et al., 2005; Tian and Burrage, 2004):

$$L_j = \min_{i \in [1, N]; v_{ij} < 0} \left[\frac{x_i}{|v_{ij}|} \right] \qquad (5.17)$$

Here the minimum is taken over only those index values i for which $\boldsymbol{v}_{ij} < 0$, and the brackets denote "greatest integer in." For example, for a reaction

$$S_1 \rightarrow S_2$$

L_j is the population of S_1. For a reaction

$$S_1 + S_2 \rightarrow S_3$$

L_j takes the smaller value between the populations of S_1 and S_2. For a reaction

$$S_1 + S_1 \rightarrow S_2$$

L_j takes the integer part of one-half of the population of S_1.

After L_j is calculated, it is compared with n_c. If $L_j < n_c$, R_j is considered a critical reaction and should be simulated by the adapted SSA part. Otherwise, R_j is noncritical and can be simulated by the tau-leaping part.

The next step is to decide how to implement the SSA part and the tau-leaping part together. To solve this problem, in every simulation step we first generate a τ' from a τ-selection procedure and a τ'' from the SSA part. If τ' is even smaller than a few fold of the expected step size of a pure SSA method, $1/a_0(\mathbf{x})$, we will stick with the pure SSA method. Otherwise, we use the tau-leaping method to simulate the noncritical reactions and the SSA method to simulate the critical reactions. The real simulation time step τ is chosen to be the smaller value between τ' and τ''. If τ'' is smaller, the critical reaction fires. Otherwise, no critical reaction should fire before τ. In both cases, the number of noncritical reaction firings is calculated using the Poisson tau-leaping method. The τ'' for the SSA part can simply follow the SSA procedure limited to only critical reactions. The τ' for the tau-leaping part is discussed in the next subsection.

4.2. The tau-selection formula

The simulation formula for the tau-leaping method is quite simple. The key point is how to select the τ value so that the leap condition is satisfied. Several tau selection formulae have been proposed in the literature. Gillespie (2001) originally proposed that the leap condition could be considered satisfied if the expected change in each propensity function $a_j(\mathbf{x})$ during the leap were bounded by $\varepsilon\, a_0(\mathbf{x})$, where ε is an error control parameter ($0 < \varepsilon \ll 1$). Later, this condition was refined by Gillespie and Petzold (2003). They showed that the largest value of ε that satisfies this requirement can be estimated as follows: First compute the $M^2 + 2M$ auxiliary quantities

$$f_{jj'}(\mathbf{x}) \equiv \sum_{i=1}^{N} \frac{\partial a_j(\mathbf{x})}{\partial x_i} v_{ij'}, \quad j,j' = 1,\ldots,M, \tag{5.18}$$

$$\mu_j(\mathbf{x}) \equiv \sum_{i=1}^{N} f_{jj'}(\mathbf{x}) a_{j'}(\mathbf{x}), \quad j,j' = 1,\ldots,M, \tag{5.19}$$

$$\sigma_j^2(\mathbf{x}) \equiv \sum_{i=1}^{N} f_{jj'}^2(\mathbf{x}) a_{j'}(\mathbf{x}), \quad j,j' = 1,\ldots,M; \tag{5.20}$$

then take

$$\tau = \min_{j \in [1,M]} \left(\frac{\varepsilon a_0(\mathbf{x})}{|\mu_j(\mathbf{x})|}, \frac{\left(\varepsilon a_0(\mathbf{x})\right)^2}{\sigma_j^2(\mathbf{x})} \right). \tag{5.20}$$

The derivation of these formulas (Gillespie and Petzold, 2003) shows that $\mu_j(\mathbf{x})\tau$ estimates the *mean* of the expected change in $a_j(\mathbf{x})$ in time τ, $\sqrt{\sigma_j^2(\mathbf{x})\tau}$ estimates the *standard deviation* of the expected change in $a_j(\mathbf{x})$ in time τ, and Eq. (5.20) essentially requires that both of those quantities be bounded by $\varepsilon\, a_0(\mathbf{x})$ for all j. We should note that Gillespie's original τ-selection formula (Gillespie, 2001) was deficient in that it lacked the σ_j^2 argument in Eq. (5.20).

The tau-selection procedure [Eq. (5.20)] seeks to set a bound on the change in each propensity function $a_j(\mathbf{x})$ during a time step τ by a small fraction ε of the *sum* $a_0(\mathbf{x})$ of all the propensity functions. Denoting the change in propensity function a_j from time t to time $t + \tau$, given $\mathbf{X}(t) = \mathbf{x}$, by $\Delta_\tau a_j(\mathbf{x})$, this requirement can be stated as

$$|\Delta_\tau a_j(\mathbf{x})| \leq \varepsilon a_0(\mathbf{x}), \quad j = 1,\ldots,M. \tag{5.21}$$

This bound is explicitly reflected in the numerators of the two fractions in the τ-selection formula [Eq. (5.20)]. Although this strategy does indeed limit the changes in the propensities during a leap as required, it does not fully accomplish the task with a proper scaling. The leap condition requires that every propensity function remains "practically constant" during a τ time period, since that is what allows the number of reaction events R_j during τ to be approximated accurately by a statistically independent Poisson random variable with mean $a_j(x)\tau$. If $a_j(x)$ for R_j reaction happens to be very small compared to $a_k(x)$ for R_k reaction, $a_j(x)$ will then be much smaller than $a_0(x)$. Equation (5.21) may allow a large *relative* change in $a_j(x)$, which could result in simulation inaccuracies.

To allow the formula for the leap condition to reflect the relative scales, we change Eq. (5.21) by

$$|\Delta_\tau a_j(\mathbf{x})| \leq \varepsilon a_j(\mathbf{x}), \; j = 1, \ldots, M. \tag{5.22}$$

However, doing this can lead to difficulties if $a_j(x)$ happens to approach zero because then Eq. (5.22) will force τ to approach zero, effectively bringing the tau-leaping process to a halt. Thus we have to make a simple modification to avoid this problem. The limit procedure described earlier is based implicitly on treating the propensity functions as continuous functions. Actually, the propensity functions change as reactions occur by discrete amounts; for every propensity function $a_j(x)$ there will always be a minimum amount by which it can change. For example, if R_j is the unimolecular reaction with propensity function $a_j(\mathbf{x}) = c_j\, x_i$, then the minimum (positive) amount by which $a_j(x)$ can change will obviously be c_j. It is not hard to show that if the propensity function of any bimolecular or trimolecular reaction R_j changes at all, it must do so by an amount greater than or equal to c_j. Since it is therefore unreasonable to require any propensity function $a_j(x)$ to change by less than c_j, we should replace the bound on the right-hand side of Eq. (5.22) with the *larger* of $\varepsilon\, a_j(x)$ and c_j:

$$\Delta_\tau a_j(\mathbf{x}) \leq \max\{\varepsilon a_j(\mathbf{x}), c_j\}, \; j = 1, \ldots, M. \tag{5.23}$$

To apply the new Eq. (5.23), we can simply replace Eq. (5.20) with

$$\tau' = \min_{j \in [1,M]} \left(\frac{\max\{\varepsilon a_j(\mathbf{x}), c_j\}}{|\mu_j(\mathbf{x})|}, \frac{\left(\max\{\varepsilon a_j(\mathbf{x}), c_j\}\right)^2}{\sigma_j^2(\mathbf{x})} \right), \tag{5.24}$$

where the definitions of μ_j and σ_j remain the same as in Eq. (5.19).

Although τ selection using Eq. (5.24) results in a more accurate simulation than τ selection using Eq. (5.20), evaluation of the functions $\mu_j(\mathbf{x})$ and

$\sigma_j^2(\mathbf{x})$ in Eqs. (5.18) and (5.19) prior to each leap tends to be very time-consuming, especially if both M and N are large. A new τ-selection formula (Cao et al., 2006) was then proposed to avoid this computational burden. Here we introduce a simplified version of this new formula. The underlying strategy of this new τ-selection procedure is to bound the relative changes in the molecular populations by a specified value ε ($0 < \varepsilon = 1$). Let

$$\Delta_\tau X_i \equiv \Delta_\tau X_i(\mathbf{x}) \triangleq X_i(t+\tau) - x_i, \text{ given } \mathbf{X}(t) = \mathbf{x}. \qquad (5.25)$$

Instead of basing the τ selection on Eq. (5.23), we base it on the condition

$$\Delta_\tau X_i \leq \max\{\varepsilon x_i, 1\}, \ \forall i \in I_{rs}, \qquad (5.26)$$

where I_{rs} denotes the set of indices of all reactant species (so $i \in I_{rs}$ if and only if x_i is an argument of at least one propensity function). Equation (26) evidently requires the relative change in X_i to be bounded by ε, except that X_i will never be required to change by an amount less than 1.

Recalling the tau-leaping formula [Eq. (5.16)], we see that the quantity defined in Eq. (5.25) will essentially be given by

$$\Delta_\tau X_i = \sum_{j=1}^{M} v_{ij} \rho\left(a_j(\mathbf{x})\tau\right), \ \forall i \in I_{rs}. \qquad (5.27)$$

Since the Poisson random variables (or the corresponding approximations) $\rho(a_j(\mathbf{x})\tau)$ on the right-hand side of Eq. (27) are statistically independent and have means and variances $a_j(\mathbf{x})\tau$, the mean and variance of that linear combination can be computed straightforwardly:

$$\langle \Delta_\tau X_i \rangle = \sum_{j=1}^{M} v_{ij}[a_j(x)\tau], \ \forall i \in I_{rs} \qquad (5.28)$$

$$\text{var}\{\Delta_\tau X_i\} = \sum_{j=1}^{M} v_{ij}^2[a_j(x)\tau], \ \forall i \in I_{rs}.$$

Using the same reasoning used in deriving the Gillespie–Petzold τ-selection procedure (Gillespie and Petzold, 2003), we may consider the bound (26) on $\Delta_\tau X_i$ to be "substantially satisfied" if it is simultaneously satisfied by the absolute mean and the standard deviation of $\Delta_\tau X_i$:

$$|\langle \Delta_\tau X_i \rangle| \leq \max\{\varepsilon x_i, 1\}, \sqrt{\text{var}\{\Delta_\tau X_i\}} \leq \max\{\varepsilon x_i, 1\}, \ \forall i \in I_{rs}, \qquad (5.29)$$

Substituting Eq. (5.28) into Eq. (5.29), we obtain the following bounds on τ:

$$\tau \leq \frac{\max\{\varepsilon_i x_i, 1\}}{|\sum_{j=1}^{M} v_{ij} a_j(\mathbf{x})|}, \quad \tau \leq \frac{\max\{\varepsilon_i x_i, 1\}^2}{\sum_{j=1}^{M} v_{ij}^2 a_j(\mathbf{x})}, \quad \forall i \in I_{rs}. \tag{5.30}$$

$$\hat{\mu}_i(\mathbf{x}) \triangleq \sum_{j-1}^{M} v_{ij} a_j(\mathbf{x}), \forall i \in I_{rs}, \tag{5.31}$$

$$\hat{\sigma}_i^2(\mathbf{x}) \triangleq \sum_{j-1}^{M} v_{ij}^2 a_j(\mathbf{x}), \forall i \in I_{rs},$$

where I_{rs} is the set of indices of all reactant species, and then taking

$$\tau = \min_{i \in I_{rs}} \left(\frac{\max\{\varepsilon x_i, 1\}}{|\hat{\mu}_i(\mathbf{x})|}, \frac{\max\{\varepsilon x_i, 1\}^2}{\hat{\sigma}_i^2(\mathbf{x})} \right) \tag{5.32}$$

The τ-selection procedure of Eqs. (5.31) and (5.32) will obviously be simpler to program and faster to execute than the τ-selection procedure of Eqs. (5.18), (5.19), and (5.24). Note in particular that the required number of computational operations increases quadratically with the number of reaction channels in the old formulas, but only linearly with the number of species in the new formulas. Since τ selection has to be performed prior to every tau leap, using these new formulas leads to substantially faster simulations when the system has many reactions and species.

Equations (5.31) and (5.32) are for the original tau-leaping method. In order to apply them to the hybrid SSA/tau-leaping method, they need a little modification. The calculation should not be extended to critical reactions, as they are handled by the adapted SSA part. Thus we let J_{ncr} denote the set of indices of the noncritical reactions. If J_{ncr} is empty (i.e., there are no noncritical reactions), we simply take $\tau = \infty$ (practically this can be a large step size, e.g., the whole simulation time interval). Otherwise, $\hat{\mu}_i$ and $\hat{\sigma}_i$ are calculated with the following formula:

$$\hat{\mu}_i(\mathbf{x}) \triangleq \sum_{j \in J_{ncr}} v_{ij} a_j(\mathbf{x}), \forall i \in I_{rs}, \tag{5.33}$$

$$\hat{\sigma}_i^2(\mathbf{x}) \triangleq \sum_{j \in J_{ncr}} v_{ij}^2 a_j(\mathbf{x}), \forall i \in I_{rs}.$$

The formula for τ remains the same as in Eq. (5.32) but the calculation of $\hat{\mu}_i$ and $\hat{\sigma}_i$ are replaced by Eq. (5.33). Note that the difference between Eqs. (5.31) and (5.33) is that only noncritical reactions are considered in Eq. (5.33), whereas all reactions are included in Eq. (5.31).

The full description of the hybrid SSA/tau-leaping method is given as follows.

Algorithm 3: The hybrid SSA/tau-leaping method

1. In state $\mathbf{x}$ at time t, identify the currently critical reactions. We calculate L_j according to Eq. (5.17). Any reaction R_j with $a_j(\mathbf{x}) > 0$ is deemed *critical* if $L_j < n_c$. Otherwise, it is *noncritical*. (We normally take $n_c = 10$ as a practical value.)
2. Let J_{ncr} denote the set of indices of the noncritical reactions. If J_{ncr} is empty, we take $\tau' = \infty$ (or the final simulation time). Otherwise, with a value chosen for ε (we normally take $\varepsilon = 0.03$), compute a candidate time leap τ' from the τ-selection Eqs. (5.32)) and (5.33)). Thus τ' tentatively estimates the time to the next noncritical reaction.
3. If τ' is less than some small multiple (which we usually take to be 10) of $1/a_0(\mathbf{x})$, abandon tau leaping temporarily, execute some modest number (which we usually take to be 100) of single-reaction SSA steps, and return to step 1. Otherwise, proceed to step 4.
4. Compute the sum $a_0^c(\mathbf{x})$ of the propensity functions of all the *critical* reactions. Generate a second candidate time leap τ'' as a sample of the exponential random variable with mean $1/a_0^c(\mathbf{x})$. As thus computed, τ'' tentatively estimates the time to the next critical reaction.
5. Take the actual time leap τ to be the smaller of τ' and τ'' and set the number of firings k_j of each reaction R_j accordingly:

    a. If $\tau' < \tau''$, take $\tau = \tau'$. For all critical reactions R_j set $k_j = 0$ (no critical reactions will occur during this leap). For all noncritical reactions R_j, generate k_j as a sample of the Poisson random variable with mean $a_j(\mathbf{x})\tau$.
    b. If $\tau'' \leq \tau'$, take $\tau = \tau''$. Generate j_c as a sample of the integer random variable with point probabilities $a_j(\mathbf{x})/a_0^c(\mathbf{x})$, where j runs over the index values of the *critical* reactions only. (The value of j_c identifies the next critical reaction, the *only* critical reaction that will occur in this leap.) Set $k_{j_c} = 1$, and for all other critical reactions R_j set $k_j = 0$. For all the noncritical reactions R_j, generate k_j as a sample of the Poisson random variable with mean $a_j(\mathbf{x})\tau$.

6. If there is a negative component in $\mathbf{x} + \Sigma_j k_j \mathbf{v}_j$, reduce τ' by half and return to step 3. Otherwise, leap by replacing $t \leftarrow t + \tau$ and $\mathbf{x} \leftarrow \mathbf{x} + \Sigma_j k_j \mathbf{v}_j$; then return to step 1 or else stop.

5. Measurement of Simulation Error

Now the only question left in the implementation of the hybrid SSA/tau-leaping method is how to select a proper error control parameter, ε. From our experience, we recommend $\varepsilon = 0.03$. We should always keep in mind that the tau-leaping method is just an approximation to the SSA method. Thus there will always be some numerical error, which depends on the error control parameter ε. The smaller ε is, the more accurate the result will be and the more time it will take to run the simulation. Thus the choice of ε is really a balance between accuracy and efficiency. We want a ε value small enough so that the overall error is acceptable. But how do we know that a particular ε is enough? There is no solid answer. When τ is fixed, it has been proved (Rathinam et al., 2005) that the errors of the mean and variance for the tau-leaping method are linear with τ. However, algorithm 3 is an adaptive method in the sense that the τ value varies in every simulation step and is directly connected to the choice of ε. Our intuition is that the errors should be linear with the value ε. However, there is no proof for that result. All we can do is run simulations on many test problems and measure the simulation errors with respect to different values of ε. The usual measurement for errors is the difference of the mean and variance between ensembles resulted from the SSA and the hybrid SSA/tau-leaping method. However, for some systems (see the Schlögl example given later), the mean and variance do not have a physical meaning. We are more interested in the error of the histograms of certain system properties (this could be the population of some species or a derived function, such as the cell cycle time, from the simulation trajectory). Thus the histogram distance is introduced to measure the simulation error of the distribution of a scalar random variable.

For a scalar random variable X, the probability density function (pdf) is defined as

$$p_X(x)dx = P(x \leq X < x + dx), \qquad (5.34)$$

for a continuous distribution. For a discrete distribution, the pdf is defined as the δ function given by

$$p_X(x) = \sum_x P(X = x)\delta(X - x). \qquad (5.35)$$

To measure the error, we need to define a distance between two probability distributions.

Suppose X and Y have probability density functions p_X and p_Y. We define the density distance between X and Y as

$$D(X, Y) = \int | p_X(s) - p_Y(s) | \, ds. \tag{5.36}$$

When X and Y are integers, Eq. (5.36) becomes

$$D(X, Y) = \sum_n \Big(| P(X = n) - P(Y = n) | \Big). \tag{5.37}$$

In many practical problems, it is difficult or impossible to obtain an analytic distribution. Instead we obtain samples from Monte Carlo simulations or observations. With those samples, the histogram of the observations is used to estimate the pdf. Let $x_1, x_2, \ldots, x_N$ be independent random variables, each having the same distribution as X. Defining the sign function

$$\kappa(x) = \begin{cases} 1 & \text{if } x \geq 0 \\ 0 & \text{if } x < 0 \end{cases}. \tag{5.38}$$

Suppose that all the sample values are bounded in the interval $I = [x_{\min}, x_{\max}]$. Let $L = x_{\max} - x_{\min}$. Divide the interval I into K subintervals and denote the subintervals by $I_i = [x_{\min} + (i-1)L/K, x_{\min} + iL/K)$. We define the characteristic function $\chi(x, I_i)$ as

$$\chi(x, I_i) = \begin{cases} 1 & \text{if } x \in I_i \\ 0 & \text{otherwise} \end{cases}. \tag{5.39}$$

Then the pdf p_X can be approximated by the histogram function h_X computed from

$$h_X(I_i) = \frac{K}{NL} \sum_{j=1}^{N} \chi(x_j, I_i). \tag{5.40}$$

The sum in Eq. (5.40) gives the number of points falling into the interval I_i. When that sum is divided by N we get the fraction of the points inside that interval, which approximates the probability of a sample point lying inside that interval. We divide this by the interval length, L/K, to approximate the probability density. Thus, $h_X(I_i)$ measures the average density function of X in the interval I_i. When K tends to infinity, the length of I_i reduces to 0. Then I_i is close to a point and h_X is close to p_X at that point.

For two groups of samples X_i and Y_j, we have the *histogram distance*

$$D_K(X, Y) = \sum_{i=1}^{K} \frac{|h_X(I_i) - h_Y(I_i)| L}{K}. \qquad (5.41)$$

Substituting Eq. (5.40) into Eq. (5.41), we obtain

$$D_K(X, Y) = \sum_{i=1}^{K} \left| \frac{\sum_{j=1}^{N} \chi(x_j, I_i)}{N} - \frac{\sum_{j=1}^{M} \chi(y_j, I_i)}{M} \right|. \qquad (5.42)$$

$D_K(X, Y)$ varies depending on the value of K. When $K = 1$ there is only one subinterval and we cannot tell the difference between X and Y. When K becomes larger we obtain more detailed information about the difference, and $D_K(X, Y)$ will increase. When K is very large we must generate a large number of samples, as otherwise there will not be enough data falling into each subinterval and there will be a large measurement error. When K, N, and M are sufficiently large, the histogram distance $D_K(X, Y)$ is close to the density distance $D(X, Y)$,

$$D_K(X, Y) \to D(X, Y) \text{ as } K, N, M \to \infty. \qquad (5.43)$$

In our numerical experiments, we use the histogram distance to measure the simulation errors.

6. SOFTWARE AND TWO NUMERICAL EXPERIMENTS

This section demonstrates application of the SSA and the hybrid SSA/tau-leaping method to two biochemical systems: a relatively simple toy model, the Schlögl model, and a realistic and far more complex model, the LacZ/LacY model. These comparison tests were performed with the software StochKit on a 1.4-GHz Pentium IV Linux workstation.

6.1. StochKit: A *stoch*astic simulation tool*kit*

Algorithm 3 has been fully implemented in the package StochKit, a software tool kit for discrete stochastic and multiscale simulation of chemically reacting systems.

StochKit is an efficient, extensible stochastic simulation toolkit developed in C++ that aims to make state-of-the-art stochastic simulation algorithms accessible to biologists and chemists, while remaining open to

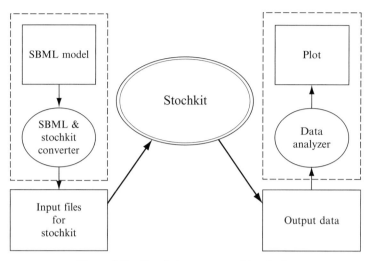

Figure 5.2 Simulation process of StochKit.

extension via new stochastic and multiscale algorithms. StochKit consists of a suite of software applications for stochastic simulation. The StochKit core implements the simulation algorithms. Additional tools are provided for the convenience of simulation and analysis. A typical simulation process of StochKit is shown in Fig. 5.2. A more detailed introduction to StochKit is given in Li *et al.* (2008). The StochKit package is freely available for download at www.engr.ucsb.edu/~cse. The user's guide is also available from that link.

There are several other algorithms in StochKit, such as the implicit tau-leaping method (Rathinam *et al.*, 2003), the trapezoidal tau-leaping method (Cao and Petzold, 2005), the slow scale SSA (Cao *et al.*, 2005a, 2005d, 2007), and its practical implementation as the multiscale SSA (Cao *et al.*, 2005b). These algorithms are still under development, while the implementations of Gillespie's SSA and the hybrid SSA/tau-leaping method are now mature enough for direct applications to biological systems. To select these two algorithms, configure the solver option in the driver file before calling the solver. There are two preset options: 1 is for the SSA and 0 is for the hybrid SSA/tau-leaping method. The code looks like

$$SolverOptions\ opt = ConfigStochRxn(1),$$

where argument 1 selects the SSA and 0 selects the hybrid SSA/tau-leaping method.

To demonstrate the application of the SSA and the hybrid SSA/tau-leaping methods, we apply both methods to the Schlögl model (Gillespie, 1992) and the LacZ/LacY model (Kierzek, 2002; Tian and Burrage, 2004).

6.2. The schlögl model

This model is famous for its bistable steady-state distribution. The reactions are

$$B_1 + 2X \underset{c_2}{\overset{c_1}{\rightleftarrows}} 3X,$$
$$B_2 \underset{c_2}{\overset{c_1}{\rightleftarrows}} X, \qquad (5.44)$$

where B_1 and B_2 denote buffered species whose respective molecular populations N_1 and N_2 are assumed to remain essentially constant over the time interval of interest. There is only one time-varying species, X; the state change vectors are $v_1 = v_3 = 1$, $v_2 = v_4 = -1$ and the propensity functions are

$$\begin{aligned}
a_1(x) &= \frac{c_1}{2} N_1 x(x-1), \\
a_2(x) &= \frac{c_2}{2} x(x-1)(x-2), \\
a_3(x) &= c_3 N_2, \\
a_4(x) &= c_4 x.
\end{aligned} \qquad (5.45)$$

For some values of the parameters this model has two stable states, which is the case for the parameter values chosen here:

$$\begin{aligned}
c_1 &= 3 \times 10^{-7}, c_2 = 10^{-4}, c_3 = 10^{-3}, c_4 = 3.5, \\
N_1 &= 10^5, N_2 = 2 \times 10^5.
\end{aligned} \qquad (5.46)$$

We made ensembles of 10^5 simulation runs from the initial state $X(0) = 250$ to time $t = 4$ using the SSA and the hybrid SSA/tau-leaping method, the latter for a range of ε values.

Figure 5.3 shows the histogram distance or "error" between the SSA ensemble and the tau-leaping ensembles as a function of ε. We can see that the errors increase roughly linearly with ε.

6.3. The LacZ/LacY model

This model was first proposed by Kierzek (2002) and later used for an efficiency test in Tian and Burrage (2004). This model has 22 reactions, 19 species, and an extremely multiscale nature. A detailed description of this model is omitted here. Interested readers can refer to the two references just given, and a list of the reaction channels and reaction rates of this model is

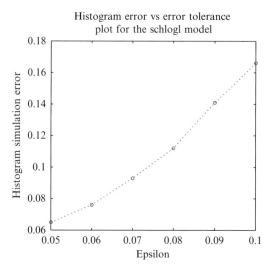

Figure 5.3 Plot of histogram distance errors corresponding to different ϵ values for the Schlögl model. Histogram distance errors are measured by 105 samples generated from the SSA method and the hybrid SSA/tau-leaping method using different τ-selection formulas.

given in Table 5.1. Tian and Burrage (2004) reported that negative populations were observed many times in their simulation using the original tau-leaping method. In our numerical experiments for this model, a single simulation from $t = 0$ to $t = 2100$ by SSA took 3359 s of CPU time. With an error tolerance of $\varepsilon = 0.03$, a single simulation by the hybrid SSA/tau-leaping method took 113.77 s of CPU time with no negative population observed during the simulation.

Since a single SSA simulation from $t = 0$ to $t = 2100$ took about 1 h on our computer, obtaining a large number of SSA samples posed a challenge. We ran the SSA from time $t = 0$ to time $t = 1000$ to obtain an "initial" state; then we made 10^5 SSA runs from time $t = 1000$ to time $t = 1001$ (which required about 3.5 h of computer time) and histogrammed the resulting populations. Finally, we made the same number of SSA/tau-leaping runs over the same time interval for a range of values for ε. Figure 5.4 shows the plot of histogram distance or "error" as a function of ε. We note again that the error increases roughly linearly with ε.

7. Conclusion

This chapter gave a review of the two most important simulation algorithms for discrete stochastic systems. Gillespie's SSA has the distinct advantage that it is an exact simulation method, but it can be slow for many

Table 5.1 A full list of reaction channels and deterministic reaction rates for LacY/LacZ model

Reaction channel	Reaction rate
PLac + RNAP → PLacRNAP	0.17
PLacRNAP → PLac + RNAP	10
PLacRNAP → TrLacZ1	1
TrLacZ1 → RbsLacZ + PLac + TrLacZ2	1
TrLacZ2 → TrLacY1	0.015
TrLacY1 → RbsLacY + TrLacY2	1
TrLacY2 → RNAP	0.36
Ribosome + RbsLacZ → RbsRibosomeLacZ	0.17
Ribosome + RbsLacY → RbsRibosomeLacY	0.17
RbsRibosomeLacZ → Ribosome + RbsLacZ	0.45
RbsRibosomeLacY → Ribosome + RbsLacY	0.45
RbsRibosomeLacZ → TrRbsLacZ + RbsLacZ	0.4
RbsRibosomeLacY → TrRbsLacY + RbsLacY	0.4
TrRbsLacZ → LacZ	0.015
TrRbsLacY → LacY	0.036
LacZ → dgrLacZ	6.42×10^{-5}
LacY → dgrLacY	6.42×10^{-5}
RbsLacZ → dgrRbsLacZ	0.3
RbsLacY → dgrRbsLacY	0.3
LacZ + lactose → LacZlactose	9.52×10^{-5}
LacZlactose → product + LacZ	431
LacY → lactose + LacY	14

practical systems. The tau-leaping method gives an approximation of the SSA that is a natural bridge connecting the discrete stochastic SSA regime at one extreme of the approximation to the continuous deterministic RRE regime at the other extreme. However, approximations of the tau-leaping method can sometimes cause unphysical results, such as negative numbers of molecules. The hybrid SSA/tau-leaping method is a practical implementation strategy for the tau-leaping method. Although it is more complex than either the SSA or the tau-leaping method, the hybrid approach combines the strengths of each method, while also avoiding the major pitfalls of each method. The three algorithms detailed in this chapter provide a recipe for implementing each of these three methods. Most importantly, the method of measuring the simulation error is given in detail. The complexity of this measure often leads to its neglect in many applications, a serious oversight for any numerical method.

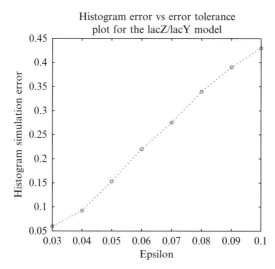

Figure 5.4 Plot of histogram distance errors corresponding to different ϵ values for the LacZ/LacY model. Histogram distance errors are measured between the population distributions of LacZlactose in 10^5 runs of the SSA and the hybrid SSA/tau-leaping method using different τ-selection formulas.

ACKNOWLEDGMENTS

This work was supported by the National Science Foundation under award CCF-0726763 and the National Institutes of Health under award GM073744.

REFERENCES

Arkin, A., Ross, J., and McAdams, H. (1998). Stochastic kinetic analysis of developmental pathway bifurcation in phage λ-infected *E. coli* cells. *Genetics* **149,** 1633–1648.

Brenan, K. E., Campbell, S. L., and Petzold, L. R. (1996). "Numerical Solution of Initial-Value Problems in Differential-Algebraic Equations." SIAM, Philadelphia, PA.

Cao, Y., Gillespie, D., and Petzold, L. (2005a). The slow-scale stochastic simulation algorithm. *J. Chem. Phys.* **122,** 014116.

Cao, Y., Gillespie, D., and Petzold, L. (2005b). Multiscale stochastic simulation algorithm with stochastic partial equilibrium assumption for chemically reacting systems. *J. Comput. Phys.* **206,** 395–411.

Cao, Y., Gillespie, D., and Petzold, L. (2005c). Avoiding negative populations in explicit tau leaping. *J. Chem. Phys.* **123,** 054104.

Cao, Y., Gillespie, D., and Petzold, L. (2005d). Accelerated stochastic simulation of the Sti_ enzyme-substrate reaction. *J. Chem. Phys.* **123,** 144917.

Cao, Y., Gillespie, D., and Petzold, L. (2006). Efficient stepsize selection for the tau-leaping method. *J. Chem. Phys.* **124,** 044109.

Cao, Y., Li, H., and Petzold, L. (2004). Efficient formulation of the stochastic simulation algorithm for chemically reacting systems. *J. Chem. Phys.* **121,** 4059–4067.

Cao, Y., and Petzold, L. (2005). Trapezoidal tau-leaping formula for the stochastic simulation of chemically reacting systems. In "Proceedings of Foundations of Systems Biology in Engineering (FOSBE 2005)," pp. 149–152.

Chatterjee, A., Vlachos, D., and Katsoulakis, M. (2005). Binomial distribution based tauleap accelerated stochastic simulation. *J. Chem. Phys.* **122**, 024112.

Cohen, S., and Hindmarsh, A. (1996). CVODE, a Stiff/Nonstiff ODE solver in C. *Comput. Phys.* **10**, 138–143.

Fedoroff, N., and Fontana, W. (2002). Small numbers of big molecules. *Science* **297**, 1129–1131.

Gibson, M., and Bruck, J. (2000). Efficient exact stochastic simulation of chemical systems with many species and many channels. *J. Phys. Chem. A* **104**, 1876.

Gillespie, D. (1976). A general method for numerically simulating the stochastic time evolution of coupled chemical reactions. *J. Comput. Phys.* **22**, 403–434.

Gillespie, D. (1977). Exact stochastic simulation of coupled chemical reactions. *J. Phys. Chem.* **81**, 2340–2361.

Gillespie, D. (1992). "Markov Processes: An Introduction for Physical Scientists." Academic Press, New York.

Gillespie, D. (2001). Approximate accelerated stochastic simulation of chemically reacting systems. *J. Chem. Phys.* **115**, 1716.

Gillespie, D., and Petzold, L. (2003). Improved leap-size selection for accelerated stochastic simulation. *J. Chem. Phys.* **119**, 8229–8234.

Gillespie, D., Petzold, L., and Cao, Y. (2007). Comment on nested stochastic simulation algorithm for chemical kinetic systems with disparate rates. *J. Chem. Phys.* **126**, 137101.

Hindmarsh, A. (1983). Scientific computing. In "ODEPACK, a Systematized Collection of ODE Solvers" (R. S. Stepleman, et al., eds.), Vol. 1, pp. 55–64. North-Holland, Amsterdam.

Kierzek, A. (2002). STOCKS: STOChastic kinetic simulations of biochemical systems with Gillespie algorithm. *Bioinformatics* **18**, 470–481.

Li, H., and Petzold, L. (2006). "Logarithmic Direct Method for Discrete Stochastic Simulation of Chemically Reacting Systems." Technical report.

Li, H., Cao, Y., Petzold, L., and Gillespie, D. (2008). Algorithms and software for stochastic simulation of biochemical reacting systems. Biotechnol. Progr. **24**, 56–61.

McAdams, H., and Arkin, A. (1997). Stochastic mechanisms in gene expression. *Proc. Natl. Acad. Sci. USA* **94**, 814–819.

McCollum, J. M., Peterson, G. D., Cox, C. D., Simpson, M. L., and Samatova, N. F. (2006). The sorting direct method for stochastic simulation of biochemical systems with varying reaction execution behavior. Comput. Biol. Chem. **30**, 39–49.

Munsky, B., and Khammash, M. (2006). The finite state projection algorithm for the solution of the chemical master equation. *J. Chem. Phys.* **124**, 044101.

Rao, C., and Arkin, A. (2003). Stochastic chemical kinetics and the quasi steady-state assumption: Application to the Gillespie algorithm. J. Chem. Phys. **118**, 4999–5010.

Rathinam, M., Petzold, L., Cao, Y., and Gillespie, D. (2003). Stiffness in stochastic chemically reacting systems: The implicit tau-leaping method. *J. Chem. Phys.* **119**, 12784–12794.

Rathinam, M., Petzold, L., Cao, Y., and Gillespie, D. (2005). Consistency and stability of tau leaping schemes for chemical reaction systems. *SIAM Multiscale Modeling* **4**, 867–895.

Tian, T., and Burrage, K. (2004). Binomial leap methods for simulating stochastic chemical kinetics. *J. Chem. Phys.* **121**, 10356–10364.

Zhang, J., and Watson, L. (2007). A modified uniformization method for the chemical master equation. In "Proc. 7th IEEE Internat. Conf. on Bioinformatics and Bioengineering, Boston, MA," pp. 1429–1433.

CHAPTER SIX

ANALYSES FOR PHYSIOLOGICAL AND BEHAVIORAL RHYTHMICITY

Harold B. Dowse

Contents

1. Introduction	142
2. Types of Biological Data and Their Acquisition	143
3. Analysis in the Time Domain	145
4. Analysis in the Frequency Domain	151
5. Time/Frequency Analysis and the Wavelet Transform	161
6. Signal Conditioning	164
7. Strength and Regularity of a Signal	169
8. Conclusions	171
References	171

Abstract

Biological data that contain cycles require specialized statistical and analytical procedures. Techniques for analysis of time series from three types of systems are considered with the intent that the choice of examples is sufficiently broad that the processes described can be generalized to most other types of physiological or behavioral work. Behavioral circadian rhythms, acoustic signals in fly mating, and the *Drosophila melanogaster* cardiac system have been picked as typical in three broad areas. Worked examples from the fly cardiac system are studied in full detail throughout. The nature of the data streams and how they are acquired is first discussed with attention paid to ensuring satisfactory subsequent statistical treatment. Analysis in the time domain, namely simple and advanced plotting of data, autocorrelation analysis, and cross-correlation, is described. The search for periodicity is conducted through examples of analysis in the frequency domain, primarily spectral analysis. Nonstationary time series pose a particular problem, and wavelet analysis of *Drosophila* mating song is described in detail as an example. Conditioning of data to improve output with digital filters, Fourier filtering, and trend removal is described. Finally, two tests for noise levels and regularity are considered.

School of Biology and Ecology and Department of Mathematics and Statistics, University of Maine, Orono, Maine

All the nonproprietary software used throughout the work is available from the author free of charge and can be specifically tailored to the needs of individual systems.

1. Introduction

Biological systems that evolve in time often do so in a rhythmic manner. Typical examples are heart beating (Bodmer *et al.*, 2004; Dowse *et al.*, 1995), circadian (Dowse, 2007), and ultradian (Dowse, 2008) biological cycles and acoustic communication, for example, in *Drosophila* mating (Kyriacou and Hall, 1982). Using objective analysis techniques to extract useful information from these time series is central to understanding and working with the systems that produce them. Digital signal analysis techniques originating with astrophysics, geophysics, and electronics have been adapted to biological series and provide critical information on any inherent periodicity, namely its frequency or period as well as its strength and regularity. The latter two may be two separate matters entirely.

The mode of data acquisition is the first concern. Often biological data are records of events as a function of time, or perhaps the number of events during a sequential series of equal time intervals or "bins." Alternatively, output may be a continuous variable, such as the titer of an enzyme or binding protein. Acquisition technique and constraints upon it may affect the outcomes of later analyses and must be taken into consideration. As part of this process, the signal must ultimately be rendered digital for computer analysis. Examples will be considered.

Initial analysis is done in the time domain and may range from something as simple as a plot of the amplitude of the process to powerful statistical techniques such as autocorrelation, which can be used for determining if significant periodicities are present. Analysis in the frequency domain, usually spectral analysis, provides information on the period or frequency of any cycles present. This usually involves one of several variants of Fourier analysis, and recent advances in that area have revealed exceptional detail in biological signals (review: Chatfield, 1989). The mating song of the fruit fly, *Drosophila melanogaster*, is rich with information, but the data stream, as is the case with many other biological systems, is irregular and variable in time. Wavelet analysis is particularly useful in this instance. Digital signals, like their analog counterparts, may be filtered to remove noise or any frequencies in other spectral ranges that can be obscuring those in the range of interest (Hamming, 1983). The strength and regularity of the biological signal are of paramount importance. Spectral analysis algorithms may be altered appropriately to provide an objective measurement of a signal-to-noise (SNR) ratio (Dowse and Ringo, 1987). A related but distinct issue is the regularity of the cycles in the signal. For example, a heart may be beating strongly, but the duration of its pacemaker duty cycle may vary considerably more than normal from beat to beat with occasional skipped beats or, conversely, may be more regular than normal. Either alteration might be a result of pathology (Glass and Mackey, 1988; Lombardi, 2000; Osaka et al., 2003). An index of rhythmic regularity is useful in this regard.

This chapter reviews modern digital techniques used to address each of these problems in turn. It uses the *Drosophila* model cardiac system extensively in this discourse, but the methods are widely applicable and other examples will be used as needed.

2. Types of Biological Data and Their Acquisition

One of the most intensively studied biological signals, of the several we shall consider, is found in the physiological and behavioral records of organisms over time. These records are commonly found to be rhythmic with periodicity in the ballpark of 24 h, the solar day. In unvarying environmental conditions, such as constant darkness (DD) or low illumination (LL), the periodicity will vary from the astronomical day, hence the term "*circadian*," or approximately daily rhythms (review: Palmer, 2002). This periodicity is the output of a biological oscillator (or oscillators), and study of this living horologue has been intense in the hopes of finding the mechanism (Dunlap, 1999; Hall, 2003). This field offers the opportunity to discuss generally applicable concepts.

Biological rhythm data take many forms, as clocks may be studied at levels ranging from intracellular fluorescence to running wheel activity. This broaches the topic of sampling. In cases of activity of an enzyme or the fluorescence level of a tag, for example, the variables are continuous and the sampling interval can be chosen arbitrarily. A primary concern here is that it be done rapidly enough to avoid "aliasing" in the periodicity region of interest. Aliasing occurs when the sampling interval is longer than the period being recorded and can be seen in old western movies when the spokes of wagon wheels seem to be going backward, a result of the interaction between the number of still frames per second and the angular velocity of the wheel (Hamming, 1983). Sampling frequency must be at least twice the frequency (half the period) of that of the sampled process. This is the Nyquist or fold-over frequency (Chatfield, 1989). A bit faster is better to be sure detail is not lost, but this is the theoretical tipping point. The tradeoff is an increasing number of data points to store and the commensurate wait for analysis programs to run if the sampling is gratuitously rapid.

The primary event in data acquisition is often an instantaneous reading of an analog signal. This may be transmembrane voltage or current in a

Xenopus oocyte clamp setup, sound levels picked up by a microphone, light intensity reported out by a photomultiplier tube, or the output of an O_2 electrode; the list is endless. In general, however, whatever is being measured, the transduction process ultimately yields a voltage. This continuous analog voltage signal needs to be converted to a format that the computer can deal with. This process is often now done by analog to digital (A/D) converters within the instruments themselves, which will have a digital computer interface capability as a matter of course. Nonetheless, research equipment must often be built from scratch in-house for a specific purpose, and here analog signals may need to be dealt with by the user. The A/D converter is a unit that assigns numbers to a given input voltage. For example, in my laboratory we monitor fly heartbeat optically (see later) (Dowse *et al.*, 1995; Johnson *et al.*, 1998). The output of the system is a voltage between −5 and +5 V, which is monitored on an oscilloscope. The computer has a DAS8, A/D 12-bit interface (Kiethly/Metrabyte) that employs 4096 0.00244 V steps, assigning proportional values between −2048 and +2048. The rate of digitization is programmable up to 1 MHz in this antique but thoroughly serviceable system. We find that for a ≈2- to 3-Hz heartbeat, 100 Hz is more than sufficient to yield excellent time resolution.

For noncontinuous data, there are other considerations. The Nyquist interval must still be factored in as a baseline for maximum sampling interval/minimum frequency (see earlier discussion), but there is a further constraint on how fast sampling may be done that has nothing to do with optimizing the number of data points to grab for computing expedience vs resolution. Common examples of this sort of data are running wheel activity in mammals (DeCoursery, 1960) and the breaking of an infrared light beam by *Drosophila* (Dowse *et al.*, 1987). Here, individual events are being registered and are summed across arbitrary intervals or "bins." The question is how well do these binned time series stand up to the sorts of analyses developed for discretely sampled continuous functions? It has been shown that bin size affects the output of time series analysis and that this effect can be profound when bin size is too small (review: Dowse and Ringo, 1994).

Over an arbitrarily short interval of the day, say a half an hour, the series of occurrences of events, such as a fly breaking a light beam in a chamber, is described by a Poisson process. There is no time structure or pattern and events occur stochastically. The probability, *P*, of *k* events occurring during the interval t, $t + 1$ is given by

$$P[N(t+1) - N(t) = k] = e^{-\lambda}\frac{\lambda^k}{k!}. \qquad (6.1)$$

The mean overall rate in events per unit time (EPUT) is given by λ (Schefler, 1969).

Over the course of a circadian day, for example, EPUT varies in a pattern, notably if the fly is behaving rhythmically. This variation can usefully be thought of as a Poisson process with a time-varying λ. In the case of running wheel data, of course, the events appear regularly spaced with a periodicity dependent on the rate of running in the apparatus, although bouts of running may be stochastically spaced throughout the active period. Nonetheless, the "amplitude" of the process remains EPUT, with the unit of time being the bin length.

Based on empirical and practical considerations, bin size much smaller than 10 minutes may cause artifact, in that perfectly good periodicities may be obscured in the presence of a lot of noise. Half-hour bins are generally small enough for good results in our experience. Longer bin lengths, for example, 1 h or longer, may act as a poorly defined low-pass digital filter, with a reduction in power transferred of about 20% at a periodicity of 2 h. Five-minute bins have a flat transfer function (the plot of power transmitted through the filter as a function of period or frequency—see detailed discussion later)(Dowse and Ringo, 1994).

3. ANALYSIS IN THE TIME DOMAIN

Time series data may be analyzed in two domains: time and frequency. They may be transformed from one to the other as needed. In the time domain, relatively simple techniques are usually used initially to visualize evolution of the system. There is also a relatively straightforward statistical analysis available, the autocorrelogram, which tests for the presence and significance of any rhythmicity. Frequency or period may be measured crudely from either plots of raw data or the autocorrelogram, and questions of phase and waveform can be addressed directly.

We shall consider the cardiac system of the fly carefully to illustrate the analyses. The heart of the insect is a simple tube that works as a peristaltic pump within an open circulatory system, moving hemolymph from the most posterior region of the abdomen forward to the brain (Curtis et al., 1999; Jones, 1977; Rizki, 1978). As air is carried to the tissues by a tracheal system, there are no pigments for gas transport, so even serious decrements in cardiac function may not necessarily be fatal (e.g., Johnson et al., 1998). Nutrients and wastes are transported by the hemolymph (Jones, 1977). Heartbeat is myogenic, arising in discrete pacemaker cells posteriorly (Dowse et al., 1995; Gu and Singh, 1995; Rizki, 1978), but as heartbeat can be retrograde, there is an alternate pacemaker near the anterior end as well (Dulcis and Levine, 2005; Wasserthal, 2007). The fly heart model is of considerable interest of late, as genes encoding heart structure and ion channels that function in the pacemaker have been shown to have analogous function in the human heart (Bodmer et al., 2004; Wolf et al., 2006).

Figure 6.1A shows a 60-s sample time series from this system, depicting the heartbeat of a wild-type *D. melanogaster* recorded optically at the P1 pupal stage. At this point in time the heart can be monitored optically, as the pupal case has not yet begun to tan and remains transparent (Ashburner, 1989). The heart is also transparent, but the nearly opaque fat bodies on either side move as the heart beats, causing a change in the amount of light passing through the animal. This is picked up by a phototransistor (FPT100) affixed in the outlet pupil of one of the eyepieces of a binocular microscope. The signal is preamplified by a 741C op amp and is further amplified by a Grass polygraph. The output voltage is digitized as described previously by a DAS8 AD converter (Kiethley/Metrabyte) at 100 Hz and recorded as a text file in a computer. The temperature of the preparation is controlled by a Sensortek100 unit and, in this instance, is maintained at 25° (cf. Dowse et al., 1995; Johnson et al., 1997, 1998). The heartbeat is very regular in this animal, although there are several gaps. Recall that this is the plot of voltage as a function of time.

While it is clear that this heart is rhythmic, it is useful to apply an objective statistical test, even in this clear example, to determine the significance of any periodicity. For example, Fig. 6.1B shows the record of a second wild-type animal's heart that is not nearly so clearly rhythmic, and Fig. 6.1C may depict a totally arrhythmic organ, also from a wild-type fly. No solid conclusion can be drawn based just on inspection of this erratic plot. An objective statistical method to determine whether a significant rhythm is present is by autocorrelation analysis (Chatfield, 1989; Levine et al., 2002). To conduct this analysis, the time series is paired with itself in register and a standard correlation analysis is done yielding the correlation coefficient, r. Since the correspondence is exactly one to one, the correlation is perfect and the resultant r is 1. The two identical series are then set out of register or "lagged" by one datum. This will cause a corresponding decrement in the correlation coefficient computed. This lagging continues one datum at a time up to about one-third of the length of the entire series. The sequential r values are plotted as a function of the lag, and this is the autocorrelogram or autocorrelation function. If the series is rhythmic, the drop in r will continue and will become negative, reaching a nadir as the peaks and valleys in the values become π antiphase. A second positive peak will occur when the peaks and valleys return to phase locking at a full 2π. The general mathematical interpretation of the coefficients in the autocorrelogram is that they are cosines of the angles between two vectors,

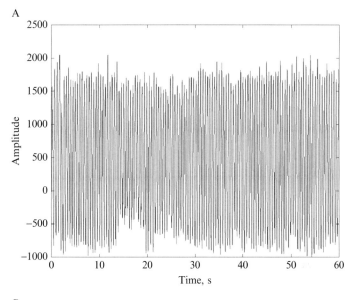

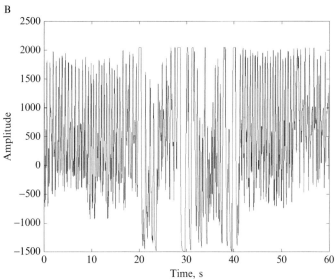

Figure 6.1 (*continued*)

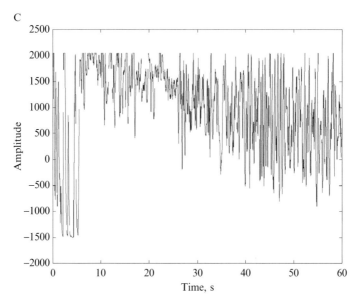

Figure 6.1 Optically acquired digital records of wild-type *Drosophila melanogaster* heartbeat. (A) Extremely regular heartbeat with few changes in amplitude or period. (B) This heart is substantially more irregular in function. There are periods during which the beat is fairly erratic interspersed with regular beating. At times, especially from about 53 s on, it can be seen that the beat is bigeminal, with weak beats alternating with the much stronger power beats. (C) Here, the heart is almost arrhythmic. It can be seen to be beating during a few intervals, notably between 45 and 50 s.

with the vectors being the original time series and that same series lagged out of register (Wiener, 1949). The process is well approximated by

$$r_k = \left(\sum_{t=1}^{N-k} (x_t - x_m)(x_{t-k} - x_m) \right) \Big/ \sum_{t=1}^{N} (x_t - x_m)^2, \quad (6.2)$$

where N is the number of samples and x_m is the mean of the series (Chatfield, 1989).

Note that the output as described earlier is normalized at each step by dividing by the variance in the entire data set (the denominator in the aforementioned equation), but need not be. If this is not done, the output is in the form of variance, and this "covariance" can be reported out instead, if this is desired, as the autocovariance function (Chatfield, 1989). Differences between and relative utilities of these two functions will become apparent when spectral analysis is considered later. Here, the normalization to get an r is useful, as it allows comparisons among experiments and the function will yield yet another useful objective statistic for comparisons, as described later.

Each time the vectors are lagged, the values on the two far ends are no longer paired and must be discarded; hence the power of the test is gradually diminished. For this reason, the usual limit of the autocorrelation computation is about $N/3$. The 95% confidence interval and hence significance of a given peak is given as $2/\sqrt{N}$, where N is the number of data points (Chatfield, 1989). Plus and minus confidence intervals are plotted as flat lines; the decrement in N as values are discarded is usually ignored. The rule of thumb interpretation of the plot is normally looking for repeated peaks equaling or exceeding the confidence interval, but a long run of peaks not quite reaching this level is usually sufficient if inspection of the raw data plot yields similar results and if the periodicity turns up in this range in the spectral analysis. Use of the autocorrelation function to provide an estimator of regularity in rhythmicity is discussed later. In the examples shown in Fig. 6.1, the heart of the third pupa is considered arrhythmic, as will be shown. Figure 6.2 depicts the autocorrelograms of data from the hearts in Fig. 6.1. As the function is symmetrical and can be lagged in either direction, data from the reverse lagging are plotted here for symmetry and ease in visual interpretation.

In the case of biological rhythm research, another way of displaying data is commonly applied. This is by way of producing a "raster plot" or actogram, in which data are broken up into 24-h segments, which are plotted one below the other sequentially, that is, "modulo" 24 h. In this way, long records may be viewed easily and the relationship between the rhythmicity and the 24-h day can be assessed (e.g., DeCoursey, 1960). However, we shall not consider this technique here. The reader is referred to the following source for a full coverage with examples (Palmer *et al.*, 1994). It is worth noting, however, that such raster plots can be very misleading. Flies bearing mutations in the *period* gene (Konopka and Benzer, 1971), considered central to the biological "clock" (Dunlap, 1999), were reported as arrhythmic based on such raster plotting and the employment of the badly flawed "spectral analysis" program erroneously called the "periodogram" (see later for a discussion)(Dowse *et al.*, 1987). By choosing a proper value for the length of the raster based on periodicity revealed by proper spectral analysis, ultradian (faster than 1/day) became clear. Even the relatively insensitive autocorrelograms showed clear, significant rhythmicity in these data (review: Dowse, 2008).

A further use of the autocorrelation algorithm can be done when it is desirable to compare the phase relationship between two time series that have similar frequencies. This may be done by way of "cross correlation." In this case, instead of comparing a time series with itself as it is lagged, a second time series is used. If they are in perfect phase, the peaks in the correlogram will be centered, but insofar as they are out of phase the central peak will be offset one way or the other. This analysis has been covered in detail elsewhere (Levine *et al.*, 2002).

One final technique can be applied in the time domain to enhance the interpretability of data; this is "time averaging." This is done commonly in electrophysiology, but has been used in circadian rhythm research as well (see, e.g., Hamblen-Coyle *et al.*, 1989). In this process, successive cycles are excised from the data stream modulo the period calculated and the peaks within are kept in phase. If this is done for a behavioral rhythm, for example, recorded in a 24-h LD cycle, then the section is simply 24 h. In electrophysiology, data sections containing individual events are excised. These data segments become rows in a matrix in register with one another. The columns produce means, which are plotted to get a composite picture

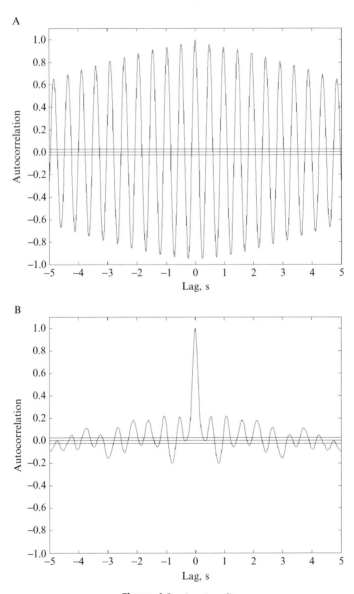

Figure 6.2 (*continued*)

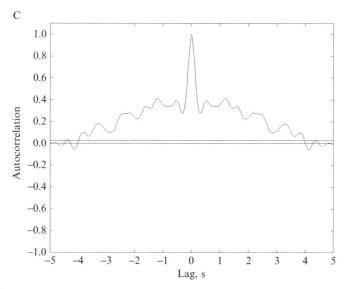

Figure 6.2 Autocorrelograms produced from the data in the previous figure, appearing in the same order. The correlogram is a time-domain analysis that allows assessment for the presence or absence of any periodicities in the data as well as their regularity (see text). The autocorrelation values, r, are without units. The horizontal lines above and below the abscissa are ± the 95% confidence interval calculated as $2/\sqrt{N}$, in this case ± 0.0258. (A) Correlogram from the data in Fig. 6.1A. This heart is exceptionally regular in its rhythmicity. The decay envelope of the function is very shallow indicating long range order and stable frequency. The height of the third peak, counting the peak at lag 0 as #1 is 0.935 and constitutes the Rhythmicity Index (RI; see section below on signal strength and regularity). (B) In keeping with the appearance of reduced regularity in Fig. 6.1B, the decay envelope is steep. The RI is 0.221. (C) The heart of this animal beats occasionally, but it is erratic in the extreme. Owing to an $RI < 0.025$, it is considered arrhythmic.

of the signal (Hille, 2001). Figure 6.3A depicts an artificially produced time series (produced by a program we have written) consisting of a square wave with a period of 25 h and 50% stochastic noise added. We shall use this as an example of circadian periodicity. For comparison, the autocorrelogram of the series is shown in Fig. 6.3B. Figure 6.3C shows the result of time averaging the series to produce a single waveform estimate.

4. ANALYSIS IN THE FREQUENCY DOMAIN

In frequency analysis, the goal is to determine either the period or the frequency of any cycles ongoing in the process. This is done by looking at signal power as a function of frequency. Power in a signal is the ensemble average of the squared values in the series (Beauchamp and Yuen, 1979). Note that if the mean is zero, this is the same as variance. The power in the signals depicted in Fig. 6.1 are 1A = 7.732 × 10⁵; 1B = 8.735 × 10⁵; and 1C = 6.813 × 10⁵. It is informative to think of it this way: the power in data is being partitioned by frequency, and the area under the curve of a spectrum, constructed as described later, is the power in the original signal.

To prepare such a spectrum, the workhorse is Fourier analysis. This begins with the remarkable observation that most functions can be

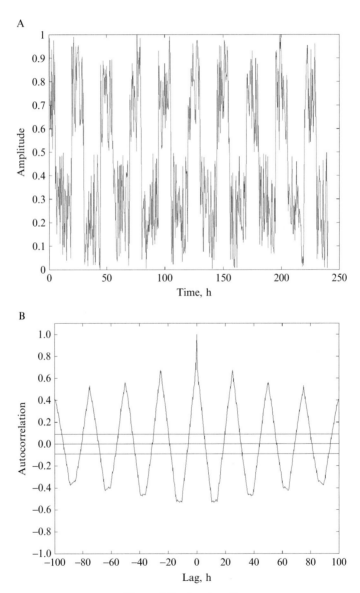

Figure 6.3 (*continued*)

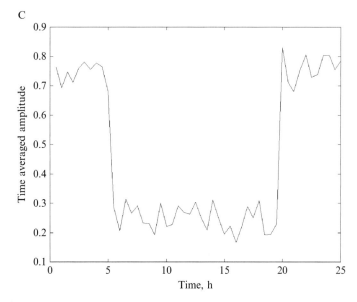

Figure 6.3 To simulate a circadian behavioral rhythm, a signal generating program was employed to produce a square wave with a period of 25 h and 50% added white noise. Data acquisition was set at one half hour intervals and 480 data points were produced with a simulated half hour sampling/binning rate. (A) The raw unconditioned signal as it was produced by the program. (B) As with the heartbeat data, the signal was analyzed with the autocorrelogram. Note the strong repeating peaks at lags of 25, 50 and 75 h. Despite the large amount of noise, given the unvarying length of the period, this is to be expected. The decay envelope is not too steep. RI for this signal is 0.561. (C) The signal was broken up into 25-h segments (50 data points each) which were inserted as rows into of a 9 X 50 (Row X Column) Matrix. Extra "odd" points were discarded. The matrix columns were summed and a mean activity was computed. This is the plot of the output of that operation, a time-averaged estimate of the underlying wave form in the presence of high frequency noise.

approximated by a series of sine and cosine terms in a process called orthogonal decomposition. Start with an arbitrary function $f(t)$ that conforms to the "Dirichlet conditions," namely that it have a finite number of maxima and minima, that it be everywhere defined, and that there be a finite number of discontinuities (Lanczos, 1956, 1966). Biological time series will almost certainly conform.

$$f(t) \cong a_0/2 + a_1 \sin t + a_2 \sin 2t + \ldots b_1 \cos t + b_2 \cos 2t + \ldots \quad (6.3)$$

The Fourier series used to approximate the function consists of pairs of sine and cosine terms that are orthogonal (Hamming, 1983). An acoustic analogy is good here. Think of the function as a guitar string, with the fundamental vibration first, followed by successive harmonics. The mathematical interpretation is a series of vectors of length R rotating in the complex plane with angular velocity ω, which is in radians/s ($\omega = 2\pi f$ or $2\pi/T$, where f is frequency and T is period). Here $R^2 = a^2 + b^2$ for each value of a and b for a given harmonic (Beauchamp and Yuen, 1979). The Fourier transform is a special case of the Fourier series, which in this form can now be used to map the series from the time domain to the frequency domain as $F(\omega)$ with the series of coefficients a and b being extracted from data (Lanczos, 1956):

$$F(\omega) = \int_{-\infty}^{\infty} f(t)e^{-i\omega t}dt, \qquad (6.4)$$

where the exponential consolidates the sine and cosine terms. A plot of R^2 calculated from the a and b coefficients extracted form the "periodogram" of the series and constitute a representation of the spectrum (Schuster, 1898). Peaks in the periodogram indicate periodicity in data at those given values. The area under the curve, as noted, is the total power in data, and for each value of R, this can be interpreted as the power in the signal at that period or frequency.

This process is not to be confused with another "periodogram" concocted some time later and used extensively in biological rhythm work. Whitaker and Robinson (1924) proposed producing a "Buys-Ballot" table for data using all possible values for frequency. This is much like the rasterizing or signal averaging techniques mentioned earlier. The rows of the series of matrices have varying length, sectioned off modulo each periodicity. The columns of the matrices are then added and means produced. For a matrix with a given set of row lengths, the variance of the column sums becomes the coefficient for the period corresponding to the length of that row. As the length varies, the variance will peak when the row equals the length of the period. The peaks and valleys will all be in register at this point as with the signal-averaged waveform discussed earlier. This method was championed for circadian rhythm studies by Enright (1965, 1990). However, it is not a mathematically sound procedure, as was demonstrated conclusively by Kendall (1946) when he noted that if there is a peak in the output, it does not mean there is any periodicity, as the variance of the column sums is independent of their order. In any event, in practice, this "periodogram" is unable to perform to the standards demanded of modern spectral analysis techniques. Historically, its widespread employment obscured important short-period (ultradian) rhythms in what appeared to be arrhythmic flies for a long time (see earlier discussion; review: Dowse, 2008). Its use is not recommended by this author. At the very least, because of Schuster's (1898) long priority, it cannot legitimately be called a periodogram. Figure 6.4 shows the Whitaker/Robinson (1924) "periodogram" for the noisy square wave depicted in Fig. 6.3A and compared further with the corresponding MESA plot in Fig. 6.6D (full discussion: Dowse and Ringo, 1989, 1991).

Figure 6.5 shows discrete Fourier analyses of the same three fly heartbeat records shown in Figs. 6.1 and 6.2. The algorithm used here is the long

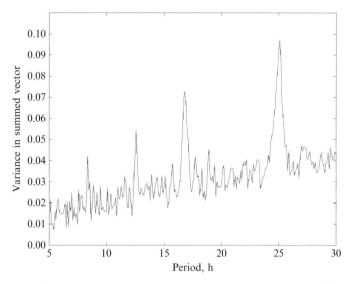

Figure 6.4 A Whitaker-Robinson "periodogram" of the data vector shown in Fig. 6.3 (A). Note the ragged, noisy output with the monotonically rising background and the multiple "harmonics".

"brute force" computational method effected by operating on the original data set. In common usage, the actual transform is done not on original data, but on either autocorrelation or autocovariance functions. If the former, the output is spectral density; if the latter, it is the true periodogram$_{SS}$. In the latter case, the area under the curve is the power in the signal, and this can be useful, whereas the normalization inherent in spectral density allows comparisons independent of the amplitude of the process across subjects (Chatfield, 1989). More computationally efficient methods, for example, the fast Fourier transform, are faster (Cooley *et al.,* 1969).

Compromises must be made in standard Fourier spectral analysis. We will consider a brief summary of the concerns here and reference more thorough coverage. Recall that as the data vector is lagged in the calculation of the autocovariance or autocorrelation function, data are lost off the "ends." As noted, this means that confidence intervals widen. Also, this imposes a limit on how long the computed function can be. To achieve the requisite number of samples to do computation of the coefficients, the function is "padded out" with zeros. Also the Fourier transform causes artifactual peaks when there are sharp discontinuities, so the step created when the autocorrelation or autocovariance function is terminated is smoothed out by a "window" function. This is a compromise in its own right, as what were perfectly good data points are altered by the smoothing operation. The practical tradeoff is between resolution and what is called side-lobe suppression (reviews: Ables, 1974; Chatfield, 1989; Kay and Marple, 1981).

In recent years, a new method for producing a spectrum that addresses these problems has become popular and, we maintain, is quite a good choice for biological time series. This technique is called "maximum entropy spectral analysis" (MESA)(Burg, 1967, 1968; Ulrych and Bishop, 1975).

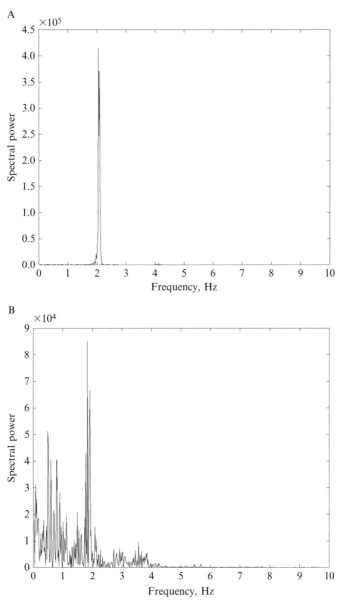

Figure 6.5 (*continued*)

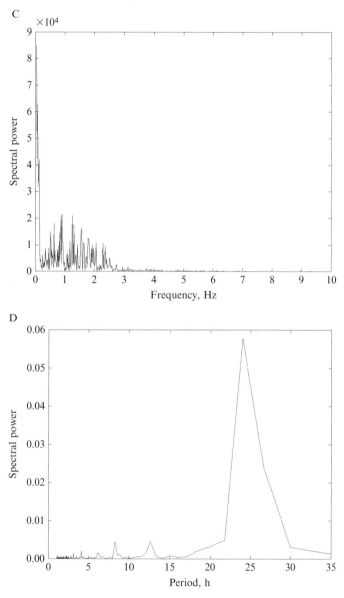

Figure 6.5 Discrete Fourier analysis of the heartbeat data shown in Fig. 6.1. (A) The very regular heart produces a clean spectral output at approximately 2 Hz, which is substantiated by the peaks in the autocorrelogram (Fig. 6.2 (A)). (B) The Fourier spectrum becomes less regular with this heartbeat which is substantially more erratic. Nonetheless, the output shows a peak at just under 2 Hz. (C) The spectrum appears as just noise in this analysis of a heart that was shown to be arrhythmic by its low RI. Note the relatively large noise component at very high frequency. (D) This analysis, when applied to the artificial square wave shown in Fig. 6.3 (A), yields a fairly broad peak at the expected 25 h.

In its most basic sense, it is a way of extending the autocorrelation out to the end in a reasonable manner, which is consistent with maximizing ignorance of the series, that is, entropy in its information sense. In choosing zeros to pad out the AC function, one is making an assumption about the process, creating values arbitrarily. It seems unlikely that if the process had continued, it would have consisted of only zeros. However, orthogonal decomposition creates no model of the system and thus cannot predict. Stochastic modeling of the system is the answer. An autoregressive (AR) function is fitted to data, which describes the evolution of the system in time. The assumption is that the system moves forward in time as a function of previous values and a random noise component. The previous values are weighted by a series of coefficients derived from known data values (Ulrych and Bishop, 1975):

$$X_t = aX_{t-1} + bX_{t-2} + cX_{t-3} + \ldots Z_t. \tag{6.5}$$

where a,b,…are the model's coefficients and Z_t is random noise. These coefficients constitute the prediction error filter (PEF). It is possible to predict values into the future, in this case functionally taking the autocorrelation function out to the needed number of values. Mathematically, it formally maximizes ignorance of the function, meaning that the values estimated are most likely based on what is known from data in hand. The spectrum is constructed from the coefficients as follows:

$$S(\omega) = P \Big/ \left|1 - \sum_{k=1}^{p} a_k e^{-i\omega k}\right|^2. \tag{6.6}$$

Maximum entropy spectral analysis has proven itself superior to ordinary Fourier analysis, as it does not produce artifacts from the various manipulations needed absent a model for the function and both resolution and sidelobe suppression are superior to standard Fourier analysis (Ables, 1974; Kay and Marple, 1981). We employ a computationally efficient algorithm described by Andersen (1974).

The number of coefficients in the prediction error filter is crucial to the output of the analysis. Too few, and resolution and important detail can be lost. If an excessive number is used, the spectrum will contain spurious peaks. In practice, an objective method has been described using the methods of Akaike (Ulrych and Bishop, 1975), based on information theory, that chooses a PEF that is consistent with the most amount of real, useful information that can be extracted. This is employed in the MESA software application demonstrated here, but we usually set a minimum filter length of about N/4 for biological rhythm analyses to ensure adequate representation of any long period cycles in the presence of considerable noise. This is not usually necessary for the heartbeat analyses.

Analyses for Rhythmicity

Figure 6.6 shows the three heartbeat records shown earlier, subjected here to MESA. Note the relationship between the sharpness of the peaks and the regularity of the rhythms. It should be pointed out that the broadness of the peak in the preceding Fourier spectrum is partially a result of the paucity of coefficients that can be computed. This number may be

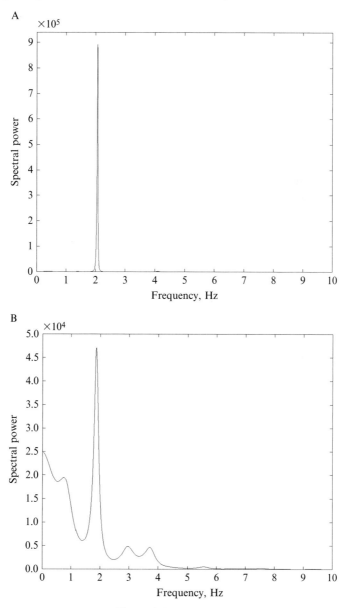

Figure 6.6 (*continued*)

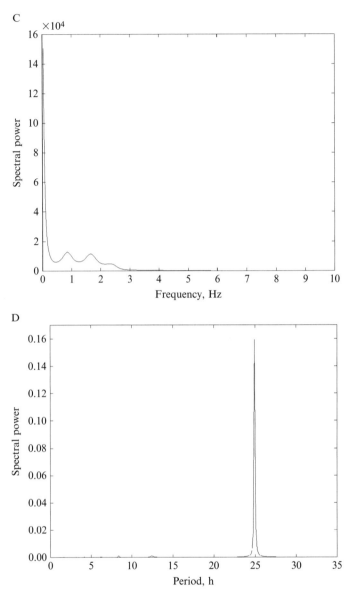

Figure 6.6 Maximum Entropy Spectral Analysis (MESA) for the four time series shown in Figs. 6.1 and 6.3. (A) The most regular heart once again produces an extremely clean plot with no noise apparent in the spectrum. The peak, taken directly by inspection of the output of the program is 2.07 Hz. (B) While less regular, with a hefty peak of noise in the high frequency range, this heart also produces a relatively clean spectral peak at 1.87 Hz, as taken from the output file as in (A). (C) This is a typical noise spectrum. The few actual beats are lost in the record. This result is common for arrhythmic hearts. (D) For the artificially produced circadian rhythm

increased if necessary (Welch, 1967) for greater resolution; however, we did not elect to do this here. It is substantially easier to increase the number of MESA coefficients to any degree needed, with a concomitant increase in computation time, but for comparison's sake, we left the number at the minimum level. For comparison, the artificially produced circadian rhythm-like signal created to demonstrate signal averaging in Fig. 6.3 has been analyzed by Fourier analysis [Fig. 6.5D, MESA (Fig. 6.6D)]. In the Whittaker–Robinson "periodogram" (Fig. 6.4), the ragged, weak 25-h peak, along with the multiple subpeaks at resonant harmonics, is striking, as is the inexorably rising background noise level. MESA produces a single peak, much sharper than in the Fourier spectrum (Fig. 6.5D) and there is little interference by the 50% of the signal that is added noise.

5. TIME/FREQUENCY ANALYSIS AND THE WAVELET TRANSFORM

The primary problem with any Fourier-based system is the fundamental assumption that the process goes on unchanged for all time, the definition of a stationary series (Chatfield, 1989). Period, phase, and amplitude are invariant. The output of the analysis degrades to the extent that the system changes with time. Biological systems are not known for being stationary. To show the effect of changing period, a "chirp" was produced. Here, the signal is generated by the following equation:

$$X_t = \cos(\omega t^{1.3}). \tag{6.7}$$

The output series is plotted in Fig. 6.7, along with a plot of a MESA done on the data. Even the redoubtable MESA is incapable of dealing with this continually moving target.

Thus it would be advantageous to follow a changing system as it evolves in time rather than looking for some consensus peak for the entire record. For example, if a heart slows down at some point, either as a result of treatment or due to alterations of the animal's internal physiology, it would be useful to be able to document this objectively and know when the change occurs. This is the role of "time-frequency analysis" and there are several methodologies (Chui, 1992). One might, for example, do short time fast Fourier transforms, meaning breaking a longer signal down into shorter

example, the MESA peak is at exactly 25 h, as would be expected. Compare the sharp, narrow peak here with the Fourier analysis of the same file depicted in Fig. 6.5(D) and the sketchy "periodogram" in Fig. 6.4. There is no evidence of the large amount of noise that was added to the signal when it was produced. This is a typical performance for this advanced signal analysis technique.

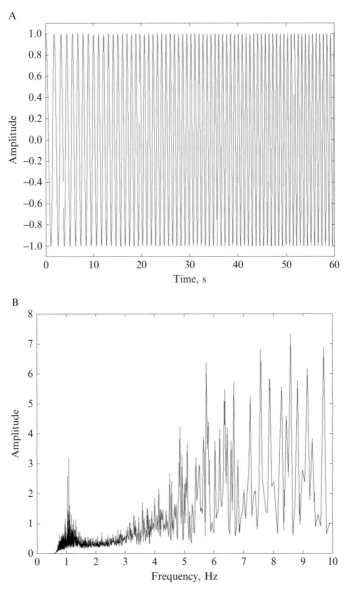

Figure 6.7 (A) This is a "chirp", or non-stationary signal artificially produced without any noise. The frequency is rising regularly and monotonically as a power of time (see text). (B) Application of MESA to this signal produces a very erratic output which is virtually uninterpretable.

segments. The loss in power of the analysis is great, however, which is unavoidable. The more you decide you want to know about frequency, the less you know about the time structure and vice versa. This relationship is rooted in quantum theory, literally the uncertainty principle (Chui, 1992; Ruskai *et al.*, 1992). It is useful to think of a plot with the frequency domain of a time series as the ordinate and the time domain as the abscissa. If one draws a rectangular box in that two-dimensional plane delimited along the ordinate by what is known reliably of frequency and on the abscissa, by knowledge of time, the uncertainty constraint means that the area of the box can never decrease below a minimum. If you want to know more about time, you contract that interval, but at the expense of widening the interval on the frequency axis, increasing uncertainty about that domain (Chui, 1992).

A very useful way to do time-frequency analysis has turned out to be to use "wavelet decomposition" (Chui, 1992; Ruskai *et al.*, 1992). Wavelets are compactly supported curves, meaning that they are nonzero only within a bounded region. They can be as simple as the Haar wavelet, which is just a single square wave, they may be derived from cardinal spline functions, or they may even be fractal (Chui, 1992; Ruskai *et al.*, 1992). This wavelet is *convolved* with the time series, meaning that each value of the wavelet is multiplied by a corresponding value of the time series and a sum of the values is computed. The wavelet is then *translated* (moved systematically) along the series, producing values from the convolution as it goes, which are the wavelet coefficients. The wavelet is then *dilated*, meaning it retains its form, but is made larger along its x axis. The process of translation is repeated and a second band of values is computed. This continues over a range of dilations. The matrix of the output is a representation of original data, but has substantially fewer points in it; this is one method of data compression. This is the *wavelet transform*. It is useful for storing signals, but more has been done than just that. For example, if you take a large matrix and waveletize it to produce a compressed "sparse" matrix, you can invert it much faster than you can invert the original. Taking the inverse wavelet transform does not restore the original matrix, rather it produces the inverted matrix, which can be useful in many applications (Chui, 1992; Ruskai *et al.*, 1992).

The critical consideration for signal analysis is that as the wavelet is dilated, it has different filtering characteristics. In this so-called "multi-resolution analysis," each wavelet band will have a different range of frequencies that have been allowed to pass the filter and, crucially, the information about the TIME when these frequencies occur is retained. Imagine, for example, a trumpet note, which is anything but constant over the course of its evolution. Fourier analysis might simply break down (as did MESA, discussed earlier, when presented with a chirp, a far simpler time series), but when wavelet analysis is done, it would show a useful breakdown of frequency as a function of time.

The utility of wavelet transform analysis for biological data is proving out in the investigation of another fly system. *D. melanogaster* males court females using a number of stylized behavioral gestures (Spieth and Ringo, 1983). Among these is the production of a "mating song." This is produced during courtship by the male extending a single wing and vibrating it, producing either a "hum," also known as a "sine song," or a "buzz," which is a series of short wing flicks called a "pulse song" (Shorey, 1962). Much information of use to the female is inherent in this signaling (Kyriacou and Hall, 1982). The sine portion is of problematic utility (Kyriacou and Hall, 1984; Talyn and Dowse, 2004; von Schilcher, 1976), but the pulse song is species specific and definitely primes the female to mate more rapidly (Kyriacou and Hall, 1982; Talyn and Dowse, 2004). There is a sinusoidal rhythm in the peak-to-peak interval, the IPI, which varies regularly with a periodicity of about a minute. Remarkably, the period of this cyclicity is, to an extent, under the control of the *period* gene, which encodes a molecule of importance in the 24-h clock mechanism (Alt *et al.*, 1998; Kyriacou and Hall, 1980). Given the staccato nature of the signal, wavelet analysis is a natural choice for time-frequency analysis and even potential automation of song analysis.

We illustrate wavelet analysis of this signal as follows (data taken by Dr. Becky Talyn in this laboratory): males and females were housed separately within 10 h after eclosion to ensure adequate mating activity. Individual males and females were aspirated into a 1-cm chamber in an "Insectavox" microphone/amplifier instrument (Gorczyca and Hall, 1987) capable of picking up sounds at this intensity. The amplified signal was collected by a computer with a Sound Blaster card digitizing at 11,025 Hz. The sound card was controlled by and data were further viewed and edited using Goldwave software. Figure 6.8A depicts a section of sine song from a typical record, while Fig. 6.8B shows pulses. This section of song was subjected to cardinal spline wavelet decomposition using a program of our devising (program written in collaboration with Dr. William Bray, Department of Mathematics and Statistics and School of Biology and Ecology, University of Maine). Figure 6.8C shows a 2-s segment of song with both sine and pulse present, along with three bands from a wavelet analysis. The central frequency of the transfer function (see later) increases from top to bottom: 141, 283, and 566 Hz. The pulses appear as sharp peaks in all bands (even the last—they are just not seen readily if the scale of the abscissa is kept the same), meaning that the wavelet sees them as "singularities," while the sine song appears in only one band, appropriate to its commonly accepted species-specific frequency of 155 Hz (Burnet *et al.*, 1977). It is hoped that the differential presence of the two components in these two bands will allow us to automate the song analyses, a project that is ongoing at this time (Dowse and Talyn, unpublished).

6. Signal Conditioning

Biological data are seldom "clean." Living systems commonly have noise associated with them, which can be a serious detriment to analysis. If the signal being acquired is analog voltage, electronic filtering can be done to remove at least a portion of this. Sixty-hertz notch filters to remove this omnipresent "hum" in electrophysiological preparations are a common example. However, we deal here with digitized signals in computers and have at hand a satisfying array of digital signal conditioning techniques for improving our analysis output once data have been recorded (Hamming, 1983).

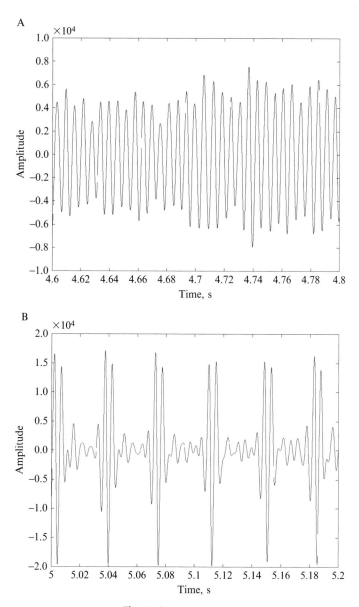

Figure 6.8 (*continued*)

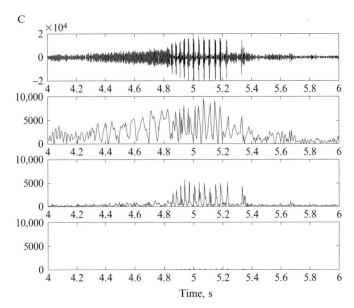

Figure 6.8 (A) Male *Drosophila melanogaster* produce mating song by vibration of their wings. The output can come in two forms, either sine song, which sounds like a hum, or pulse song which makes a staccato buzz. (A) short segment of sine song and (B) pulse song. (C) Top panel: two seconds of song from which the above segments were excised as examples. Panel 2: Output of a cardinal spline wavelet analysis with a central frequency pass at 141 Hz, near the commonly reported species norm of ~150 Hz (see text). The ordinates of this panel and those below is power rather than amplitude, so the output can be thought of as a time-frequency analysis, showing how much power is present at a given time rather than a way of looking at all frequencies present across all time. Note that both the pulses and the sine pass power through the operation. Panel 3: Here, the pulses continue to pass power, as they are seen by the wavelet as singularities, while the sine disappears. The central frequency of the band is 283 Hz. Panel 4: There is nothing in the signal that passes through at this central frequency of 566 Hz with the exception of pulses that are barely visible at this scale. Bands higher or lower than the ones shown are similar.

The first sort of problem is very common in the study of biological behavioral rhythms. This is trend. An animal may be consistently rhythmic throughout a month-long running-wheel experiment, for example, but the level of its total activity may increase or decrease through that period of time. There may also be long-period fluctuations as well. This is also commonly seen in experiments where rhythmic enzymatic activity is being recorded and the substrate is depleted throughout the time period, leaving a decay envelope superimposed over the fluctuations of interest. There are numerous examples that could be cited. One exemplary problem is the search for rhythms as amplitude declines monotonically (review: Levine *et al.*, 2002). The first way to tackle this is to remove any linear trend. In doing this, it is also sound practice to remove the mean from the series. This is equivalent to removing the "direct current" or DC part of the signal (Chatfield, 1989). There may be a strong oscillation in a parameter superimposed on a large DC offset, which will detract from the analysis. Removing the mean will leave only the "alternating current" or AC portion. The technique is extremely straightforward. One fits a regression line to data and subtracts that regression line from original data point by point (Dowse, 2007).

More problematic is a situation where there are nonlinear trends, for example, if one is looking for ultradian periodicity, and have a strong circadian period upon which it is superimposed. There are several ways to combat this. We will consider two here. First, "Fourier filtering" can be done to remove long-period rhythmicity. The discrete Fourier transform is first taken, and the Fourier coefficients are computed. Recall that these coefficients are orthogonal, meaning that they are totally independent of one another. What one does in one area of the spectrum does not affect actions taken in another. Hence, one simply zeroes out the coefficients for periodicities or frequencies one wishes to eliminate and then does the inverse Fourier transform. The resultant reconstructed time series then no longer has those frequencies and there is no disturbance of the other periodicities of interest (Dowse, 2007; Levine et al., 2002; Lindsley et al., 1999).

This is by way of doing a "high pass filter," meaning that the higher frequencies pass with a removal of the longer. Other filtering techniques are available, and digital filters of many sorts can be applied. We shall consider these in the context of "low pass" filters as the techniques are similar. The function of any filter is characterized by its transfer function, which is a plot of power transmitted as a function of frequency (Hamming, 1983).

The more common problem with biological signals in physiology is high frequency noise. This can arise within the electronics of the data acquisition systems themselves or be part of the actual process. Either way, this "static" can be highly detrimental to analyzing the longer frequency periodicities of interest. The amount and power in the noise portion of the spectrum can be computed for reasons relating to understanding the process itself, which is the subject of the next section. Here, we deal with ways of removing the noise from data to strengthen analysis, for example, to be certain frequency estimates are as accurate as may be obtained from data in hand.

We have already discussed several techniques for minimizing noise. Signal averaging is one done in the time domain, yielding more accurate waveform approximation. Noise is distributed stochastically throughout the spectrum and hence cancels out when multiple cycles are superimposed. The signal itself, however, reinforces itself continually cycle after cycle. Binning of unary data also removes substantial noise as this process acts as a low pass filter in and of itself (Dowse and Ringo, 1994). However, we now begin work with digital filters per se, beyond the Fourier filtering discussed briefly immediately preceding.

The simplest sort of digital filter is a moving average. This may be nothing more than a three-point process: $Y_t = (X_{t-1} + X_t + X_{t+1})/3$,

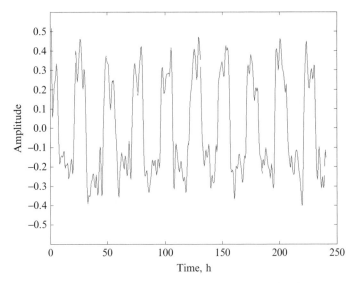

Figure 6.9 The signal shown in Fig. 6.3 (A) was filtered with a low-pass recursive Butterworth filter (see text) with a 3 db attenuation at a period of 4 h. The 50% added noise is substantially reduced, and the underlying signal is far easier to see. There is a slight phase lag introduced compared to the original.

where $X_t = X_1, X_2, \ldots, X_N$ is the original series, and Y_t is the filtered version. This is surprisingly effective and can be all that is needed. However, far more sophisticated filters are available and there are freeware programs available to compute the coefficients. Chebyshev and Butterworth recursive filters are well-known examples (review: Hamming, 1983). We consider here the Butterworth. This is called recursive because not only are original time series data incorporated into the moving filtering process, but previously filtered values are used as well. Butterworth filters are highly accurate in their frequency stopping characteristics, and the cutoff can be made quite sharp. The cutoff is usually expressed in decibels and is a quantification of the decrement in amplitude at a particular frequency. The cutoff frequency is the value at which the decrement, for example, 3 dB, is specified. In Fig. 6.9, the artificially produced square wave depicted first in Fig. 6.3A is shown after filtering with a two-pole low pass Butterworth filter with a $\approx$3-dB amplitude decrement at the cutoff period of 4 h (Hamming, 1983). The filter recursion is

$$Y_t = (X_t + 2X_{t-1} + X_{t-2} + AY_{t-1} + BY_{t-2})/C, \qquad (6.8)$$

where X_t is the original data series and Y_t is the output. A and B are filter coefficients: A = 9.656 and B = −3.4142. C is the "gain" of the filter, meaning the amplitude change in the filtered output and C = 10.2426. Note again that both original data points and points from the output vector are combined. There is a net phase delay in the series produced by the filter, which should be kept in mind if phase is something of interest, as is often the case in biological rhythm work. Note also that the mean has been subtracted out, as was described in an earlier section. One important warning, it is highly inadvisable to run a filter more than once to achieve further smoothing as this will result in multiplication of error in the signal (Hamming, 1983).

7. Strength and Regularity of a Signal

The amount of noise in a biological signal is more than just an annoyance when it comes to analysis. Assuming that the noise arises in the system being monitored and not in the acquisition hardware, it may be part of the process itself and thus will be of extreme interest. How strong and regular is the signal? These central questions can reflect on values of parameters in the oscillating system, for example, activity of enzymes, or, in the heart pacemaker, conductivity and kinetics of the component ion channels. Note that strength and regularity are not the same thing. A wild-type fly heart beating at 2 Hz may have a much greater regularity than a mutant heart beating at the same rate and amplitude, all else being equal (Johnson *et al.*, 1989). In the heart, "noise" may be an artifact resulting from interference by unrelated systems, but might also derive from poorly functioning ion channels (Ashcroft, 2000; Dowse *et al.*, 1995). Thus quantification of noise and irregularity can be enlightening and useful (Glass and Mackey, 1988).

One standard way to quantify noise in a system in engineering is by the signal-to-noise ratio (Beauchamp and Yuen, 1979). As noted earlier, power in digital signals is the ensemble average of the squared values of the vector. Note that a noiseless, DC signal would still have power by this definition, while its variance would be zero. If all the noise in the signal is constrained to one region of the spectrum, while the signal is in another, SNR computation from the power spectrum would be simple. This is not usually the case, however, and given the erratic waveforms in biological signals, it is necessary to use alternate strategies.

To characterize such signals, we developed an algorithm based on MESA that allows waveform-independent calculation of SNRs (Dowse and Ringo, 1987, 1989). In this method, we fit the autoregressive function to the vector in the usual manner; however, we use the coefficients calculated *not* in the prediction error filter to compute a spectrum, but plug them into the actual AR model and use this equation, thus fleshed out with numbers derived from data, to predict upcoming values from past ones. A new series is thus generated one datum at a time from the previous values of the original. If, say, there are 30 coefficients, value Y_t in a series of predicted values is generated by the previous 30 values of X (X_{t-1},

$X_{t-2}, \ldots X_{t-30}$), the original series, operated on by the filter. This forms a predicted output vector Y in parallel to the original. This process continues, working through the time series one value at a time until the entire predicted series is developed. The power in this series (as defined earlier) is the "signal" as it reflects the output of the model underlying the original series. The generated series Y is subtracted point by point from the values of X, and the difference series is the "noise." The ratio of the power in the generated series to that of the noise series is the SNR. The SNRs of the three heartbeat records are A = 2134, B = 667, and C = 488.

Noise in the system obscuring the output is not the only variation. The output of the oscillator generating the periodicity may also be irregular in another way, namely its period may vary from cycle to cycle. In the heart pacemaker, this may be a result of chaos (Glass and Mackey, 1988). The variation is not stochastic, but rather derives from a deterministic process, which has no perfect repeating orbit in phase space, and is considered "pseudoperiodic." This is considered quite normal and even necessary in a healthy heart, but in excess, chaos can be life-threatening. Fibrillation is the worst-case scenario, but irregular heartbeat plagues millions of patients and may often be fatal; analysis of the regularity of heartbeat may prove to be a useful predictive tool (Lombardi, 2000).

To assess this beat-to-beat variability, one may go to the length of recording these intervals from raw data and looking at the variance. Alternatively, difficult algorithms can be used to compute how chaotic a system might be (Glass and Mackey, 1988). We have chosen a simple way to characterize this phenomenon in physiological oscillators based on the autocorrelation function (review: Levine *et al.*, 2002). We measure the height of the third peak in the autocorrelogram, counting the first peak as the peak at lag zero, which is termed the rhythmicity index (RI). The decay envelope of the autocorrelogram is a function of the long-range regularity in the signal (Chatfield, 1989). If there is a lot of variation between beats, plus possible beat-to-beat decrement in amplitude, the function will decay more rapidly than in a regular series. With a perfect sinusoid, it will not decay at all. We have a program employing a sequential bubble sort algorithm that automatically retrieves this value from the output files of the autocorrelation program. Values of RI for the heights of the third peaks of the autocorrelograms depicted in Fig. 6.2 are A = 0.935, B = 0.221, and C = arrhythmic (no significant third peak, in fact, no third peak at all). Note that as in the autocorrelogram itself, from which this statistic is derived, values are normalized to the unit circle. These values are distributed normally and can be compared statistically. While this is not sophisticated enough to be used diagnostically, it is of more than sufficient resolution for us to compare heartbeat among strains with mutations affecting heart function (see, e.g., Sanyal *et al.*, 2005).

8. Conclusions

A complete suite of programs for the analysis of biological time series has been described, capable of dealing with a wide range of signals ranging from behavioral rhythmicity in the range of 24-h periods to fly sine song in the courtship ritual, commonly in the range of 150 Hz in *D. melanogaster*. Frequencies outside this range can be dealt with easily and are limited only by data acquisition systems. It should be possible to pick and choose among the various techniques to assemble a subset for almost any situation. All the programs demonstrated here, other than the proprietary MATLAB used extensively for plotting output and the Goldwave program used in programming the SoundBlaster card, are available from the author free of charge. They may be requested in executable form or in FORTRAN source code from which they may be translated into other programming languages. For the former, a computer capable of running DOS window applications is required, but we have tested all applications in operating systems up to and including Windows XP.

REFERENCES

Ables, J. G. (1974). Maximum entropy spectral analysis. *Astron. Astrophys. Suppl. Ser.* **15**, 383–393.
Alt, S., Ringo, J., Talyn, B., Bray, W., and Dowse, H. (1998). The period gene controls courtship song cycles in *Drosophila melanogaster*. *Anim. Behav.* **56**, 87–97.
Andersen, N. (1974). On the calculation of filter coefficients for maximum entropy spectral analysis. *Geophysics* **39**, 69–72.
Ashburner, M. (1989). "*Drosophila*, a Laboratory Manual." Cold Spring Harbor Press, Cold Spring Harbor, NY.
Ashcroft, F. (2000). "Ion Channels and Disease." Academic Press, New York.
Bodmer, R., Wessels, R. J., Johnson, E., and Dowse, H. (2004). Heart development and function. *In* "Comprehensive Molecular Insect Science" (L. I. Gilbert, K. Iatrou, and S. Gill, eds.), Vol. 2. Elsevier, New York.
Beauchamp, K., and Yuen, C. K. (1970). "Digital Methods for Signal Analysis." Allen & Unwin, London.
Burg, J. P. (1967). Maximum entropy spectral analysis. *In* "Modern Spectrum Analysis" (1978) (D. G. Childers, ed.), pp. 34–41. Wiley, New York.
Burg, J. P. (1968). A new analysis technique for time series data. *In* "Modern Spectrum Analysis" (1978) (D. G. Childers, ed.), pp. 42–48. Wiley, New York.
Burnet, B., Eastwood, L., and Connolly, K. (1977). The courtship song of male *Drosophila* lacking aristae. *Anim. Behav.* **25**, 460–464.
Chatfield, C. (1989). "The Analysis of Time Series: An Introduction." Chapman and Hall, London.
Chui, C. K. (1992). "An Introduction to Wavelets." Academic Press, New York.
Cooley, J. W., Lewis, P. A. W., and Welch, P. D. (1969). Historical notes on the fast Fourier transform. *IEE Trans. Aud. Elect.* **AU15**, 76–79.

Curtis, N., Ringo, J., and Dowse, H. B. (1999). Morphology of the pupal heart, adult heart, and associated tissues in the fruit fly, *Drosophila melanogaster. J. Morphol.* **240,** 225–235.

DeCoursey, P. (1960). Phase control of activity in a rodent. *In* "Cold Spring Harbor Symposia on Quantitative Biology XXV" (A. Chovnik, ed.), pp. 49–55.

Dowse, H. (2007). Statistical analysis of biological rhythm data. *In* "Methods in Molecular Biology: Circadian Rhythms" (E. Rosato, ed.), Vol. 362, pp. 29–45. Humana Press.

Dowse, H. B. (2008). Mid-range ultradian rhythms in *Drosophila* and the circadian clock problem. *In* "Ultradian Rhythmicity in Biological Systems: An Inquiry into Fundamental Principles" (D. L. Lloyd and E. Rossi, eds.). Springer Verlag, Berlin, in press.

Dowse, H., and Ringo, J. (1987). Further evidence that the circadian clock in *Drosophila* is a population of coupled ultradian oscillators. *J. Biol. Rhythms* **2,** 65–76.

Dowse, H. B., Hall, J. C., and Ringo, J. M. (1987). Circadian and ultradian rhythms in *period* mutants of *Drosophila melanogaster. Behav. Genet.* **17,** 19–35.

Dowse, H. B., and Ringo, J. M. (1989). The search for hidden periodicities in biological time series revisited. *J. Theor. Biol.* **139,** 487–515.

Dowse, H. B., and Ringo, J. M. (1991). Comparisons between "periodograms" and spectral analysis: Apples are apples after all. *J. Theor. Biol.* **148,** 139–144.

Dowse, H. B., and Ringo, J. M. (1994). Summing locomotor activity into "bins": How to avoid artifact in spectral analysis. *Biol. Rhythm Res.* **25,** 2–14.

Dowse, H. B., Ringo, J. M., Power, J., Johnson, E., Kinney, K., and White, L. (1995). A congenital heart defect in *Drosophila* caused by an action potential mutation. *J. Neurogenet.* **10,** 153–168.

Dulcis, D., and Levine, R. (2005). Innervation of the heart of the adult fruit fly, *Drosophila melanogaster. J. Comp. Neurol.* **465,** 560–578.

Dunlap, J. (1999). Molecular bases for circadian clocks. *Cell* **96,** 271–290.

Enright, J. T. (1965). The search for rhythmicity in biological time-series. *J. Theor. Biol.* **8,** 662–666.

Enright, J. T. (1990). A comparison of periodograms and spectral analysis: Don't expect apples to taste like oranges. *J. Theor. Biol.* **143,** 425–430.

Glass, L., and Mackey, M. C. (1988). "From Clocks to Chaos: The Rhythms of Life." Princeton Univ. Press, Princeton, NJ.

Gorczyca, M., and Hall, J. C. (1987). The INSECTAVOX, and integrated device for recording and amplifying courtship songs. *Drosophila Inform. Serv.* **66,** 157–160.

Gu, G.-G., and Singh, S. (1995). Pharmacological analysis of heartbeat in *Drosophila. J. Neurobiol.* **28,** 269–280.

Hall, J. (2003). "Genetics and Molecular Biology of Rhythms in *Drosophila* and Other Insects." Academic Press, New York.

Hamblen-Coyle, M., Konopka, R. R., Zwiebel, L. J., Colot, H. V., Dowse, H. B., Rosbash, M. R., and Hall, J. C. (1989). A new mutation at the *period* locus of *Drosophila melanogaster* with some novel effects on circadian rhythms. *J. Neurogenet.* **5,** 229–256.

Hamming, R. W. (1983). "Digital Filters." Prentice-Hall, New York.

Hille, B. (2001). "Ion Channels of Excitable Membranes." Sinauer, Sunderland, MA.

Johnson, E., Ringo, J., Bray, N., and Dowse, H. (1998). Genetic and pharmacological identification of ion channels central to *Drosophila's* cardiac pacemaker. *J. Neurogenet.* **12,** 1–24.

Johnson, E., Ringo, J., and Dowse, H. (1997). Modulation of *Drosophila* heartbeat by neurotransmitters. *J. Comp. Physiol. B* **167,** 89–97.

Jones, J. (1977). "The Circulatory System of Insects." Charles C. Thomas, Springfield, IL.

Kay, S. M., and Marple, S. G. Jr. (1981). Spectrum analysis, a modern perspective. *IEEE Proc.* **69,** 1380–1419.

Kendall, M. G. (1946). "Contributions to the study of oscillatory time series." Cambridge Univ. Press, Cambridge.

Konopka, R. J., and Benzer, S. (1971). Clock mutants of *Drosophila melanogaster*. *Proc. Natl. Acad. Sci. USA* **68,** 2112–2116.

Kyriacou, C. P., and Hall, J. C. (1980). Circadian rhythm mutation in *Drosophila melanogaster* affects short-term fluctuations in the male's courtship song. *Proc. Natl. Acad. Sci. USA* **77,** 6729–6733.

Kyriacou, C. P., and Hall, J. C. (1982). The function of courtship song rhythms in *Drosophila*. *Anim. Behav.* **30,** 794–801.

Kyriacou, C. P., and Hall, J. C. (1984). Learning and memory mutations impair acoustic priming of mating behaviour in *Drosophila*. *Nature* **308,** 62–64.

Lanczos, C. (1956). "Applied Analysis." Prentice Hall, New York.

Lanczos, C. (1966). "Discourse on Fourier Series." Oliver and Boyd, Edinburgh.

Levine, J., Funes, P., Dowse, H., and Hall, J. (2002). Signal Analysis of Behavioral and Molecular Cycles. *Biomed. Central. Neurosci.* **3,** 1.

Lindsley, G., Dowse, H., Burgoon, P., Kilka, M., and Stephenson, L. (1999). A persistent circhoral ultradian rhythm is identified in human core temperature. *Chronobiol. Int.* **16,** 69–78.

Lombardi, F. (2000). Chaos theory, heart rate variability, and arrhythmic mortality. *Circulation* **101,** 8–10.

Osaka, M., Kumagai, H., Katsufumi, S., Onami, T., Chon, K., and Wantanabe, M. (2003). Low-order chaos in sympatheic nerve activity and scaling of heartbeat intervals. *Phys. Rev. E* **67,** 1–4104915.

Palmer, J. D. (2002). "The Living Clock." Oxford Univ. Press, London.

Palmer, J. D., Williams, B. G., and Dowse, H. B. (1994). The statistical analysis of tidal rhythms: Tests of the relative effectiveness of five methods using model simulations and actual data. *Mar. Behav. Physiol.* **24,** 165–182.

Rizki, T. (1978). The circulatory system and associated cells and tissues. *In* "The Genetics and Biology of *Drosophila*" (M. Ashburner and T. R. F. Wright, eds.), pp. 1839–1845. Academic Press, New York.

Ruskai, M., Beylkin, G., Coifman, R., Daubechies, I., Mallat, S., Meyer, Y., and Raphael, L. (eds.) (1992). *In* "Wavelets and Their Applications." Jones & Bartlett, Boston.

Sanyal, S., Jennings, T., Dowse, H. B., and Ramaswami, M. (2005). Conditional mutations in SERCA, the sarco-endoplasmic reticulum Ca^{2+}-ATPase, alter heart rate and rhythmicity in *Drosophila*. *J. Comp. Physiol. B* **176,** 253–263.

Schefler, W. (1969). "Statistics for the Biological Sciences." Addison Wesley, Reading, MA.

Schuster, A. (1898). On the investigation of hidden periodicities with application to a supposed 26-day period of meterological phenomena. *Terrestrial Magn. Atmos. Electr.* **3,** 13–41.

Shorey, H. H. (1962). Nature of sound produced by *Drosophila* during courtship. *Science* **137,** 677–678.

Spieth, H., and Ringo, J. M. (1983). Mating behavior and sexual isolation in *Drosophila*. *In* "The Genetics and Biology of *Drosophila*" (M. Ashburner, H. Carson, and J. Thompson, eds.). Academic Press, New York.

Talyn, B., and Dowse, H. (2004). The role of courtship song in sexual selection and species recognition by female *Drosophila melanogaster*. *Anim. Behav.* **68,** 1165–1180.

Ulrych, T., and Bishop, T. (1975). Maximum entropy spectral analysis and autoregressive decomposition. *Rev. Geophys. Space Phys.* **13,** 183–300.

von Schilcher, F. (1976). The function of pulse song and sine song in the courtship of *Drosophila melanogaster*. *Anim. Behav.* **24,** 622–625.

Wasserthal, L. (2007). *Drosophila* flies combine periodic heartbeat reversal with a circulation in the anterior body mediated by a newly discovered anterior pair of ostial valves and 'venous'channels. *J. Exp. Biol.* **210,** 3703–3719.

Welch, P. (1967). The use of Fast Fourier Transform for the estimation of power spectra: A method based on time averaging over short, modified periodograms. *IEEE Trans. Audio Electroacoust.* **AU-15,** 70–73.

Whitaker, E., and Robinson, G. (1924). "The Calculus of Observations." Blackie, Glasgow.

Wiener, N. (1949). "Extrapolation, Interpolation, and Smoothing of Stationary Time-Series." MIT Press, Cambridge, MA.

Wolf, M., Amrein, H., Izatt, J., Choma, M., Reedy, M., and Rockman, H. (2006). *Drosphila* as a model for the identification of genes causing adult human heart disease. *Proc. Nat. Acad. Sci. USA* **103,** 1394–1399.

CHAPTER SEVEN

A Computational Approach for the Rational Design of Stable Proteins and Enzymes: Optimization of Surface Charge–Charge Interactions

Katrina L. Schweiker*,† and George I. Makhatadze*

Contents

1. Introduction	176
2. Computational Design of Surface Charge–Charge Interactions	183
2.1. Calculating pair wise charge–charge interaction energies	183
2.2. Optimization of surface charges using the genetic algorithm	188
3. Experimental Verification of Computational Predictions	190
3.1. Single site substitutions — proof of concept and test of robustness of the TK-SA model	191
3.2. Rational design of surface charge–charge interactions using a genetic algorithm	197
3.3. Effects of stabilization on enzymatic activity	200
4. Closing Remarks	202
Acknowledgments	204
References	204

Abstract

The design of stable proteins and enzymes is not only of particular biotechnological importance, but also addresses some important fundamental questions. While there are a number of different options available for designing or engineering stable proteins, the field of computational design provides fast and universal methods for stabilizing proteins of interest. One of the successful computational design strategies focuses on stabilizing proteins through the optimization of charge–charge interactions on the protein surface. By optimizing surface interactions, it is possible to alleviate some of the challenges that

---

* Department of Biology and Center for Biotechnology and Interdisciplinary Studies, Rensselaer Polytechnic Institute, Troy, New York
† Department of Biochemistry and Molecular Biology, Penn State University College of Medicine, Hershey, Pennsylvania

accompany efforts to redesign the protein core. The rational design of surface charge–charge interactions also allows one to optimize only the interactions that are distant from binding sites or active sites, making it possible to increase stability without adversely affecting activity. The optimization of surface charge–charge interactions is discussed in detail along with the experimental evidence to demonstrate that this is a robust and universal approach to designing proteins with enhanced stability.

1. Introduction

The problem of how to design a stabilized protein encompasses many fundamental questions. How does the amino acid sequence of the protein dictate its fold and stability? How do the physical–chemical intramolecular interactions in the native state govern the stability of this conformation? What role does the unfolded state play in protein stability? To what extent are stability and function intertwined; is it possible to stabilize a protein without decreasing its activity? While there is still much to be learned about how a given amino acid sequence dictates the fold of the protein and what interactions will be present in the native sate, much has been learned about the contributions of intramolecular forces to stability in the native state (Dill, 1990; Makhatadze and Privalov, 1995). To this end, an understanding of the contributions of these forces to stability makes it possible to design stable proteins from first principles.

The term "protein stability" can take on different definitions depending on how one is interested in characterizing the protein. These include the Gibbs free energy of unfolding, the melting temperature, *in vivo* lifetime, resistance to proteolysis, folding/unfolding rates, and temperature or rates of inactivation. For the purposes of this chapter, we are interested in the measures of thermodynamic stability, the Gibbs free energy of unfolding (ΔG) and thermostability (T_m). Due to the manner in which ΔG and T_m are related, it is possible to alter the *thermodynamic stability* (ΔG) of a protein without affecting its *thermostability* (T_m)(Loladze et al., 2001) and vice versa (Makhatadze et al., 2004). One can also affect both ΔG and T_m simultaneously (Beadle et al., 1999; Grattinger et al., 1998; Rees and Robertson, 2001). These possibilities are a direct result of the relationship among ΔG, T_m, and the other thermodynamic parameters: enthalpy (ΔH), entropy (ΔS), and the change in heat capacity upon unfolding (ΔC_P). The thermodynamic stability (ΔG) for a protein that undergoes two-state unfolding ($N \Leftrightarrow U$) can be described by the equilibrium constant, K_{eq}, which is the ratio of the fraction of unfolded protein (F_U) to folded protein (F_N), such that at a given temperature:

$$\Delta G(T) = -RT \cdot \ln(K_{eq}) = -RT \cdot \ln\left(\frac{F_U}{F_N}\right). \quad (7.1)$$

The $\Delta G(T)$ of a protein can also be related to its thermostability (T_m) and the other thermodynamic parameters mentioned earlier via the Gibbs–Helmholtz relationship:

$$\Delta G(T) = \Delta H(T_m) + \Delta C_P \cdot (T - T_m) - T \cdot \left[\frac{\Delta H(T_m)}{T_m} + \Delta C_P \cdot \ln\left(\frac{T}{T_m}\right)\right] \quad (7.2)$$

where T_m is the transition temperature, or the temperature where 50% of the protein molecules are unfolded; $\Delta H(T_m)$ is the enthalpy of unfolding at the transition temperature; and ΔC_P is the change in heat capacity upon unfolding, which characterizes the temperature dependence of both enthalpy and entropy functions. The form of the Gibbs–Helmholtz relationship given by Eq. (7.2) assumes that ΔC_P is not a temperature-dependent variable. Under this assumption, the enthalpy and entropy functions at a given temperature, T, are defined by $\Delta H(T) = \Delta H(T_m) + \Delta C_p(T - T_m)$ and $\Delta S(T) = \Delta H(T_m)/T_m + \Delta C_p \cdot \ln(T/T_m)$, respectively. Changes in the heat capacity, enthalpy, and entropy of unfolding due to substitutions in the amino acid sequence of the protein will define the thermodynamic mechanisms by which T_m and/or ΔG are altered.

The stability function for a typical globular protein defined by Eq. (7.2) is a bell-shaped curve (Fig. 7.1A). The thermodynamic stability of a protein is zero when half of the molecules are folded and half are unfolded, which occurs at two temperatures: T_m (transition temperature for heat denaturation) and T_c (transition temperature for cold denaturation)(Griko Iu and Privalov, 1986; Griko Yu et al., 1989; Griko et al., 1988; Ibarra-Molero et al., 1999b; Privalov, 1990). The stability function has a maximum (ΔG_{max}) at the temperature where the entropy is equal to zero (T_{max}). The changes in ΔH, ΔS, and ΔC_P due to substitutions within the protein will define the thermodynamic mechanism that stabilizes the protein. Figure 7.1 shows three of the possible mechanisms, and the extreme versions of each have been modeled to illustrate the differences among them more clearly. However, one should remember that in practice, it is often more appropriate to explain experimental observations using combinations of these models (see Deutschman and Dahlquist, 2001). If a protein is

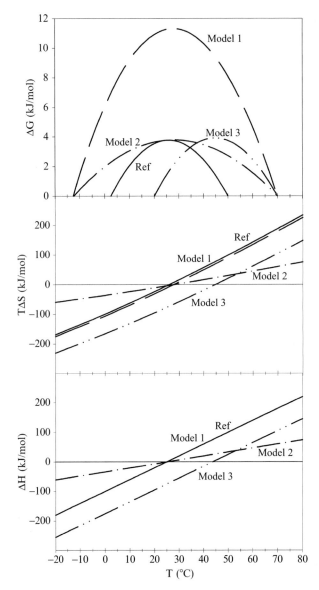

Figure 7.1 Three thermodynamic mechanisms of thermostabilization (Makhatadze et al., 2004). To highlight the differences more clearly, these model functions represent extreme examples of each mechanism of stabilization. In each panel, different thermodynamic models are represented by the following lines: solid, reference model; dashed, model 1; dash-dot-dashed, model 2; and dash-dot-dot-dashed, model 3. (A) Gibbs free energy as a function of temperature. (B) Temperature dependence of the entropic term. (C) Temperature dependence of the enthalpy function.

stabilized via the first mechanism, a large increase in both the maximum stability (ΔG_{max}) and the thermostability (T_m)(model 1, Fig. 7.1A) will be observed (Beadle et al., 1999; Grattinger et al., 1998). This is caused by a small decrease in the entropy function (Fig. 7.1B), while the enthalpy function (Fig. 7.1C) and the change in heat capacity upon unfolding are unchanged relative to the reference model. In the second model, a dramatic decrease in ΔC_P upon substitution creates a ΔG function with a shallower temperature dependence (Guzman-Casado et al., 2003; Hollien and Marqusee, 1999). This results in an increase in the thermostability of the protein without affecting T_{max} or the absolute value of ΔG_{max}. The third model results in the entire stability function shifting to higher temperatures (Makhatadze et al., 2004). This is caused by a large decrease in both entropy and enthalpy functions, without changing their temperature dependencies (i.e., no change in ΔC_P). As a result, T_c, T_m, and T_{max} increase, while the absolute value of ΔG_{max} is not affected. In each of these three models, the stability of the protein at room temperature (ΔG_{RT}) is affected differently. In both the first and the third mechanisms, ΔG_{RT} changes relative to the reference model — it increases if the protein is stabilized via model 1, but decreases if the protein is stabilized by model 3 (Fig. 7.1A). The second model demonstrates that it is possible to increase the thermostability of a protein without affecting ΔG_{RT}. With an understanding of the underlying thermodynamic mechanisms of stabilization, one should be able to design proteins to meet any desired thermodynamic criteria.

The contributions of different intramolecular interactions to the stability of the native state of globular proteins have been studied extensively. The core of a globular protein typically contains a large number of nonpolar residues and is stabilized by the hydrophobic interactions between them (Dill, 1990; Kauzmann, 1959; Loladze et al., 2002; Matthews, 1995). The high energetic cost of desolvation of polar residues means that the burial of polar residues is usually very unfavorable (Makhatadze and Privalov, 1995; Serrano et al., 1992). However, this energetic penalty can be offset through the formation of hydrogen bonds with other polar groups or buried water molecules (Makhatadze and Privalov, 1995; Serrano et al., 1992). All buried residues are also stabilized by van der Waals (packing) interactions (Desjarlais and Handel, 1995a; Lazar et al., 1997; Loladze et al., 2002; Serrano et al., 1992). In fact, hydrogen bonding and packing interactions in the protein core have been demonstrated to be as important for protein stability as hydrophobic interactions (Griko et al., 1994; Loladze and Makhatadze, 2002; Loladze et al., 2002; Makhatadze et al., 1993, 1994; Pace et al., 1996; Wintrode et al., 1994; Yu et al., 1994).

The results of some early studies on the forces that govern protein stability suggested that residues on the surface of the protein do not provide significant contributions to stability. In one example, a systematic set of mutations were made in T4 lysozyme, and it was observed that substitutions at many of the positions that were highly flexible and/or exposed to a solvent did not have a significant effect on the stability of this protein (Matthews, 1995; Sun et al., 1991). Another example studied the interactions between charged surface residues in barnase (Sali et al., 1991) and found that most interactions between the solvent-exposed charged residues had only weak contributions to the stability of the protein. This set a precedent in the field of protein stability as the observations could be rationalized easily by the idea that the residues on the surface of a protein are exposed to solvent in both the native and the unfolded states and their environments do not change significantly upon unfolding, and therefore their relative contributions to ΔG would be smaller than residues in the core. As a result of this hypothesis, the computational protein design field began to focus on optimizing interactions in the protein core (Dantas et al., 2003; DeGrado et al., 1999; Desjarlais and Handel, 1995a; Hurley et al., 1992; Korkegian et al., 2005; Pokala and Handel, 2001; Street and Mayo, 1999). However, attempts to stabilize proteins by redesigning the protein core have had mixed success (Desjarlais and Handel, 1995a; Hurley et al., 1992; Korkegian et al., 2005; Lazar et al., 1997). In general, core substitutions that fill cavities will enhance packing interactions and are therefore stabilizing (Hurley et al., 1992; Korkegian et al., 2005), whereas core substitutions that create cavities and thus decrease the packing interactions are destabilizing (Desjarlais and Handel, 1995a; Hurley et al., 1992; Lazar et al., 1997; Loladze et al., 2001, 2002). However, one should proceed with caution when making cavity filling substitutions because large, hydrophobic residues can also be destabilizing due to steric clashes within a tightly packed protein core. *De novo* attempts to redesign the protein core demonstrate how difficult it can be to model which substitutions will be stabilizing and which will be destabilizing (Desjarlais and Handel, 1995a,b; Lazar et al., 1997). One explanation is that the core of the protein is packed very tightly, suggesting that the intramolecular interactions within the protein core are already optimized. In order to further improve the interactions within the core, one would need extremely precise modeling of the positions of the side chains. Another issue was that most early core redesign methods were modeling interactions using a fixed backbone (Desjarlais and Handel, 1995a; Korkegian et al., 2005; Kuhlman and Baker, 2000). While this assumption was necessary to minimize the search space, and reduce computation time, it has been demonstrated that the backbone does indeed shift to accommodate substitutions within the protein core (Baldwin and Matthews, 1994; Hurley et al., 1992). However, one of the early attempts to incorporate backbone flexibility into a core design algorithm did not show significant improvements over fixed backbone methods in predicting the effects of core substitutions on protein stability (Desjarlais and Handel, 1999).

One alternative to the computational redesign of the protein core is to focus on redesigning interactions on the protein surface. Although it had been argued that surface residues were not important for stability, support for this idea comes from a few sources. For example, when the differences between mesophilic and thermophilic proteins from the same family were examined, it was observed that differences in stability appear to come primarily from an increase in electrostatic interactions (Huang and Zhou, 2006; Makhatadze et al., 2003, 2004; Motono et al., 2008; Perl et al., 1998; Sanchez-Ruiz and Makhatadze, 2001), which are more likely to be found on the protein surface than in the core. Similar results came from attempts to stabilize proteins using directed evolution, where it was observed that the stabilizing substitutions are often found on the surface of the protein (Hamamatsu et al., 2006; Morawski et al., 2001; Wunderlich and Schmid, 2006; Wunderlich et al., 2005a,b). The idea that surface residues can be important for stability was also supported by a theoretical study on the physical origin of stability (Berezovsky and Shakhnovich, 2005). It was suggested that as a response to evolutionary pressure, mesophilic proteins can evolve high thermostability by increasing the number of charged residues (Berezovsky and Shakhnovich, 2005), and once again, charged residues are much more likely to be found on the surface than in the core of the protein. Finally, experimental observations have shown that, in some cases, the interactions between charged surface residues have relatively large contributions to stability (Anderson et al., 1990; Horovitz et al., 1990; Serrano et al., 1990). A number of studies have exploited this information and shown that it is possible to modulate the stability of a number of proteins through altering the charge–charge interactions on the protein surface (Fernandez et al., 2000; Grimsley et al., 1999; Loladze et al., 1999; Makhatadze et al., 2004; Nagi and Regan, 1997; Perl et al., 2000; Predki et al., 1996; Spector et al., 2000; Strickler et al., 2006). Indeed, the optimization of interactions on the protein surface can provide similar increases in stability to what can be obtained through the optimization of the protein core (Table 7.1). One of the advantages of redesigning the surface of the protein is that surface residues have greater conformational flexibility than those in the core. As a result, the modeling of surface side chains does not have to be as precise as core side chain modeling in order to obtain a good description of the energetics of interactions. In addition, the flexibility of surface side chains means that they are generally more tolerant to substitutions than residues in the core. For these reasons, the optimization of surface interactions should be considered a viable alternative approach to stabilizing proteins. The remainder of this review discusses one of the successful surface redesign approaches and highlights some of the important experimental verifications of this method.

Table 7.1 Comparison of different computational design approaches[a]

Design method	Protein name	# Mutations (total residues)	ΔT_m and/or $\Delta\Delta G$	Location of substitutions
Rosetta	Procarboxypeptidase (Dantas et al., 2003)	48 (71)	30 kJ/mol	Surface and core
ORBIT	Engrailed homeodomain (Zollars et al., 2006)	24 (51)	33 °C	Surface and core
ORBIT	Thioredoxin (Bolon et al., 2003)	3 (104)	10 kJ/mol	Core
Rosetta	Yeast cytosine deaminase (Korkegian et al., 2005)	3 (158)	10 °C	Core
Rosetta	Procarboxypeptidase (Dantas et al., 2007)	4 (71)	16 kJ/mol	Surface
TK-SA	Fyn, Ubq, U1A, Procarb, Acp, CDC42, Ten (Schweiker et al., 2007; Strickler et al., 2006)	4–7 (72–190)	4–12 °C 4–18 kJ/mol	Surface
Poisson–Boltzmann	psbd41 (Spector et al., 2000)	1 (41)	9–12 °C 3 kJ/mol	Surface
Altered Coulombic interactions	RnaseT1, RnaseSa (Grimsley et al., 1999)	1 (96–104)	2–7 °C 2–5 kJ/mol	Surface
Sequence-based design	CspB-Bs (Perl et al., 2000)	2 (67)	13–21 °C 8.8–14 kJ/mol	Surface

[a] A comparison of different protein design approaches. The first two rows highlight instances where the Rosetta and ORBIT algorithms were used to increase the stability of proteins by much more than what has been demonstrated to be possible by stabilizing proteins using only surface interactions. However, in both of these instances, over half of the protein sequence was subjected to substitution, resulting in optimization of both core and surface interactions. When Rosetta and ORBIT were used to make only a small number of substitutions in the protein core, similar increases in both thermodynamic stability (ΔG) and thermostability (T_m) relative to the surface redesign approaches were observed. Results demonstrate that surface interactions can be as important as interactions in the protein core for modulating the stability of proteins.

2. Computational Design of Surface Charge–Charge Interactions

The computational approach described here is the rational design of surface charge–charge interactions. The first step to stabilizing proteins by this method is to calculate the pair wise charge–charge interaction energies in the wild-type protein. Second, the interaction energies are optimized using a genetic algorithm. The genetic algorithm will identify many sequences that have increased stabilities relative to the wild type. Since it is not possible to test all of these sequences experimentally, only a few are selected for characterization. Structures for the selected sequences are created using homology modeling, and the charge–charge interaction energies are calculated for the designed variants to better understand the details of how the substitutions are predicted to affect the stability. Finally, the stabilities of the selected sequences are characterized experimentally.

2.1. Calculating pair wise charge–charge interaction energies

The interaction energies between pairs of charges are calculated using the Tanford–Kirkwood model, corrected for solvent accessibility (TK-SA) (Matthew and Gurd, 1986a,b; Matthew et al., 1985; Tanford and Kirkwood, 1957). In the TK-SA model, the protein is represented by a low dielectric sphere that is impenetrable to solvent (Fig. 7.2). The charged groups in the protein are represented by point charges that occupy fixed positions in the protein sphere. It is assumed that the interactions between charges are the only type of interaction between the groups (Tanford and Kirkwood, 1957). The energy of the charge–charge interactions between two residues on the protein surface, i and j, is

$$E_{ij} = e^2 \left(\frac{A_{ij} - B_{ij}}{2b} - \frac{C_{ij}}{2a} \right) \cdot (1 - SA_{ij}), \qquad (7.3)$$

where e is the unit charge, b is the radius of the protein sphere and is related to the specific volume of the protein, and a is the radius of the ion exclusion boundary. The terms A_{ij}, B_{ij}, and C_{ij} have been defined previously by Tanford and Kirkwood (1957). A_{ij} represents the energy between the charges in the low dielectric environment of the protein interior and is a function of the protein dielectric constant ($\varepsilon_P = 4$ for the interior of the protein) and the distance between the charges, r_{ij}. B_{ij} reflects the contributions from both the low dielectric environment of the protein and the high dielectric environment of the solvent that surrounds the protein and is a function of the solvent dielectric constant ($\varepsilon_s = 78.5$ for water), protein

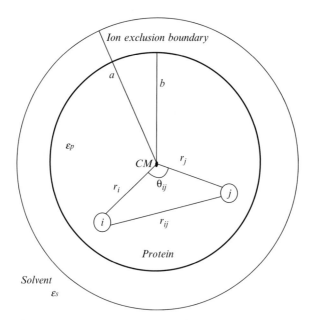

Figure 7.2 Schematic representation of the Tanford–Kirkwood model of interactions between charged residues. The protein is represented by a hard sphere with low dielectric (ε_P) of radius b from the center of mass (CM). It is surrounded by a larger sphere of radius a, which is the ion exclusion boundary. These spheres are immersed in water, represented by high dielectric (ε_S). The other parameters in the TK-SA model are the identity of charges on the protein surface, represented by the small spheres i and j and the distance between them, r_{ij}.

dielectric constant, ε_P, and the relative positions of the charges on the protein surface, which are defined by r_i, r_j, and θ_{ij}. C_{ij} is a function of the ionic strength of the solvent and the positions of the charges. The average solvent accessibility of residues i and j is represented by the term SA_{ij} and is calculated by the method of Richmond as described previously (Ibarra-Molero et al., 1999a; Richmond, 1984).

The contribution of these pair wise charge–charge interaction energies to the Gibbs free energy of unfolding of the protein is determined from the pK_a shifts of the ionizable residues in the protein relative to model compounds. At a given pH, charges on the ionizable residues can be represented by a protonation state, χ. The energy of this protonation state for the folded protein molecule is

$$\Delta G_N(\chi) = -RT \cdot (\ln 10) \sum_{i=1}^{n} (q_i + x_i) \cdot pK_{\text{int},i}$$
$$+ \frac{1}{2} \sum_{i,j=1}^{n} E_{ij}(q_i + x_i) \cdot (q_j + x_j), \quad (7.4)$$

where R is the universal gas constant; T is the temperature in Kelvin (298K is the standard temperature for these calculations); x_i and x_j represent the protonation of groups i and j and will have a value of 0 or 1; q_i and q_j are the charges of groups i and j in the unprotonated state and have a value of -1 or 0; and $pK_{int,i}$ is the intrinsic pK_a of group i if all other groups in the protein have zero charge. In this approach, intrinsic pK_a values for the ionizable groups of proteins are determined from model compounds and are Asp, 4.0; Glu, 4.5; His, 6.3; Lys, 10.6; Arg, 12.0; N-ter, 7.7; and C-ter, 3.6. There have been several reports that specific interactions between charged residues occur in the unfolded state and need to be considered to predict the thermodynamic stability of proteins accurately (Anil et al., 2006; Cho and Raleigh, 2005; Elcock and McCammon, 1998; Kuhlman et al., 1999; Oliveberg et al., 1995; Pace et al., 1990, 2000; Swint-Kruse and Robertson, 1995; Whitten and Garcia-Moreno, 2000; Zhou, 2002a,b). However, the contributions of these interactions are small ($\approx$2 kJ/mol for a pK_a shift of 0.4 units compared to $\approx$20 kJ/mol for the total ΔG of unfolding for a protein) and are expected to be even smaller when comparing the unfolded state contributions to $\Delta\Delta G$ (the difference in stability between wild-type and designed protein variants). Therefore, we assume that there are no residual charge–charge interactions in the unfolded state of the protein, and as such the energy of the protonation state, χ, in the unfolded state is

$$\Delta G_U(\chi) = -RT \cdot (\ln 10) \sum_{i=1}^{n} (q_i + x_i) \cdot pK_{int,i}. \qquad (7.5)$$

These energy functions can then be used to define partition functions for the native (Z_N) and unfolded (Z_U) states of the protein:

$$Z_N = \sum_{\chi} \exp\left(-\frac{\Delta G_N(\chi)}{RT} - v(x) \cdot (\ln 10) \cdot pH\right) \qquad (7.6)$$

$$Z_U = \sum_{\chi} \exp\left(-\frac{\Delta G_U(\chi)}{RT} - v(x) \cdot (\ln 10) \cdot pH\right), \qquad (7.7)$$

where $v(\chi)$ is the number of protonated ionizable groups in the χ protonation state. By using the neutral forms of the native [$\Delta G_N(\chi) = 0$] and unfolded [$\Delta G_U(\chi) = 0$] states of the protein as reference states for both Z_N and Z_U, the overall contribution of the charge–charge interactions to the Gibbs free energy of unfolding can be described as

$$\Delta G_{qq} = -RT \cdot \ln\left(\frac{Z_U}{Z_N}\right). \tag{7.8}$$

In order to calculate the charge–charge interaction energies, one must know the distances between the charged residues, which can be determined from a high-resolution three-dimensional structural representation of the protein obtained through either X-ray crystallography or nuclear magnetic resonance (NMR). Structures obtained from X-ray crystallography represent static snapshots of one possible configuration of the positions of side chains in a protein molecule. However, in solution, the surface side chains will have a certain degree conformational freedom, which could alter the relative positions of the ionizable groups compared to the crystal structure. To account for the flexibility of the side chains, an ensemble of structures is created by homology modeling using the Modeller software package (Marti-Renom et al., 2000). However, because NMR structure determination experiments are performed in solution, the conformational flexibility of surface side chains is, to some degree, already accounted for in an ensemble of NMR structures so no homology modeling is needed in these cases. Once the structural ensemble has been generated through homology modeling or obtained from NMR experiments, TK-SA calculations are performed on each individual structure, and results are then averaged over the entire ensemble. The flexibility of surface residues also makes it possible to use homology modeling to generate structural representations of proteins that have not yet had their structures solved, provided that they have a high degree of sequence similarity to known structures. Indeed, this approach was used to redesign human acylphosphatase (Strickler et al., 2006).

Figure 7.3A shows results of TK-SA calculations for ubiquitin (Strickler et al., 2006). The value of ΔG_{qq} for a given residue represents the total energy of the interactions between that residue and every other charged residue in the protein. Unfavorable interactions are represented by positive values of ΔG_{qq}, whereas favorable interactions are indicated by negative values of ΔG_{qq}. Note that ubiquitin has several charged residues that participate in unfavorable interactions and therefore provide unfavorable contributions to stability. This general trend has been observed in all proteins redesigned by this approach so far (Lee et al., 2005; Loladze et al., 1999; Makhatadze et al., 2003, 2004; Permyakov et al., 2005; Sanchez-Ruiz and Makhatadze, 2001; Schweiker et al., 2007; Strickler et al., 2006) and leads to the idea that it should be possible to increase the stability of these proteins by neutralizing or reversing the charges of the residues that participate in unfavorable interactions. In particular, one would expect that the reversal of an existing

A Computational Approach for the Rational Design of Stable Proteins and Enzymes 187

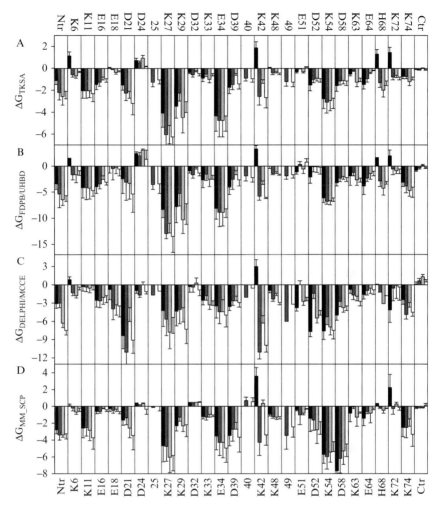

Figure 7.3 Surface charge–charge interaction energies (ΔG_{qq}) for wild-type and designed variants of ubiquitin at pH 5.5, reprinted with permission from Strickler *et al.* (2006). Copyright 2006 by the American Chemical Society. The charge–charge interactions were calculated using four different models: (A) TK-SA, (B) FDPB-UHBD (C) MMCE, and (D) MM.SCP. Each bar represents the total energy of charge–charge interactions of the corresponding residue with every other residue in the protein, averaged over an ensemble of 11 structures. Positive values of ΔG_{qq} are indicative of unfavorable interactions, whereas negative values correspond to favorable interactions. Black bars, wild-type ubiquitin; dark gray bars, UBQ-GA#1; light gray bars, UBQ-GA#2; and white bars, UBQ-GA#3 represent designed sequences identified by the genetic algorithm. UBQ-GA#1 and UBQ-GA#3 included uncharged polar residues in the optimization, whereas UBQ-GA#2 did not.

charge that participates in unfavorable interactions should yield greater increases in stability than neutralization. In fact, this behavior has been observed experimentally (Grimsley et al., 1999; Loladze et al., 1999; Spector et al., 2000).

2.2. Optimization of surface charges using the genetic algorithm

While it is possible to make single or double substitutions and observe a significant increase in the stability of a given protein (Lee et al., 2005; Loladze et al., 1999; Permyakov et al., 2005; Strickler et al., 2006), those variants are often not representative of the most favorable charge distribution for the protein. The ideal approach for identifying the optimal charge distribution for a protein, given its sequence, would be to use an exhaustive search algorithm to calculate the energies of every possible ionization state. However, this approach is computationally prohibitive for all but the smallest of peptides. For a protein with n charged positions on the surface, there are three possible charged states at each position (−1, 0, +1), and therefore, it would take 3^n calculations to identify all possible charge distributions. Even for a relatively small protein, such as ubiquitin, with only 23 charged surface residues, $3^{23} \approx 10^{11}$ calculations would need to be performed. Assuming that one processor can perform 100 TK-SA calculations per second, it would take over 31 years to perform the exhaustive search! An excellent alternative to exhaustive calculations is the genetic algorithm (Godoy-Ruiz et al., 2005; Ibarra-Molero and Sanchez-Ruiz, 2002; Strickler et al., 2006). The genetic algorithm is faster than exhaustive calculations because it does not seek to find each and every one of the best charge distributions, but instead identifies some of the sequences that are among the most optimal. For a protein like ubiquitin, only around 5×10^4 calculations are required to identify some of the best sequences using the genetic algorithm. Once again, assuming that one processor can perform 100 TK-SA calculations per second, it would only take a little over 8 min to identify optimized sequences using the genetic algorithm. This significant reduction in computation time makes it possible for optimization by a genetic algorithm to be performed on a standard desktop personal computer.

Although the genetic algorithm and its implementation have been described in great detail elsewhere (Godoy-Ruiz et al., 2005; Ibarra-Molero and Sanchez-Ruiz, 2002; Strickler et al., 2006), a conceptual overview of how it works is important for understanding the computational design approach discussed here (see Fig. 7.4). In our implementation of the genetic algorithm, only residues with greater than 50% solvent accessibility are included in this optimization. The surface charge distributions available to a protein, given its sequence, are represented *in silico* by a "chromosome."

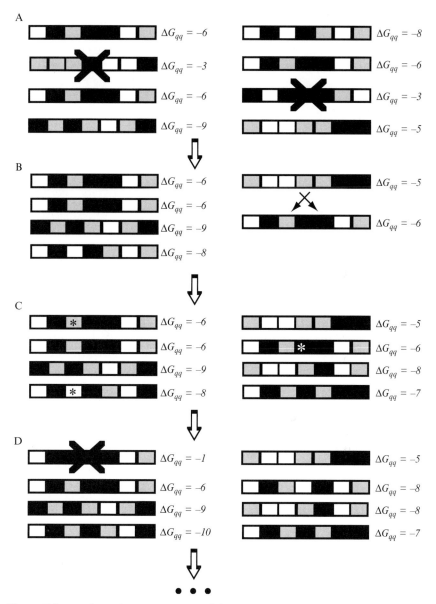

Figure 7.4 A schematic representation of the genetic algorithm. The steps of the algorithm are described in detail in the text. In each "chromosome," black boxes are representative of positive charges, gray boxes are negative charges, and white boxes indicate neutral residues. The ΔG_{qq} values are defined by Eq. 7.8. The large "X" in (A) and (D) indicates sequences whose energies were above the cutoff (−5 kJ/mol in this example) and were therefore not kept for the $n+1$ generation. (B) Black arrows show the "crossover" events used to finish populating the $n+1$ generation. (C) Stars indicate "point mutations" used to introduce more diversity into the population. (D) Chromosomes have energies below the predetermined cutoff that will be kept for the next generation; these steps will be repeated iteratively until the sequences in population of "chromosomes" have reached convergence.

The elements of these "chromosomes" are the charged states of the amino acid residues on the surface of the protein. An initial population of "chromosomes" is generated that contains a certain number of wild-type charge distributions and a certain number of randomly generated distributions (Fig. 7.4A). The "chromosomes" are scored based on their total charge–charge interaction energies, which are calculated by the TK-SA model described in the previous section. The lowest energy "chromosomes" are kept for the next generation where "crossover" events are used to finish populating the $n + 1$ generation (Fig. 7.4B). Once this generation has been populated, "point mutations" are used to introduce more diversity into the population (Fig. 7.4C). An energetic penalty helps minimize the number of energetically neutral or weak "mutations" and makes the "crossover" events essential for proper sampling of the available charge distributions. This process is repeated iteratively until the lowest energy "chromosomes" have remained identical for a predetermined number of cycles (Fig. 7.4D).

Results of the genetic algorithm for ubiquitin are shown in Figure 7.5 and serve to demonstrate the effectiveness of the genetic algorithm to appropriately sample the entire sequence space available to a given protein (Strickler et al., 2006). By examining the charge–charge interaction energies as a function of the number of amino acid substitutions relative to the wild-type sequence, one can see that, in general, an increasing number of substitutions leads to a significant increase in favorable charge–charge interactions. However, after a certain number of substitutions (8 to 10 for ubiquitin), the increase in favorable energy obtained per additional substitution begins to level off. The observation that a large increase in stability can be obtained with a small number of substitutions also holds for other proteins (Schweiker et al., 2007), suggesting that it is possible to increase protein stability via the optimization of surface charge–charge interactions with only a few substitutions.

3. Experimental Verification of Computational Predictions

One of the most important facets of computational design is to test the predictions experimentally. The experimental characterization of the stabilities of the designed variants serves two important purposes. First, it is the only way to know if the physical model being used to make the predictions is appropriate or if it is lacking in some of the fundamental aspects. Second, only by testing this approach on a number of proteins with different sizes, secondary structures, and three-dimensional topologies can one determine how universal this approach is and what improvements, if any, should be made. This section highlights the results of some of the key experiments that validated the TK-SA computational design method.

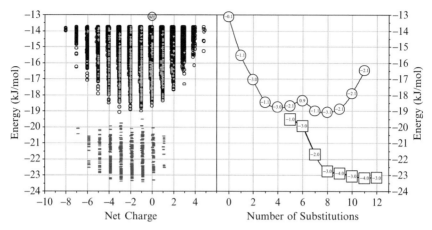

Figure 7.5 Analysis of the ability of the genetic algorithm to find the optimal sequence of ubiquitin, reprinted with permission from Strickler et al. (2006). Copyright 2006 by the American Chemical Society. The charge–charge interaction energies were calculated at pH 5.5 using the TK-SA model. Each sequence is characterized by the energy, net charge, and number of substitutions relative to the wild-type protein. (A) The ability of the genetic algorithm to effectively sample the sequence space searched by more exhaustive calculations is assessed. Open black circles represent results of exhaustive calculations. The grey crosshair represents the genetic algorithm. The light gray hash marks (energies below −19 kJ/mol) represent results of the genetic algorithm when previously uncharged surface residues were also included in optimization. (B) The relationship between the number of substitutions in the sequence and the energy of the lowest sequence with that number of substitutions. Open circles correspond to the grey crosshairs in A, whereas open squares represent the gray hash marks in A. Numbers within the symbol are the net charges of those sequences at pH 5.5.

3.1. Single site substitutions — proof of concept and test of robustness of the TK-SA model

The first test of the hypothesis that optimizing surface charge–charge interactions will increase the stability of a protein was performed using ubiquitin as a model system (Loladze et al., 1999). In this test, three single site substitutions were made to neutralize the charges at positions predicted to contribute unfavorably to stability (K6Q, H68Q, and R72Q), three single substitutions were made to reverse charges at unfavorable positions (K6E, R42E, and H68E), and three single substitutions were made to neutralize charges at favorable positions (K27Q, K29Q, and K29N) to serve as controls for these experiments. The stabilities of these nine ubiquitin variants were measured by monitoring changes in secondary structure as a function of denaturant concentration using far-UV circular dichroism spectroscopy (CD)(Pace, 1986). As predicted, the neutralization of unfavorable charges was stabilizing. Furthermore, the reversal of unfavorable charges resulted in larger increases in stability (≈ 1 kJ/mol) than charge neutralization.

In addition, the neutralization of charges predicted to contribute favorably to the stability of ubiquitin (K27Q, K29Q, and K29N) resulted in variants with significantly decreased stability relative to the wild type, suggesting that the TK-SA model can qualitatively predict the effects of surface substitutions on the stability of ubiquitin.

One of the advantages of computational design methods over other approaches used to stabilize proteins, such as directed evolution or sequence-based design, is that it is universal. In other words, because computational design approaches model the energetics of the intramolecular interactions, one should be able to use the same algorithm to redesign many different proteins without developing different selection criteria for each protein. After it had been demonstrated that the TK-SA approach could be used to successfully stabilize one model protein, the next important step was to test the robustness of this model. This was done using several proteins, and the results of each test are described in this section. Initially, the calculated values of ΔG_{qq} were compared to the experimental stabilities reported in the literature for ubiquitin (Loladze et al., 1999), the bacterial cold-shock protein (CspB)(Frankenberg et al., 1999; Perl et al., 2000), RNaseSA (Grimsley et al., 1999), the peripheral subunit binding domain (psbd41)(Spector et al., 2000), rubredoxin (Strop and Mayo, 2000), barnase (Horovitz et al., 1990; Sali et al., 1991; Serrano et al., 1990; Tissot et al., 1996), λ-repressor (Marqusee and Sauer, 1994), T4 lysozyme (Nicholson et al., 1991), the B1 domain of protein G (GB1)(Lassila et al., 2002; Merkel et al., 1999), and the zinc-finger domain (Blasie and Berg, 1997). The changes in both experimentally measured thermostabilites (ΔT_m) and stabilities ($\Delta\Delta G_{exp}$) of the variants relative to their wild-type proteins were compared to the changes in the charge–charge interaction energy ($\Delta\Delta G_{qq}$) expected from the substitutions (Makhatadze et al., 2003; Sanchez-Ruiz and Makhatadze, 2001). It was observed in all cases that changes in both thermostability and stability for these proteins could be predicted qualitatively based on the calculated changes in $\Delta\Delta G_{qq}$ (Makhatadze et al., 2003; Sanchez-Ruiz and Makhatadze, 2001). These results provided the first indication for the robustness of the TK-SA design strategy. More extensive testing was performed by making many substitutions in three different model systems: α-lactalbumin (Permyakov et al., 2005), ribosomal protein L30e (Lee et al., 2005), and bacterial cold shock protein (CspB)(Gribenko and Makhatadze, 2007; Makhatadze et al., 2004).

α-Lactalbumin is a small calcium-binding protein that has been observed to bind electrostatically to highly basic proteins and histones. The *apo* form of the protein was predicted to have many unfavorable surface charge–charge interactions (Permyakov et al., 2005). While the presence of calcium does create favorable interactions for the residues involved in metal binding, residues far from the binding loop maintained unfavorable interactions. However, the TK-SA approach was able to successfully predict the effects of single site substitutions on the stability of α-lactalbumin. It was also observed that changes in the thermostability of α-lactalbumin are in direct correlation with changes in calcium affinity (Permyakov *et al.*, 2005).

In order to learn more about the extent to which surface charge–charge interactions affect stability, the *Thermococcus celer* ribosomal protein L30e was used as a model system (Lee *et al.*, 2005). In this study, the TK-SA model was used to predict the effects of charge to alanine substitutions at all 26 charged positions of this protein. In addition to eliminating the charges at these positions, the alanine substitutions alter other important intramolecular interactions, such as hydrophobicity, secondary structure propensity, and side chain packing interactions. If these other types of interactions contribute more to stability than charge–charge interactions at these positions, then one would expect the calculated values of $\Delta\Delta G_{qq}$ to predict the experimentally observed changes in stability incorrectly. However, the experimentally measured changes in stability were predicted correctly for 20 of the 26 positions studied. The remaining 6 positions were all located at either the N or the C termini of α helices and thus are likely to participate in specific interactions at the ends of the helix; it has been shown previously that the identity of helix-capping residues is very important for thermodynamic stability (Aurora and Rose, 1998; Bang *et al.*, 2006; Chakrabartty *et al.*, 1993; Doig and Baldwin, 1995; Ermolenko *et al.*, 2002; Gong *et al.*, 1995; Makhatadze, 2005; Marshall *et al.*, 2002; Thomas *et al.*, 2001; Thomas and Makhatadze, 2000; Viguera and Serrano, 1995). Results of the L30e experiments suggest that the nonelectrostatic interactions important for the helix-capping motifs contribute more to stability than the charge–charge interactions at these positions.

The bacterial cold shock protein CspB was used as a model system to gain a better understanding of the possible thermodynamic mechanism of stabilization through the rational design of surface charges. The surface charge–charge interaction energies were calculated and compared for the CspB proteins from the mesophilic bacterium *Bacillus subtilis* (CspB-Bs), the thermophilic *Bacillus caldolyticus* (CspB-Bc), and the hyperthermophilic *Thermotoga maritima* (CspB-Tm)(Makhatadze *et al.*, 2004; Sanchez-Ruiz and Makhatadze, 2001). Although the sequences of these three variants of CspB are highly homologous, the distributions of the surface charges are very different (Fig. 7.6). CspB-Bs has the greatest number of unfavorable charge–charge interactions (Fig. 7.6A), whereas CspB-Tm has the most favorable charge–charge interactions (Fig. 7.6C). This trend correlates with the relative thermostabilities of these proteins. To determine whether the high stability of CspB-Tm did indeed come from the increased number of favorable surface charge–charge interactions, a cold shock protein (CspB-TB) was engineered to have the same core residues as CspB-Bs and the same surface charge distribution of CspB-Tm (Fig. 7.6D). The thermal stabilities of CspB-Bs and CspB-TB were measured using far-UV CD and it was found

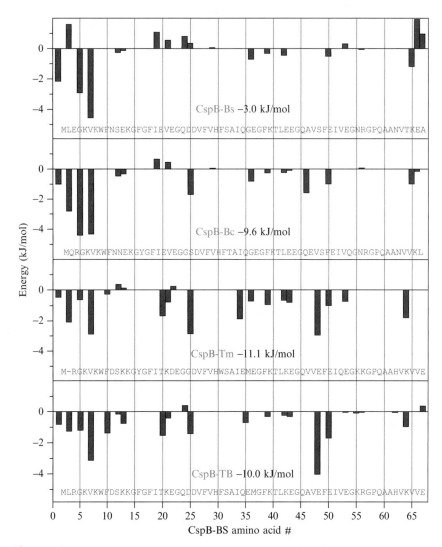

Figure 7.6 Comparison of charge–charge interaction energies for CspB-Bs, CspB-Bc, CspB-Tm, and CspB-TB, reprinted from Makhatadze *et al.* (2004). Each bar represents the total energy of charge–charge interactions of the corresponding residue with every other charged residue in the protein, averaged over an ensemble of 11 structures. Positive values of ΔG_{qq} are indicative of unfavorable interactions, whereas negative values correspond to favorable interactions. Of the wild-type proteins, CspB-Bs has the largest number of unfavorable interactions, whereas CspB-Tm has the greatest number of favorable interactions. CspB-TB was engineered to have the same core as CspB-Bs, but a similar surface charge distribution to CspB-Tm.

that CspB-TB had an increase in thermostability of 20 °C relative to the CspB-Bs protein (Makhatadze et al., 2004). This result further supported the idea that the rational design of surface charge–charge interactions could be a more effective way to stabilize proteins than making single substitutions at unfavorable positions.

CspB-Bs and CspB-TB also provided a special opportunity to address two important questions. This is because CspB-Bs and CspB-TB are structurally similar, yet have dramatically different surface charge distributions. The first question was whether charge–charge interactions in the unfolded state provide significant contributions to stability. To answer this question, a number of single substitutions were made in each of the proteins (Gribenko and Makhatadze, 2007). For most of the substitutions, the TK-SA approach was able to semiquantitatively (relative rank order) predict the effects of the substitutions on the stability of each protein. Since most of the positions that were predicted incorrectly were located in a β hairpin, it is possible that residual charge–charge interactions in the unfolded state could affect the overall contributions of these residues to the stability of the native state. However, when the Gaussian chain model of the unfolded state (Zhou, 2002a) was incorporated into the calculations, no significant improvement in the correlation between calculated and experimental stabilities was observed. Furthermore, when the putative unfolded state structure was disrupted, by destabilizing the $\beta 2$–$\beta 3$ hairpin, there was also no significant improvement between calculations and experiments (Gribenko and Makhatadze, 2007). Rather, it was found that these residues were part of a complex network of charge–charge interactions that, when disrupted, led to markedly better agreement between calculated and experimentally measured stabilities (Gribenko and Makhatadze, 2007).

At first, the observation that the unfolded state of CspB did not have a significant effect on the predictions of the TK-SA model seemed to be in conflict with previous observations that consideration of the unfolded state was necessary to accurately predict experimentally measured stabilities of a number of different proteins (Anil et al., 2006; Cho and Raleigh, 2005; Elcock and McCammon, 1998; Kuhlman et al., 1999; Oliveberg et al., 1995; Pace et al., 1990, 2000; Swint-Kruse and Robertson, 1995; Whitten and Garcia-Moreno, 2000; Zhou, 2002a). In most of these examples, it seems that there were specific nonelectrostatic interactions in the unfolded state of the proteins that affected the predictions of the thermodynamic stability (ΔG). However, the TK-SA approach predicts $\Delta \Delta G_{qq}$ or the difference in charge–charge interaction energies between a wild-type protein and its designed variant. In the absence of specific interactions in the unfolded state, the high dielectric of the solvent, to which most of the protein is exposed, will screen out interactions between charges separated by more than 10 residues. Even for residues that are much closer in sequence, the calculated charge–charge interaction energies in the unfolded state are smaller than for interactions in the native state of the protein. Since the unfolded state contributions of nonspecific interactions are small, even in terms of ΔG_{qq}, they would be expected to have even smaller contributions to $\Delta\Delta G_{qq}$. This idea, combined with CspB results, would suggest that including the unfolded state charge–charge interactions is not always necessary to improve the correlation between predicted and experimental stabilities.

The second question that the CspB proteins were uniquely suited to address is whether the surface substitutions affect interactions other than the charge–charge interactions. Ideally, to analyze only the effects of altered charge–charge interactions on protein stability, one would make substitutions that perturbed only the charge of the side chain, without affecting other factors, such as size, hydrophobicity, and/or packing interactions (Spector et al., 2000). Incorporating nonnatural amino acids into the protein sequence is an effective way to accomplish this goal (Spector et al., 2000), but it is only experimentally possible for small proteins. Furthermore, the relatively small number of naturally occurring amino acids offers only limited options for reversing or neutralizing charges in proteins. As such, it is often easier to use the natural amino acids lysine or glutamic acid for charge reversals and glutamine for charge neutralizations. Although this approach simplifies the design process, it is also possible that observed changes in stability are actually due to changes in other important properties, such as hydrogen bonding patterns, hydrophobicity, secondary structure propensity, or packing. It is also possible that the charge reversals/neutralizations could alter short-range (i.e., salt bridges) rather than long-range charge–charge interactions. Short-range and long-range interactions are affected differently by changes in the ionic strength of a solution — long-range interactions tend to get weaker with increasing salt concentrations, whereas short-range interactions tend to persist (Dominy et al., 2002, 2004; Gvritishvili et al., 2008; Kao et al., 2000; Lee et al., 2002), making it possible to determine which interactions contribute more to the observed increases in stability. To address this issue, the same substitutions were made at the same surface positions in the different electrostatic environments of CspB-Bs and CspB-TB. If the substitutions affect primarily long-range charge–charge interactions, then one would expect an inverse correlation between changes in stability and changes in the halophilicity of the proteins (Perl and Schmid, 2001; Perl et al., 1998, 2000). Indeed, for most of the substitutions, this behavior was observed (Gribenko and Makhatadze, 2007). The surface substitutions that did not display an inverse correlation between stability and halophilicity occurred at the same position: V20 in CspB-Bs and K20 in CspB-TB. The introduction of charge at V20 in CspB-Bs results in a protein that is both less stable and less halophilic, whereas introduction of a hydrophobic residue at K20 in CspB-Tb results in a protein with increased stability and halophilicity, suggesting that hydrophobic interactions are much more important than charge–charge interactions at this position (Gribenko and Makhatadze, 2007).

3.2. Rational design of surface charge–charge interactions using a genetic algorithm

Studies on the proteins described earlier provided strong evidence that the rational design of surface charge–charge interactions could be successful for many different proteins. Furthermore, studies with the engineered CspB-TB protein not only gave important insights into the nature of how proteins are stabilized by this approach, but also led to the idea that it should be possible to computationally optimize the entire surface charge distribution for any given protein. To optimize the surface charge–charge interactions, a genetic algorithm (GA) was used. The TKSA-GA approach to stabilize proteins was tested using seven model proteins: ubiquitin, procarboxypeptidase, the Fyn SH3 domain, acylphosphatase, tenascin, U1A, and CDC42. These proteins all have different sizes, three-dimensional topologies, secondary structural composition, and surface charge distributions (Gribenko et al., 2008; Schweiker et al., 2007; Strickler et al., 2006) (Fig. 7.7). The surface charge–charge interaction energies were calculated for each protein using the TK-SA model, and then the optimal surface charge distributions were identified using the genetic algorithm. Results discussed in this section are summarized in Table 7.2.

One of the first proteins redesigned by this approach was ubiquitin (Fig. 7.7A). In this study, the stabilities of two variants with single substitutions at unfavorable positions (Ubq-6, K6E, and Ubq-72, R74E) and one variant with charge reversals at both positions (Ubq-6/72, K6E/R74E) were characterized as a reference for the magnitude of $\Delta\Delta G$ expected when charge–charge interactions were optimized using the genetic algorithm. It was observed that both Ubq-6 and Ubq-72 were more stable than wild-type ubiquitin ($\Delta\Delta G_{Des-WT} = 3.3$ and 1.7 kJ/mol, respectively) and that Ubq-6/72 was more stable than either single variant ($\Delta\Delta G_{Des-WT} = 5.2$ kJ/mol)(Strickler et al., 2006). Once again, this demonstrates that reversing the charges at unfavorable positions can lead to significant increases in stability. The next step was to see if the optimization of surface charges using the genetic algorithm would provide even larger increases in stability. Figure 7.5 shows an analysis of the results of the genetic algorithm for ubiquitin. In general, there is an increase in favorable charge–charge interaction energy with an increasing number of substitutions. However, after 8 to 10 substitutions, the increase in favorable charge–charge interaction energy gained per additional substitution begins to level off (Fig. 7.5B). As a result, it should be possible to obtain significant increases in stability with just a few mutations. Three of the sequences that were predicted to increase the stability of ubiquitin were selected for further characterization.

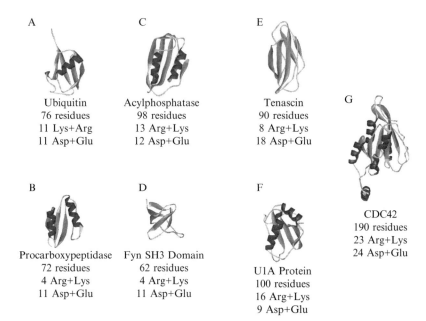

Figure 7.7 Structures of the seven proteins redesigned using the TK-SA model. The surface charge–charge distribution of each protein was optimized using the genetic algorithm. The different sizes, shapes, secondary structures, and three-dimensional topologies of these proteins provide a good test of the robustness of this rational design approach. PDB codes for the structures are (A) 1UBQ (Vijay-Kumar *et al.*, 1987)(ubiquitin), (B) 1AYE (Garcia-Saez *et al.*, 1997)(activation domain of human procarboxypeptidase), (C) 2ACY (Thunnissen *et al.*, 1997)(acylphosphatase), (D) 1FYN (Musacchio *et al.*, 1994)(Fyn SH3 domain), (E) 1TEN (Leahy *et al.*, 1992)(tenascin), (F) 1URN (Oubridge *et al.*, 1994)(ribosomal U1A protein), and (G) 1A4R (Rudolph *et al.*, 1999)(CDC42).

Figure 7.3A provides a comparison of the results of the TK-SA calculations for wild-type and designed variants of ubiquitin. One of the variants only optimized existing charges (Ubq-GA2 — four substitutions, 5.3% of the sequence), whereas the other two variants also allowed for neutral polar residues on the surface to be included in the optimization (Ubq-GA1 — five substitutions, 6.6% of the sequence; Ubq-GA3 — 6 substitutions, 7.9% of the sequence). Figure 7.3A shows that several positions that have unfavorable contributions to stability in the wild-type protein are now predicted to contribute favorably in each of the designed variants. Indeed, when the stabilities of these three designed sequences were characterized using urea-induced unfolding, it was found that not only were they much more stable than the wild type ($\Delta\Delta G_{UbqGA1-WT} = 13.2$, $\Delta\Delta G_{UbqGA2-WT} = 18.4$, and $\Delta\Delta G_{UbqGA3-WT} = 17.7$ kJ/mol), but all three variants also had much larger increases in stability than one obtains by focusing only on one or two unfavorable positions (Strickler *et al.*, 2006).

Table 7.2 TK-SA/GA results

Protein name	Number of substitutions (% of sequence)	T_m or $\Delta\Delta G$
Ubiquitin		
GA1	5/76 (6.6%)	13.2 kJ/mol
GA2	4/76 (5.3%)	18.4 kJ/mol
GA3	6/76 (7.9%)	17.7 kJ/mol
Procarboxypeptidase		
GA1	5/72 (6.9%)	4.1 kJ/mol
GA2	7/72 (9.7%)	10.7 kJ/mol
Acylphosphatase		
GA1	4/98 (4.1%)	7 kJ/mol
GA2	5/98 (5.1%)	11 °C
Fyn SH3 domain		
Fyn1	1/62 (1.6%)	4.7 kJ/mol
Fyn2	2/62 (3.2%)	2.3 kJ/mol
Fyn3	3/62 (4.8%)	6.7 kJ/mol
Fyn5	5/62 (8%)	7.1 kJ/mol
Tenascin		
GA1	4/90 (4.4%)	5.4 kJ/mol
U1A protein		
GA1	4/100 (4%)	4.1 kJ/mol
CDC42		
GA1	8/190 (4.2%)	10 °C

Two of the sequences of procarboxypeptidase (Fig. 7.7B) predicted to be more stable than the wild type were also selected for experimental characterization. One sequence contained five substitutions (6.9% of the sequence) and one contained seven substitutions (9.7% of the sequence). Both designed sequences had significantly increased stabilities relative to the wild-type protein ($\Delta\Delta G_{Des-WT}$ = 4.1 and 10.7 kJ/mol)(Strickler et al., 2006). Two designed variants of acylphosphatase (Fig. 7.7C) were also studied. The first variant (Acp-GA1) contained four substitutions (4.1% of the sequence) and was stabilized by 7.0 kJ/mol relative to the wild type (Strickler et al., 2006). The second variant (Acp-GA2) contained five substitutions (5.1% of the sequence) and was stabilized by 11 °C relative to the wild type (Gribenko et al., 2008). Four sequences of the FynSH3 domain (Fig. 7.7D) were selected for characterization to understand the stepwise effects of the optimization of charge–charge interactions on protein stability

(Schweiker et al., 2007). One of the variants contained five substitutions (Fyn5; 8% of the sequence) and the others contained one (Fyn1; 1.6% of the sequence), two (Fyn2; 3.2% of the sequence), or three (Fyn3; 4.8% of the sequence) of those five substitutions in their sequences. Each of the four variants was more stable than the wild type ($\Delta\Delta G_{Fyn1} = 4.7$ kJ/mol, $\Delta\Delta G_{Fyn2} = 2.3$ kJ/mol, $\Delta\Delta G_{Fyn3} = 6.7$ kJ/mol, $\Delta\Delta G_{Fyn5} = 7.1$ kJ/mol) (Schweiker et al., 2007). Importantly, the TK-SA model was able to predict the relative rank order of the stabilities of these variants (Fig. 7.8). The ubiquitin, procarboxypeptidase, acylphosphatase, and FynSH3 results demonstrate how the flexibility of this approach allows for large increases in stability to be obtained with two or three different designed sequences. This was an important observation because it suggested that it should be possible to choose sequences that do not make substitutions in or near the active/binding site of proteins, thus providing a way to increase their stability without affecting their function significantly.

Only one optimized sequence of tenascin (Fig. 7.7E) and U1A (Fig. 7.7F) were selected for further characterization (Strickler et al., 2006). The designed sequence for each of these proteins had four substitutions ($\approx$4% of each sequence). Once again, for each of these proteins, the designed variants were significantly more stable than the wild type ($\Delta\Delta G_{Des-WT} = 5.4$ and 4.1 kJ/mol for tenascin and U1A, respectively). Moreover, this was the first successful stabilization of the U1A ribosomal protein. The largest protein redesigned by this approach was CDC42 (190 amino acids; Fig. 7.7G). With only eight amino acid substitutions (4.2% of the sequence), it was possible to thermostabilize CDC42 by 10 °C (Gribenko et al., 2008), which is quite remarkable for such a large protein. These results highlight how the rational design of surface charge–charge interactions is a universal approach for stabilizing proteins of different sizes and structures.

3.3. Effects of stabilization on enzymatic activity

In order for any design approach to be useful for practical applications, the protein must retain its activity. To determine if this was indeed true for proteins stabilized by optimizing the surface charges, activity assays were performed on three of the proteins: CspB, acylphosphatase, and CDC42. The activities of each designed variant were compared to their respective wild-type proteins.

CspB-Bs is expressed by B. subtilis when it is exposed to cold temperatures and protects cells from these conditions by acting as an RNA chaperone (Ermolenko and Makhatadze, 2002). CspB can also bind polypyrimidine single-stranded DNA (ssDNA) sequences (Lopez and Makhatadze, 2000; Lopez et al., 1999, 2001). Interactions between both CspB-Bs and CspB-TB (23° more thermostable than CspB-Bs) with ssDNA templates were measured using fluorescence spectroscopy. Not

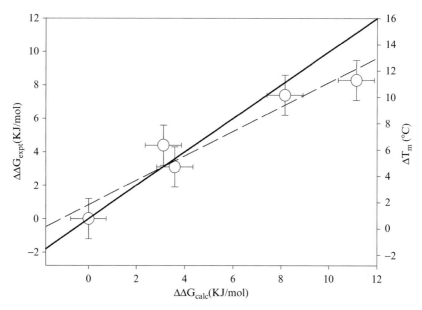

Figure 7.8 Comparison of the changes in stability of the designed Fyn variants calculated from the TK-SA model ($\Delta\Delta G_{calc}$) to changes in stability observed experimentally, reprinted with permission from Schweiker et al. (2007). The dashed line represents a perfect correlation (($\Delta\Delta G_{calc} = \Delta\Delta G_{expt}$), whereas the solid line represents the linear regression of data (slope = 0.73, $R^2 = 0.96$). The TK-SA model was able to correctly predict the relative rank order of both changes in thermodynamic stability (($\Delta\Delta G_{expt}$) and changes in thermostability (ΔT_m).

only could CspB-TB bind ssDNA at higher temperatures (37 °C) than CspB-Bs, but it also bound ssDNA with a higher affinity than CspB-Bs at lower temperatures (25 °C) (Makhatadze et al., 2004). Based on the findings of these functional studies, the structure of CspB in complex with pT7 was solved (Max et al., 2006).

Acylphosphatase is an enzyme that binds its charged substrate, acylphosphate, and catalyzes the hydrolysis to produce carboxylate and inorganic phosphate. The hydrolysis of benzoylphosphate by the acylphosphatase variants was measured using a continuous UV absorption assay (Ramponi et al., 1966, 1967). The Acp-GA1 variant was found to be inactive, and examination of its sequence revealed that one of the stabilizing substitutions was in the active site of the enzyme. As a result, a sequence was selected that contained only substitutions that were distant from the active site (Acp-GA2). The designed Acp-GA2 enzyme was able to maintain its catalytic activity at higher temperatures than the wild-type enzyme, while also remaining active at lower temperatures (Gribenko et al., 2008).

CDC42 is an important cell signaling protein that binds GTP and catalyzes the hydrolysis of GTP to GDP + P_i. This activity can be monitored using a colorimetric assay to detect the amount of free phosphate released upon hydrolysis. Since the thermal inactivation/denaturation of CDC42 is not reversible, the functional properties were characterized as the residual activity after incubation at high temperatures ($\approx 10\,^\circ$C higher than the T_m of wild-type CDC42). Once again, it was observed that the designed variant was not only able to retain activity at temperatures where the wild-type enzyme was inactivated, but it was also as active as the wild type at lower temperatures (Gribenko et al., 2008).

The functional studies of the three proteins discussed in this section support two important conclusions. First, it is possible to stabilize proteins through the rational design of surface charge–charge interactions without perturbing activity. Second, it provides a strong argument against the idea that a dichotomy exists between protein stability and function. Historically, it was believed that if a protein was thermostable, then it must be rigid at lower temperatures. Because proteins and enzymes need some flexibility to function properly, an idea developed that if a protein were stabilized, it would become more rigid and, therefore, less active at lower temperatures. This idea was supported by observations that proteins isolated from thermophilic organisms were not as active at lower temperatures as they were at higher temperatures (Giver et al., 1998; Meiering et al., 1992; Roca et al., 2007; Shoichet et al., 1995; Varley and Pain, 1991; Zhang et al., 1995). However, proteins that exist in thermophilic organisms are under no evolutionary pressure to function at decreased temperatures. This does not mean per se that stability and flexibility/function should be mutually exclusive. Indeed, the results presented here show that by optimizing surface charge interactions at regions of the protein that are far from the active site, it is possible to increase stability and maintain activity at both higher and lower temperatures.

4. Closing Remarks

As evidenced in this chapter, the TK-SA method provides a simple model that is still effective for determining the qualitative changes in charge–charge interactions on the protein surface. However, a few assumptions that go into the model must be taken into consideration when it is applied to the rational design of stable proteins. First, the model assumes that the protein is spherical. For globular proteins, this assumption does not appear to have adverse effects, even when the shape of the protein deviates somewhat from a sphere. For example, tenascin (Fig. 7.7D) is more cylindrical than spherical, but this approach was still able to predict stabilizing substitutions successfully (Strickler et al., 2006). Second, the model assumes that the interactions between charges are the only electrostatic interactions that occur in the native state. This assumption could pose the biggest challenge for accurate predictions, as it is known that hydrogen bonding and partial dipoles also provide significant contributions to electrostatic interaction energies. To ameliorate this issue, surface side chains involved in intramolecular hydrogen bonds are not included in the optimization procedure described here. Third, it has been reported that this model ignores important parameters, such as self-energy and solvation (Garcia-Moreno and Fitch, 2004; Schutz and Warshel, 2001; Warshel, 2003). While this is true, when results of the TK-SA calculations for surface residues are compared to results of calculations on surface residues using other continuum electrostatic models, such as the finite difference solution of the Poisson–Boltzmann equation (FDPB/UHBD)(Fig. 7.7.3B)(Antosiewicz et al., 1994, 1996), the Multi-Conformer Continuum Electrostatic model (MCCE)(Fig. 7.3C)(Alexov and Gunner, 1997; Georgescu et al., 2002; Honig and Nicholls, 1995; Xiao and Honig, 1999; Yang et al., 1993), the Microenvironment Modulated Screened Coulomb Potential model (MM_SCP) (Fig. 7.3D)(Mehler and Guarnieri, 1999), or the Langevin dipole model (PDLD)(Lee et al., 1993; Schutz and Warshel, 2001; Sham et al., 1998), they are qualitatively similar for all models. The only advantage of TK-SA over the continuum models is that it is less computationally demanding.

Another potential pitfall to the rational design of surface charges is that it is currently not possible to predict protein stabilities quantitatively. This is largely due to the fact that only interactions between charges in the native state of the protein are being considered. While it has been shown that unfolded state effects do not seem to be significant for the proteins redesigned by this approach, unfolded state effects have been demonstrated to be important for other proteins (Cho and Raleigh, 2005; Kuhlman et al., 1999; Oliveberg et al., 1995; Trefethen et al., 2005), and we do not consider them here. In addition, other important factors for protein stability, such as side chain hydrophobicity, secondary structure propensity, hydrogen bonding, packing interactions, and helix capping interactions, are not considered (Baldwin and Matthews, 1994; Ermolenko et al., 2002; Fernandez et al., 2000; Fersht and Winter, 1992; Loladze and Makhatadze, 2005; Loladze et al., 2002; Makhatadze, 2005; Makhatadze and Privalov, 1995; Pace et al., 1996; Serrano et al., 1992; Thomas and Makhatadze, 2000; Thomas et al., 2001). Nevertheless, the TK-SA model does provide excellent qualitative predictions of protein stability. In order to obtain quantitative predictions, the other factors mentioned here will need to be included in the computational optimization approach. Important questions that will need to be addressed in the development of a quantitative algorithm are as follow: which of these interactions are the most important for modulating stability and how quantitative should the algorithm be to be practical? It seems likely that incorporating just a few of the factors mentioned here will give the algorithm the ability to predict the stability of designed sequences within the errors of experimental techniques.

ACKNOWLEDGMENTS

The experimental and computational work on charge–charge interactions in proteins in the G.I.M. laboratory is supported by a grant from the National Science Foundation (MCB-0110396). Helpful discussions and comments by Dr. Werner Streicher are greatly appreciated.

REFERENCES

Alexov, E. G., and Gunner, M. R. (1997). Incorporating protein conformational flexibility into the calculation of pH-dependent protein properties. *Biophys. J.* **72**, 2075–2093.
Anderson, D. E., Becktel, W. J., and Dahlquist, F. W. (1990). pH-induced denaturation of proteins: A single salt bridge contributes 3-5 kcal/mol to the free energy of folding of T4 lysozyme. *Biochemistry* **29**, 2403–2408.
Anil, B., Craig-Schapiro, R., and Raleigh, D. P. (2006). Design of a hyperstable protein by rational consideration of unfolded state interactions. *J. Am. Chem. Soc.* **128**, 3144–3145.
Antosiewicz, J., McCammon, J. A., and Gilson, M. K. (1994). Prediction of pH-dependent properties of proteins. *J. Mol. Biol.* **238**, 415–436.
Antosiewicz, J., McCammon, J. A., and Gilson, M. K. (1996). The determinants of pKas in proteins. *Biochemistry* **35**, 7819–7833.
Aurora, R., and Rose, G. D. (1998). Helix capping. *Protein Sci.* **7**, 21–38.
Baldwin, E. P., and Matthews, B. W. (1994). Core-packing constraints, hydrophobicity and protein design. *Curr. Opin. Biotechnol.* **5**, 396–402.
Bang, D., Gribenko, A. V., Tereshko, V., Kossiakoff, A. A., Kent, S. B., and Makhatadze, G. I. (2006). Dissecting the energetics of protein alpha-helix C-cap termination through chemical protein synthesis. *Nat. Chem. Biol.* **2**, 139–143.
Beadle, B. M., Baase, W. A., Wilson, D. B., Gilkes, N. R., and Shoichet, B. K. (1999). Comparing the thermodynamic stabilities of a related thermophilic and mesophilic enzyme. *Biochemistry* **38**, 2570–2576.
Berezovsky, I. N., and Shakhnovich, E. I. (2005). Physics and evolution of thermophilic adaptation. *Proc. Natl. Acad. Sci. USA* **102**, 12742–12747.
Blasie, C. A., and Berg, J. M. (1997). Electrostatic interactions across a beta-sheet. *Biochemistry* **36**, 6218–6222.
Bolon, D. N., Marcus, J. S., Ross, S. A., and Mayo, S. L. (2003). Prudent modeling of core polar residues in computational protein design. *J. Mol. Biol.* **329**, 611–622.
Chakrabartty, A., Doig, A. J., and Baldwin, R. L. (1993). Helix capping propensities in peptides parallel those in proteins. *Proc. Natl. Acad. Sci. USA* **90**, 11332–11336.
Cho, J. H., and Raleigh, D. P. (2005). Mutational analysis demonstrates that specific electrostatic interactions can play a key role in the denatured state ensemble of proteins. *J. Mol. Biol.* **353**, 174–185.
Dantas, G., Corrent, C., Reichow, S. L., Havranek, J. J., Eletr, Z. M., Isern, N. G., Kuhlman, B., Varani, G., Merritt, E. A., and Baker, D. (2007). High-resolution structural and thermodynamic analysis of extreme stabilization of human procarboxypeptidase by computational protein design. *J. Mol. Biol.* **366**, 1209–1221.

Dantas, G., Kuhlman, B., Callender, D., Wong, M., and Baker, D. (2003). A large scale test of computational protein design: Folding and stability of nine completely redesigned globular proteins. *J. Mol. Biol.* **332,** 449–460.

DeGrado, W. F., Summa, C. M., Pavone, V., Nastri, F., and Lombardi, A. (1999). De novo design and structural characterization of proteins and metalloproteins. *Annu. Rev. Biochem.* **68,** 779–819.

Desjarlais, J. R., and Handel, T. M. (1995a). De novo design of the hydrophobic cores of proteins. *Protein Sci.* **4,** 2006–2018.

Desjarlais, J. R., and Handel, T. M. (1995b). New strategies in protein design. *Curr. Opin. Biotechnol.* **6,** 460–466.

Desjarlais, J. R., and Handel, T. M. (1999). Side-chain and backbone flexibility in protein core design. *J. Mol. Biol.* **290,** 305–318.

Deutschman, W. A., and Dahlquist, F. W. (2001). Thermodynamic basis for the increased thermostability of CheY from the hyperthermophile *Thermotoga maritima*. *Biochemistry* **40,** 13107–13113.

Dill, K. A. (1990). Dominant forces in protein folding. *Biochemistry* **29,** 7133–7155.

Doig, A. J., and Baldwin, R. L. (1995). N- and C-capping preferences for all 20 amino acids in alpha-helical peptides. *Protein Sci.* **4,** 1325–1336.

Dominy, B. N., Minoux, H., and Brooks, C. L. (2004). An electrostatic basis for the stability of thermophilic proteins. *Proteins* **57,** 128–141.

Dominy, B. N., Perl, D., Schmid, F. X., and Brooks, C. L. 3rd, (2002). The effects of ionic strength on protein stability: The cold shock protein family. *J. Mol. Biol.* **319,** 541–554.

Elcock, A. H., and McCammon, J. A. (1998). Electrostatic contributions to the stability of halophilic proteins. *J. Mol. Biol.* **280,** 731–748.

Ermolenko, D. N., and Makhatadze, G. I. (2002). Bacterial cold-shock proteins. *Cell Mol. Life Sci.* **59,** 1902–1913.

Ermolenko, D. N., Thomas, S. T., Aurora, R., Gronenborn, A. M., and Makhatadze, G. I. (2002). Hydrophobic interactions at the Ccap position of the C-capping motif of alpha-helices. *J. Mol. Biol.* **322,** 123–135.

Fernandez, A. M., Villegas, V., Martinez, J. C., Van Nuland, N. A., Conejero-Lara, F., Aviles, F. X., Serrano, L., Filimonov, V. V., and Mateo, P. L. (2000). Thermodynamic analysis of helix-engineered forms of the activation domain of human procarboxypeptidase A2. *Eur. J. Biochem.* **267,** 5891–5899.

Fersht, A., and Winter, G. (1992). Protein engineering. *Trends Biochem. Sci.* **17,** 292–295.

Frankenberg, N., Welker, C., and Jaenicke, R. (1999). Does the elimination of ion pairs affect the thermal stability of cold shock protein from the hyperthermophilic bacterium *Thermotoga maritima*? *FEBS Lett.* **454,** 299–302.

Garcia-Moreno, E. B., and Fitch, C. A. (2004). Structural interpretation of pH and salt-dependent processes in proteins with computational methods. *Methods Enzymol.* **380,** 20–51.

Garcia-Saez, I., Reverter, D., Vendrell, J., Aviles, F. X., and Coll, M. (1997). The three-dimensional structure of human procarboxypeptidase A2: Deciphering the basis of the inhibition, activation and intrinsic activity of the zymogen. *EMBO J.* **16,** 6906–6913.

Georgescu, R. E., Alexov, E. G., and Gunner, M. R. (2002). Combining conformational flexibility and continuum electrostatics for calculating pK(a)s in proteins. *Biophys. J.* **83,** 1731–1748.

Giver, L., Gershenson, A., Freskgard, P. O., and Arnold, F. H. (1998). Directed evolution of a thermostable esterase. *Proc. Natl. Acad. Sci. USA* **95,** 12809–12813.

Godoy-Ruiz, R., Perez-Jimenez, R., Garcia-Mira, M. M., Plaza del Pino, I. M., and Sanchez-Ruiz, J. M. (2005). Empirical parametrization of pK values for carboxylic acids in proteins using a genetic algorithm. *Biophys. Chem.* **115,** 263–266.

Gong, Y., Zhou, H. X., Guo, M., and Kallenbach, N. R. (1995). Structural analysis of the N- and C-termini in a peptide with consensus sequence. *Protein Sci.* **4,** 1446–1456.

Grattinger, M., Dankesreiter, A., Schurig, H., and Jaenicke, R. (1998). Recombinant phosphoglycerate kinase from the hyperthermophilic bacterium *Thermotoga maritima*: Catalytic, spectral and thermodynamic properties. *J. Mol. Biol.* **280,** 525–533.

Gribenko, A. V., and Makhatadze, G. I. (2007). Role of the charge–charge interactions in defining stability and halophilicity of the CspB proteins. *J. Mol. Biol.* **366,** 842–856.

Gribenko, A. V., Patel, M. M., and Makhatadze, G. I. (2008). Rational redesign of enzyme stability. Submitted for publication.

Griko Iu, V., and Privalov, P. L. (1986). Cold denaturation of myoglobin in alkaline solutions. *Dokl. Akad. Nauk SSSR* **291,** 709–711.

Griko Yu, V., Venyaminov, S., and Privalov, P. L. (1989). Heat and cold denaturation of phosphoglycerate kinase (interaction of domains). *FEBS Lett.* **244,** 276–278.

Griko, Y. V., Makhatadze, G. I., Privalov, P. L., and Hartley, R. W. (1994). Thermodynamics of barnase unfolding. *Protein Sci.* **3,** 669–676.

Griko, Y. V., Privalov, P. L., Sturtevant, J. M., and Venyaminov, S. (1988). Cold denaturation of staphylococcal nuclease. *Proc. Natl. Acad. Sci. USA* **85,** 3343–3347.

Grimsley, G. R., Shaw, K. L., Fee, L. R., Alston, R. W., Huyghues-Despointes, B. M., Thurlkill, R. L., Scholtz, J. M., and Pace, C. N. (1999). Increasing protein stability by altering long-range coulombic interactions. *Protein Sci.* **8,** 1843–1849.

Guzman-Casado, M., Parody-Morreale, A., Robic, S., Marqusee, S., and Sanchez-Ruiz, J. M. (2003). Energetic evidence for formation of a pH-dependent hydrophobic cluster in the denatured state of Thermus thermophilus ribonuclease H. *J. Mol. Biol.* **329,** 731–743.

Gvritishvili, A. G., Gribenko, A. V., and Makhatadze, G. I. (2008). Cooperativity of complex salt bridges. *Protein Sci.* **17,** 1285–1290.

Hamamatsu, N., Nomiya, Y., Aita, T., Nakajima, M., Husimi, Y., and Shibanaka, Y. (2006). Directed evolution by accumulating tailored mutations: Thermostabilization of lactate oxidase with less trade-off with catalytic activity. *Protein Eng. Des. Sel.* **19,** 483–489.

Hollien, J., and Marqusee, S. (1999). A thermodynamic comparison of mesophilic and thermophilic ribonucleases H. *Biochemistry* **38,** 3831–3836.

Honig, B., and Nicholls, A. (1995). Classical electrostatics in biology and chemistry. *Science* **268,** 1144–1149.

Horovitz, A., Serrano, L., Avron, B., Bycroft, M., and Fersht, A. R. (1990). Strength and co-operativity of contributions of surface salt bridges to protein stability. *J. Mol. Biol.* **216,** 1031–1044.

Huang, X., and Zhou, H. X. (2006). Similarity and difference in the unfolding of thermophilic and mesophilic cold shock proteins studied by molecular dynamics simulations. *Biophys. J.* **91,** 2451–2463.

Hurley, J. H., Baase, W. A., and Matthews, B. W. (1992). Design and structural analysis of alternative hydrophobic core packing arrangements in bacteriophage T4 lysozyme. *J. Mol. Biol.* **224,** 1143–1159.

Ibarra-Molero, B., Loladze, V. V., Makhatadze, G. I., and Sanchez-Ruiz, J. M. (1999a). Thermal versus guanidine-induced unfolding of ubiquitin: An analysis in terms of the contributions from charge-charge interactions to protein stability. *Biochemistry* **38,** 8138–8149.

Ibarra-Molero, B., Makhatadze, G. I., and Sanchez-Ruiz, J. M. (1999b). Cold denaturation of ubiquitin. *Biochim. Biophys. Acta* **1429,** 384–390.

Ibarra-Molero, B., and Sanchez-Ruiz, J. M. (2002). Genetic algorithm to design stabilizing surface-charge distributions in proteins. *J. Phys. Chem. B* **106,** 6609–6613.

Kao, Y. H., Fitch, C. A., Bhattacharya, S., Sarkisian, C. J., Lecomte, J. T., and Garcia-Moreno, E. B. (2000). Salt effects on ionization equilibria of histidines in myoglobin. *Biophys. J.* **79,** 1637–1654.

Kauzmann, W. (1959). Some factors in the interpretation of protein denaturation. *Adv. Protein Chem.* **14,** 1–63.

Korkegian, A., Black, M. E., Baker, D., and Stoddard, B. L. (2005). Computational thermostabilization of an enzyme. *Science* **308,** 857–860.

Kuhlman, B., and Baker, D. (2000). Native protein sequences are close to optimal for their structures. *Proc. Natl. Acad. Sci. USA* **97,** 10383–10388.

Kuhlman, B., Luisi, D. L., Young, P., and Raleigh, D. P. (1999). pKa values and the pH dependent stability of the N-terminal domain of L9 as probes of electrostatic interactions in the denatured state: Differentiation between local and nonlocal interactions. *Biochemistry* **38,** 4896–4903.

Lassila, K. S., Datta, D., and Mayo, S. L. (2002). Evaluation of the energetic contribution of an ionic network to beta-sheet stability. *Protein Sci.* **11,** 688–690.

Lazar, G. A., Desjarlais, J. R., and Handel, T. M. (1997). De novo design of the hydrophobic core of ubiquitin. *Protein Sci.* **6,** 1167–1178.

Leahy, D. J., Hendrickson, W. A., Aukhil, I., and Erickson, H. P. (1992). Structure of a fibronectin type III domain from tenascin phased by MAD analysis of the selenomethionyl protein. *Science* **258,** 987–991.

Lee, C. F., Makhatadze, G. I., and Wong, K. B. (2005). Effects of charge-to-alanine substitutions on the stability of ribosomal protein L30e from *Thermococcus celer. Biochemistry* **44,** 16817–16825.

Lee, K. K., Fitch, C. A., and Garcia-Moreno, E. B. (2002). Distance dependence and salt sensitivity of pairwise, coulombic interactions in a protein. *Protein Sci.* **11,** 1004–1016.

Lee, P. S., Chu, Z. T., and Warshel, A. (1993). Microscopic and semimicroscopic calculations of electrostatic energies in proteins by the POLARIS and ENZYMIX programs. *J. Comput. Chem.* **14,** 161–185.

Loladze, V. V., Ermolenko, D. N., and Makhatadze, G. I. (2001). Heat capacity changes upon burial of polar and nonpolar groups in proteins. *Protein Sci.* **10,** 1343–1352.

Loladze, V. V., Ermolenko, D. N., and Makhatadze, G. I. (2002). Thermodynamic consequences of burial of polar and non-polar amino acid residues in the protein interior. *J. Mol. Biol.* **320,** 343–357.

Loladze, V. V., Ibarra-Molero, B., Sanchez-Ruiz, J. M., and Makhatadze, G. I. (1999). Engineering a thermostable protein via optimization of charge-charge interactions on the protein surface. *Biochemistry* **38,** 16419–16423.

Loladze, V. V., and Makhatadze, G. I. (2002). Removal of surface charge-charge interactions from ubiquitin leaves the protein folded and very stable. *Protein Sci.* **11,** 174–177.

Loladze, V. V., and Makhatadze, G. I. (2005). Both helical propensity and side-chain hydrophobicity at a partially exposed site in alpha-helix contribute to the thermodynamic stability of ubiquitin. *Proteins* **58,** 1–6.

Lopez, M. M., and Makhatadze, G. I. (2000). Major cold shock proteins, CspA from *Escherichia coli* and CspB from *Bacillus subtilis*, interact differently with single-stranded DNA templates. *Biochim. Biophys. Acta* **1479,** 196–202.

Lopez, M. M., Yutani, K., and Makhatadze, G. I. (1999). Interactions of the major cold shock protein of *Bacillus subtilis* CspB with single-stranded DNA templates of different base composition. *J. Biol. Chem.* **274,** 33601–33608.

Lopez, M. M., Yutani, K., and Makhatadze, G. I. (2001). Interactions of the cold shock protein CspB from *Bacillus subtilis* with single-stranded DNA: Importance of the T base content and position within the template. *J. Biol. Chem.* **276,** 15511–15518.

Makhatadze, G. I. (2005). Thermodynamics of alpha-helix formation. *Adv. Protein Chem.* **72,** 199–226.

Makhatadze, G. I., Clore, G. M., Gronenborn, A. M., and Privalov, P. L. (1994). Thermodynamics of unfolding of the all beta-sheet protein interleukin-1 beta. *Biochemistry* **33,** 9327–9332.

Makhatadze, G. I., Kim, K. S., Woodward, C., and Privalov, P. L. (1993). Thermodynamics of BPTI folding. *Protein Sci.* **2,** 2028–2036.

Makhatadze, G. I., Loladze, V. V., Ermolenko, D. N., Chen, X., and Thomas, S. T. (2003). Contribution of surface salt bridges to protein stability: Guidelines for protein engineering. *J. Mol. Biol.* **327,** 1135–1148.

Makhatadze, G. I., Loladze, V. V., Gribenko, A. V., and Lopez, M. M. (2004). Mechanism of thermostabilization in a designed cold shock protein with optimized surface electrostatic interactions. *J. Mol. Biol.* **336,** 929–942.

Makhatadze, G. I., and Privalov, P. L. (1995). Energetics of protein structure. *Adv. Protein Chem.* **47,** 307–425.

Marqusee, S., and Sauer, R. T. (1994). Contributions of a hydrogen bond/salt bridge network to the stability of secondary and tertiary structure in lambda repressor. *Protein Sci.* **3,** 2217–2225.

Marshall, S. A., Morgan, C. S., and Mayo, S. L. (2002). Electrostatics significantly affect the stability of designed homeodomain variants. *J. Mol. Biol.* **316,** 189–199.

Marti-Renom, M. A., Stuart, A. C., Fiser, A., Sanchez, R., Melo, F., and Sali, A. (2000). Comparative protein structure modeling of genes and genomes. *Annu. Rev. Biophys. Biomol. Struct.* **29,** 291–325.

Matthew, J. B., and Gurd, F. R. (1986a). Calculation of electrostatic interactions in proteins. *Methods Enzymol.* **130,** 413–436.

Matthew, J. B., and Gurd, F. R. (1986b). Stabilization and destabilization of protein structure by charge interactions. *Methods Enzymol.* **130,** 437–453.

Matthew, J. B., Gurd, F. R., Garcia-Moreno, B., Flanagan, M. A., March, K. L., and Shire, S. J. (1985). pH-dependent processes in proteins. *CRC Crit. Rev. Biochem.* **18,** 91–197.

Matthews, B. W. (1995). Studies on protein stability with T4 lysozyme. *Adv. Protein Chem.* **46,** 249–278.

Max, K. E., Zeeb, M., Bienert, R., Balbach, J., and Heinemann, U. (2006). T-rich DNA single strands bind to a preformed site on the bacterial cold shock protein Bs-CspB. *J. Mol. Biol.* **360,** 702–714.

Mehler, E. L., and Guarnieri, F. (1999). A self-consistent, microenvironment modulated screened coulomb potential approximation to calculate pH-dependent electrostatic effects in proteins. *Biophys. J.* **77,** 3–22.

Meiering, E. M., Serrano, L., and Fersht, A. R. (1992). Effect of active site residues in barnase on activity and stability. *J. Mol. Biol.* **225,** 585–589.

Merkel, J. S., Sturtevant, J. M., and Regan, L. (1999). Sidechain interactions in parallel beta sheets: The energetics of cross-strand pairings. *Structure* **7,** 1333–1343.

Morawski, B., Quan, S., and Arnold, F. H. (2001). Functional expression and stabilization of horseradish peroxidase by directed evolution in *Saccharomyces cerevisiae*. *Biotechnol. Bioeng.* **76,** 99–107.

Motono, C., Gromiha, M. M., and Kumar, S. (2008). Thermodynamic and kinetic determinants of Thermotoga maritima cold shock protein stability: A structural and dynamic analysis. *Proteins* **71,** 655–669.

Musacchio, A., Saraste, M., and Wilmanns, M. (1994). High-resolution crystal structures of tyrosine kinase SH3 domains complexed with proline-rich peptides. *Nat. Struct. Biol.* **1,** 546–551.

Nagi, A. D., and Regan, L. (1997). An inverse correlation between loop length and stability in a four-helix-bundle protein. *Fold Des.* **2,** 67–75.

Nicholson, H., Anderson, D. E., Dao-pin, S., and Matthews, B. W. (1991). Analysis of the interaction between charged side chains and the alpha-helix dipole using designed thermostable mutants of phage T4 lysozyme. *Biochemistry* **30,** 9816–9828.

Oliveberg, M., Arcus, V. L., and Fersht, A. R. (1995). pKA values of carboxyl groups in the native and denatured states of barnase: The pKA values of the denatured state are on average 0.4 units lower than those of model compounds. *Biochemistry* **34,** 9424–9433.

Oubridge, C., Ito, N., Evans, P. R., Teo, C. H., and Nagai, K. (1994). Crystal structure at 1.92 A resolution of the RNA-binding domain of the U1A spliceosomal protein complexed with an RNA hairpin. *Nature* **372,** 432–438.

Pace, C. N. (1986). Determination and analysis of urea and guanidine hydrochloride denaturation curves. *Methods Enzymol.* **131,** 266–280.

Pace, C. N., Alston, R. W., and Shaw, K. L. (2000). Charge-charge interactions influence the denatured state ensemble and contribute to protein stability. *Protein Sci.* **9,** 1395–1398.

Pace, C. N., Laurents, D. V., and Thomson, J. A. (1990). pH dependence of the urea and guanidine hydrochloride denaturation of ribonuclease A and ribonuclease T1. *Biochemistry* **29,** 2564–2572.

Pace, C. N., Shirley, B. A., McNutt, M., and Gajiwala, K. (1996). Forces contributing to the conformational stability of proteins. *FASEB J.* **10,** 75–83.

Perl, D., Mueller, U., Heinemann, U., and Schmid, F. X. (2000). Two exposed amino acid residues confer thermostability on a cold shock protein. *Nat. Struct. Biol.* **7,** 380–383.

Perl, D., and Schmid, F. X. (2001). Electrostatic stabilization of a thermophilic cold shock protein. *J. Mol. Biol.* **313,** 343–357.

Perl, D., Welker, C., Schindler, T., Schroder, K., Marahiel, M. A., Jaenicke, R., and Schmid, F. X. (1998). Conservation of rapid two-state folding in mesophilic, thermophilic and hyperthermophilic cold shock proteins. *Nat. Struct. Biol.* **5,** 229–235.

Permyakov, S. E., Makhatadze, G. I., Owenius, R., Uversky, V. N., Brooks, C. L., Permyakov, E. A., and Berliner, L. J. (2005). How to improve nature: Study of the electrostatic properties of the surface of alpha-lactalbumin. *Protein Eng. Des. Sel.* **18,** 425–433.

Pokala, N., and Handel, T. M. (2001). Review: Protein design — where we were, where we are, where we're going. *J. Struct. Biol.* **134,** 269–281.

Predki, P. F., Agrawal, V., Brunger, A. T., and Regan, L. (1996). Amino-acid substitutions in a surface turn modulate protein stability. *Nat. Struct. Biol.* **3,** 54–58.

Privalov, P. L. (1990). Cold denaturation of proteins. *Crit. Rev. Biochem. Mol. Biol.* **25,** 281–305.

Ramponi, G., Treves, C., and Guerritore, A. (1967). Hydrolytic activity of muscle acyl phosphatase on 3-phosphoglyceryl phosphate. *Experientia* **23,** 1019–1020.

Ramponi, G., Treves, C., and Guerritore, A. A. (1966). Aromatic acyl phosphates as substrates of acyl phosphatase. *Arch. Biochem. Biophys.* **115,** 129–135.

Rees, D. C., and Robertson, A. D. (2001). Some thermodynamic implications for the thermostability of proteins. *Protein Sci.* **10,** 1187–1194.

Richmond, T. J. (1984). Solvent accessible surface area and excluded volume in proteins: Analytical equations for overlapping spheres and implications for the hydrophobic effect. *J. Mol. Biol.* **178,** 63–89.

Roca, M., Liu, H., Messer, B., and Warshel, A. (2007). On the relationship between thermal stability and catalytic power of enzymes. *Biochemistry* **46,** 15076–15088.

Rudolph, M. G., Wittinghofer, A., and Vetter, I. R. (1999). Nucleotide binding to the G12V-mutant of Cdc42 investigated by X-ray diffraction and fluorescence spectroscopy: Two different nucleotide states in one crystal. *Protein Sci.* **8,** 778–787.

Sali, D., Bycroft, M., and Fersht, A. R. (1991). Surface electrostatic interactions contribute little of stability of barnase. *J. Mol. Biol.* **220,** 779–788.

Sanchez-Ruiz, J. M., and Makhatadze, G. I. (2001). To charge or not to charge? *Trends Biotechnol.* **19,** 132–135.

Schutz, C. N., and Warshel, A. (2001). What are the dielectric "constants" of proteins and how to validate electrostatic models? *Proteins* **44,** 400–417.

Schweiker, K. L., Zarrine-Afsar, A., Davidson, A. R., and Makhatadze, G. I. (2007). Computational design of the Fyn SH3 domain with increased stability through optimization of surface charge charge interactions. *Protein Sci.* **16,** 2694–2702.

Serrano, L., Horovitz, A., Avron, B., Bycroft, M., and Fersht, A. R. (1990). Estimating the contribution of engineered surface electrostatic interactions to protein stability by using double-mutant cycles. *Biochemistry* **29,** 9343–9352.

Serrano, L., Kellis, J. T. Jr.,, Cann, P., Matouschek, A., and Fersht, A. R. (1992). The folding of an enzyme. II. Substructure of barnase and the contribution of different interactions to protein stability. *J. Mol. Biol.* **224,** 783–804.

Sham, Y. Y., Muegge, I., and Warshel, A. (1998). The effect of protein relaxation on charge-charge interactions and dielectric constants of proteins. *Biophys. J.* **74,** 1744–1753.

Shoichet, B. K., Baase, W. A., Kuroki, R., and Matthews, B. W. (1995). A relationship between protein stability and protein function. *Proc. Natl. Acad. Sci. USA* **92,** 452–456.

Spector, S., Wang, M., Carp, S. A., Robblee, J., Hendsch, Z. S., Fairman, R., Tidor, B., and Raleigh, D. P. (2000). Rational modification of protein stability by the mutation of charged surface residues. *Biochemistry* **39,** 872–879.

Street, A. G., and Mayo, S. L. (1999). Computational protein design. *Structure* **7,** R105–R109.

Strickler, S. S., Gribenko, A. V., Gribenko, A. V., Keiffer, T. R., Tomlinson, J., Reihle, T., Loladze, V. V., and Makhatadze, G. I. (2006). Protein stability and surface electrostatics: A charged relationship. *Biochemistry* **45,** 2761–2766.

Strop, P., and Mayo, S. L. (2000). Contribution of surface salt bridges to protein stability. *Biochemistry* **39,** 1251–1255.

Sun, D. P., Soderlind, E., Baase, W. A., Wozniak, J. A., Sauer, U., and Matthews, B. W. (1991). Cumulative site-directed charge-change replacements in bacteriophage T4 lysozyme suggest that long-range electrostatic interactions contribute little to protein stability. *J. Mol. Biol.* **221,** 873–887.

Swint-Kruse, L., and Robertson, A. D. (1995). Hydrogen bonds and the pH dependence of ovomucoid third domain stability. *Biochemistry* **34,** 4724–4732.

Tanford, C., and Kirkwood, J. G. (1957). Theory of protein titration curves. I. General equations for impenetrable spheres. *J. Am. Chem. Soc.* **79,** 5333–5339.

Thomas, S. T., Loladze, V. V., and Makhatadze, G. I. (2001). Hydration of the peptide backbone largely defines the thermodynamic propensity scale of residues at the C$'$ position of the C-capping box of alpha-helices. *Proc. Natl. Acad. Sci. USA* **98,** 10670–10675.

Thomas, S. T., and Makhatadze, G. I. (2000). Contribution of the 30/36 hydrophobic contact at the C-terminus of the alpha-helix to the stability of the ubiquitin molecule. *Biochemistry* **39,** 10275–10283.

Thunnissen, M. M., Taddei, N., Liguri, G., Ramponi, G., and Nordlund, P. (1997). Crystal structure of common type acylphosphatase from bovine testis. *Structure* **5,** 69–79.

Tissot, A. C., Vuilleumier, S., and Fersht, A. R. (1996). Importance of two buried salt bridges in the stability and folding pathway of barnase. *Biochemistry* **35,** 6786–6794.

Trefethen, J. M., Pace, C. N., Scholtz, J. M., and Brems, D. N. (2005). Charge-charge interactions in the denatured state influence the folding kinetics of ribonuclease Sa. *Protein Sci.* **14,** 1934–1938.

Varley, P. G., and Pain, R. H. (1991). Relation between stability, dynamics and enzyme activity in 3-phosphoglycerate kinases from yeast and *Thermus thermophilus*. *J. Mol. Biol.* **220,** 531–538.

Viguera, A. R., and Serrano, L. (1995). Experimental analysis of the Schellman motif. *J. Mol. Biol.* **251,** 150–160.

Vijay-Kumar, S., Bugg, C. E., and Cook, W. J. (1987). Structure of ubiquitin refined at 1.8 A resolution. *J. Mol. Biol.* **194,** 531–544.

Warshel, A. (2003). Computer simulations of enzyme catalysis: Methods, progress, and insights. *Annu. Rev. Biophys. Biomol. Struct.* **32,** 425–443.

Whitten, S. T., and Garcia-Moreno, E. B. (2000). pH dependence of stability of staphylococcal nuclease: Evidence of substantial electrostatic interactions in the denatured state. *Biochemistry* **39,** 14292–14304.

Wintrode, P. L., Makhatadze, G. I., and Privalov, P. L. (1994). Thermodynamics of ubiquitin unfolding. *Proteins* **18,** 246–253.

Wunderlich, M., Martin, A., and Schmid, F. X. (2005a). Stabilization of the cold shock protein CspB from *Bacillus subtilis* by evolutionary optimization of Coulombic interactions. *J. Mol. Biol.* **347,** 1063–1076.

Wunderlich, M., Martin, A., Staab, C. A., and Schmid, F. X. (2005b). Evolutionary protein stabilization in comparison with computational design. *J. Mol. Biol.* **351,** 1160–1168.

Wunderlich, M., and Schmid, F. X. (2006). *In vitro* evolution of a hyperstable Gbeta1 variant. *J. Mol. Biol.* **363,** 545–557.

Xiao, L., and Honig, B. (1999). Electrostatic contributions to the stability of hyperthermophilic proteins. *J. Mol. Biol.* **289,** 1435–1444.

Yang, A. S., Gunner, M. R., Sampogna, R., Sharp, K., and Honig, B. (1993). On the calculation of pKas in proteins. *Proteins* **15,** 252–265.

Yu, Y., Makhatadze, G. I., Pace, C. N., and Privalov, P. L. (1994). Energetics of ribonuclease T1 structure. *Biochemistry* **33,** 3312–3319.

Zhang, X. J., Baase, W. A., Shoichet, B. K., Wilson, K. P., and Matthews, B. W. (1995). Enhancement of protein stability by the combination of point mutations in T4 lysozyme is additive. *Protein Eng.* **8,** 1017–1022.

Zhou, H. X. (2002a). A Gaussian-chain model for treating residual charge-charge interactions in the unfolded state of proteins. *Proc. Natl. Acad. Sci. USA* **99,** 3569–3574.

Zhou, H. X. (2002b). Residual electrostatic effects in the unfolded state of the N-terminal domain of L9 can be attributed to nonspecific nonlocal charge-charge interactions. *Biochemistry* **41,** 6533–6538.

Zollars, E. S., Marshall, S. A., and Mayo, S. L. (2006). Simple electrostatic model improves designed protein sequences. *Protein Sci.* **15,** 2014–2018.

CHAPTER EIGHT

Efficient Computation of Confidence Intervals for Bayesian Model Predictions Based on Multidimensional Parameter Space

Amber D. Smith,* Alan Genz,[†] David M. Freiberger,* Gregory Belenky,* *and* Hans P. A. Van Dongen*

Contents

1. Introduction	214
2. Height of the Probability Density Function at the Boundary of the Smallest Multidimensional Confidence Region	215
3. Approximating a One-Dimensional Slice of the Probability Density Function by Means of Normal Curve Spline Pieces	217
4. Locating the Boundary of the Smallest Multidimensional Confidence Region	219
5. Finding the Minimum and Maximum of the Prediction Model over the Confidence Region	220
6. An Application: Bayesian Forecasting of Cognitive Performance Impairment during Sleep Deprivation	221
7. 95% Confidence Intervals for Bayesian Predictions of Cognitive Performance Impairment during Sleep Deprivation	223
8. Conclusion	227
Acknowledgments	229
Appendix	229
References	230

Abstract

A new algorithm is introduced to efficiently estimate confidence intervals for Bayesian model predictions based on multidimensional parameter space. The algorithm locates the boundary of the smallest confidence region in the multidimensional probability density function (pdf) for the model predictions by approximating a one-dimensional slice through the mode of the pdf with splines made of pieces of normal curve with continuous z values. This computationally

* Sleep and Performance Research Center, Washington State University, Spokane, Washington
[†] Department of Mathematics, Washington State University, Pullman, Washington

efficient process (of order N) reduces estimation of the lower and upper bounds of the confidence interval to a multidimensional constrained nonlinear optimization problem, which can be solved with standard numerical procedures (of order N^2 or less). Application of the new algorithm is illustrated with a five-dimensional example involving the computation of 95% confidence intervals for predictions made with a Bayesian forecasting model for cognitive performance deficits of sleep-deprived individuals.

1. INTRODUCTION

Bayesian methods have found widespread use in forecasting applications such as time series prediction. A broad class of Bayesian modeling and prediction problems can be formulated in terms of a given prediction model $P(\Theta)$, which is a function of an n-dimensional free parameter space $\Theta = \{\theta_1, \theta_2, \ldots, \theta_n\}$, where Θ is described by a continuous probability density function (pdf) denoted as $f(\Theta)$. In case the Bayesian model $P(\Theta)$ is derived from experimental data Y by means of maximum-likelihood regression, the pdf $f(\Theta)$ is typically composed of a likelihood function $L(Y,\Theta)$ and priors $p_k(\theta_k)$ for a subset of the parameters, as follows:

$$f(\Theta) = L(\Theta, Y) \prod_{k=1}^{n'} p_k(\theta_k),$$

where $n' \leq n$. The prediction model is often formulated such that the priors are each assumed to be normally distributed[1] with zero mean: $p_k(\theta_k) = p_N(\theta_k; 0, \omega_k^2)$, where the right-hand term indicates a univariate normal distribution over θ_k with mean 0 and standard deviation ω_k. Note that the narrower the prior distributions are, the more the overall pdf $f(\Theta)$ tends to resemble a multivariate normal distribution.

Finding the optimal parameter values Θ for the prediction model is relatively straightforward as it merely involves finding the mode of the pdf, which is done by maximizing $f(\Theta)$ with respect to the parameters Θ.[2] With the optimal parameter values Θ assumed to be in hand, Bayesian predictions $P(\Theta)$ can be made. However, deriving a confidence interval

---

[1] If a model parameter θ must remain strictly positive or strictly negative, the assumption of a normal distribution is commonly replaced by the assumption of a lognormal distribution. However, the model can then be reformulated in terms of a new parameter θ', which is distributed normally. For instance, consider a time series model of the form $P(\theta, t) = (\gamma + \theta)\sin(2\pi t - \varphi)$, where φ and γ (with $\gamma > 0$) are fixed parameters and t denotes time. To keep the amplitude $(\gamma + \theta)$ strictly positive, the model may be reformulated as $P(\theta, t) = \gamma e^{\theta'} \sin(2\pi t - \varphi)$, where θ' is normally distributed with zero mean.

[2] This is generally done by minimizing $-2 \log f(\Theta)$ over Θ, which yields the same result but is numerically easier to do.

(such as the frequently featured 95% confidence interval) for a prediction $P(\Theta)$ based on the uncertainty in the parameter values Θ as represented by $f(\Theta)$ is not so straightforward. It first involves finding the smallest region of the multidimensional parameter space Θ that captures the desired portion (e.g., 95%) of the (hyper)area under the pdf curve. It then involves finding the minimum and maximum taken on by $P(\Theta)$ over the identified smallest region to determine the lower and upper bounds of the confidence interval.

These two steps could be performed numerically through an extensive parameter grid search, but this process would be of computational order N^n, where N denotes the number of grid points in each dimension of the pdf and n is the number of parameters (dimensions) in Θ. For more than a handful of parameters, this makes the problem computationally burdensome. The objective of this chapter is to introduce a more efficient numerical algorithm to compute confidence intervals for Bayesian model predictions based on multidimensional parameter space.

2. HEIGHT OF THE PROBABILITY DENSITY FUNCTION AT THE BOUNDARY OF THE SMALLEST MULTIDIMENSIONAL CONFIDENCE REGION

Our first goal is to find the smallest region in the multidimensional parameter space Θ that captures the required percentage (e.g., 95%) of the (hyper)area under the pdf curve $f(\Theta)$. For instance, to find the 95% confidence interval of a Bayesian prediction based on a univariate pdf, this involves finding the smallest interval (or set of intervals) in the one-dimensional domain such that the area under the curve captured by that interval equals 95% of the total area under the curve (which, by definition, equals 1). To find the 95% confidence interval of a Bayesian prediction determined by a bivariate pdf, this involves finding the smallest area (or set of areas) in the two-dimensional domain such that the volume under the curve captured by that region is 95% of the total volume under the pdf curve (which, again, equals 1). And so on.

As illustrated in Fig. 8.1, it turns out that the boundary delineating the smallest confidence region in a multidimensional, continuous pdf always projects to a level (i.e., fixed-height) contour on the surface of the pdf (Box and Tao, 1992). A mathematical proof is given in the appendix and is illustrated with a one-dimensional example in Fig. 8.2. We shall see that knowing that the value (height) of $f(\Theta)$ on the boundary of the confidence region is a fixed constant is helpful in reducing the dimensionality of the computations needed to pinpoint the boundary.

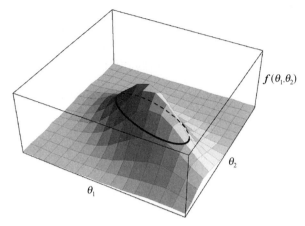

Figure 8.1 Illustration of the boundary of the confidence region of a two-dimensional pdf. For any multidimensional, continuous pdf, the projection of the boundary of the smallest region in the domain that is associated with a given probability (say, 95%) onto the surface of the pdf constitutes a level contour. In other words, the projected boundary (black curve) is of constant height.

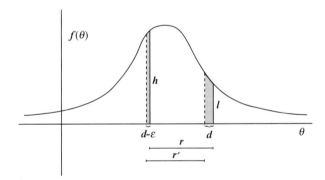

Figure 8.2 One-dimensional illustration of the proof that the projected boundary of the confidence region for a continuous pdf is a level contour. Suppose r is believed to be the smallest region in the domain that captures a given area under the curve. Let the boundary of r have unequal heights h and l. Then there exists an interval d inside the boundary, and an interval $d - \varepsilon$ outside the boundary (where $\varepsilon > 0$), such that the areas under the curve above these intervals are equal (gray areas). Thus, the region $r' = r - \varepsilon$ is smaller than r, but it captures the same area under the curve. Therefore, r is not actually the smallest region. This inconsistency can only be resolved by making $h = l$. It follows that the projection of the boundary of the smallest region onto the pdf curve must have identical height.

3. APPROXIMATING A ONE-DIMENSIONAL SLICE OF THE PROBABILITY DENSITY FUNCTION BY MEANS OF NORMAL CURVE SPLINE PIECES

The observation that the boundary of the smallest multidimensional confidence region projects to a level contour on the pdf curve—that is, the height of the pdf at the boundary is the same throughout—implies that, without loss of generality, an equation for the boundary can be derived from a one-dimensional slice of the parameter space through the mode of the pdf. The shape of this one-dimensional slice can be conveniently approximated by means of a series of spline pieces. This forms the basis for a new, efficient algorithm for locating the boundary.

Let $\theta = \theta_1$ be the parameter of choice over which the pdf is sliced, fixing the remaining parameters θ_k ($k > 1$) at the values corresponding to the mode of the pdf. Let the resulting one-dimensional slice be denoted as $g(\theta)$. It is assumed here that $g(\theta)$ is unimodal, differentiable, and monotonically decreasing from the mode [with $\lim_{\theta \to \infty} g(\theta) = 0$]. Let $\theta_{(0)}$ denote the location of the maximum of $g(\theta)$. This serves as a starting point for approximating $g(\theta)$ using splines with known statistical properties—namely, pieces of normal curve.

To construct the splines, we increase θ incrementally from $\theta_{(0)}$ in fixed, small intervals d, thereby constructing a series of knot points on the θ axis:

$$\vec{\theta} = \{\theta_{(0)}, \theta_{(1)}, \theta_{(2)}, \ldots, \theta_{(N)}\}$$
$$= \{\theta_{(0)}, \theta_{(0)} + d, \theta_{(0)} + 2d, \ldots, \theta_{(0)} + Nd\},$$

where N is the number of spline pieces.[3] Evaluating $g(\theta)$ at these knot points yields the series $\vec{g} = \{g_{(0)}, g_{(1)}, g_{(2)}, \ldots, g_{(N)}\} = \{g(\theta_{(0)}), g(\theta_{(1)}), g(\theta_{(2)}), \ldots, g(\theta_{(N)})\}$. We transform $\vec{\theta}$ by subtracting $\theta_{(0)}$ from each value, which yields $\vec{\theta}_T = \{\theta_{T(0)}, \theta_{T(1)}, \theta_{T(2)}, \ldots, \theta_{T(N)}\} = \{0, d, 2d, \ldots, Nd\}$. We also transform $\vec{g}$, dividing each value by $g_{(0)}$, which yields $\vec{g}_T = \{1, h_1, h_2, \ldots, h_N\}$.

The knots ($\vec{\theta}_T, \vec{g}_T$) describe the pdf slice $g(\theta)$ in standardized form so that the maximum occurs at coordinates (0,1). We approximate the standardized curve with a series of normal curve spline pieces, connected at their boundaries (i.e., at the knots) and, importantly, with continuous z values. Here z is the number of standard deviations away from the mean of the normal curve used in the spline piece at hand. This procedure is illustrated in Fig. 8.3.

---

[3] Equal spacing of the knot points is not required, but tends to facilitate the numerical process on a computer.

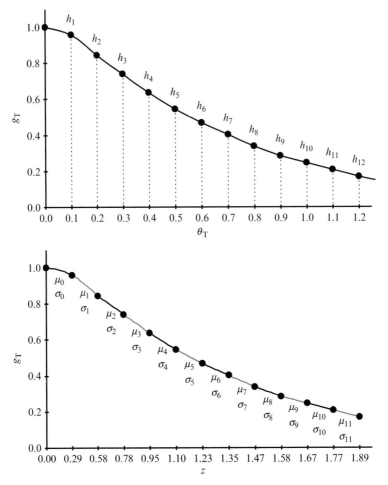

Figure 8.3 Construction of splines for the approximation of a one-dimensional pdf slice. (Top) The pdf slice after transformation to shifted θ_T and standardized g_T coordinates. Thus, the maximum occurs at coordinates (0, 1). Knots are defined at θ_T points $\{0, d, 2d, \ldots\}$, where in this example $d = 0.1$. At the knot points, the standardized pdf slice takes on the heights $\{1, h_1, h_2, \ldots\}$. (Bottom) Approximation of the pdf slice by normal curve spline pieces; the different spline pieces are alternately marked black and gray. Each normal curve spline piece is described by a different mean μ_j and standard deviation σ_j as indicated below the composite curve. The spline pieces are defined such that they have continuity in the z values (the number of standard deviations away from the mean of the spline piece at hand) across the knots. The z values at the knot points are indicated on the abscissa. Note that although they are continuous, they are not equidistant.

Although each normal curve spline piece may involve a different mean and a different standard deviation, between the knots where the spline piece is defined and approximates the standardized pdf slice, the incremental area under the curve captured by the spline piece is a close approximation of the incremental area under the curve captured by the pdf slice. As such, the z continuity requirement allows us to keep track of the (approximated) percentage of the area under the pdf slice captured between the maximum [at $\theta_{T(0)} = 0$] and z. The smaller the spline pieces, the better the agreement between the splines and the pdf slice. Thus, as is desirable for a numerical approximation algorithm, the accuracy of the procedure can be improved by decreasing the size of the spline pieces d and thereby increasing the number of spline pieces N.

Let the splines be labeled such that spline piece j is defined between the knots $(\theta_{T(j)}, g_{T(j)})$ and $(\theta_{T(j+1)}, g_{T(j+1)})$. Each of the normal curve spline pieces can be characterized by a mean μ_j and a standard deviation σ_j. For the first spline piece $j = 0$, which describes the pdf slice between the maximum $(0,1)$ and the subsequent knot (d, h_1), we define $z_0 = 0$, and require that $\mu_0 = 0$. To find σ_0, we solve $e^{\frac{-(d-\mu_0)^2}{2\sigma_0^2}} = h_1$, where the left-hand term is simply the normal distribution standardized to have a maximum of 1. It follows that $\sigma_0 = \frac{d}{\sqrt{-2 \ln(h_1)}}$. We may now determine the z value associated with the ending knot of the $j = 0$ spline piece: $z_1 = \frac{d-\mu_0}{\sigma_0}$.

In order to keep continuity in z, the ending knot of the spline piece for $j = 0$ determines the z value at the starting knot of the normal curve spline piece for $j = 1$. This constrains the mean and standard deviation μ_1 and σ_1 for the $j = 1$ spline piece: $z_1 = \frac{d-\mu_1}{\sigma_1}$. Furthermore, μ_1 and σ_1 are also determined by the ending knot of the $j = 1$ spline piece: $e^{\frac{-(2d-\mu_1)^2}{2\sigma_1^2}} = h_2$. Substitution of the previously computed z_1 yields $e^{-\left(z_1+\frac{d}{\sigma_1}\right)^2/2} = h_2$, where the only unknown is σ_1. It follows that $\sigma_1 = \frac{d}{\sqrt{-2 \ln(h_2)}-z_1}$. To find μ_1, we solve the constraint from the starting knot of the spline piece, $z_1 = \frac{d-\mu_1}{\sigma_1}$, which yields $\mu_1 = d - z_1\sigma_1$. We may now determine the z value associated with the ending knot of the spline piece for $j = 1$: $z_2 = \frac{2d-\mu_1}{\sigma_1}$. This process is iterated, where it is found that $\sigma_j = \frac{d}{\sqrt{-2 \ln(h_{j+1})}-z_j}$, $\mu_j = jd - z_j\sigma_j$, and $z_{j+1} = \frac{(j+1)d - \mu_j}{\sigma_j}$.

4. LOCATING THE BOUNDARY OF THE SMALLEST MULTIDIMENSIONAL CONFIDENCE REGION

If the original multidimensional pdf were a standard multivariate normal distribution over n parameters, then the location of the boundary of the smallest multidimensional confidence region capturing a given percentage (e.g., 95%) of the hyperarea under the curve would be known. The boundary can be shown to occur at the z value where a standard χ distribution with n degrees of freedom reaches the required percentage of its area under the curve (Evans et al., 1993). A practical way of obtaining this critical value $z_\star$ is to look up the critical value in a standard cumulative χ^2 distribution table for n degrees of freedom and take the square root of the value from the table.

Recall that we could locate the boundary of the smallest confidence region associated with the required percentage of hyperarea under the multidimensional pdf curve by considering a one-dimensional slice through the mode. Recall further that we approximated the one-dimensional slice with normal curve spline pieces for which we required continuity in the z values in order to keep track of the percentage of area under the pdf slice, as if the slice were a standard normal curve and the original multidimensional pdf were a standard multivariate normal distribution. Given these premises, to locate the boundary we must iteratively compute the splines until the spline piece is encountered that encompasses the critical value $z_\star$ for the χ_n distribution (as obtained by taking the square root of the critical value found in a χ_n^2 table).

Let this spline piece be number j, such that $z_j \leq z_\star < z_{j+1}$. The z values in spline piece number j are governed by the equation $z = \frac{\theta_T - \mu_j}{\sigma_j}$. Using this equation and reversing the previously made transformation of θ to θ_T, we convert $z_\star$ to the corresponding critical value for θ: $\theta_\star = \theta_{(0)} + \mu_j + \sigma_j z_\star$. It follows that in the original pdf slice, the boundary is located at $\theta_\star$, where it has a height $g(\theta_\star)$. This implies that the boundary of the smallest confidence region, which projects to a fixed height on the surface of the multidimensional pdf, is completely described by $f(\Theta) = g(\theta_\star)$.

5. Finding the Minimum and Maximum of the Prediction Model over the Confidence Region

The aforementioned equation for the boundary of the smallest multidimensional confidence region forms a constraint for finding the minimum and maximum that the prediction model $P(\Theta)$ can have over the confidence region, that is, the lower and upper bounds of the confidence interval for the prediction. In general, the minimum and maximum may occur either on or within the boundary of the confidence region so that the constraint condition is $f(\Theta) \geq g(\theta_\star)$. Finding the minimum and maximum for $P(\Theta)$ subject to this boundary constraint is a standard multidimensional constrained nonlinear optimization problem, which can be solved with the method of Lagrange multipliers (Nocedal and Wright, 1999). Readily available routines such as the "fmincon" procedure in MATLAB (The MathWorks, Inc.) can be used to find the solutions numerically. If the confidence region is contiguous (i.e., the boundary is a continuous curve), then the solutions are what we are looking for: the lower and upper bounds of the confidence interval for the prediction $P(\Theta)$ based on the uncertainty in the parameter values Θ as represented by $f(\Theta)$.

6. AN APPLICATION: BAYESIAN FORECASTING OF COGNITIVE PERFORMANCE IMPAIRMENT DURING SLEEP DEPRIVATION

To illustrate the use of the new algorithm introduced earlier, we consider a Bayesian forecasting model for the prediction of deficits of individuals in cognitive performance during periods of sleep deprivation (Van Dongen et al., 2007). The magnitude of such deficits is proportional to the pressure for sleep, which builds up in a saturating exponential manner across time spent awake and is modulated across the 24 h of the day by a "circadian" oscillation (Van Dongen and Dinges, 2005). Thus, the performance deficits vary as a function of time, but there are also large interindividual differences (Van Dongen et al., 2004). Information about the distribution of interindividual differences in the population is used in the form of Bayesian priors to improve the accuracy of predictions for the magnitude of performance deficits over time in a specific individual (Van Dongen et al., 2007). Our present goal is to compute 95% confidence intervals for performance predictions made with this Bayesian forecasting model.

For a given individual i, the prediction model for the level of performance impairment during a period of acute total sleep deprivation is as follows (from Van Dongen et al., 2007):

$$P_i(\xi_i, \varphi_i, \nu_i, \eta_i, \lambda_i, t) = \xi_i e^{-\rho e^{\nu_i}(t-t_0)}$$
$$+ \gamma e^{\eta_i} \sum_{q=1}^{5} a_q \sin(2q\pi(t - \varphi_i)/24) + \kappa + \lambda_i,$$

where t denotes time (in hours) and t_0 is the start time (i.e., time of awakening). This model contains a number of fixed parameters: ρ is the population-average buildup rate of sleep pressure across time awake; γ is the population-average amplitude of the circadian oscillation; κ determines the population-average basal performance capability; and the coefficients a_q reflect the relative amplitudes of harmonics of the circadian oscillation (see Borbély and Achermann, 1999). The five unknown

Table 8.1 Parameters in the Bayesian forecasting model of cognitive performance impairment during sleep deprivation (from Van Dongen et al., 2007)

Parameter	Value	Description
t_0	7.5	Time of awakening
ρ	0.0350	Population-average buildup rate of sleep pressure
γ	4.30	Population-average amplitude of circadian oscillation
κ	29.7	Population-average basal performance capability
ξ_i	Unknown	Initial sleep pressure for individual i
φ_i	Unknown	Temporal alignment of circadian oscillation for individual i
v_i	Unknown	Adjustment of buildup rate of sleep pressure for individual i
η_i	Unknown	Adjustment of amplitude of circadian oscillation for individual i
λ_i	Unknown	Adjustment of basal performance capability for individual i
ω_v^2	1.15	Variance of v_i across individuals in the population
ω_η^2	0.294	Variance of η_i across individuals in the population
ω_λ^2	36.2	Variance of λ_i across individuals in the population
σ^2	77.6	Error variance

parameters in this model are $\Theta = \{\xi_i, \varphi_i, v_i, \eta_i, \lambda_i\}$, where ξ_i represents the specific individual's initial sleep pressure from prior sleep loss, φ_i determines the temporal alignment of the individual's circadian oscillation, v_i adjusts the buildup rate of sleep pressure across time awake for the individual, η_i adjusts the amplitude of the circadian oscillation for the individual, and λ_i adjusts the basal performance capability of the individual at hand. Table 8.1 summarizes the key parameters in the model and shows their values if known a priori (from Van Dongen et al., 2007).[4]

In the absence of any information about the performance of individual i, a reasonable assumption would be that the person is well represented by the population average, which means $\xi_i = -28.0$ and $\varphi_i = 0.6$ (Van Dongen et al., 2007) and, as necessitated by the way the prediction model is formulated, $v_i = \eta_i = \lambda_i = 0$. However, if we have performance measurements y_ℓ at past time points t_ℓ (where $\ell = 1, 2, \ldots$) for the individual i, then better estimates for the five free parameters can be obtained by maximizing the likelihood function

---

[4] These values express cognitive performance in terms of the number of lapses (reaction times exceeding 500 ms) on a 10-min psychomotor vigilance task (Dorrian et al., 2005).

$$l_i(\xi_i, \varphi_i, \nu_i, \eta_i, \lambda_i) \propto \prod_{\ell=1}^{m} p_N\left(y_\ell; P_i(\xi_i, \varphi_i, \nu_i, \eta_i, \lambda_i, t_\ell), \sigma^2\right),$$

where σ^2 is the error variance (see Table 8.1).

Moreover, from a completed laboratory study (see Van Dongen and Dinges, 2005) we know the population variance describing the interindividual differences in ν_i, η_i, and λ_i. Thus, Bayesian priors can be formulated for these parameters: $p_\nu(\nu_i) = p_N(\nu_i; 0, \omega_\nu^2)$, $p_\eta(\eta_i) = p_N\left(\eta_i; 0, \omega_\eta^2\right)$, and $p_\lambda(\lambda_i) = p_N(\lambda_i; 0, \omega_\lambda^2)$.[5] Values for the variances ω_ν^2, ω_η^2, and ω_λ^2 are shown in Table 8.1. Using the priors for ν_i, η_i, and λ_i, we can now formulate $f(\Theta)$ for the Bayesian forecasting model of cognitive performance deficits during sleep deprivation:

$$f(\xi_i, \varphi_i, \nu_i, \eta_i, \lambda_i) = c\, l_i(\xi_i, \varphi_i, \nu_i, \eta_i, \lambda_i) p_\nu(\nu_i) p_\eta(\eta_i) p_\lambda(\lambda_i),$$

where c is a normalization constant.

For a given individual i, a set of 23 past performance measurements $\{(t_1, y_1), (t_2, y_2), \ldots, (t_{23}, y_{23})\}$ was obtained during 51.5 h of total sleep deprivation.[6] We use these 23 data points to define $f(\xi_i, \varphi_i, \nu_i, \eta_i, \lambda_i)$. The mode (i.e., maximum) of $f(\xi_i, \varphi_i, \nu_i, \eta_i, \lambda_i)$ is at $(\xi_i, \varphi_i, \nu_i, \eta_i, \lambda_i) = (-31.5, 5.99, 0.502, -0.181, 7.22)$.[7] This is the starting point for the application of our new algorithm for finding the boundary of the 95% confidence region.

7. 95% CONFIDENCE INTERVALS FOR BAYESIAN PREDICTIONS OF COGNITIVE PERFORMANCE IMPAIRMENT DURING SLEEP DEPRIVATION

One-dimensional slices through the mode of $f(\xi_i, \varphi_i, \nu_i, \eta_i, \lambda_i)$ along each of the parameter axes are shown in Fig. 8.4. The slices over ξ_i and λ_i are shaped exactly like normal curves, and the slices over ν_i and η_i are shaped approximately like normal curves. The slice over φ_i is different—the

---

[5] Note that across different individuals i, the term $\kappa + \lambda_i$ forms a normal distribution determining the basal performance level, but that the factors ρe^{ν_i} and γe^{η_i} in the prediction model constitute lognormal distributions for the buildup rate of sleep pressure and the amplitude of circadian oscillation, respectively.

[6] These data points are shown as closed circles in the bottom panel for Subject C in Figure 2 in Van Dongen et al. (2007).

[7] In Van Dongen et al. (2007), parameter estimates are reported that were derived from the marginal pdfs. For each parameter, this involved collapsing the pdf over the other four parameters by means of integration. This procedure yielded different results than would be obtained when using the location of the mode of the five-dimensional pdf to estimate the parameter values. As in this chapter, however, Van Dongen et al. (2007) did use the mode of the five-dimensional pdf for making performance predictions.

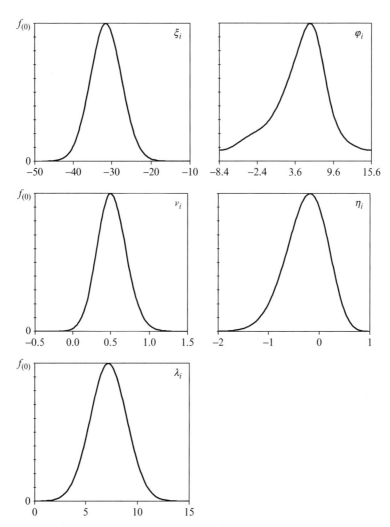

Figure 8.4 One-dimensional slices through the mode of $f(\xi_i, \varphi_i, v_i, \eta_i, \lambda_i)$ over each of the five free parameters, given a set of 23 past performance measurements (t_ℓ, y_ℓ) for the individual i. Slices are scaled to the maximum value of the pdf, labeled as $f_{(0)}$.

sinusoidal term in $P_i(\xi_i, \varphi_i, v_i, \eta_i, \lambda_i, t)$ causes oscillations in $f(\xi_i, \varphi_i, v_i, \eta_i, \lambda_i)$ across φ_i so that it does not actually meet the standard pdf criteria of $\lim_{\varphi_i \to -\infty} f_\varphi(\varphi_i) = 0$ and $\lim_{\varphi_i \to \infty} f_\varphi(\varphi_i) = 0$. However, with sufficient data points m contributing to the likelihood function $l_i(\xi_i, \varphi_i, v_i, \eta_i, \lambda_i)$, the slice over φ_i between two successive minima becomes unimodal and begins to resemble a normal curve as well (Fig. 8.4, upper right). Still, in the present case, the height of the slice over φ_i at the minima is approximately

0.079 times the height of the mode of $f(\xi_i, \varphi_i, \nu_i, \eta_i, \lambda_i)$. This is higher than what we expect to find for the height of the boundary of the 95% confidence region, and so using a slice over φ_i would not be a good choice for locating the boundary of the confidence region. Using one of the normal curve-shaped slices, over ξ_i or over λ_i, would be the easiest way to proceed, as our approximation of these slices would be exact with even a single spline piece. We will, however, illustrate the use of our spline-based algorithm by focusing on the more skewed left-hand (negative) tail of the slice $g(\eta_i)$ over η_i (Fig. 8.4, center right).

Let $\eta_{(0)}$ denote the location of the maximum of $g(\eta_i)$. To construct the splines, we decrease η_i incrementally from $\eta_{(0)}$ in intervals of $d = 0.1$, thereby constructing a series of knot points on the η_i axis: $\vec{\eta}_i = \{\eta_{(0)}, \eta_{(1)}, \eta_{(2)}, \ldots, \eta_{(N)}\} = \{\eta_{(0)}, \eta_{(0)} - d, \eta_{(0)} - 2d, \ldots, \eta_{(0)} - Nd\}$. Evaluating $g(\eta_i)$ at these knot points yields the series $\vec{g} = \{g_{(0)}, g_{(1)}, g_{(2)}, \ldots, g_{(N)}\} = \{g(\eta_{(0)}), g(\eta_{(1)}), g(\eta_{(2)}), \ldots, g(\eta_{(N)})\}$. We transform $\vec{\eta}_i$ by subtracting each value from $\eta_{(0)}$, which yields $\vec{\eta}_T = \{0, d, 2d, \ldots, Nd\}$. We also transform $\vec{g}$, dividing each value by the maximum $g_{(0)}$, which yields $\vec{g}_T = \{1, h_1, h_2, \ldots, h_N\}$.

Starting from the maximum at $\eta_{T(0)}$, we now approximate the (transformed) slice using normal curve splines with continuous z values. Figure 8.5 shows the characteristic z_j, μ_j, and σ_j of each of the spline pieces j. The critical value $z_\star$ for finding the 95% confidence region, derived from the 95% probability column of the cumulative χ^2 distribution table with 5 degrees of freedom, is $z_\star = \sqrt{11.0705} = 3.327$. This critical value occurs in the spline piece for $j = 16$ (see Fig. 8.5). It follows that the boundary to the left of the maximum of the slice is located at $\eta_\star = \eta_{(0)} - \mu_{16} - \sigma_{16} z_\star = -0.181 + 0.156 - 0.531 \cdot 3.327 = -1.791$. The corresponding height of the boundary is $g(\eta_\star) = 3.945 \cdot 10^{-3} g_{(0)}$. Reducing the spline knot interval to $d = 0.001$ yields negligible further precision.

We can check this result for the height of the boundary of the 95% confidence region using Monte Carlo simulation (Fishman, 1996). Figure 8.6 compares the boundary height results of the spline algorithm and the Monte Carlo algorithm as a function of N. There is good agreement between the estimates obtained with the two methods, even though the spline algorithm is computationally much more efficient.

Given a time t and a corresponding performance prediction $P_i(\xi_i, \varphi_i, \nu_i, \eta_i, \lambda_i, t)$, the lower and upper bounds of the 95% confidence interval for the performance prediction can be determined by finding the minimum or maximum of $P_i(\xi_i, \varphi_i, \nu_i, \eta_i, \lambda_i, t)$ subject to the boundary constraint $f(\xi_i, \varphi_i, \nu_i, \eta_i, \lambda_i) \geq g(\eta_\star)$. Note that $f(\xi_i, \varphi_i, \nu_i, \eta_i, \lambda_i)$ is not a function of t; therefore, the boundary constraint proper needs to be assessed

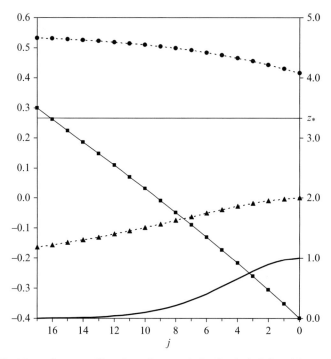

Figure 8.5 Normal curve spline piece characteristics for the left-hand (negative) tail of the η_i slice shown in Fig. 8.4. As a function of the spline piece knot number j (in reverse order), the graph shows the left-hand portion of the transformed slice (solid curve), the means (triangles) and standard deviations (circles) for the spline pieces, and the corresponding z values (boxes). Also shown is the level of the critical value $z_\star$ (thin line). The vertical scale for the dashed curves is on the left axis; for the solid curves it is on the right axis.

only once to determine the 95% confidence intervals for performance predictions at different times t [as long as the prior performance measurements continue to be the same set $\{(t_1, y_1), (t_2, y_2), \ldots, (t_{23}, y_{23})\}$ and no new performance measurements are added]. However, depending on the prediction time t the prediction model $P_i(\xi_i, \varphi_i, \nu_i, \eta_i, \lambda_i, t)$ may take on different minima and maxima over the 95% confidence region, and so the multidimensional constrained nonlinear optimization process to determine the minimum and maximum needs to be repeated for each different prediction time t.

We complete the example by computing performance predictions from the time of the last available measurement t_{23} (at 51.5 h awake) to 24 h later in steps of 1 h and estimating the lower and upper bounds of the 95% confidence interval for each of these predictions. The procedure "fmincon" in MATLAB is used to solve the multidimensional constrained nonlinear optimization problem for each of the predictions. The resulting confidence

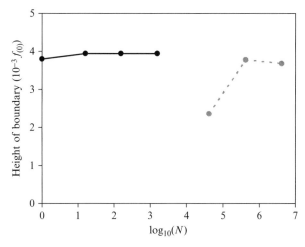

Figure 8.6 Height of the boundary of the 95% confidence region of $f(\xi_i, \varphi_i, v_i, \eta_i, \lambda_i)$. The estimated height of the boundary located with the new spline-based algorithm is shown as a function of the logarithm of the number of spline pieces N (black curve). The estimated height determined with Monte Carlo simulation is also shown as a function of the logarithm of the number of data points N (gray dashed curve). The boundary height is expressed as a fraction of the mode (maximum) of the pdf, labeled as $f_{(0)}$.

intervals around the predictions are shown in Fig. 8.7. Results confirm that the efficient algorithm outlined in this chapter yields accurate estimates of the confidence intervals.

8. Conclusion

This chapter introduced a two-step algorithm to efficiently estimate confidence intervals for Bayesian model predictions that can be formulated in terms of a model $P(\Theta)$ and an n-dimensional free parameter space Θ described by a continuous pdf $f(\Theta)$. We assume that the location of the mode (maximum) of $f(\Theta)$ is already known.

The first step of the new algorithm deals with finding the boundary of the smallest region of Θ that captures the desired percentage of the (hyper)area under the pdf curve. This boundary projects to a level (i.e., fixed-height) contour on the surface of the pdf. Therefore, the boundary can be located in a one-dimensional slice over one of the parameters in Θ. We assume that the slice is unimodal, differentiable, and monotonically decreasing from the mode. It can then be approximated with splines constructed using pieces of normal curve, connected at their boundaries and with continuous z values. Locating the boundary requires iteratively

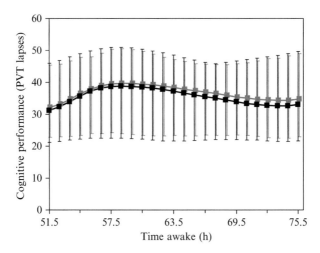

Figure 8.7 Estimated 95% confidence intervals for hourly predictions of cognitive performance during sleep deprivation. The predictions and confidence intervals computed with the new algorithm are shown in black.[4] For comparison, predictions and confidence intervals found previously by means of a parameter grid search are shown in gray (redrawn from the bottom panel for Subject C in Figure 2 in Van Dongen et al., 2007). The confidence intervals obtained with the new algorithm are slightly bigger because the estimation of the lower and upper bounds is not limited to discrete points in the domain as for the grid search algorithm. PVT, psychomotor vigilance task.

computing the splines until the spline piece is encountered that encompasses the critical value $z_\star$ for the required percentage of area under the curve, as determined for a χ_n distribution. Having located the boundary, we then determine the height of the boundary.

The second step of the new algorithm deals with assessing the minimum and maximum of $P(\Theta)$ over the confidence region for which the height of the boundary has been determined. If the confidence region is contiguous (i.e., the boundary is a continuous curve), then this is a standard optimization problem that can be solved with the method of Lagrange multipliers. This yields the estimated lower and upper bounds of the confidence interval for the Bayesian model prediction $P(\Theta)$ based on the uncertainty in the parameter values Θ as represented by $f(\Theta)$.

Our previously implemented method of estimating confidence intervals for Bayesian model predictions (Van Dongen et al., 2007) involved an n-dimensional parameter grid search, which was of computational order N^n and thus inefficient. For $n > 2$, our new algorithm has much greater computational efficiency. The spline procedure for finding the boundary of the smallest multidimensional confidence region operates in only one dimension and is a linear process; thus, this procedure is of order N. If a critical assumption (e.g., unimodality) is violated, the boundary can instead be found using Monte Carlo simulation, which is of order N^2 (Fishman,

1996) and thus still more efficient than a grid search (although Monte Carlo simulation requires much greater N to be reliable). Further, the optimization step to find the lower and upper bounds of $P(\Theta)$ over the confidence region is of order N^2 or less, depending on the algorithm used (Press et al., 2007). All in all, therefore, when compared to the parameter grid search method, the new algorithm described in this chapter constitutes a major improvement in computational efficiency (for $n > 2$).

For our five-dimensional example of the 95% confidence intervals for Bayesian predictions of cognitive performance impairment during sleep deprivation for a given individual, total computation times on a typical present-day desktop computer are in the order of a second for the spline-based algorithm, tens of seconds when using Monte Carlo simulation (up to $N = 10^6$), and in the order of a minute for performing a parameter grid search (as described in Van Dongen et al., 2007). Thus, we have successfully developed a considerably more efficient numerical algorithm for computing the confidence intervals of Bayesian model predictions based on multidimensional parameter space. Across a variety of suitable applications, this new algorithm will make it possible to provide confidence intervals for Bayesian model predictions in real time.

ACKNOWLEDGMENTS

We thank Christopher Mott, David Goering, Kevin Smith, Ron Gentle, Peter McCauley, Domien Beersma, David Terman, Rudy Gunawan, and David Dinges for their contributions. This work was funded by Air Force Office of Scientific Research Grant FA9550-06-1-0055, Defense University Research Instrumentation Program Grant FA9550-06-1-0281, and U.S. Army Medical Research and Materiel Command Award W81XWH-05-1-0099.

APPENDIX PROOF THAT THE BOUNDARY OF THE CONFIDENCE REGION FOR A MULTIDIMENSIONAL, CONTINUOUS PDF IS A LEVEL CONTOUR

Theorem: Consider a pdf $f(\Theta)$, where Θ is in $\mathbb{R}^n$, which is the surface of an $n + 1$-dimensional hypervolume. Then for a given hypervolume V' under the pdf curve, where $V' = V(A^n) = \int_{A^n} f(\Theta) d\Theta$, there exists a hyperarea domain $A^n \subseteq \mathbb{R}^n$ (where A^n is not necessarily unique), which is the smallest hyperarea associated with the hypervolume V'. The projection of the boundary of this smallest hyperarea A^n onto the surface $f(\Theta)$ is a level contour.

Proof: Suppose the projection of the boundary of the smallest domain A^n associated with the given volume V' is not a level contour. This implies that at least one maximum point and one minimum point exist on the boundary.

Furthermore, there exists a compact region $X_1 \subseteq A^n$ such that X_1 is connected to the boundary of A^n at the minimum point, where $f(\Theta)$ has a maximum height h_1. There also exists a compact region X_2 outside the interior of A^n such that X_2 is connected to the boundary of A^n at the maximum point, where $f(\Theta)$ has a minimum height h_2, with $h_2 > h_1$. Denoting the areas of X_1 and X_2 as $\alpha(X_1)$ and $\alpha(X_2)$, there is one of three possible relationships in size between $\alpha(X_1)$ and $\alpha(X_2)$.

If $\alpha(X_1) = \alpha(X_2)$, then the volumes over the regions X_1 and X_2 are such that $V(X_1) \leq h_1 \cdot \alpha(X_1) < h_2 \cdot \alpha(X_2) \leq V(X_2)$. We can adjust X_2 by removing a portion $\varepsilon \subseteq X_2$ (i.e., the adjusted area is $X_2 \backslash \varepsilon$), such that $h_1 \cdot \alpha(X_1) = h_2 \cdot \alpha(\hat{X}_2 \backslash \varepsilon)$. We may define a new domain $(A^n)' = (A^n \backslash X_1) \cup (X_2 \backslash \varepsilon)$, that is the original domain A^n with X_1 removed and $X_2 \backslash \varepsilon$ added, for which $V((A^n)') = V(A^n)$ but $\alpha((A^n)') < \alpha(A^n)$. Thus, we have found a smaller domain that captures the same volume V'.

If $\alpha(X_1) < \alpha(X_2)$, then there exists a smaller compact region $X_3 \subseteq X_2$ over which h_2 is a lower bound for the heights of that region, and X_3 is connected to the boundary of A^n such that $\alpha(X_1) = \alpha(X_3)$. This leads back to the first case.

If $\alpha(X_1) > \alpha(X_2)$, then there exists a smaller compact region $X_4 \subseteq X_1$ over which h_1 is an upper bound for the heights of that region, and X_4 is connected to the boundary of A^n such that $\alpha(X_4) = \alpha(X_2)$. This also leads back to the first case.

In all three cases, if the boundary is not a level contour, then the original domain A^n is not the smallest hyperarea domain associated with the hypervolume V', which contradicts our supposition. It follows that the projection of the boundary of the smallest hyperarea onto the surface $f(\Theta)$ must be a level contour.

REFERENCES

Borbély, A. A., and Achermann, P. (1999). Sleep homeostasis and models of sleep regulation. *J. Biol. Rhythms* **14,** 557–568.

Box, G. E. P., and Tiao, G. C. (1992). "Bayesian Inference in Statistical Analysis." Wiley-Interscience, New York.

Evans, M., Hastings, N., and Peacock, B. (1993). "Statistical Distributions," 2nd Ed. Wiley-Interscience, New York.

Fishman, G. S. (1996). "Monte Carlo: Concepts, Algorithms, and Applications." Springer-Verlag, New York.

Nocedal, J., and Wright, S. J. (1999). "Numerical Optimization." Springer-Verlag, New York.

Press, W. H., Teukolsky, S. A., Vetterling, W. T., and Flannery, B. P. (2007). "Numerical Recipes," 3rd Ed. Cambridge Univ. Press, New York.

Van Dongen, H. P. A., Baynard, M. D., Maislin, G., and Dinges, D. F. (2004). Systematic interindividual differences in neurobehavioral impairment from sleep loss: Evidence of trait-like differential vulnerability. *Sleep* **27,** 423–433.

Van Dongen, H. P. A., and Dinges, D. F. (2005). Sleep, circadian rhythms, and psychomotor vigilance. *Clin. Sports Med.* **24,** 237–249.

Van Dongen, H. P. A., Mott, C. G., Huang, J.-K., Mollicone, D. J., McKenzie, F. D., and Dinges, D. F. (2007). Optimization of biomathematical model predictions for cognitive performance impairment in individuals: Accounting for unknown traits and uncertain states in homeostatic and circadian processes. *Sleep* **30,** 1129–1143.

CHAPTER NINE

Analyzing Enzymatic pH Activity Profiles and Protein Titration Curves Using Structure-Based pK_a Calculations and Titration Curve Fitting

Jens Erik Nielsen

Contents

1. Introduction	234
2. Calculating the pH dependence of protein characteristics	235
2.1. Equations describing protein ionization reactions	236
2.2. Theory of protein pK_a calculation methods	237
2.3. Calculating electrostatic energies from protein structures	239
2.4. A note on structure relaxation	243
3. Setting up and Running a pK_a Calculation	243
3.1. Modeling protein conformational change	243
4. Analyzing the Results of a pK_a Calculation	244
4.1. Extracting pK_a values from calculated titration curves	244
4.2. Accuracy of protein pK_a calculation methods	245
5. How Reliable Are Calculated pK_a Values?	246
5.1. Sensitivity analysis	247
6. Predicting pH Activity Profiles	248
6.1. More than one CCPS	249
7. Decomposition Analysis	249
7.1. Redesigning protein pK_a values	250
8. Predicting Protein Stability Profiles	251
8.1. Predicting protein–ligand-binding pH profiles	252
9. Fitting pH Titration Curves, pH Activity Profiles, and pH Stability Profiles	252

School of Biomolecular and Biomedical Science, Centre Synthesis and Chemical Biology, UCD Conway Institute, University College Dublin, Belfield, Dublin, Ireland

Methods in Enzymology, Volume 454
ISSN 0076-6879, DOI: 10.1016/S0076-6879(08)03809-3

© 2009 Elsevier Inc.
All rights reserved.

9.1. Classic equations for fitting pH-dependent characteristics
of proteins 253
9.2. Statistical mechanical fitting (GloFTE) 255
10. Conclusion 255
References 256

Abstract

The pH dependence of protein biophysical characteristics is often analyzed to gain an improved understanding of protein stability, enzyme activity, and protein–ligand-binding processes. Indeed, much of our understanding of the catalytic mechanisms of enzymes derives from studies of the pH dependence of catalytic activity, and the ability to redesign the pH-dependent properties of enzymes continues to be of high relevance for both industrial and medical applications of proteins. This chapter discusses current theoretical methods for calculating protein pK_a values and illustrates how one can analyze protein pK_a calculation results to study calculation accuracy, pH stability profiles, and enzymatic pH activity profiles. A description of how one can analyze the importance of individual titratable groups is presented along with details on methods for redesigning protein pK_a values and enzymatic pH activity profiles. Finally, I discuss novel methods for fitting experimental nuclear magnetic resonance titration curves and enzymatic pH activity profiles that can be used to derive information on electrostatic interaction energies in proteins.

1. INTRODUCTION

An important question that an experimental biologist or biochemist must ask him or herself when planning an experiment is "Which buffer do I use?" Tight control of pH remains one of the most important features of any biological experiment, and a poor choice of buffer or unexpected fluctuation in pH is probably one of the most frequent reasons that experimental results are not reproducible.

So why is control of pH important when dealing with biological systems? The primary reasons are the omnipresence of water in biological systems, the high number of titratable groups (groups that can titrate, i.e., change protonation state with pH) in biomolecules, and the consequential importance of electrostatic effects for biomolecular stability. pH effects and electrostatic effects have been studied extensively using theoretical methods since the pioneering work of Linderstrøm-Lang (1924) [for a review, see Warshel *et al.* (2006)], and numerous theoretical and experimental studies are being carried out on projects ranging from studies on the importance of protein electrostatics for protein stability (Baldwin, 2007; Strickler *et al.*, 2006) to studies of electrostatic effects in membrane proteins (Cymes and Grosman, 2008; Pisliakov *et al.*, 2008).

pH is important because it changes the properties of the biomolecules that constitute a biological system. Specifically, a change in pH will cause titratable groups to lose or gain protons, thus altering the charge of the titratable group and the net charge on the protein, often with significant effects on the function of the biomolecule. The classic example of regulation of function using pH is hemoglobin, where a change in pH leads to altered affinity for oxygen, thus ensuring efficient delivery of O_2 to tissues that produce high levels of CO_2. The pH dependence of the catalytic activity of many enzymes, such as trypsin, chymotrypsin, and lysozyme, are other well-known examples of dramatic pH-dependent effects in biomolecules.

This chapter discusses how current theoretical methods can be used to study the pH-dependent properties of biomolecules. The methods described are, in principle, applicable to all types of biomolecules, but research in the field has focused almost exclusively on the titratable properties of proteins. The theoretical methods discussed here have therefore been tested and validated extensively only on protein molecules.

The chapter starts by describing the theory behind protein pK_a calculations, continues by discussing how to perform and analyze a protein pK_a calculation, and concludes by discussing how to predict and analyze the pH dependence of protein biophysical characteristics.

2. CALCULATING THE pH DEPENDENCE OF PROTEIN CHARACTERISTICS

Before embarking on a detailed account of how to perform a pK_a calculation and analyze the results, it is worth reviewing the procedures available to study the pH dependence of protein characteristics. Figure 9.1 displays an information flowchart where the individual procedures available are shown as boxes. Thus a researcher performing a study of some pH-dependent characteristic will start by identifying a protein structure appropriate for the desired conditions and then proceed by performing a pK_a calculation on that protein structure. Once the pK_a calculation has been completed, one can proceed with validation tests, prediction of pH activity profiles, pH stability profiles, and pK_a value redesign. However, the accuracy of the predictions and analyses following a pK_a calculation will always depend on data produced in the pK_a calculations: the intrinsic pK_a values and the site–site interaction energy matrix. These calculated energies are highly dependent on the choice of protein structure, force field, pK_a calculation parameters, and procedures used in the pK_a calculation method itself, thus making it important to correctly choose, prepare, and set up the structures used with the pK_a calculation software.

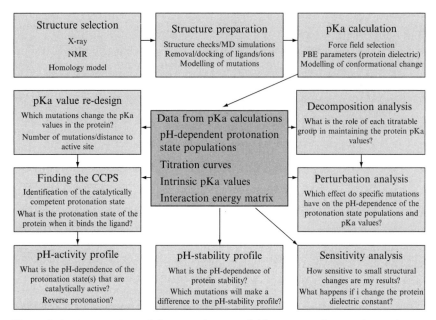

Figure 9.1 An overview of the procedures available to researchers studying the pH dependence of protein biophysical characteristics. Arrows indicate dependencies between the procedures, that is, one has to perform a pK_a calculation before one can predict an enzymatic pH activity profile.

It is furthermore important to familiarize oneself with the limitations of the pK_a calculation method used. This chapter discusses a pK_a calculation methodology that models changes in the hydrogen bond network explicitly but all other protein dynamics implicitly (via a higher protein dielectric constant). This method will do well on relatively rigid systems, but will fail for protein systems with large conformational transitions. Therefore, with protein pK_a calculations, as with all other structure-based energy calculation methods, it is important to know the characteristics of each method well and to apply common sense when analyzing the results. If something seems implausible, then it most likely is!

2.1. Equations describing protein ionization reactions

The pH-dependent ionization of a titratable group is classically described using the Henderson–Hasselbalch equation:

$$pH = pK_a + \log \frac{[A^-]}{[HA]} \quad (9.1)$$

where

$$pH = -\log[H_3O^+] \qquad (9.2)$$

$$pK_a = -\log K_a \qquad (9.3)$$

$$K_a = \frac{[H_3O^+][A^-]}{[HA]} \qquad (9.4)$$

and [HA] and [A$^-$] are the concentrations of the acid and corresponding base, respectively. Since a proton is lost in the conversion from HA to A$^-$, the charge on the titratable group changes during the process. In proteins we define bases as titratable groups that convert from positive to neutral and acids as titratable groups that exist in neutral and negative form.

If we can calculate the energy difference between the neutral and the charged form of a titratable group then we can, using some kind of sampling or simulation technique, calculate its pK_a value. Most present-day protein pK_a calculation methods do exactly that, although the details of the computational procedures used are slightly more complicated. The following gives a brief account of the theory behind protein pK_a calculation methods based on continuum methods and discusses the uses of protein pK_a calculation methods while paying particular attention to the challenges encountered when describing the protein structures.

2.2. Theory of protein pK_a calculation methods

A protein consisting of N titratable group can generally exist in 2^N distinct protonation states, as each titratable group can occupy two states (a charged and a neutral state). Each of the 2^N protonation states can be associated with a specific free energy, and once the pH dependence of these energies is known, it is possible to evaluate the partition function

$$p_i = \frac{e^{-\frac{E_i}{kT}}}{\sum_{j=1}^{2^N} e^{-\frac{E_j}{kT}}} \qquad (9.5)$$

at each pH value of interest. In this way we obtain a relative population for each protonation state as a function of pH. If the number of titratable groups is significantly larger than 30, it is not possible to evaluate the partition function [Eq. (9.5)] explicitly; in these cases a Monte Carlo sampling procedure can be used successfully to obtain the relative population of protonation states (Beroza et al., 1991). Other methods (Myers et al., 2006;

Yang et al., 1993) have been shown to be able to solve this problem by evaluating the partition function explicitly only for strongly interacting clusters of titratable groups while treating weak interactions between clusters with more approximate methods (Tanford and Roxby, 1972).

Figure 9.2 (top) shows a plot of the pH dependence of the population of the protonation states for a hypothetical three-group system obtained by evaluation of the partition function using the pKaTool program (Nielsen, 2006a,b). To construct titration curves for each titratable group from such a plot we now simply add the population of the protonation states that have a specific group in its charged state. For example, if we have eight (2^3) protonation states and states 1, 3, 4, and 7 constitute the states where a specific Asp is charged, then the fractional charge of that Asp at a given pH value is the sum of the populations of these protonation states at that pH. This calculation is repeated for each pH value of interest, and results are plotted as a function of pH to produce titration curves like those shown in Fig. 9.2 (bottom).

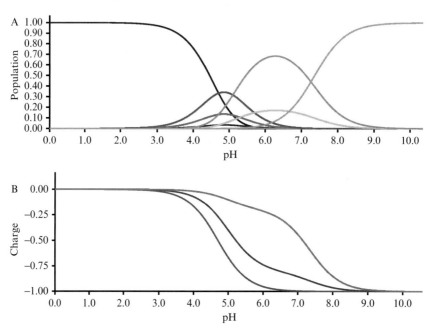

Figure 9.2 pH dependence of the relative populations of the eight possible protonation states for a system consisting of three titratable acids: I (black titration curve) intrinsic pK_a 4.7, II (blue titration curve) intrinsic pK_a value 5.1, and III (red titration curve) intrinsic pK_a value 5.7. II and III interact with an energy of 3.6 kT. All other pair wise interaction energies are zero. (Top) pH dependence of population of protonation states. The first set of bell-shaped curves corresponds to one deprotonated group, whereas the next set of bell-shaped curves corresponds to the protonation states with two groups deprotonated. The single curves at high and low pH correspond to protonation states with all groups protonated (low pH) and all groups deprotonated (high pH). (Bottom) titration curves. (See Color Insert.)

Titration curves describing a simple two-state titration can be fitted using the Henderson–Hasselbalch equation [Eq. (9.1)] to extract a pK_a value for the group in question. However, more complicated titration curves, such as the two titration curves displaying the highest pK_a values in Fig. 9.2 (bottom), obviously cannot be described by the Henderson–Hasselbalch equation; indeed, such titration curves represent both problems and sources of extra information, as discussed later.

We now turn our attention to the methods for calculating the pH dependence of the energy for each protonation state. Often it is assumed that the energy of a protonation state can be described using Eq.(9.6):

$$E_x = \sum_{i=1}^{N} \gamma_i \ln(10) kT(pH - pK_{a, \text{int}, i}) + \sum_{i=1}^{N} \sum_{j=1}^{N} \gamma_i \gamma_j E(i,j) \quad (9.6)$$

where the first term describes the ionization of a single titratable group in the absence of all other titratable groups. This artificial pK_a value (p$K_{a, \text{int}}$) is called the "intrinsic pK_a value" and is assumed to be pH independent. Interactions between titratable groups are modeled by the second term, and it is assumed that the energy of any protonation state of the protein can be modeled by adding pair wise interaction energies between two titratable groups.

2.3. Calculating electrostatic energies from protein structures

Equation (9.6) relates intrinsic pK_a values and pair wise interaction energies to the energy of a specific protonation state at a particular pH; this knowledge allows us to calculate titration curves from which we can extract pK_a values. However, we have yet to address the issue of how to calculate the intrinsic pK_a values and pair wise interaction energies from a protein structure. This is the crucial step that allows the calculation of titration curves using three-dimensional structural information.

Electrostatic interaction energies can be calculated from protein structures using a variety of methods. The following discussion focuses on methods that model the solvent as a continuum and, in particular, discusses the use of Poisson–Boltzmann equation (PBE)-based methods. PBE solvers for biological macromolecules are readily available (Baker et al., 2001; Madura et al., 1995; Nicholls and Honig, 1991) and arguably provide for the most convenient way of calculating the electrostatic potential in and around a macromolecule. Using a series of PBE calculation schemes [described in Yang et al. (1993)], it is possible to calculate desolvation energies, background interaction energies, and pair wise interaction energies between titratable groups (from here on referred to as "site–site" interaction energies). Desolvation energies describe the energy penalty associated with transferring a charged residue from a solvent to its position in the protein, background interaction energies describe the interaction with permanent protein dipoles, and site–site interaction energies describe the electrostatic interaction between the charges on titratable groups.

The input for a PBE solver consists of a protein structure expressed as a set of point charges, each associated with a radius, and a set of calculation parameters, including the infamous protein dielectric constant (see later).

It turns out that the PBE parameters and the preparation of a protein structure have a rather large effect on the electrostatic energies from the PBE run, and therefore also have a large effect on the calculated pK_a values. The following describes the current "best practice" for performing pK_a calculations with the WHAT IF pK_a calculation routines. The set of rules are derived mostly from studies with the WHAT IF pK_a calculation routines (Nielsen and Vriend, 2001), but should apply generally to most pK_a calculation programs.

The structure preparation process consists of the following three steps.

1. Choosing a set of protein structures.
2. Examining the protein structures.
3. Adapting the protein structures to the experimental conditions.

2.3.1. Choosing a set of protein structures

In general, pK_a calculation programs perform best when they are used with the type of protein structures that they were benchmarked on. Because pK_a calculation programs typically are benchmarked using X-ray crystal structures, it is no surprise that calculations on nuclear magnetic resonance (NMR) structures are more inaccurate than calculations on X-ray structures. The high importance of desolvation energies in determining protein pK_a values means that the details of atom packing are of particular importance. Because atoms are not observed directly in NMR data, atom packing in NMR structures generally is of lower quality than in X-ray structures. The first rule of pK_a calculations is therefore: do not use NMR structures if you can avoid it.

The same reasoning goes for homology models/comparative models of proteins; although no detailed studies have been carried out on the accuracy of pK_a values calculated from homology models, it is fair to assume that the accuracy is going to be much lower than those calculated from real X-ray structures.

2.3.2. Examining the Protein Structures

Once a suitable set of crystal structures has been identified, it is time to examine these to establish which structures will be most suitable for the calculation. Common structure descriptors such as resolution and protein structure check scores do not appear to correlate significantly with the quality of the pK_a calculation results (Nielsen and McCammon, 2003), but it is clear that very low resolution structures should be avoided. Furthermore, because protein crystallographers often adjust buffer conditions, pH, and temperature to grow crystals, it is important to examine which conditions each X-ray structure has been produced under. Often, specific buffer components will be present in the X-ray structure and it is important to (1) establish which structural rearrangements the buffer component has caused and (2) to remove the molecule before starting the pK_a calculation. Similarly, it is important to examine the protein structure for completeness, as some atoms can be missing from the X-ray structure because no electron density was observed for them. If only a few heavy atoms are missing, then these can generally be reconstructed using tools such as PDB2PQR (Dolinsky et al., 2004, 2007) or WHAT IF (Vriend, 1990), but if larger stretches of amino acids are missing, then it is best to avoid using the structure. If the use of the structure is unavoidable, then make sure that the missing set of residues is replaced by a stretch of alanine residues that make few contacts with the protein. The insertion of poly alanine ensures that the protein does not contain an extra set of termini, while at the same time avoiding the difficult process of modeling a missing loop onto the structure. Once the pK_a calculation has been competed for the poly-ala structure, one can examine the impact of modeling a loop with the correct sequence and, in this way, estimate the pK_a changes that the missing loop is able to induce.

2.3.2.1. pH effects The X-ray structure should furthermore be examined for possible pH-induced effects; in general, one should choose the X-ray structure solved at a pH value closest to the pK_a value that one is interested in. This is important, as any pK_a calculation in general, will calculate the fractional charges on the protein most accurately at the pH of the crystal. If the protein is known to undergo pH-induced conformational transitions, then it is important to realize that the pK_a calculations will be valid only for the conformation represented in the X-ray structure. The pK_a calculation package will most likely not be able to predict the structural transition correctly, although molecular dynamics-based pK_a calculation techniques are attempting to address this question (Baptista et al., 2002; Khandogin and Brooks, 2005; Mongan et al., 2004).

2.3.2.2. Crystal contacts For protein X-ray structures it is well known that the crystal environment influences the structure for a subset of surface residues; in a systematic study of calculated pK_a values on HEWL X-ray structures, it was shown that crystal contacts can have quite dramatic effects on the calculated pK_a values (Nielsen and McCammon, 2003). Specifically, it was found that crystal contacts affecting the structure of Asp 48 in HEWL influence the hydrogen bond network in such a way that the pK_a value of Glu 35 (16 Å from Asp 48) is calculated with an error of up to 2 units. In some cases the error in the calculated pK_a value is so large that the identity of the HEWL proton donor is predicted incorrectly.

2.3.2.3. Water molecules Most protein X-ray structures include a number of water molecules, and it is often questioned whether such water molecules should be retained in the PDB structure or simply deleted. There are good arguments for retaining water molecules that are tightly bound to the protein, but the overriding argument is that most pK_a calculation methods have been benchmarked on structures where all water molecules have been removed. This approach has been used because it is the "cleanest" way to treat water molecules, as it avoids complicated arguments on which water molecules to retain.

2.3.3. Adapting the protein structure to the experimental conditions

Before starting the pK_a calculation process itself, it is worth checking if there is anything that can be done to make the X-ray structure resemble the experimental conditions more. It is best to avoid extensive modeling exercises, and it is essential to always check the impact of your modeling steps on the final pK_a values. However, some steps can be completed without introducing significant errors in the structure, and some steps must be completed to get realistic results. In summary, for the WHAT IF pK_a calculation package, one must

1. Remove all water molecules.
2. Remove all ions that are crystallization artifacts.
3. Identify regions with many crystal contacts and correct these if possible.
4. Model any missing atoms as discussed earlier.
5. Mutate residues that are different between the protein wild type sequence and the experimental X-ray structure.

It is always important to check how large changes in the pK_a values these modeling steps introduce. If a specific step introduces unexpectedly large changes in the pK_a values, then one must assess critically whether the current X-ray structure is the best choice for the current experiment by manually examining the structural differences introduced.

2.4. A note on structure relaxation

When making changes to a protein structure it is often customary to allow the structure to adapt to the changes by performing a short molecular dynamics (MD) simulation and/or an energy minimization. In this context it is worth remembering that current molecular dynamics simulation methods almost invariably produce structures that are less accurate than the starting X-ray structure. One should therefore not resort to a full energy minimization or MD run when having to make one or more changes to an X-ray structure. Each modification should be examined, and any structure clashes should be resolved locally by rotamer optimization of the mutated residue or possibly by optimizing the rotamers for a small subset of residues close to the mutated residue. In general, the fewer changes to the atom coordinates one makes, the more accurate the structure will be. In summary: if optimization is needed, keep it local and minimal.

3. Setting up and Running a pK_a Calculation

Once a protein structure has been identified and prepared, it is time to run the pK_a calculation. Most pK_a calculation packages will perform a pK_a calculation without any manual intervention and deliver a set of files containing titration curves and pK_a values. There are thus few choices that a researcher has to make in this step except for choosing which pK_a calculation package to use. There has been much debate on which value or the protein dielectric constant ($\varepsilon_{protein}$) to use with PBE-based pK_a calculations, and for the nonexpert it is often hard to decide which $\varepsilon_{protein}$ to use. In general, one should use the $\varepsilon_{protein}$ that was used when the pK_a calculation method was benchmarked. It is important to realize that the absolute magnitude of the $\varepsilon_{protein}$ used has little physical meaning and merely reflects how each pK_a calculation package models protein conformational change. The more explicit a package models protein conformational change, the lower an $\varepsilon_{protein}$ (Sham et al., 1998) that package will tend to use.

3.1. Modeling protein conformational change

While colorful pictures of X-ray structures tend to give the impression that proteins are structures, it is well known that protein structures are highly dynamic and can occupy a large number of conformations under a given set of conditions. pH affects the structure of a protein, and it is therefore relevant to ask how one can calculate pK_a values using a static protein structure.

Early PBE-based methods for calculating pK_a values simply assumed that there was no structural change accompanying protonation/deprotonation events. These methods modeled the effect of pH simply by changing the charge on titratable groups and calculating the relevant electrostatic energies as discussed earlier.

This kind of modeling, while simple, clearly does not capture the structural changes that accompany pH changes and spurred the development of methods that model the change in protein structure by optimizing the hydrogen bond network for each protonation state used with the PBE solver in the pK_a calculations (Alexov and Gunner, 1997; Nielsen and Vriend, 2001). Such methods allow the protein to respond to changes in the protonation state by rearranging dipoles to minimize the energy of each protein structure. Allowing for relaxation of the hydrogen bond network generally gives improved results, especially for buried sites, as the only relaxation possible for buried titratable sites typically is the movement of hydrogen bond-mediated dipoles.

For surface residues, the improvement obtained by hydrogen bond optimization is less convincing, and in some cases the agreement with experimental data becomes significantly worse. This phenomenon presents itself because surface residues generally change their conformation more than buried sites when a deprotonation reaction occurs. Whereas buried sites are "locked in" by the protein interior and the only change in environment comes from reorientation of hydrogen bond dipoles, surface residues can move more freely and even change rotamer states. This knowledge was exploited (Alexov and Gunner, 1999; Georgescu et al., 2002) to construct an algorithm that includes the optimization of rotamer states for each protonation state.

4. Analyzing the Results of a pK_A Calculation

Completion of a pK_a calculation marks the end of the first stage of our protein analysis. We now have calculated pK_a values that have been derived from calculated titration curves and are ready to start the next stage of our pK_a calculation analysis: an assessment of the reliability of the calculated pK_a values.

4.1. Extracting pK_a values from calculated titration curves

In a standard pK_a calculation program, a protein structure, a set of charges and radii, and a set of calculation parameters are converted into titration curves. The final step in such a pK_a calculation procedure is to extract pK_a values from the titration curves. This extraction can be done in a number ways, with most methods assuming the pK_a value of a titration curve to be the $pK½$ value. The $pK½$ value is the lowest pH value at which the group is half-protonated.

The practice of using pK½ values is, of course, something that would never be accepted when measuring pK_a values experimentally. However, this method has been adopted in many pK_a calculation methods because the distinct non-Henderson–Hasselbalch (non-HH) titration curves observed with standard pK_a calculation methods prevent the extraction of classic pK_a values using the HH equation. The rationale for using pK½ values is that it is better to have a pK_a value than an oddly shaped titration curve that cannot be compared to experimental pK_a values.

Although some pK_a calculation packages report both pK½ values and pK_a values obtained from fits to the HH equation (Demchuk and Wade, 1996; Nielsen, 2006a), it is largely up to the user of a pK_a calculation package to check the titration curves manually. This task can be made easier using graphical interfaces such as pKaTool (Nielsen, 2006a,b) or Web-based solutions such as the H++ server (Gordon et al., 2005). In summary, when having completed a pK_a calculation, one must check the titration curves for all residues that one is interested in and identify titration curves that cannot be described by a single pK_a value. In subsequent analyses and conclusions, these pK_a values should be discounted and conclusions for those residues must be based on a direct analysis the titration curves.

Furthermore, some pK_a calculation packages (Li et al., 2005) calculate pK_a values directly without calculating titration curves. These methods thus circumvent the problem of extracting pK_a values from titration curves, but consequently suffer from the inability to calculate titration curves for tightly coupled systems.

It should be mentioned that calculated non-HH titration curves have found use in detecting functional sites in proteins (Ondrechen et al., 2001) because they are produced by strong electrostatic interaction energies, which are normally found in enzyme active sites. In a related study it was found that unfavorable electrostatic interaction energies also can be used to detect protein functional sites (Elcock, 2001).

4.2. Accuracy of protein pK_a calculation methods

How accurate can one expect a calculated pK_a value to be? The answer to that question depends on which structure and pK_a calculation method you are using and which residues you are interested in. Most pK_a calculation packages have been optimized to give the lowest overall root mean square deviation for a set of experimentally measured pK_a values such as those found in the PPD database (Toseland et al., 2006), but there is a big difference in how well different packages calculate pK_a values for surface and buried residues.

Generally, the more physically realistic a pK_a calculation method is, the better it will be at accurately calculating pK_a values for residues in unusual environments. However, such a pK_a calculation program will also be more sensitive to structural errors, will require more computational resources, and will typically perform worse for large sets of pK_a values. More coarse-grained and fast methods, however, tend to perform slightly better for the average titratable group at the expense of getting pK_a values wrong for residues in unusual environments.

Most pK_a-related research is directed at studying unusual parts of proteins, such as functional sites such as binding clefts and active sites; before embarking on a pK_a calculation project it is therefore essential to investigate if the pK_a calculation package being used is known to perform well for the environment that you are interested in. Typically a pK_a calculation package can be expected to calculate pK_a values correctly within 1.0 units, but the performance of individual pK_a calculation programs can vary considerably from this average value. Even when you have chosen the appropriate pK_a calculation package you must still perform a number of control calculations to assess the reliability of your results as explained in the next section.

5. How Reliable Are Calculated pK_A Values?

We have now calculated a set of pK_a values for our protein of interest and are interested in concluding something about our protein from the results. Let us assume that we are interested in identifying a general acid in the active site of enzyme X. A general acid is often an Asp or a Glu with an elevated pK_a value, and it is hoped that we can find an acidic residue that has a high calculated pK_a value.

Our hypothetical calculation does indeed show an Asp with a calculated pK_a value of 6.5, and the question is now how confident we can be that this residue is the general acid in the catalytic mechanism. First it is important to test if the position of the identified residue is consistent with the assumed binding mode of the substrate and the stereochemistry of the catalytic mechanism. If no structural considerations preclude the residue from acting as the general acid, it becomes important to test the reliability of the calculated pK_a values themselves. Often the PDB contains structures of similar or identical enzymes, and pK_a calculations on such structures can often hint at the robustness of the conclusions; if the same residue has a high pK_a value in the majority of the structures, then we can be more confident of our prediction. If there is no consensus in the calculated pK_a values, then we have to be more cautious with our conclusions. If there are few or no alternative X-ray structures available, the aforementioned analyses become impossible to conduct. In these cases, and to further validate one's results, it is often useful to conduct a sensitivity analysis of the calculated titration curves, as described next.

5.1. Sensitivity analysis

It is possible to test how sensitive the calculated pK_a values are to the uncertainties in atom positions that are inherent in any protein structure. The following discussion assumes that the protein structure used already has been examined for crystal contacts that could influence the result, and we therefore address only the question of how wrong the pK_a calculations potentially could be if the structure is correct within the limits specified by the resolution. To address this issue, we calculate how much desolvation energies, background interaction energies, and site–site interaction energies vary when we perturb the protein structure slightly. Such an analysis shows that for an average structure, the desolvation energies and background interaction energies vary by up to 50%, whereas site–site interaction energies typically vary by up to 20% (Nielsen, 2006a).

The aforementioned changes to the desolvation and background interaction energies are quite dramatic and are a result of the high importance of the packing quality for the calculation of these energies. Site–site interaction energies are less dependent on atom positions, as they, on average, reflect longer range interactions. However, for strongly interacting ion pairs (short distances), one can also expect changes in the order of 50%.

Given that any energy we calculate could be wrong by ±50%, can we still trust the conclusions from our pK_a calculations? Let us return to the hypothetical example involving the general acid in enzyme X. We now recalculate the titration curves for protein X by systematically varying the intrinsic pK_a values and site–site interaction energies according to the aforementioned criteria and thus obtain a picture of how large the changes in the pK_a value of the active site titratable groups are. If the changes are so large that they frequently lead to a different group being identified as the proton donor (or indeed if a proton donor frequently cannot be identified), then it puts our initial conclusion in doubt.

However, if the pK_a value of our Asp consistently is the highest in the active site, possibly with a few extreme perturbations yielding a lower pK_a value, then we can be relatively certain that our conclusion is valid, even when taking into account the uncertainty of PBE-based pK_a calculations. The aforementioned analysis is, of course, somewhat tedious and time-consuming to carry out by hand, but the analysis can conveniently be carried out for a subsystem of titratable groups in pKaTool (Nielsen, 2006a,b).

The following is a summary of how to validate conclusions from a pK_a calculation study.

1. Establish rigorous criteria for making a conclusion from the pK_a calculation results. E.g. pK_a value of proton donor has to be at least 6.0 and at least 1.5 units higher than the pK_a of any other acidic residue in the active site.

2. Perform pK_a calculations on highly similar proteins and enzymes. Do these calculations show the same trends as the first calculation?
3. Perturb the intrinsic pK_a values and site–site interaction energies as described earlier and recalculate titration curves. Use the variation in the recalculated titration curves to estimate the potential error in the pK_a calculation. Reexamine the conclusions.
4. pH activity profiles and the pH dependence of protein stability. Do these agree with experimental measurements?
5. Use any other experimental data, such as FT-IR titration curves and mutation studies, to validate the calculated pK_a values.

6. Predicting pH Activity Profiles

Often we calculate pK_a values in the active site of an enzyme in order to predict the pH activity profile of the enzyme. Before being able to predict the pH activity profile it is necessary to identify the catalytically competent protonation state (CCPS) for the enzyme.

Typically one or two groups are required to be in a specific protonation state so as to provide the protons or charges needed in the catalytic mechanism. A good example of this is found for hen egg white lysozyme, where the general acid/base (Glu 35) donates a proton to the glycosidic bond, whereas the catalytic nucleophile (Asp 52) attacks the C1 of the sugar substrate (Fig. 9.3). However, although these two groups are the only catalytically active groups, it is necessary to examine the binding cleft carefully to see if other groups must be in a specific protonation state for the substrate to bind or for catalysis to occur. An example where a third group must be in a specific protonation state comes from the N35D mutant of *Bacillus circulans* xylanase (BCX), where Glu 172 (the proton donor), Glu 78 (the nucleophile), and a third residue (Asp 35) are required to be in a specific protonation state (Joshi *et al.*, 2000). Asp 35 is situated next to Glu 172, and only when Asp 35 is protonated and donates a hydrogen bond to Glu 172 can catalysis occur.

Figure 9.3 The predicted pH dependence of the stability of hen egg white lysozyme (PDB ID 2lzt). The black curve shows the pH dependence of ΔG_{fold}, whereas colored areas show decomposition of the total charge difference. Figure produced with pKaTool (Nielsen, 2006a). (See Color Insert.)

Although BCX N35D is the best studied case where more than two residues are required to be in a specific protonation state, it is likely that many enzyme active sites contain two or more groups that must be in a specific protonation state for catalysis to occur. The catalytic groups themselves are easy to identify, but in addition to those, several groups often must be in a specific protonation state to bind the substrate. Indeed it is well known that several residues change protonation state upon ligand binding, and in X-ray structures of enzyme–substrate complexes, titratable groups often make important contacts with the substrate or provide hydrogen bonds to support the ionization state of the catalytic residues.

Once the CCPS (or CCPSs) has been identified, the pH activity profile can be predicted simply by extracting the pH dependence of the population of the CCPS. Note that only in the trivial case of perfect Henderson–Hasselbalch titration curves for the catalytic groups is it possible to correctly predict the pH activity profile from the pK_a values or the titration curves alone. In all other cases the pH activity profile is predicted incorrectly from pK_a values or titration curves. With many pK_a calculation packages, a bit of extra effort is involved in obtaining the pH dependence of the CCPS populations, and sometimes manual recalculation of a select set of protonation state populations is the easiest approach. Such a calculation can be performed with little effort in pKaTool, which also can predict pH activity profiles without manual recalculation from a set of WHAT IF pK_a calculation files.

6.1. More than one CCPS

Catalytic mechanisms tend to convey the impression that only a single protonation state is catalytically active. A single CCPS will always give a pH activity profile with one or two transitions. Many enzymes are known to have pH-activity profiles with "shoulders" or "tails", and it is therefore clear that multiple active site protonation states can be catalytically competent. When predicting the pH activity profile for an enzyme it is therefore important to examine if that several protonation states can be catalytically active for your enzyme.

7. Decomposition Analysis

It is commonly of interest to ascribe roles to different amino acids in enzyme active sites. In pK_a calculations it is similarly of interest to examine the effect that a titratable group has on the pK_a values of all other titratable groups. One can examine this question in two different ways: in the context of the active site or in isolation. If we are measuring the effects of group A on group B, then we can remove group A from the active site and recalculate all titration curves in the system and observe how much the pK_a value of group B has changed. Alternatively, we can calculate the effect of group A on the pK_a value of group B by calculating titration curves for a system consisting only of groups A and B. We compare the titration curve of group B in this system with the titration curve of group B when it does not interact with any other groups (its intrinsic titration).

This kind of analysis yields two ΔpK_a values for each pair of residues: an in-system ΔpK_a and an ex-system ΔpK_a. For weakly coupled systems these two ΔpK_a values are identical, whereas significant differences are seen for tightly coupled systems. The difference in ΔpK_a values essentially means that the tightly coupled system behaves as a single entity and that it cannot be viewed as a collection of individual titratable groups. Table 9.1 shows an example of a theoretical decomposition analysis for three active site groups in *B. circulans*. It is clearly seen that the $\Delta\Delta$pK_a (the difference between the in-system ΔpK_a and the ex-system ΔpK_a) is nonzero for both Glu 78 and Arg 112.

7.1. Redesigning protein pK_a values

As described previously, the titrational behavior of active site groups determines the pH activity profile of an enzyme. Consequently, if the active site pK_a values are changed then the pH activity profile will change with them. The decomposition analysis described in the previous section can be performed interactively in pKaTool and gives a fast analysis of how the pK_a values and the pH activity profile change upon the removal of titratable groups in the active site. However, this knowledge is of little use in protein engineering, since mutations in the active site of an enzyme generally completely abolish or significantly reduce the catalytic activity of the enzyme. It is therefore of interest to examine if the pK_a values of the active site residues can be changed in some other way. The simplest way to answer

Table 9.1 Decomposition analysis for Glu 172 in the active site of *Bacillus circulans* xylanase[a]

Residue removed	Δp$K_{a\text{in.system}}$, Glu 172	Δp$K_{a\text{wx.system}}$ Glu 172	$\Delta\Delta$pK_a Glu 172
Glu 78	−1.74	−1.62	0.12
Arg 112	0.71	0.83	0.12

[a] Glu 78 and Arg 112 are predicted to have a context-specific effect on the titrational behavior of Glu 172.

this question is to produce a protein model containing a large number of point mutations outside the active site and recalculate the active site pK_a values with this new protein model. However, this procedure becomes very time-consuming when studying the effect of a large number of point mutations, since all intrinsic pK_a values and pair wise interaction energies have to be recalculated for all titratable groups in the mutant structure.

We have therefore developed a fast ΔpK_a calculation algorithm that can be used for redesigning protein pK_a values(Tynan-Connolly and Nielsen, 2006, 2007). The method relies on the assumption that the introduction of a mutation does not change any pair wise electrostatic interaction energies other than those involving the mutated residue itself. Furthermore, we assume that the intrinsic pK_a values of all other titratable groups remain unperturbed. The first assumption is quite reasonable since the structure of the protein (and thus the dielectric boundary) is not altered significantly by a single mutation. The second assumption (that the intrinsic pK_a values stay constant) is more problematic and is initially resolved by only allowing mutations that are further away than 10 Å from a group whose pK_a shift we want to predict. A subsequent modification to the code allows for recalculation of intrinsic pK_a values within a specific distance cutoff from each mutation. Initially we precalculate the pair wise electrostatic interaction energies and intrinsic pK_a values for each mutation and use these energies to modify the wild-type interaction energy matrix and intrinsic pK_a value list. Typically we are interested in observing the effect of a mutation on the pK_a values for a small set of residues. We therefore only have to calculate titration curves for all titratable groups for a few pH values around the wild-type pK_a values for those residues in order to get a quick estimate of the mutant pK_a values. The speedup in this step (typically a factor 10–20), combined with the speedup associated with not recalculating all wild-type energies (this speedup is very size dependent), allows pK_a values for the mutated protein to be calculated quickly, thus enabling us to study the effect of many mutations in a very short amount of time.

The aforementioned methodology has been implemented in pK_d routines and is available for download and via a Web server (Tynan-Connolly and Nielsen, 2006), but essentially the methodology can be incorporated in any PBE-based pK_a calculation methodology to allow for the redesign of protein pK_a values.

8. Predicting Protein Stability Profiles

It is well known that the stability of proteins varies with pH, and in many cases it is of interest to predict or reengineer the pH stability profile of a protein. It is straightforward to calculate the variation of protein stability with pH once one has calculated the titration curves for all titratable groups in the protein. According to Tanford (1970), the pH dependence of protein stability can be found by evaluating the integral

$$\Delta G_{fold} = \int_{pH\_ref}^{pH} \ln(10) kT \Delta Q_{fold} \quad (9.7)$$

for various values of pH. In Eq. (9.7), pH_ref is a reference pH value where the ΔG of protein folding is assumed to be zero, k is Boltzmann's constant, T is the absolute temperature, and ΔQ_{fold} is the difference in the total charge between unfolded and folded states ($\Delta Q_{fold} = Q_{folded} - Q_{unfolded}$). The evaluation of this integral for calculated pK_a values of hen egg white lysozyme produces the plot shown in Fig. 9.4. Please note that Eq. (9.7) only describes the relative difference in stability between folded and unfolded states. The absolute value of ΔG_{fold} cannot be predicted using this equation and has to be established by an independent method at a specific pH value.

8.1. Predicting protein–ligand-binding pH profiles

The pH dependence of protein–ligand-binding energies can be predicted in a similar way. The only difference is that the folded and unfolded states refer to the holo and apo states of the receptor protein.

9. Fitting pH Titration Curves, pH Activity Profiles, and pH Stability Profiles

Analysis of experimental data remains an important step in studying the pH-dependent properties of proteins. Ultimately, theoretical calculations aim to replace experiments as the main source of information in the biosciences, and it is therefore of the utmost importance to compare calculated results to experimental results. To be able to compare theoretical data to experimental data we must know how to predict experimental data, but equally we must know how to analyze experimental data to reduce risks of misinterpretations that could have important consequences for our understanding of pH-dependent phenomena of proteins. Furthermore, a detailed analysis of experimental data can often lead to measurements of biophysical characteristics that can be compared directly to calculated results. An example of such an analysis is discussed in this section, but before we turn out attention to advanced analysis of pH-dependent experimental

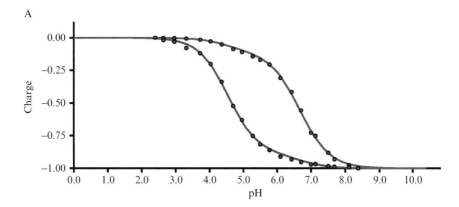

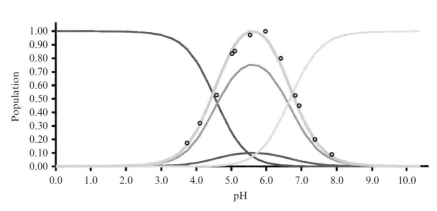

Figure 9.4 Fits of titration curves (top) and the pH activity profile (bottom) for *Bacillus circulans* xylanase using Eq. (9.6). Circles indicate experimental data points from McIntosh *et al.* (1996), whereas lines indicate functions fitted using GloFTE/pKaTool (Søndergaard *et al.*, 2008). (Top) Blue curve, Glu 172; black curve, Glu 78. (Bottom) black curve, Glu78H, Glu172H; red curve, Glu78-, Glu172H; blue curve, Glu78H, Glu172-; cyan curve, Glu78-, Glu172-; green curve is the red curve (the population catalytically competent protonation state) normalized. (See Color Insert.)

data, it is useful to review some of the classic equations for the pH dependence of biophysical characteristics of proteins.

9.1. Classic equations for fitting pH-dependent characteristics of proteins

Nuclear magnetic resonance spectroscopy is the single most used technique for studying protein residue titration curves. Typically, two-dimensional NMR spectra are acquired at a range of pH values, and the NMR chemical shift for specific atoms is tracked as a function of pH. Such experiments provide information on the titration process of a titratable group; in the simplest case, an NMR a titration curve is fitted to an equation describing a two-state titration:

$$\delta_{obs} = \Delta\delta_{group} \frac{1}{10^{(pK_a - pH)} + 1} + \delta_{min} \qquad (9.8)$$

where δ_{min} and $\delta_{min} + \Delta\delta_{group}$ correspond to the end-point chemical shifts of a nucleus in the presence of the conjugate acid and base forms of a titratable group, respectively.

In the case of NMR titration curves displaying one or more titrations, the curve can be fitted using Eq. (9.9):

$$\delta_{obs} = \frac{\Delta\delta_{group1}}{10^{(pK_{a1} - pH)} + 1} + \frac{\Delta\delta_{group2}}{10^{(pK_{a2} - pH)} + 1} + \delta_{min} \qquad (9.9)$$

Equation (9.9) assumes two completely independent titrations. Equations (9.8) and (9.9) can trivially be extended to deal with three or more titrations, and it is also possible to derive equations for coupled titrations.

Why do we need equations for both coupled and uncoupled titrations? The reason lies in the high sensitivity of the NMR chemical shift and in the ability for strongly coupled groups to have distinct non-HH-shaped titration curves. The high sensitivity of the NMR chemical shift to the chemical environment of atoms means that the chemical shift will report even minimal changes in protein structure or the electric field. Thus, if we monitor the chemical shift of an atom in a titratable residue, the NMR chemical shift will be affected not only by the titration of that residue, but also by small changes in the protein structure and the electric field. An NMR pH titration curve for a residue whose charge titrates with a perfect HH curve can thus be bi- or triphasic because the NMR chemical shift senses titrations from distant residues in addition to titration from the residue itself. The following discussion refers to these "sensed" titrations as "ghost" titrations, as they in effect are ghosts of other real charge titrations.

The charge titration of a titratable group can also be non-HH shaped if the group is part of a strongly coupled system, as shown in Fig. 9.2. That is, the loss of charge on that single residue occurs in a way that is distinctly non two-state. It is impossible to distinguish between a ghost titration and a real non-HH charge titration using the NMR chemical shift information alone. It is therefore up to the experimental researcher to decide if a charge titration curve is HH shaped or non-HH shaped. This decision will have an effect on the pK_a values ascribed to individual titratable groups and adds an extra level of uncertainty to pK_a values determined from titration curves that are not monophasic.

9.2. Statistical mechanical fitting (GloFTE)

Instead of fitting NMR pH titration curves to Eqs. (9.8) and (9.9) and extensions of these, it should also be possible to fit NMR pH titration curves to titration curves produced by the evaluation of Eq. (9.6). This requires the assumption that the change in chemical shift is directly correlated to the charge titration and thus that there are no ghost titrations in the titration curve. We can be most confident of this since we measure the chemical shift very close to the titratable group since the chemical shift of the atoms close to the titration is dominated by the charge titration curve. We explored the possibility of applying Eq. (9.6) using data on the titration of the active site residues of *B. circulans* xylanase (Joshi *et al.*, 2000; McIntosh *et al.*, 1996) and found that it was indeed possible to fit non-HH titration curves very accurately with this method (Søndergaard *et al.*, 2008). Because Eq. (9.6) also gives information on the pH dependence of the individual protonation state populations, it is possible to fit the pH activity profile of the enzyme simultaneously with the NMR titration curves (see Fig. 9.4). An additional benefit of using Eq. (9.6) is that, for non-HH-shaped titration curves, we obtain accurate information on the pair wise electrostatic interaction energies for the system. These interaction energies can be used to validate the pair wise interaction energies used in pK_a calculations and thus present a route to improve pK_a calculation predictions.

Additionally, we have demonstrated that if a given charge titration curve is very complex, then it can be used to derive information on the titrational behavior of other nearby groups. Thus by using only the titration curve of Glu 172 in the N35D mutant of *B. circulans* xylanase, it is possible to predict the titration curves of the neighboring residues Asp 35 and Glu 78 (Søndergaard *et al.*, 2008). By using the pair wise interaction energies from this prediction, it is also possible to uniquely identify these three residues in the X-ray structure of BCX N35D, thus raising the possibility of using non-HH-shaped NMR pH titration curves to assign resonances in an NMR spectrum.

10. Conclusion

Theoretical pK_a calculation methods can be used to accurately predict protein charge titration curves, enzymatic pH activity profiles, and pH stability profiles. They can also yield valuable information on the roles of individual residues in determining the pK_a values of other titratable groups and provide a basis for calculations aimed at redesigning protein pK_a values. However, it is essential to be very careful when selecting and preparing protein structures for use with pK_a calculations; additionally, one must test the validity of the calculations by examining the sensitivity of the calculations to small structural perturbations. Calculations on a large number of X-ray structures can often provide additional support for pK_a calculation-based conclusions and have revealed that crystal contacts can have profound effects on the calculation pK_a values. Finally, it has been shown that experimentally measured NMR pH titration curves and pH activity profiles can be analyzed to yield pair wise electrostatic interaction energies that can be used to validate the calculated energies used in pK_a calculations.

In summary, in the last decade, PBE-based pK_a calculation techniques have been refined and extended to address a wide range of problems. With careful execution and analysis, these techniques can be very helpful in analyzing the pH dependence of protein biophysical characteristics.

This work was supported by a Science Foundation Ireland PIYR award (04/YI1/M537).

REFERENCES

Alexov, E. G., and Gunner, M. R. (1997). Incorporating protein conformational flexibility into the calculation of pH-dependent protein properties. *Biophys. J.* **72,** 2075–2093.

Alexov, E. G., and Gunner, M. R. (1999). Calculated protein and proton motions coupled to electron transfer: Electron transfer from QA- to QB in bacterial photosynthetic reaction centers. *Biochemistry* **38,** 8253–8270.

Baker, N. A., Sept, D., Joseph, S., Holst, M. J., and McCammon, J. A. (2001). Electrostatics of nanosystems: Application to microtubules and the ribosome. *Proc. Natl. Acad. Sci. USA* **98,** 10037–10041.

Baldwin, R. L. (2007). Energetics of protein folding. *J. Mol. Biol.* **371,** 283–301.

Baptista, A. M., Teixeira, V. H., and Soares, C. M. (2002). Constant-pH molecular dynamics using stochastic titration. *J. Chem. Phys.* **117,** 4184–4200.

Beroza, P., Fredkin, D. R., Okamura, M. Y., and Feher, G. (1991). Protonation of interacting residues in a protein by a Monte Carlo method: Application to lysozyme and the photosynthetic reaction center of Rhodobacter sphaeroides. *Proc. Natl. Acad. Sci. USA* **88,** 5804–5808.

Cymes, G. D., and Grosman, C. (2008). Pore-opening mechanism of the nicotinic acetylcholine receptor evinced by proton transfer. *Nat. Struct. Mol. Biol.* **15,** 389–396.

Demchuk, E., and Wade, R. C. (1996). Improving the continuum dielectric approach to calculating pKas of ionizable groups in proteins. *J. Phys. Chem.* **100,** 17373–17387.

Dolinsky, T. J., Czodrowski, P., Li, H., Nielsen, J. E., Jensen, J. H., Klebe, G., and Baker, N. A. (2007). PDB2PQR: Expanding and upgrading automated preparation of biomolecular structures for molecular simulations. *Nucleic Acids Res.* **35,** W522–W525.

Dolinsky, T. J., Nielsen, J. E., McCammon, J. A., and Baker, N. A. (2004). PDB2PQR: An automated pipeline for the setup of Poisson-Boltzmann electrostatics calculations. *Nucleic Acids Res.* **32,** W665–W667.

Elcock, A. H. (2001). Prediction of functionally important residues based solely on the computed energetics of protein structure. *J. Mol. Biol.* **312,** 885–896.

Georgescu, R. E., Alexov, E. G., and Gunner, M. R. (2002). Combining conformational flexibility and continuum electrostatics for calculating pK(a)s in proteins. *Biophys. J.* **83,** 1731–1748.

Gordon, J. C., Myers, J. B., Folta, T., Shoja, V., Heath, L. S., and Onufriev, A. (2005). H++: A server for estimating pKas and adding missing hydrogens to macromolecules. *Nucleic Acids Res.* **33,** W368–W371.

Joshi, M. D., Sidhu, G., Pot, I., Brayer, G. D., Withers, S. G., and McIntosh, L. P. (2000). Hydrogen bonding and catalysis: A novel explanation for how a single amino acid substitution can change the pH optimum of a glycosidase. *J. Mol. Biol.* **299,** 255–279.

Khandogin, J., and Brooks, C. L., 3rd (2005). Constant pH molecular dynamics with proton tautomerism. *Biophys. J.* **89,** 141–157.

Li, H., Robertson, A. D., and Jensen, J. H. (2005). Very fast empirical prediction and rationalization of protein pKa values. *Proteins* **61,** 704–721.

Linderstrom-Lang, K. (1924). Om proteinstoffernes ionisation. *Compt. zrend. ztrav. Lab. Carlsberg* **15,** 1–29.

Madura, J. D., Briggs, J. M., Wade, R. C., Davis, M. E., Luty, B. A., Ilin, A., Antosiewicz, J., Gilso, M. K., Bagheri, B., Scott, L. R., and McCammon, J. A. (1995). Electrostatics and diffusion of molecules in solution: Simulations with the University of Houston Brownian Dynamics Program. *Comput. Phys. Commun.* **91,** 57–95.

McIntosh, L. P., Hand, G., Johnson, P. E., Joshi, M. D., Körner, M., Plesniak, L. A., Ziser, L., Wakarchuk, W. W., and Withers, S. G. (1996). The pKa of the general acid/base carboxyl group of a glycosidase cycles during catalysis: A 13C-NMR study of Bacillus circulans xylanase. *Biochemistry* **35,** 9958–9966.

Mongan, J., Case, D. A., and McCammon, J. A. (2004). Constant pH molecular dynamics in generalized Born implicit solvent. *J. Comput. Chem.* **25,** 2038–2048.

Myers, J., Grothaus, G., Narayanan, S., and Onufriev, A. (2006). A simple clustering algorithm can be accurate enough for use in calculations of pKs in macromolecules. *Proteins* **63,** 928–938.

Nicholls, A., and Honig, B. (1991). A rapid finite difference algorithm, utilizing successive over-relaxation to solve the Poisson-Boltzmann equation. *J. Comp. Chem.* **12,** 435–445.

Nielsen, J. E. (2006a). Analysing the pH-dependent properties of proteins using pKa calculations. *J. Mol. Graph.* **25,** 691–699.

Nielsen, J. E. (2006b). pKaTool. University College Dublin, Dublin, Ireland, http://enzyme.ucd.ie/Science/pKa/pKaTool.

Nielsen, J. E., and McCammon, J. A. (2003). On the evaluation and optimisation of protein X-ray structures for pKa calculations. *Protein Sci.* **12,** 313–326.

Nielsen, J. E., and Vriend, G. (2001). Optimizing the hydrogen-bond network in Poisson-Boltzmann equation-based pK(a) calculations. *Proteins* **43,** 403–412.

Ondrechen, M. J., Clifton, J. G., and Ringe, D. (2001). THEMATICS: A simple computational predictor of enzyme function from structure. *Proc. Natl. Acad. Sci. USA* **98,** 12473–12478.

Pisliakov, A. V., Sharma, P. K., Chu, Z. T., Haranczyk, M., and Warshel, A. (2008). Electrostatic basis for the unidirectionality of the primary proton transfer in cytochrome c oxidase. *Proc. Natl. Acad. Sci. USA* **105,** 7726–7731.

Sham, Y. Y., Muegge, I., and Warshel, A. (1998). The effect of protein relaxation on charge-charge interactions and dielectric constants of proteins. *Biophys. J.* **74,** 1744–1753.

Søndergaard, C. R., McIntosh, L. P., Pollastri, G., and Nielsen, J. E. (2008). Determination of electrostatic interaction energies and protonation state populations in enzyme active sites by global fits of NMR-titration data and pH-activity profiles. *J. Mol. Biol.* **376,** 269–287.

Strickler, S. S., Gribenko, A. V., Gribenko, A. V., Keiffer, T. R., Tomlinson, J., Reihle, T., Loladze, V. V., and Makhatadze, G. I. (2006). Protein stability and surface electrostatics: A charged relationship. *Biochemistry* **45,** 2761–2766.

Tanford, C. (1970). Protein denaturation. Part C. Theoretical models for the mechanism of denaturation. . *Adv. Protein Chem.* **25,** 1–95.

Tanford, C., and Roxby, R. (1972). Interpretation of protein titration curves: Application to lysozyme. *Biochemistry* **11,** 2193–2198.

Toseland, C. P., McSparron, H., Davies, M. N., and Flower, D. R. (2006). PPD v1.0–an integrated, web-accessible database of experimentally determined protein pKa values. *Nucleic Acids Res.* **34,** D199–D203.

Tynan-Connolly, B., and Nielsen, J. E. (2006). pKD: Re-designing protein pKa values. *Nucleic Acids Res.* **34,** W48–W51.

Tynan-Connolly, B. M., and Nielsen, J. E. (2007). Re-designing protein pKa values. *Protein Sci.* **16,** 239–249.

Vriend, G. (1990). WHAT IF: A molecular modeling and drug design program. *J. Mol. Graph* **8,** 52–6, 29.

Warshel, A., Sharma, P. K., Kato, M., and Parson, W. W. (2006). Modeling electrostatic effects in proteins. *Biochim. Biophys. Acta* **1764,** 1647–1676.

Yang, A. S., Gunner, M. R., Sampogna, R., Sharp, K., and Honig, B. (1993). On the calculation of pKas in proteins. *Proteins* **15,** 252–265.

CHAPTER TEN

Least Squares in Calibration: Weights, Nonlinearity, and Other Nuisances

Joel Tellinghuisen

Contents

1. Introduction	260
2. Review of Least Squares	263
2.1. Formal equations for linear least squares	263
2.2. Simple illustrations	266
2.3. Extensions and practical concerns	268
3. Experiment Design Using V_{prior} — Numerical Illustrations	269
3.1. Characterizing a nonlinear response function	269
3.2. Uncertainty in the unknown	274
3.3. Should I try a data transformation?	276
3.4. Variance function estimation	277
3.5. When χ^2 is too large	281
4. Conclusion	282
References	283

Abstract

In the least-squares fitting of data, there is a unique answer to the question of how the data should be weighted: inversely as their variance. Any other weighting gives less than optimal efficiency and leads to unreliable estimates of the parameter uncertainties. In calibration, knowledge of the data variance permits exact prediction of the precision of calibration, empowering the analyst to critically examine different response functions and different data structures. These points are illustrated here with a nonlinear response function that exhibits a type of saturation curvature at large signal like that observed in a number of detection methods. Exact error propagation is used to compute the uncertainty in the fitted response function and to treat common data transformations designed to reduce or eliminate the effects of data heteroscedasticity. Data variance functions can be estimated with adequate reliability from

Department of Chemistry, Vanderbilt University, Nashville, Tennessee remarkably small data sets, illustrated here with three replicates at each of seven calibration values. As a quantitative goodness-of-fit indicator, χ^2 is better than the widely used R^2; in one application it shows clearly that the dominant source of uncertainty is not the measurement but the preparation of the calibration samples, forcing the conclusion that the calibration regression should be reversed.

1. INTRODUCTION

Calibration is the operation of determining an unknown quantity by comparison measurements on samples for which the targeted quantity is presumed to be known. In analytical work the unknown quantity is typically a concentration or amount, and the calibration equation is achieved by least-squares (LS) fitting the measured property to a smooth function of the known concentrations or amounts of the calibration samples. The bulk of analytical work involves a single measured property of a single targeted quantity, giving a calibration response function of form $y = f(x)$ and called *univariate calibration*. Most such work also assumes a linear response function, $f(x) = a + bx$, and tacitly assumes the data are *homoscedastic* (i.e., have constant uncertainty), hence employs unweighted or "ordinary" LS (OLS). In this chapter I will confine my attention to univariate calibration. However, linearity of response and homoscedasticity are both at some level approximations, so I will deal specifically with deviations from these assumptions. In so doing, I will describe and illustrate techniques that are minor but important extensions of the routine methods long in use in analytical work. These techniques focus on the statistical fundamentals of calibration and should become standard tools in every modern analyst's bag of tricks.

The major emphasis of the present work centers on three simple statistical realities that seem to be underappreciated by analysts:

- OLS is a special case of weighted LS (WLS) where the data weights w_i are all the same.
- Minimum-variance estimation of the calibration parameters is achieved by employing weights

$$w_i \propto 1/\sigma_i^2, \qquad (10.1)$$

where σ_i^2 is the variance of y_i.
- Knowledge of the x structure of the data and σ_i^2 permits *exact* prediction of calibration precision in linear LS (LLS) and near exact in nonlinear LS (NLS), irrespective of actual data.

Regarding the first of these, some workers understand WLS in a more specific sense, as the use of replicate-based estimates s_i^2 of σ_i^2 to compute the weights. This is actually *not* a sound choice when the number of replicates m is small; for example, the large uncertainty in such estimates ensures that the results will be worse than OLS when the data are homoscedastic or nearly so (Jacquez and Norusis, 1973; Tellinghuisen, 2007). Regarding the last, LLS refers to the LS equations, not to the response function (RF), so it includes many nonlinear RFs, such as polynomials in x. Concerning the second, the proportionality suffices to obtain correct estimates of the adjustable parameters, but absolute knowledge of σ_i^2 is needed for *experiment design* — the a priori prediction of the precision of calibration.

These points have become increasingly relevant in bioanalytical work in recent years, as analysts have come to recognize that their data can be strongly *heteroscedastic* (of varying σ_i), especially for chromatographic techniques (Almeida *et al.*, 2002; Baumann and Wätzig, 1995; Burns *et al.*, 2004; Johnson *et al.*, 1988; Karnes and March, 1991; Kiser and Dolan, 2004; Nagaraja *et al.*, 1999; Sadray *et al.*, 2003; Zhou *et al.*, 2003). However, the methods for handling such data often seem to be a mysterious trial-and-error mix of experimenting with various weighting formulas, such as $1/x$, $1/\sqrt{x}$, $1/x^2$ and the same functions of y; and variable transformations, such as $\sqrt{y}$ vs $\sqrt{x}$ and $\ln(y)$ vs $\ln(x)$. The choice of a proper weighting formula has sometimes been based on statistically flawed tests such as "quality coefficients" (Tellinghuisen, 2008c). All such attempts can be seen as indirect methods of dealing with data heteroscedasticity and/or nonlinearity of response. Such roundabout approaches are not only statistically questionable, they are also more time demanding than approaches that focus directly on the data variance (Hwang, 1994). For example, I have recently shown that one can estimate a two-parameter variance function for use with a linear response function, with only minor ($\approx 10\%$) loss of precision for as few as three replicates at six calibration values (Tellinghuisen, 2008a, 2008d). The needed 18 measurements are fewer than employed in many bioanalytical protocols, and the derived data variance function is both more complex and more physically reasonable than the single parameter choices that are the focus of much trial-and-error effort.

To add further emphasis to the significance of Eq. (10.1), in most situations "minimum variance" means the best possible, and any weighting other than inversely as the data variance can only do worse. Further, the standard equations used to estimate parameter uncertainties become unreliable when the weighting is incorrect. The magnitudes of these problems depend on the range of weights for the data and the placement of the unknown relative to the knowns. For example, neglect of weights has relatively minor consequences when the variances for the calibration samples vary by less than a factor of 10 over the calibration range; and one-shot calibration can be achieved with a negligible loss of precision and a complete neglect of weights when the unknown can be approximately centered in the range of the calibration samples (Tellinghuisen, 2007). However, bioanalytical work often requires comprehensive calibration over several orders of magnitude of the independent variable (x), for use with multiple unknowns. The strong heteroscedasticity already noted for chromatographic data is usually dominated by proportional error ($\sigma \propto y$) over typical working ranges (Zeng et al., 2008), and in this situation proper weighting can be very important.

Dedication to the linear response function in analytical work approaches religion in its firmness of establishment, with the quest for the linear RF or the "region of linear response" sometimes seeming obsessive. This linearity emphasis perhaps stems from the need to estimate statistical uncertainties for unknowns in calibration and the availability of a mathematical expression for this purpose when the RF is linear (Danzer and Currie, 1998; Miller, 1991),

$$\sigma_{x0}^2 = \frac{1}{b^2}\left[\frac{\sigma_0^2}{m} + \frac{1}{\Sigma \sigma_i^{-2}} + \frac{(y_0 - \bar{y}_w)^2}{b^2 \Sigma (x_i - \bar{x}_w)^2 / \sigma_i^2}\right]. \qquad (10.2)$$

Here y_0 is the average response from m measurements for the unknown, $x_0 = (y_0 - a)/b$, having a priori known variance σ_0^2, with the weights taken as $w_i = 1/\sigma_i^2$, also known. The subscript "w" designates weighted average, e.g., $\bar{y}_w = \Sigma w_i y_i / \Sigma w_i$. If the data variances are known in only a relative sense, Eq. (10.2) is replaced by its a posteriori version (see later),

$$s_{x0}^2 = \frac{s_1^2}{b^2}\left[\frac{1}{mw_0} + \frac{1}{\Sigma w_i} + \frac{(y_0 - \bar{y}_w)^2}{b^2 \Sigma w_i (x_i - \bar{x}_w)^2}\right], \qquad (10.3)$$

where again it is assumed that all m measurements of y_0 have the same weight w_0 and where $s_1^2 = S_w/(n-2)$, with $S_w = \Sigma w_i (y_i - a - b x_i)^2$. Equation (10.2) represents an "exact" result permitting assignment of confidence limits using the normal distribution [but with limitations due to the nonlinear nature of the estimator of x_0 (Draper and Smith, 1998; Shukla, 1972; Tellinghuisen 2000a)], whereas s_1^2 and hence s_{x0}^2 in Eq. (10.3) represent estimates and hence require use of the t distribution in assigning confidence limits.

The formal equivalent of Eqs. (10.2) and (10.3) for RFs other than linear is "messy," a difficulty that has probably been partly responsible for the aversion to other RFs. However, when treated as a numerical problem, the computation of σ_{x0}^2 (s_{x0}^2) can be accomplished easily for any RF and any weighting through a nonlinear algorithm that makes the error propagation computations behind Eqs. (10.2) and (10.3) an automatic part of the output (Tellinghuisen, 2000c, 2005a). This facility makes it easy for the analyst to experiment with various RFs to achieve a statistically optimal calibration formula, with assurance of statistically correct uncertainty estimates for the unknown.

In following sections I will first briefly review the essential LS relations and then illustrate the power of knowledge of the data variance in experiment design for a case requiring a nonlinear RF. Importantly, the predictions are valid for both correct and incorrect RFs, permitting one to predict biases for reasonable modifications of the RF. Closely related is the estimation of unknowns and their uncertainties, for which purpose I will illustrate the new algorithm. I will then treat the estimation of the data variance function (VF) from replicate data. Throughout, I will emphasize the use of the $c2$ statistic as a figure of merit for the fits, over the widely used, equivalent, but hard-to read $R2$. I will close with a realistic example that leads to the conclusion that the presumed independent variable (x) is actually more uncertain than y, and describe how to handle such cases. Throughout, I will use the KaleidaGraph program (Synergy Software) to illustrate the computations (Tellinghuisen, 2000b), but will provide sufficient detail to facilitate verification by readers using other data analysis tools.

2. Review of Least Squares

2.1. Formal equations for linear least squares

The working least-squares relations are available from many sources (e.g., Albritton *et al.*, 1976; Bevington, 1969; Draper and Smith, 1998; Hamilton, 1964; Press *et al.*, 1986), including in matrix notation in an earlier contribution by me in this series (Tellinghuisen, 2004); accordingly, only the necessary essentials are given here. The equations are obtained by minimizing the sum S,

$$S = \Sigma w_i \delta_i^2, \qquad (10.4)$$

with respect to a set of adjustable parameters $\boldsymbol{\beta}$, where δ_i is the residual (observed–calculated mismatch) for the ith point. $\boldsymbol{\beta}$ is a column vector containing one element for each of the p adjustable parameters. The LS problem is *linear* if the measured values of the dependent variable (y) can be related to those of the independent variable(s) ($x, u, \ldots$) and the adjustable parameters through

$$\mathbf{y} = \mathbf{X}\boldsymbol{\beta} + \boldsymbol{\delta}, \qquad (10.5)$$

where $\mathbf{y}$ and $\boldsymbol{\delta}$ are column vectors containing n elements (for the n measured values), and the *design matrix* $\mathbf{X}$ has n rows and p columns and depends only on the values of the independent variable(s) and not on $\boldsymbol{\beta}$ or $\mathbf{y}$. In this case, the LS equations are

$$\mathbf{X}^T \mathbf{W} \mathbf{X} \boldsymbol{\beta} \equiv \mathbf{A}\boldsymbol{\beta} = \mathbf{X}^T \mathbf{W} \mathbf{y}, \qquad (10.6)$$

where $\mathbf{X}^T$ is the transpose of $\mathbf{X}$ and the weight matrix $\mathbf{W}$ is diagonal, with n elements $W_{ii} = w_i$. The solution of these equations is

$$\boldsymbol{\beta} = \mathbf{A}^{-1}\mathbf{X}^T\mathbf{W}\,\mathbf{y}, \quad (10.7)$$

where $\mathbf{A}^{-1}$ is the inverse of $\mathbf{A}$. Knowledge of the parameters permits calculation of the residuals $\boldsymbol{\delta}$ from Eq. (10.5) and thence S (= $\boldsymbol{\delta}^T\mathbf{W}\boldsymbol{\delta}$).

While these equations may seem intimidating to the novice, they are very powerful for computational purposes and can be implemented directly in symbolic programs such as Mathematica and MathCad. Most important for present purposes, the parameter variances and covariances are proportional to the diagonal and off-diagonal elements of $\mathbf{A}^{-1}$, respectively. In particular, when the data variances are known and we take $w_i = 1/\sigma_i^2$, we obtain the a priori variance–covariance matrix,

$$\mathbf{V}_{\text{prior}} = \mathbf{A}^{-1}. \quad (10.8)$$

Alternatively, if the σ_i are known only in only a relative sense, we obtain the a posteriori $\mathbf{V}$ from

$$\mathbf{V}_{\text{post}} = \frac{S}{\nu}\mathbf{A}^{-1}, \quad (10.9)$$

where ν is the number of statistical degrees of freedom, $n - p$.

It is useful to emphasize here the assumptions behind these equations and their implications. First, all uncertainty is assumed to be random and to reside in the dependent variable, which must have *finite variance*. If the data are *unbiased*, the parameter estimates $\boldsymbol{\beta}$ will be unbiased. If the data error is distributed normally, the parameters will be similarly normal (Gaussian) in distribution, with variances given *exactly* by $\mathbf{V}_{\text{prior}}$ and approximately by $\mathbf{V}_{\text{post}}$. Any weighting other than $w_i \propto \sigma_i^{-2}$ will not yield minimum-variance estimates of the parameters and will render the $\mathbf{V}$ matrices unreliable for prediction purposes; however, incorrect weighting leaves LLS parameter estimates unbiased. These last results apply especially to heteroscedastic data analyzed by OLS.

In OLS, one uses $w_i = 1$ for homoscedastic data. Then the prefactor S/ν in Eq. (10.9) becomes an estimate s^2 of the (constant) data variance σ^2. For heteroscedastic data of unknown scale, S/ν is the estimated variance for data of unit weight, s_1^2. When the data variances are known and $w_i = \sigma_i^{-2}$, S becomes an estimate of χ^2 and S/ν of the reduced χ^2 (χ_ν^2). The former has average value ν and variance 2ν, so χ_ν^2 has average unity and variance $2/\nu$. Importantly these results mean that sampling estimates of s^2 and s both have statistical uncertainty proportional to their magnitudes, or proportional error. The probability distribution of χ_ν^2 is given by

$$P(z)dz = Cz^{(\nu-2)/2}\exp(-\nu z/2)dz, \quad (10.10)$$

where $z = \chi_\nu^2$ and C is a normalization constant. $P(z)$ is strongly asymmetrical for small ν (exponential for $\nu = 2$, for example) but becomes Gaussian in the limit of large ν.

Analytical chemists commonly use R^2 (R) to judge the quality of calibration fits. R^2 and S are related by

$$R^2 = 1 - \frac{S}{\Sigma w_i (y_i - \bar{y}_w)^2}. \qquad (10.11)$$

While it is customary to look for the value of R^2 closest to unity for the "best" fit, the theoretical value of R^2 for a properly weighted fit should be less than 1, since the average of S ($= \chi^2$) should equal ν. Through Eq. (10.11) R^2 and S contain equivalent information about the quality of the fit, and if nothing is known a priori about the data error, either can be used as a figure of merit: minimizing S (i.e., minimizing the estimated variance) is equivalent to making R^2 as close as possible to 1. However, significant changes in S can be squeezed into surprisingly small ranges in R^2, making the latter harder to "read" in this context. When the data error is known a priori and the weights w_i are taken as σ_i^{-2}, χ^2 is a better tool for quantitatively assessing the fit quality.

Another property of χ^2 that is useful in ad hoc LS fitting is that any adjustable parameter whose magnitude is less than its standard error can be set to zero (dropped from the fit model) with a resultant reduction in χ_ν^2 (for known σ_i) or in the estimated variance S/ν (unknown σ_i)(Press et al., 1986; Tellinghuisen, 2005b). In particular, this test can often be used to eliminate the intercept in a calibration RF when there is no a priori hypothesis for including it. This can greatly improve the precision of calibration at small x.

$\mathbf{V}_{\text{prior}}$ requires no actual data for its evaluation; thus various RFs can be examined for assumed data error structures in experiment design applications with exactly fitting data. $\mathbf{V}_{\text{prior}}$ also remains valid when the number of data points equals the number of parameters ($\nu = 0$). In both of these situations it functions as a method of error propagation (see later), mapping data uncertainty into parameter uncertainty. In Monte Carlo (MC) simulations, the data error is known by assignment, so $\mathbf{V}_{\text{prior}}$ applies and we expect, for example, 95% of the estimated values of β_1 to fall within $\pm 1.96\, V_{11,\text{prior}}^{1/2}$ of the true value. Conversely, a significant deviation from this prediction indicates a flaw in the MC procedures. Note further that $\mathbf{A}$ scales with σ_y^{-2} and with the number of data points n, so that $\mathbf{V}$ scales with σ_y^2 and with $1/n$. Thus, the parameter standard errors go as σ_y and as $n^{-1/2}$. For example, if the data are assumed to have 5% uncertainty, all parameter standard errors (SEs) double if this is increased to 10%, while the SEs decrease by a factor of 2 if four replicates are averaged at each point.

To calculate the uncertainty in some function f of uncertain variables $\boldsymbol{\beta}$, we must employ *error propagation*. The widely used textbook expression for this purpose,

$$\sigma_f^2 = \Sigma \left(\frac{\partial f}{\partial \beta_j}\right)^2 \sigma_{\beta_j}^2, \tag{10.12}$$

is valid only for independent and uncorrelated variables. Here we are especially interested in the uncertainty in the calibration function itself and the unknown x_0. Since the parameters from an LS fit are typically correlated, we must use the full expression,

$$\sigma_f^2 = \mathbf{g}^T \mathbf{V} \mathbf{g}, \tag{10.13}$$

in which the jth element of the vector $\mathbf{g}$ is $\partial f/\partial \beta_j$. This expression is rigorously correct for functions f that are linear in variables β_j that are themselves normal variates (Tellinghuisen, 2001). Equation (10.13) degenerates into Eq. (10.12) when the off-diagonal elements of $\mathbf{V}$ vanish, which is the condition for zero correlation among the parameters.

In many cases the computation of σ_f can be facilitated by simply redefining the fit function so that f is one of the adjustable parameters of the fit. For example, consider fitting a set of data to the quadratic function $y = a + bz + cz^2/2$, where $z = (x - x_1)$. Fits to this function are statistically equivalent for all x_1; and σ_a, σ_b, and σ_c can be seen to be the standard errors in f and its first two derivatives at $x = x_1$. Thus, one can generate values for these quantities and their errors at all desired x_1 by simply repeating the fit for each x_1 of interest.

2.2. Simple illustrations

Consider first perhaps the simplest application of these results, the fitting of heteroscedastic data to a weighted average using the response function $y = a$. The $\mathbf{X}$ matrix contains a single column of n 1's, giving $\mathbf{A} = \Sigma w_i$ and $\mathbf{A}^{-1}$ the reciprocal of this. The rhs of Eq. (10.6) is $\Sigma w_i y_i$, so $a = \Sigma w_i y_i / \Sigma w_i$ [as given after Eq. (10.2)]. If the σ_i are known, $\mathbf{V}_{\text{prior}} = \mathbf{A}^{-1}$; if they are all the same, $\mathbf{V}_{\text{prior}} = \sigma^2/n$, from which $\sigma_a = \sigma/\sqrt{n}$. If the scale of the σ_i is not known, we have $\mathbf{V}_{\text{post}} = \Sigma w_i \delta_i^2 / [(n-1) \Sigma w_i]$; if all w_i are the same, $\mathbf{V}_{\text{post}} = \Sigma \delta_i^2/[(n-1) n]$, which can be recognized as the expression for the sampling estimate of the variance in the mean. These last results illustrate two important points about LS: (1) The parameter standard errors are truly standard errors, meaning that they correctly reflect the increasing precision with increasing n and (2) the results for the parameters (and their a posteriori standard errors) are independent of scale factors in the w_i, which always cancel.

As a variation on this exercise, consider the fit of homoscedastic data to $y = ax$. The $\mathbf{X}$ matrix contains a single column of values x_i, giving $\mathbf{A} = \Sigma w_i x_i^2$. The rhs of Eq. (10.6) is $\Sigma w_i x_i y_i$, so $a = \Sigma w_i x_i y_i / \Sigma w_i x_i^2$. If σ is known,

$\mathbf{V}_{\text{prior}}$ ($= \sigma_a^2$) $= \sigma^2/\Sigma\, x_i^2$; if σ is not known, we set $w_i = 1$ (OLS) and obtain $\mathbf{V}_{\text{post}} = (S/\nu)/\Sigma\, x_i^2$, which differs from $\mathbf{V}_{\text{prior}}$ through replacement of the true variance σ^2 by its sampling estimate.

Suppose we recognize that each value of y yields an estimate of a, $a_i = y_i/x_i$ and choose to obtain our best estimate of a by averaging these. This must be a weighted average, as the variable transformation has altered the weights of the different values. Letting $z_i = y_i/x_i$, the LS estimate of a should employ weights $w_{zi} \propto 1/\sigma_{zi}^2$. We obtain σ_{zi}^2 by applying error propagation, Eq. (10.12), in which we remember that by assumption x_i has no uncertainty. We find $w_{zi} \propto x_i^2$ and the reader may verify that all results of this approach are identical to those from the original fit to $y = ax$ — a general result for LLS under such linear (in the parameters) transformations that preserve data normality.

In the last example, suppose the original data had proportional error, $\sigma_i \propto y_i$. Since we assume that y is proportional to x, we also have $\sigma_i \propto x_i$ and we see that the transformation to z renders the original weighted fit into a simple unweighted average, which is generally the point of such variable transformations. The two treatments again yield identical results for any data set, provided we actually evaluate the weights in the original fit to $y = ax$ using x_i instead of y_i. If we use the latter, the weights become subject to the random error in y_i, and the results become slightly inequivalent. This problem can be reduced by assessing the weights in y but using the LS estimated values, $y_{\text{calc},i} = ax_i$. However, this approach requires that the w_i be reassessed iteratively in the LS fit, and it still fails to yield results identical to those for the unweighted average. It is these problems that cause analysts to prefer weighting schemes based on functions of x instead of y. However, in a fundamental sense, it is y that is uncertain and it is reasonable to expect its error to depend on it directly. For example, in spectrophotometry, the statistical data error can be related to the wavelength and the absorbance in a universal way (Ingle and Crouch, 1988; Tellinghuisen, 2000d), whereas expression in terms of the concentration of the absorbing analyte (x) would perforce be different for each analyte that has a different absorption spectrum.

We move now to the case of the standard linear response function, $y = a + bx$. $\mathbf{X}$ contains two columns — the first all 1's, the second the n x_i values ($X_{i2} = x_i$). $\mathbf{A}$ is 2×2 with elements $A_{11} = \Sigma\, w_i$, $A_{12} = A_{21} = \Sigma\, w_i x_i$, and $A_{22} = \Sigma\, w_i x_i^2$. $\mathbf{A}^{-1}$ has elements $A^{-1}_{11} = A_{22}/D$, $A^{-1}_{12} = A^{-1}_{21} = -A_{12}/D$, and $A^{-1}_{22} = A_{11}/D$, where $D = \det(\mathbf{A}) = A_{11}A_{22} - A_{12}^2$. (The matrix novice should verify that $\mathbf{A}\mathbf{A}^{-1} = \mathbf{A}^{-1}\mathbf{A} = \mathbf{I}$, the identity matrix, with 1's on the diagonal and 0 off.) The rhs of Eq. (10.6) is a two-element column vector containing $\Sigma\, w_i\, y_i$ in the first position and $\Sigma\, w_i\, x_i\, y_i$ in the second. The solutions to these equations are readily available (e.g., Draper and Smith, 1998), and the reader should verify that Eq. (10.7) yields these results. Note again that use of $w_i = 1$ yields the OLS results.

Applying error propagation, Eq. (10.13), to the response function itself, $f = a + bx$, we find $g_1 = 1$, $g_2 = x$, and $\sigma_f^2 = V_{11} + 2\,V_{12}\,x + V_{22}\,x^2$ from

V_{prior} or $s_f^2 = V_{11} + 2 V_{12} x + V_{22} x^2$ from V_{post}, where σ_f^2 and s_f^2 represent the true and estimated variance in f, respectively. Since $V_{11} = \sigma_a^2$ (s_a^2) and $V_{22} = \sigma_b^2$ (s_b^2), we see that this result is the same as that from Eq. (10.12), plus the correlation term, $2 V_{12} x$.

If we fit to a line through the origin with a quadratic correction, $y = ax + bx^2$, we find $X_{i1} = x_i$, and $X_{i2} = x_i^2$, giving $A_{11} = \Sigma w_i x_i^2$, $A_{12} = A_{21} = \Sigma w_i x_i^3$, and $A_{22} = \Sigma w_i x_i^4$. The rhs of Eq. (10.6) has elements $\Sigma w_i x_i y_i$ and $\Sigma w_i x_i^2 y_i$. Alternatively, we might transform to $z = y/x$, as in the earlier example of the straight line through the origin. This transformation will again yield identical results when weights are properly taken into account; and if the original data have proportional error, the transformed data can be fitted using OLS.

2.3. Extensions and practical concerns

Most of the key equations given earlier have assumed that the RF is linear in the adjustable parameters [Eq. (10.5)] giving linear LS. This assumption is applicable to many RFs of interest, including polynomials in x, but not to all. By way of example, a simple case of nonlinear LS is the fit of data to a constant in the form $y = 1/A$: While the value of A obtained in such a fit is exactly the reciprocal of the value obtained in the linear fit to $y = a$, the former fit is nonlinear. The equations for NLS are not reviewed here, but the key expressions for V_{prior} and V_{post} are the same (although the elements of A are different)(Johnson and Frasier, 1985; Press et al., 1986; Johnson and Faunt, 1992). The chief differences are that NLS fits must be solved iteratively, and the elements of A and V may include dependence on the adjustable parameters and on y. One can still define an "exact" V_{prior} for exactly fitting data; however, nonlinear estimators are inherently nonnormal and indeed may not even have finite variance (e.g., A in the foregoing simple example). Thus the predictive properties for nonlinear parameters become approximate only, but are typically sufficiently reliable when the parameter relative standard errors are smaller than 0.1 (10%)(Tellinghuisen, 2000e). Interestingly, the important parameter x_0 in calibration is such a nonlinear estimator (Shukla, 1972). Data analysis programs such as KaleidaGraph, Igor, Origin, and SigmaPlot make it easy to carry out NLS fits, so some such applications are discussed below.

Users of commercial data analysis programs like those just named should be aware that these programs do not always make clear which V is being used to produce the parameter standard errors. For example, recent versions of KaleidaGraph use V_{post} in unweighted fits but V_{prior} in weighted fits. Thus in cases where weights are known in only a relative sense (e.g., from a data transformation of homoscedastic data of unknown σ), the user must scale the parameter error estimates by the factor $(S/v)^{1/2}$ to obtain the correct a posteriori values. KaleidaGraph also requires that the user provide data σ values, from which the w_i are computed in the program. Origin offers the user a choice of ways of inputting weights and an option for including the factor $(S/\nu)^{1/2}$ to convert to a posteriori. (In the KaleidaGraph program, the quantity called "Chisq" in the output box is just S, which is χ^2 only when the input σ_i values are valid in an absolute sense.)

3. Experiment Design Using V_{PRIOR} — Numerical Illustrations

3.1. Characterizing a nonlinear response function

In this section I illustrate how knowledge of the data variance can facilitate experiment design. I consider a three-parameter model consisting of a straight line with finite intercept exhibiting saturation curvature at large x, accommodated by a cubic term in the model. Such behavior occurs in several analytical techniques, including fluorescence. Bioanalytical workers frequently use a geometric data structure, so I have chosen seven x_i values of this sort, spanning two orders of magnitude: 0.1, 0.2, 0.5, 1, 2, 5, 10. For comparison, I have also used seven evenly spaced x_i covering the same region. Both data structures are investigated for the two common extremes of data error: constant and proportional. For the latter, I have assumed 5% of y; for the former, I take $\sigma = 0.6$, chosen to give about the same SE in the slope for the model.

The model is illustrated in Fig. 10.1, which also shows results obtained for fitting to an altered function with the cubic term replaced by a quadratic

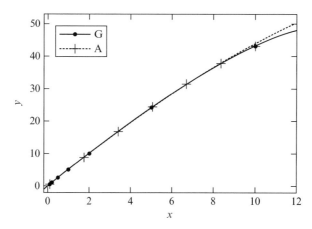

Figure 10.1 Model response function, $y = 0.1 + 5x - 0.007 x^3$, showing geometric x-data structure (G) and arithmetic structure (A). The dashed curve shows results of fitting the A data to a modified response function in which the cubic correction term is replaced by $-cx^2$.

term. The curvature in this model is strong enough to permit most workers to recognize it in the presence of the assumed data error, but many would reasonably choose the quadratic correction term, so it is of interest to examine the effects of this model error. Results of all calculations, done with KaleidaGraph, are summarized in Table 10.1.

From the first four result lines in Table 10.1, all parameters are statistically significant to more than 2 σ except the intercept for constant error, which would fail to survive most tests of significance, even if it were hypothesized to be present. In ad hoc fitting we would always drop this term, a choice that has significant advantages for the overall calibration precision, as discussed later. Recalling that the parameter SEs scale with the data error, we see that a would become justified in the constant-error model only if the data error were reduced to ≈ 0.1. Dependence on the data structure (G or A) is small, so the common practice of choosing the geometric structure is not supported. Here it is worth mentioning that for the linear response function and constant error, the optimal placement of the calibration points is half at each end of the range, whereas for a quadratic RF, a three-value scheme is best (Francois et al., 2004); however, when the form of the RF is not known in advance, better sampling of the calibration range seems advisable.

The next four lines give results for fitting to the alternative quadratic response function. Again the intercept is insignificant for the constant-error models. These results include the model error, in the form of the increase in χ^2. Since the average value of χ^2 is 4 in this case, these increments would not be obvious for individual data sets, especially for the large spread of reasonable values of $\chi^2 \left[\sigma(\chi^2) = \sqrt{8} \right]$. An analyst comparing the two response functions would opt for the quadratic about as often as the cubic on the basis of comparing χ^2 values for the two fits.

Because the intercept is insignificant for constant error, we delete it from the response function and refit both data models to both response functions, obtaining the results in the last four lines. The $\Delta \chi^2$ values remain small, and in this case they result in a reduction in χ_v^2 and in the a posteriori estimated data variance, since v increases from 4 to 5. More importantly, forcing the fit to go through the origin has significantly reduced the SEs for the other two parameters, greatly improving the precision of calibration at small x.

To appreciate this last point, we must investigate the statistical error in the response function. This information is also important for the uncertainty in an unknown to be determined with this RF, because the total uncertainty for the unknown x_0 from measured y_0 contains contributions from both y_0 and the calibration function at $x = x_0$. At the same time, to check effects of picking the wrong RF we must examine the bias, which is the difference between the fitted incorrect RF and the true RF. The variance with bias can be taken as the sum of the squared bias and σ_f^2 from Eq. (10.13).

Table 10.1 Statistical properties of test calibration response function: $y = a + bx - cx^3$[a]

Model[a]	a	σ_a	b	σ_b	c	σ_c
GP	0.1	0.03470	5	0.1536	0.007	0.002749
GC	0.1	0.3360	5	0.1856	0.007	0.001818
AP	0.1	0.03536	5	0.1671	0.007	0.002829
AC	0.1	0.5090	5	0.1645	0.007	0.001564
GPQ ($\Delta\chi^2 = 0.28$)	0.08646	0.03751	5.1124	0.1876	0.07719	0.03099
GCQ ($\Delta\chi^2 = 0.22$)	−0.0333	0.3557	5.3303	0.2692	0.10142	0.02653
APQ ($\Delta\chi^2 = 0.23$)	0.07985	0.03970	5.2067	0.2436	0.08500	0.03502
ACQ ($\Delta\chi^2 = 0.23$)	−0.1278	0.5434	5.4022	0.2524	0.10605	0.02405
GC2 ($\Delta\chi^2 = 0.09$)			5.0368	0.1384	0.007274	0.001568
AC2 ($\Delta\chi^2 = 0.04$)			5.0262	0.0960	0.007178	0.001274
GCQ2 ($\Delta\chi^2 = 0.23$)			5.3129	0.1953	0.09998	0.02162
ACQ2 ($\Delta\chi^2 = 0.65$)			5.3549	0.1528	0.10239	0.01834

[a] Seven x_i values disposed geometrically (G; 0.1, 0.2, 0.5, 1, 2, 5, 10) or arithmetically (A, 0.1–10, evenly spaced), having proportional data error (P, $\sigma_i = 0.05\, y_i$) or constant (C, $\sigma_i = 0.6$). The label "Q" indicates results for replacing x^3 in the response function by x^2, whereas "2" designates two-parameter response functions with intercept omitted.

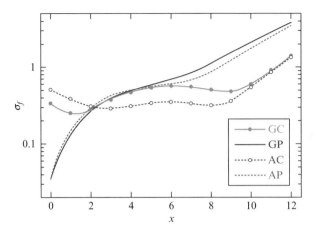

Figure 10.2 Statistical error in the model response function for both data structures and both weightings: solid curves for G, dashed for A, points for constant weighting (C), and none for proportional weighting (P).

Figure 10.2 shows the statistical error in the model RF for the two data structures and two weightings; results are also presented in Table 10.2 to facilitate numerical checks by readers. Results for the two weightings are comparable only because of my choice of magnitudes for the constant and proportional error. However, effects of x structure are correctly compared for each weighting, and we see that for both weightings the G structure performs better only in the small x region, and then only marginally so. This further calls into question the widespread use of the G structure in bioanalytical work. The observation that for proportional error the A structure yields smaller uncertainty than G at large x is particularly noteworthy, because from Table 10.1, all three parameters are actually less precise for the A structure. Thus the smaller σ_f at large x is a manifestation of stronger interparameter correlation for the A structure.

Figure 10.3 shows the effects of dropping the intercept from the constant-error model for the A structure. The statistical error in the modified RF is almost everywhere lower than that in the original RF, but especially so at small x, where it vanishes rigorously at $x = 0$ — the real virtue of an intercept-free RF. While the bias is prominent at $x = 0$, it is quickly subsumed into the total error, which remains comparable to that in the original RF at larger x. Of course the bias occurs here precisely because we built it into the original RF; in the real world the analyst would know only that the fitted intercept was statistically insignificant, hence should be dropped from the RF. The present results show that even with the bias, the total uncertainty remains much smaller at small x when the RF is made to pass through the origin. In fact, this comparison favors deleting the intercept even more when the RFs in question are strictly linear.

Table 10.2 Statistical error in response functions and in selected x_0 determinations[a]

x	σ_f					σ_{x0}		
	GC	GP	AC	AP	AC	AP	ACQ2
---	---	---	---	---	---	---	---
0.0	0.3360	0.03470	0.5090	0.03536			
0.1					0.15566	0.0084835	0.11258
0.2					0.15403	0.012856	0.11309
0.5					0.14954	0.029650	0.11497
1.0	0.2535	0.1310	0.3881	0.1490	0.14352	0.059259	0.11912
2.0	0.2821	0.2713	0.3089	0.2989	0.13728	0.11888	0.12971
3.0	0.3765	0.3944	0.2891	0.4242			
4.0	0.4730	0.4967	0.3124	0.5153	0.14504	0.23790	0.15120
5.0	0.5425	0.5858	0.3423	0.5710			
6.0	0.5717	0.6873	0.3538	0.6085	0.16414	0.36609	0.16508
7.0	0.5565	0.8471	0.3396	0.6794			
8.0	0.5090	1.1171	0.3201	0.8658	0.18604	0.55281	0.18148
9.0	0.4849	1.5341	0.3651	1.2280			
10.0	0.5999	2.1182	0.5478	1.7824	0.28026	0.96468	0.23622
11.0	0.9135	2.8822	0.8692	2.5323			
12.0	1.4030	3.8386	1.3150	3.4846	0.73204	2.1430	0.34367

[a] Data structures and error identified as in Table 10.1: G and A, geometric and arithmetic structures; C and P, constant and proportional data error.

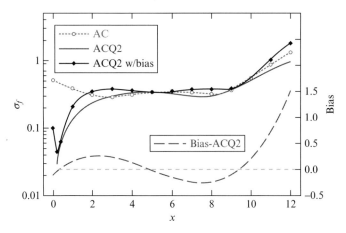

Figure 10.3 Statistical error and bias for arithmetic data structure, showing effects of dropping the statistically insignificant intercept for the constant error model. The bias (linear scale to right) does not exceed 0.3 in magnitude within the range of the data; it is included in the "w/ bias" curve by taking the total variance to be a sum of the statistical variance and (bias)^2.

In closing this section it is worth emphasizing that the results given for incorrect fit models are, like those for the correct function, exact. That is to say, Monte Carlo calculations run on incorrect linear models confirm that the bias, the parameter precisions, and $\Delta\chi^2$ are as observed for the exact data belonging to the correct model, and the biased parameters are normal in their distribution.

3.2. Uncertainty in the unknown

The unknown x_0 is obtained from the measured response y_0 and the calibration function by solving the equation $y_0 = f(x_0)$, which gives the familiar result in the case of the linear RF. The uncertainty in x_0 can then be calculated from the two contributing sources — y_0 and the RF — using error propagation, Eq. (10.12). These two sources add independently and each contributes a similar variance term to σ_{x0}^2, giving

$$\sigma_{x0}^2 = \frac{\sigma_f^2(x_0) + \sigma_{y0}^2}{(df/dx)_0^2}, \qquad (10.14)$$

where the derivatives and variances are evaluated at corresponding y_0 and x_0 values. This expression is relatively easy to evaluate for response functions other than linear, so there is little reason for analysts to continue to rely so heavily on linear RFs only. Moreover, through my nonlinear algorithm, this error propagation is done automatically, at the same time that the calibration data are fitted (Tellinghuisen, 2000c, 2005a). In this algorithm the y_0 value is added to the data set and provided with a dummy x value that can be used to treat it differently in the fit. In the present case, a single y_0 value is fitted exactly (even in the case of real data) to

$$y_0 = a + bx_0 - cx_0^3, \qquad (10.15)$$

in which x_0 is a fourth adjustable parameter in the fit, the output of which thus includes its uncertainty. KaleidaGraph can be used for this fit by defining the fit function as, e.g.,

$$(x > 10)?(a + b^*d - c^*d^\wedge 3) : (a + b^*x - c^*x^\wedge 3); \qquad (10.16)$$

in which the dummy x value is some value greater than the largest x_i value in the calibration set and d is x_0. C programmers will recognize the syntax of the statement in Eq. (10.16); the initial test determines whether the quantity in the first set of parentheses is evaluated (for y_0) or the second (the calibration data, all having $x_i \leq 10$). The y_0 value can be given any suitable σ value. Thus single measurements would be expected to have the same σ as the calibration data for the same y, whereas values obtained by averaging m measurements would have σ smaller by the factor $\sqrt{m}$. Alternatively, if y_0 is assigned a very small σ, its contribution to the numerator of Eq. (10.14) vanishes and one can use the result to evaluate σ_f at $x = x_0$.

Results for σ_{x_0} resemble those for σ_f, but it is instructive to examine the results for the constant-error model with deleted intercept. Fig. 10.4 shows

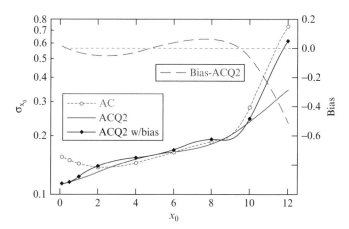

Figure 10.4 Statistical error and bias in x_0 as determined for the same data structure and RFs illustrated in Fig. 10.3. The bias (observed − true) is treated the same as in Fig. 10.3 in assessing the total σ with bias.

that the extra precision of calibration at small x is not fully realized in the determination of the unknown. The reason is that y_0 still has its full uncertainty, and this dominates in the numerator of Eq. (10.14). Thus, to reap the precision rewards of intercept-free calibration, it is necessary to reduce the uncertainty in y_0, e.g., by taking replicates. (In general, best precision in x_0 is obtained by splitting the total measurements equally between the calibration data and y_0.) Some numerical results for σ_{x0} are included in Table 10.2.

3.3. Should I try a data transformation?

As noted earlier, the analyst can sometimes use a data transformation to simplify the analysis. There are two reasons one might take this path: (1) to simplify the response function and (2) to simplify the data weighting. In my opinion, the first of these is seldom valid (see later) and in particular I have just illustrated how one can straightforwardly extract both x_0 and its uncertainty from any response function. There is some validity to the second reason, as strongly heteroscedastic and noisy data raise the question of just how the data error and hence the weights are to be evaluated. For example, if the data have proportional error, the weights themselves will be afflicted with experimental error if evaluated using $\sigma_i \propto y_i$. In the aforementioned computations we had the advantage of working with exact data, ensuring correct weights for the assumed proportional error model. With real data, one can use $\sigma_i \propto y_{\text{calc},i}$, but this requires iterative adjustment of the weights, since $y_{\text{calc},i}$ is not known at the outset. The use of weighting functions of the (error-free) x_i values avoids this problem but raises others and denies the physical reality that the data uncertainty is fundamentally a property of the dependent variable y and not of the x that produces this y.

In the present case we might attempt to transform the proportional error model into one with constant error by dividing through by x, as in the simple examples given earlier. Because of the way y is related to x through the model, this transformation does not produce constant error, although the remaining heteroscedasticity is small enough to be ignored in most applications. Also, the transformed response function, $z = a/x + b - cx^2$, varies strongly near $x = 0$ in a way that is not desirable for calibration. (This would not be the case if a were negligible.) However, with the conversion of weights through $\sigma_z = \sigma_y/x$, we do obtain parameter SEs in this fit to z that are identical to those given in Table 10.1 for the GP and AP models, as predicted in the earlier illustrations.

A more promising transformation is $z = \ln(y)$. By error propagation, $\sigma_z = \sigma_y/y$, so with $\sigma_y = 0.05\, y$, $\sigma_z = 0.05$. "Exact" equivalence is obtained by now defining the response function as $z = \ln(a + bx - cx^3)$. This is a nonlinear fit, which is nonetheless carried out easily with data analysis programs such as KaleidaGraph [e.g., by just entering `ln (a + b*x - c*x^3)` in the "Define Fit" box and providing initial estimates for a, b, and c]. While this approach does eliminate the ambiguity in calculating the weights of the original model and yields identical "exact" parameter SEs, the actual parameter errors now vary somewhat for different real data sets due to the nonlinearity. In any event, this is not the approach typically taken. Rather $z = \ln(y)$ is fitted to a simple function of $\ln(x)$. If we try polynomials in $\ln(x)$, we indeed achieve a good fit with a constant, a linear term, and a cubic term, giving a modest $\Delta\chi^2 = 0.13$. In contrast, replacing the cubic term by a quadratic one yields a significantly poorer fit, with $\Delta\chi^2 = 2.04$.

Some detection methods have inherently nonlinear response functions that can be made linear through a variable transformation. For example, methods that utilize an adsorption isotherm, fluorescence quenching, or enzyme kinetics have a hyperbolic dependence of form $f(x) = a/(1 + bx)$ or $ax/(1 + bx)$ and have frequently been handled by inversion. If the data have proportional error, they will still have proportional error on inversion, but it is nonetheless better to fit original (normal) data using NLS than their inverses (nonnormal) using LLS. This recommendation becomes even stronger if the original data have constant uncertainty, because by Eq. (10.12) the w_i for fitting $z = 1/y$ should be $\propto y^4$, so neglect of weights following such a transformation can have disastrous consequences for the calibration precision. Even when the weights are properly taken into account, the parameter estimates are biased and even *inconsistent*, which means that they retain bias in the limit of an infinite number of data points (Tellinghuisen, 2000f).

3.4. Variance function estimation

Up until now we have had the luxury of knowing the data error structure, being able to assign weights by assumption. Unfortunately, nature seldom hands the analyst this information. [An exception is the use of counting instruments, which often follow Poisson statistics closely enough to permit taking the data variance equal to the count; except for very small counts, Poisson data are also nearly Gaussian in distribution (Bevington, 1969).] From the experimental side we can never have perfect a priori information about σ_i. However, there is good evidence that when the data variance depends smoothly on the experimental parameters, it is possible to estimate the VF reliably enough to treat σ_i as known, even from relatively few total measurements. Importantly, this information permits one to use the χ^2 statistic as a quantitative indicator of *goodness of fit* and to estimate confidence intervals using the normal distribution rather than the t distribution.

In connection with the last statement, it is worth recalling the statistical properties of χ^2 and their implications for the precision of sampling estimates of σ^2 and σ. Since the variance in $\left(\chi_\nu^2\right)$ is $2/\nu$, the relative standard deviation in the sampling estimate s^2 is $\sqrt{2/\nu}$. From error propagation, the relative standard deviation in s is half that in s^2 or $\sqrt{1/(2\nu)}$. For $\nu = 200$, this translates into a nominal 5% standard deviation in s and hence also in the a posteriori parameter SE estimates ($V_{jj}^{1/2}$), but it is a whopping 50% when $\nu = 2$. Further, the distribution of s^2 is far from Gaussian for small ν — exponential for $\nu = 2$ [see Eq. (10.10)]. Still, such data can be fitted reliably using LS (Tellinghuisen, 2008a), and I will now illustrate how variance function estimation (VFE) can be accomplished for replicate synthetic data generated using our sample RF.

A simple and instructive way to generate random normal variates from which to produce synthetic data is to just sum 12 random numbers as produced by many data analysis programs, including KaleidaGraph (Tellinghuisen, 2008b). Individual random numbers follow the uniform distribution, usually over the default interval 0–1. The average of this distribution is 0.5 and the variance 1/12, so a sum of 12 such numbers has mean 6 and $\sigma = 1$. Importantly, and an impressive illustration of the central limit theorem of statistics, the distribution of this sum is very nearly Gaussian. This approach has been used to generate three replicate data sets for both the G and the A data structures of our sample RF, using the VF,

$$\sigma^2 = a^2 + (by)^2 \qquad (10.17)$$

with $a = 0.1$ and $b = 0.01$. This function is physically reasonable for bridging the range between very low signal, where constant noise generally dominates instrumental statistics, and high signal, where proportional error seems prominent in many techniques, including spectrophotometry, isothermal titration calorimetry, and high-performance liquid chromatography (HPLC)(Ingle and Crouch, 1988; Tellinghuisen, 2000d, 2005c; Zeng et al., 2008). The values of a and b in Eq. (10.17) were chosen to make both contributions to the VF significant in part of the x range of the data and to keep the RF intercept significant in most data sets. The synthetic data are given in Table 10.3. The means and sample variances for the three replicate values are also included.

As was noted earlier, sampling estimates of variances have proportional error. For $\nu = 2$, the proportionality constant is $\sqrt{2/\nu} = 1$. To fit the s^2 values directly to our variance function, we should weight inversely as σ^4, but because of the large uncertainty in the s^2 estimates themselves, we cannot approximate σ^4 with s^4 for this purpose. Rather, we must use the calculated VF from the fit, which means that we must iteratively adjust the weights, refit, and then readjust the weights. This approach works surprisingly well, even for as few as three replicates, with their concomitant large uncertainty and strongly nonnormal distribution (Tellinghuisen, 2008a).

There is an alternative approach that avoids this iteration. As was noted earlier, for proportional error, a logarithmic transformation yields constant weights. Here they are all unity, since we have three replicates for each

Table 10.3 Synthetic replicate data for test calibration model[a,b]

x	y_1	y_2	y_3	Mean	s^2
0.10	0.565	0.536	0.520	0.54033	0.00052033
0.20	1.102	1.031	1.078	1.0703	0.0013043
0.50	2.585	2.614	2.566	2.5883	0.00058434
1.00	5.029	5.046	5.014	5.0297	0.00025633
2.00	9.943	9.914	10.140	9.9990	0.015121
5.00	23.664	24.199	24.386	24.083	0.14041
10.00	43.422	43.104	42.777	43.101	0.10401
0.10	0.635	0.591	0.470	0.56533	0.0073003
1.75	8.661	8.986	9.005	8.8840	0.037387
3.40	16.929	17.309	16.967	17.068	0.043801
5.05	23.921	24.629	25.011	24.520	0.30588
6.70	31.634	31.692	31.392	31.573	0.025321
8.35	37.193	37.441	37.337	37.324	0.015509
10.00	43.387	44.255	43.068	43.570	0.37736

[a] Geometric data first, followed by arithmetic data.
[b] VFE analysis shown in Fig. 10.5. Resulting fit of G data to RF = $a + bx - cx^3$ yielded $a = 0.0615 \pm 0.0101$, $b = 4.9944 \pm 0.0182$, $c = 0.006937 \pm 0.000284$, and $\chi^2 = 6.96$. The fit of A data to the same function gave $a = 0.1098 \pm 0.0657$, $b = 5.0219 \pm 0.0297$, $c = 0.007170 \pm 0.000366$, and $\chi^2 = 15.70$. Replacement of x^3 by x^2 in the RF gave $a = -0.02276 \pm 0.06909$, $b = 5.3474 \pm 0.0446$, $c = 0.09967 \pm 0.00505$, and $\chi^2 = 10.69$. Deletion of a in this model led to $b = 5.3370 \pm 0.0313$, $c = 0.09875 \pm 0.00422$, and $\chi^2 = 10.80$. All stated parameter SEs are a priori.

value. Fitting to our VF requires a nonlinear algorithm, but this is easy with programs like KaleidaGraph, where this example is analogous to that mentioned earlier in connection with data transformations: just enter ln (a^2 + (b*x)^2) in the "Define Fit" box and supply initial estimates for a and b. Results are shown in Fig. 10.5 for both data structures.

Both replicate sets have yielded reasonable estimates for b and for χ^2, but the G structure has produced an anomalously small a. Using exact data, the reader can verify that the exact parameter SEs here are 0.0266 and 0.0443 for a in the G and A structures, respectively, and 0.00371 and 0.00261 for b for G and A, respectively. The differences reflect the different data densities in the two structures. Also, calculations have shown that the log transformation yields significant negative bias to the estimates of the individual VF parameters and overshoots on χ^2 (Tellinghuisen, manuscript in preparation); however, it is the b/a ratio that determines the correctness of weights for the fit of data to the RF, and this ratio is biased only slightly.

We next use the fitted VF to weight the data in the fit to the RF. Because our VF applies to individual points, we can fit all 21 points of the combined sets using this function or we can fit the average values using weights $3/\sigma^2(y_{avg})$ to account for the averaging. Results for the parameters

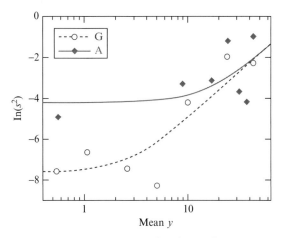

Figure 10.5 Estimation of variance function, $\sigma^2 = a^2 + (by)^2$, from unweighted fits of the logarithm of the sampling estimates s^2 obtained from the replicate data in Table 10.3. For the G data structure, the fit yields $a = 0.0225 \pm 0.0101$, $b = 0.00817 \pm 0.00273$, and $\chi^2 = 7.71$; for the A structure, $a = 0.123 \pm 0.069$, $b = 0.00786 \pm 0.00353$, and $\chi^2 = 9.97$. The true values of a and b used to generate the synthetic data were 0.1 and 0.01, respectively.

and their a priori SEs will be identical, but the χ^2 values will not be, as we have $v = 18$ for the combined points but $v = 4$ for the averages. The curvature is clear for data this precise, so initially we fit to our RF and to modifications with the cubic term replaced by a quadratic or quartic.

For the G structure we find that the cubic is preferred overwhelmingly, using χ^2 as our figure of merit (6.93 vs >18 for the alternative functions). At the same time, the R values all start with four 9s, showing how the useful information is compressed into small differences between 1 and R or R^2 when these quantities are used. For the A structure, the quadratic modification clearly outperforms the other functions ($\chi^2 = 10.6$ cf 15.7 for the true RF). Furthermore, the intercept in this fit is insignificant — $a = -0.023 \pm 0.069$ — so we drop it and refit, obtaining a modest increase of χ^2 to 10.8, but a significant reduction in χ_v^2. Results for both data structures are included in Table 10.3.

In these two cases, use of χ^2 for goodness of fit has yielded unambiguous answers for the RF among the test functions (albeit not the true function in one case). Of course we have not examined every possible function having the proper shape, and it is likely that others — like a constant minus a declining exponential (Tellinghuisen, 2000c) — might perform as well or better. In many cases, the calibration needs will be satisfied more than adequately by any such function. In cases where subtle differences might be important, it is easy to simply determine the unknown(s) with the two or more comparable RFs and include the differences (model uncertainty) in the assessment of the overall uncertainty of x_0. In the interest of conservative safety of estimation, it is also reasonable to quote the larger of the $\mathbf{V}_{\text{prior}}$- or $\mathbf{V}_{\text{post}}$-based uncertainty estimates unless there is high confidence in the a priori values.

3.5. When χ^2 is too large

With allowance for the bias effects, the χ^2 values returned by the best fits to data in Table 10.3 are reasonable. For example, the larger value 10.8 leads to a reduced χ^2 of 2.16, which occurs more than 5% of the time for $v = 5$; however, this estimate is biased positively by $\approx 70\%$, correction for which raises the probability to $\approx 25\%$ (Bevington, 1969). An extremely small χ^2 value usually just means that the data error has been grossly overestimated and is seldom a source of lost sleep. However, excessively large values imply that there is something wrong with the model, data, or both (which again might be just a misestimation of the data error). In a case study of extensive HPLC data for acetaldehyde analyzed in the manner illustrated here, Zeng *et al.* (2008) observed a limiting χ_v^2 of ≈ 400 for a high-order polynomial. Examination of the residuals showed essentially random scatter, so it was clear that further modifications of the RF could not help. The effective data error was simply 20 times larger than we thought on the basis of our analysis of the VF.

Our replicate data were designed to characterize the instrumental data variance function, which we accomplished by just injecting multiple samples of each prepared calibration solution. It was immediately clear that our overall data error was dominated by something other than the instrument error. Since solution preparation was the only other experimental activity, this meant that the uncertainty was primarily in our solution concentrations, not our measured quantity (the HPLC peak area). Recalling the fundamental postulates of LS, it is clear that we should have taken the HPLC average peak area as independent variable and concentration as dependent variable, which we then did.

With the remarkable precision of many modern analytical instruments, it is likely that this situation will be found to occur frequently, but it required attention to χ^2 for us to recognize it and would not be evident to an analyst using R^2 as a figure of merit (the fit yielded better than three nines). There is actually a nice upside to this situation: The estimator of the unknown x_0 is now linear (since it is obtained by just inserting the error-free y_0 into the calibration formula) and it will often have only the uncertainty of calibration associated with it.

There is also a downside: Getting true replicate data may entail much more effort, and it can be hard to ensure their randomness. In our case study, the instrumental replicates were of no value in determining the effective variance function for the reversed regression. However, our residuals did appear to be random, and we had enough data points to use another approach — generalized least squares (GLS)(Baumann and Wätzig, 1995; Carroll and Ruppert, 1988; Davidian and Haaland, 1990; Hwang, 1994). In this method, the LS residuals themselves are used to estimate the VF in an iterative computation in which the data are initially fitted to the RF, after which the residuals are fitted to the VF, which is then used to estimate new weights for the data in the next fit to the RF. This procedure requires fitting the squares or absolute values of the residuals and has the same concerns about weighting in the estimation of the VF, since the δ_i^2 values have the statistical properties of χ^2. Also, because the estimation of the VF is inextricably tied to the estimation of the RF, it is more difficult to examine different functional forms for both of these. The bootstrapping computation coverges adequately in typically four to six cycles, and test computations show that GLS is actually slightly more efficient than VFE from replicates when the forms of both the VF and the RF are known (Tellinghuisen, 2008d). However the efficiency edge for GLS is not great, so when there are questions about the functional forms of the VF or RF, it is better to record true replicates, so as to permit estimation of the VF independent of the RF. Accordingly, GLS will not be discussed further here.

4. Conclusion

Knowledge of the data variance greatly empowers the data analyst, as it facilitates reliable prediction of uncertainties and confidence intervals for linear and nonlinear LS fits of the data in question. In calibration, such experiment design can be used to optimize the data structure to achieve desired goals. In the present illustrations with a realistic nonlinear calibration function, results showed strong dependence on the data heteroscedasticity but weak dependence on the data structure — in particular, little value in the geometric structure preferred in much bioanalytical work. If an analysis protocol calls for a demanding percent reliability at small x, this would have to be accomplished through an unbalanced replicate scheme if the data are homoscedastic. However, for the proportional error often exhibited by chromatographic data, there is a natural proclivity to satisfy a percent reliability criterion at all x (Tellinghuisen, 2008c).

Since statistically optimal weights should be taken proportional to the inverse data variance, analysts should focus their efforts on determining this property instead of on indirect methods that attempt to select a weighting scheme by trial and error. Several studies have shown that the numbers of replicates often taken in bioanalytical protocols suffice to determine the data variance function with adequate precision to yield near-optimal calibration functions. However, more effort may be required to ascertain the true functional form of the variance function. The form of Eq. (10.17), used here to illustrate VFE, has been found to be a realistic approximation for the instrumental variance for several techniques. Although this form is more general than the single-parameter choices often used for weighting functions, there is no guarantee that it will be found to apply in any particular analytical method. Also, it is likely that error sources other than instrumental will be found to dominate in many protocols. In such cases it may prove difficult to obtain statistically reliable replicates for assessing the effective data error.

Regarding these last points, it is highly desirable to have reliable information about the instrumental data error, and facility with χ^2 as a goodness-of-fit indicator, in order to be able to recognize when the dominant error is other than instrumental. In such cases, reversed regression should be used in calibration. This approach has some clear statistical advantages that at least partially offset the just-noted drawback about getting reliable information on the effective variance.

REFERENCES

Albritton, D. L., Schmeltekopf, A. L., and Zare, R.N (1976). An introduction to the least-squares fitting of spectroscopic data. *In* "Molecular Spectroscopy: Modern Research II" (K. Narahari Rao, ed.), pp. 1–67. Academic Press, New York.

Almeida, A. M., Castel-Branco, M. M., and Falcão, A. C. (2002). Linear regression for calibration lines revisited: Weighting schemes for bioanalytical methods. *J. Chromatogr. B* **774,** 215–222.

Baumann, K., and Wätzig, H. (1995). Appropriate calibration functions for capillary electrophoresis II. Heteroscedasticity and its consequences. *J. Chromatogr. A* **700,** 9–20.

Bevington, P. R. (1969). "Data Reduction and Error Analysis for the Physical Sciences." McGraw-Hill, New York.

Burns, M. J., Valdivia, H., and Harris, N. (2004). Analysis and interpretation of data from real-time PCR trace detection methods using quantitation of GM soya as a model system. *Anal. Bioanal. Chem.* **378,** 1616–1623.

Carroll, R. J., and Ruppert, D. (1988). "Transformation and Weighting in Regression." Chapman and Hall, New York.

Danzer, K., and Currie, L. A. (1998). Guideline for calibration in analytical chemistry. Part 1. Fundamentals and single component calibration. *Pure Appl. Chem.* **70,** 993–1014.

Davidian, M., and Haaland, P. D. (1990). Regression and calibration with nonconstant error variance. *Chemom. Intell. Lab. Syst.* **9,** 231–248.

Draper, R. N., and Smith, H. (1998). "Applied Regression Analysis," 3rd Ed. Wiley, New York.

Francois, N., Govaerts, B., and Boulanger, B. (2004). Optimal designs for inverse prediction in univariate nonlinear calibration models. *Chemom. Intell. Lab. Syst.* **74,** 283–292.

Hamilton, W. C. (1964). "Statistics in Physical Science: Estimation, Hypothesis Testing, and Least Squares." Ronald Press Co., New York.

Hwang, L.-J. (1994). Impact of variance function estimation in regression and calibration. *Methods Enzymol.* **240,** 150–170.

Ingle, J. D. Jr., and Crouch, S. R. (1988). "Spectrochemical Analysis." Prentice-Hall, Englewood Cliffs, NJ.

Jacquez, J. A., and Norusis, M. (1973). Sampling experiments on the estimation of parameters in heteroscedastic linear regression. *Biometrics* **29,** 771–779.

Johnson, E. L., Reynolds, D. L., Wright, D. S., and Pachla, L. A. (1988). Biological sample preparation and data reduction concepts in pharmaceutical analysis. *J. Chromatogr. Sci.* **26,** 372–379.

Johnson, M. L., and Frasier, S. G. (1985). Nonlinear least-squares analysis. *Methods Enzymol.* **117,** 301–342.

Johnson, M. L., and Faunt, L. M. (1992). Parameter estimation by least-squares methods. *Methods Enzymol.* **210,** 1–37.

Karnes, H. T., and March, C. (1991). Calibration and validation of linearity in chromatographic biopharmaceutical analysis. *J. Pharm. Biomed. Anal.* **9,** 911–918.

Kiser, M. K., and Dolan, J. W. (2004). Selecting the best curve fit: Which curve-fitting function should be used? *LC-GC North Am.* **22,** 112–115.

Miller, J. N. (1991). Basic statistical methods for analytical chemistry. Part 2. Calibration and regression methods: A review. *Analyst* **116,** 3–14.

Nagaraja, N. V., Paliwal, J. K., and Gupta, R. C. (1999). Choosing the calibration model in assay validation. *J. Pharm. Biomed. Anal.* **20,** 433–438.

Press, W. H., Flannery, B. P., Teukolsky, S. A., and Vetterling, W. T. (1986). "Numerical Recipes." Cambridge Univ. Press, Cambridge, UK.

Sadray, S., Rezaee, S., and Rezakhah, S. (2003). Non-linear heteroscedastic regression model for determination of methotrexate in human plasma by high-performance liquid chromatography. *J. Chromatogr. B* **787,** 293–302.

Shukla, G. K. (1972). On the problem of calibration. *Technometrics* **14,** 547–553.

Tellinghuisen, J. (2000a). Inverse vs. classical calibration for small data sets. *Fresenius J. Anal. Chem.* **368,** 585–588.

Tellinghuisen, J. (2000b). Nonlinear least squares using microcomputer data analysis programs: KaleidaGraph in the physical chemistry teaching laboratory. *J. Chem. Educ.* **77,** 1233–1239.

Tellinghuisen, J. (2000c). A simple, all-purpose nonlinear algorithm for univariate calibration. *Analyst* **125,** 1045–1048.

Tellinghuisen, J. (2000d). Statistical error calibration in UV-visible spectrophotometry. *Appl. Spectrosc.* **54,** 431–437.

Tellinghuisen, J. (2000e). A Monte Carlo study of precision, bias, inconsistency, and non-Gaussian distributions in nonlinear least squares. *J. Phys. Chem. A* **104,** 2834–2844.

Tellinghuisen, J. (2000f). Bias and Inconsistency in linear regression. *J. Phys. Chem. A* **104,** 11829–11835.

Tellinghuisen, J. (2001). Statistical error propagation. *J. Phys. Chem. A* **105,** 3917–3921.

Tellinghuisen, J. (2004). Statistical error in isothermal titration calorimetry. *Methods. Enzymol.* **383,** 245–282.

Tellinghuisen, J. (2005a). Simple algorithms for nonlinear calibration by the classical and standard additions methods. *Analyst* **130,** 370–378.

Tellinghuisen, J. (2005b). Understanding least squares through Monte Carlo calculations. *J. Chem. Educ.* **82,** 157–166.

Tellinghuisen, J. (2005c). Statistical error in isothermal titration calorimetry: Variance function estimation from generalized least squares. *Anal. Biochem.* **343,** 106–115.

Tellinghuisen, J. (2007). Weighted least squares in calibration: What difference does it make? *Analyst* **132,** 536–543.

Tellinghuisen, J. (2008a). Least squares with non-normal data: Estimating experimental variance functions. *Analyst* **133,** 161–166.

Tellinghuisen, J. (2008b). Stupid statistics! *Methods Cell Biol.* **84,** 739–780.

Tellinghuisen, J. (2008c). Weighted least squares in calibration: The problem with using "quality coefficients" to select weighting formulas. *J. Chromatogr. B* **872,** 162–166.

Tellinghuisen, J. (2008d). Variance function estimation by replicate analysis and generalized least squares: A Monte Carlo comparison, to be published.

Zeng, Q. C., Zhang, E., and Tellinghuisen, J. (2008). Univariate calibration by reversed regression of heteroscedastic data: A case study. *Analyst* **133,** 1649–1655.

Zeng, Q. C., Zhang, E., Dong, H., and Tellinghuisen, J. (2008). Weighted least squares in calibration: Estimating data variance functions in HPLC. *J. Chromatogr. A* **1206,** 147–152.

Zhou, M., Wei, G., Liu, Y., Sun, Y., Xiao, S., Lu, L., Liu, C., and Zhong, D. (2003). Determination of vertilmicin in rat serum by high-performance liquid chromatography using 1-fluoro-2,4-dinitrobenzene derivatization. *J. Chromatogr. B* **798,** 43–48.

CHAPTER ELEVEN

Evaluation and Comparison of Computational Models

Jay I. Myung, Yun Tang, *and* Mark A. Pitt

Contents

1. Introduction	288
2. Conceptual Overview of Model Evaluation and Comparison	288
3. Model Comparison Methods	292
3.1. Akaike Information Criterion and Bayesian Information Criterion	292
3.2. Cross-Validation and Accumulative Prediction Error	293
3.3. Bayesian Model Selection and Stochastic Complexity	295
4. Model Comparison at Work: Choosing between Protein Folding Models	297
5. Conclusions	301
Acknowledgments	303
References	303

Abstract

Computational models are powerful tools that can enhance the understanding of scientific phenomena. The enterprise of modeling is most productive when the reasons underlying the adequacy of a model, and possibly its superiority to other models, are understood. This chapter begins with an overview of the main criteria that must be considered in model evaluation and selection, in particular explaining why generalizability is the preferred criterion for model selection. This is followed by a review of measures of generalizability. The final section demonstrates the use of five versatile and easy-to-use selection methods for choosing between two mathematical models of protein folding.

Department of Psychology, Ohio State University, Columbus, Ohio

1. Introduction

How does one evaluate the quality of a computational model of enzyme kinetics? The answer to this question is important and complicated. It is important because mathematics makes it possible to formalize the reaction, providing a precise description of how the factors affecting it interact. Study of the model can lead to significant understanding of the reaction, so much so that the model can serve not merely as a description of the reaction, but can contribute to explaining its role in metabolism. Model evaluation is complicated because it involves subjectivity, which can be difficult to quantify.

This chapter begins with a conceptual overview of some of the central issues in model evaluation and selection, with an emphasis on those pertinent to the comparison of two or more models. This is followed by a selective survey of model comparison methods and then an application example that demonstrates the use of five simple yet informative model comparison methods.

Criteria on which models are evaluated can be grouped into those that are difficult to quantify and those for which it is easier to do so (Jacobs and Grainger, 1994). Criteria such as *explanatory adequacy* (whether the theoretical account of the model helps make sense of observed data) and *interpretability* (whether the components of the model, especially its parameters, are understandable and are linked to known processes) rely on the knowledge, experience, and preferences of the modeler. Although the use of these criteria may favor one model over another, they do not lend themselves to quantification because of their complexity and qualitative properties. Model evaluation criteria for which there are quantitative measures include *descriptive adequacy* (whether the model fits the observed data), *complexity* or *simplicity* (whether the model's description of observed data is achieved in the simplest possible manner), and *generalizability* (whether the model provides a good predictor of future observations). Although each criterion identifies a property of a model that can be evaluated on its own, in practice they are rarely independent of one another. Consideration of all three simultaneously is necessary to assess fully the adequacy of a model.

2. Conceptual Overview of Model Evaluation and Comparison

Before discussing the three quantitative criteria in more depth, we highlight some of the key challenges of modeling. Models are mathematical representations of the phenomenon under study. They are meant to capture patterns or regularities in empirical data by altering parameters that correspond to variables that are thought to affect the phenomenon. Model specification is difficult because our knowledge about the phenomenon being modeled is rarely complete. That is, the empirical data obtained from studying the phenomenon are limited, providing only partial information (i.e., snapshots) about its properties and the variables that influence it. With limited information, it is next to impossible to construct the "true" model. Furthermore, with only partial information, it is likely that multiple models are plausible; more than one model can provide a good account of the data. Given this situation, it is most productive to view models as approximations, which one seeks to improve through repeated testing.

Another reason models can be only approximations is that data are inherently noisy. There is always measurement error, however small, and there may also be other sources of uncontrolled variation introduced during the data collection process that amplifies this error. Error clouds the regularity in the data, increasing the difficulty of modeling. Because noise cannot be removed from the data, the researcher must be careful that the model is capturing the meaningful trends in the data and not error variation. As explained later, one reason why generalizability has become the preferred method of model comparison is how it tackles the problem of noise in data.

The descriptive adequacy of a model is assessed by measuring how well it fits a set of empirical data. A number of goodness-of-fit (GOF) measures are in use, including sum of squared errors (SSE), percent variance accounted for, and maximum likelihood (ML; e.g., Myung, 2003). Although their origins differ, they measure the discrepancy between the empirical data and the ability of a model to reproduce those data. GOF measures are popular because they are relatively easy to compute and the measures are versatile, being applicable to many types of models and types of data. Perhaps most of all, a good fit is an almost irresistible piece of evidence in favor of the adequacy of a model. The model appears to do just what one wants it to—mimic the process that generated the data. This reasoning is often taken a step further by suggesting that the better the fit, the more accurate the model. When comparing competing models, then, the one that provides the best fit should be preferred.

Goodness of fit would be suitable for model evaluation and comparison if it were not for the fact that data are noisy. As described earlier, a data set contains the regularity that is presumed to reflect the phenomenon of interest plus noise. GOF does not distinguish between the two, providing a single measure of a model's fit to both (i.e., GOF = fit to regularity + fit to noise). As this conceptual equation shows, a good fit can be achieved for the wrong reasons, by fitting noise well instead of the regularity. In fact, the better a model is at fitting noise, the more likely it will provide a superior fit than a competing model, possibly resulting in the selection of a model that in actuality bears little resemblance to the process being modeled. GOF alone is a poor criterion for model selection because of the potential to yield misleading information.

This is not to say that GOF should be abandoned. On the contrary, a model's fit to data is a crucial piece of information. Data are the only link to the process being modeled, and a good fit can indicate that the model mimics the process well. Rather, what is needed is a means of ensuring that a model does not provide a good fit for the wrong reason.

What allows a model to fit noisy data better than its competitors is that it is the most complex. *Complexity* refers to the inherent flexibility of a model that allows it to fit diverse data patterns (Myung and Pitt, 1997). By varying the values of its parameters, a model will produce different data patterns. What distinguishes a simple model from a complex one is the sensitivity of the model to parameter variation. For a simple model, parameter variation will produce small and gradual changes in model performance. For a complex model, small parameter changes can result in dramatically different data patterns. It is this flexibility in producing a wide range of data patterns that makes a model complex. For example, the cubic model $y = ax^2 + bx + c$ is more complex than the linear model $y = ax + b$. As shown in the next section, model selection methods such as AIC and BIC include terms that penalize model complexity, thereby neutralizing complexity differences among models.

Underlying the introduction of these more sophisticated methods is an important conceptual shift in the goal of model selection. Instead of choosing the model that provides the best fit to a single set of data, choose the model that, *with its parameters held constant*, provides the best fit to the data if the experiment were repeated again and again. That is, choose the model that generalizes best to replications of the same experiment. Across replications, the noise in the data will change, but the regularity of interest should not. The more noise that the model captures when fit to the first data set, the poorer its measure of fit will be when fitting the data in replications of that experiment because the noise will have changed. If a model captures mostly the regularity, then its fits will be consistently good across replications. The problem of distinguishing regularity from noise is solved by focusing on generalizability. A model is of questionable worth if it does not have good predictive accuracy in the same experimental setting. Generalizability evaluates exactly this, and it is why many consider generalizability to be the best criterion on which models should be compared (Grunwald et al., 2005).

The graphs in Fig. 11.1 summarize the relationship among the three quantitative criteria of model evaluation and selection: GOF, complexity, and generalizability. Model complexity is along the x axis and model fit along the y axis. GOF and generalizability are represented as curves whose performance can be compared as a function of complexity. The three smaller graphs contain the same data set (dots) and the fits to these data by increasingly more complex models (lines). The left-most model in Fig. 11.1

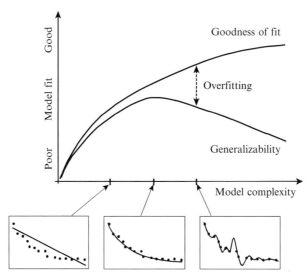

Figure 11.1 An illustration of the relationship between goodness of fit and generalizability as a function of model complexity. The y axis represents any fit index, where a larger value indicates a better fit (e.g., maximum likelihood). The three smaller graphs provide a concrete example of how fit improves as complexity increases. In the left graph, the model (line) is not complex enough to match the complexity of data (dots). The two are well matched in complexity in the middle graph, which is why this occurs at the peak of the generalizability function. In the right graph, the model is more complex than data, capturing microvariation due to random error. Reprinted from Pitt and Myung (2002).

underfits the data. Data are curvilinear, whereas the model is linear. In this case, GOF and generalizability produce similar outcomes because the model is not complex enough to capture the bowed shape of the data. The model in the middle graph of Fig. 11.1 is a bit more complex and does a good job of fitting only the regularity in the data. Because of this, the GOF and generalizability measures are higher and also similar. Where the two functions diverge is when the model is more complex than is necessary to capture the main trend. The model in the right-most graph of Fig. 11.1 captures the experiment-specific noise, fitting every data point perfectly. GOF rewards this behavior by yielding an even higher fit score, whereas generalizability does just the opposite, penalizing the model for its excess complexity.

The problem of overfitting is the scourge of GOF. It is easy to see when overfitting occurs in Fig. 11.1, but in practice it is difficult to know when and by how much a model overfits a data set, which is why generalizability is the preferred means of model evaluation and comparison. By using generalizability, we evaluate a model based on how well it predicts the statistics of future samples from the same underlying processes that generated an observed data sample.

3. Model Comparison Methods

This section reviews measures of generalizability currently in use, touching on their theoretical foundations and discussing the advantages and disadvantages of their implementation. Readers interested in more detailed presentations are directed to Myung *et al.* (2000) and Wagenmakers and Waldorp (2006).

3.1. Akaike Information Criterion and Bayesian Information Criterion

As illustrated in Fig. 11.1, good generalizability is achieved by trading off GOF with model complexity. This idea can be formalized to derive model comparison criteria. That is, one way of estimating the generalizability of a model is by appropriately discounting the model's goodness of fit relative to its complexity. In so doing, the aim is to identify the model that is sufficiently complex to capture the underlying regularities in the data but not unnecessarily complex to capitalize on random noise in the data, thereby formalizing the principle of Occam's razor.

The Akaike information criterion (AIC; Akaike, 1973; Bozdogan, 2000), its variation called the second-order AIC (AICc; Burnham and Anderson, 2002; Sugiura, 1978), and the Bayesian information criterion (BIC; Schwartz, 1978) exemplify this approach and are defined as

$$AIC = -2\ln f(y|w^*) + 2k$$

$$AICc = -2\ln f(y|w^*) + 2k + \frac{2k(k+1)}{n-k-1} \quad (11.1)$$

$$BIC = -2\ln f(y|w^*) + k\ln(n)$$

where y denotes the observed data vector, $\ln f(y|w^*)$ is the natural logarithm of the model's maximized likelihood calculated at the parameter vector w^*, k is the number of parameters of the model, and n is the sample size. The first term of each comparison criterion represents a model's lack of fit measure (i.e., inverse GOF), with the remaining terms representing the model's complexity measure. Combined, they estimate the model's generalizability such that the lower the criterion value, the better the model is expected to generalize.

The AIC is derived as an asymptotic (i.e., large sample size) approximation to an information theoretic distance between two probability distributions, one representing the model under consideration and the other representing the "true" model (i.e., data-generating model). As such, the smaller the AIC value, the closer the model is to the "truth." AICc represents a small sample size version of AIC and is recommended for data with relatively small n with respect to k, say $n/k < 40$ (Burnham and Anderson, 2002). BIC, which is a Bayesian criterion, as the name implies, is derived as an asymptotic expression of the minus two log marginal likelihood, which is described later in this chapter.

The aforementioned three criteria differ from one another in how model complexity is conceptualized and measured. The complexity term in AIC depends on only the number of parameters, k, whereas both AICc and BIC consider the sample size (n) as well, although in different ways. These two dimensions of a model are not the only ones relevant to complexity, however. Functional form, which refers to the way the parameters are entered in a model's equation, is another dimension of complexity that can also affect the fitting capability of a model (Myung and Pitt, 1997). For example, two models, $y = ax^b + e$ and $y = ax + b + e$, with a normal error e of constant variance, are likely to differ in complexity, despite the fact that they both assume the same number of parameters. For models such as these, the aforementioned criteria are not recommended because they are insensitive to the functional form dimension of complexity. Instead, we recommend the use of the comparison methods, described next, which are sensitive to all three dimensions of complexity.

3.2. Cross-Validation and Accumulative Prediction Error

Cross-validation (CV; Browne, 2000; Stone, 1974) and the accumulative prediction error (APE; Dawid, 1984; Wagenmakers, Grunwald and Steyvers, 2006) are sampling-based methods for estimating generalizability from the data, without relying on explicit, complexity-based penalty terms as in AIC and BIC. This is done by simulating the data collection and prediction steps artificially using the observed data in the experiment.

Cross-validation and APE are applied by following a three-step procedure: (1) divide the observed data into two subsamples, the calibration sample, y_{cal}, simulating the "current" observations and the validation sample, y_{val}, simulating "future" observations; (2) fit the model to y_{cal} and obtain the best-fitting parameter values, denoted by $w^*(y_{cal})$; and (3) with the parameter values fixed, the model is fitted to y_{val}. The resulting prediction error is taken as the model's generalizability estimate.

The two comparison methods differ from each other in how the data are divided into calibration and validation samples. In CV, each set of $n - 1$ observations in a data set serves as the calibration sample, with the remaining observation treated as the validation sample on which the prediction error is calculated. Generalizability is estimated as the average of n such prediction errors, each calculated according to the aforementioned three-step procedure. This particular method of splitting the data into calibration and validation samples is known as the leave-one-out CV in statistics. Other methods of splitting data into two subsamples can also be used. For example, the data can be split into two equal halves or into two subsamples of different sizes. In the remainder of this chapter, CV refers to the leave-one-out cross validation procedure.

In contrast to CV, in APE the size of the calibration sample increases successively by one observation at a time for each calculation of prediction error. To illustrate, consider a model with k parameters. We would use the first $k + 1$ observations as the calibration sample so as to make the model identifiable, and the $(k + 2)$-th observation as the validation sample, with the remaining observations not being used. The prediction error for the validation sample is then calculated following the three-step procedure. This process is then repeated by expanding the calibration sample to include the $(k + 2)$-th observation, with the validation sample now being the $(k + 3)$-th observation, and so on. Generalizability is estimated as the average prediction error over the $(n - k - 1)$ validation samples. Time series data are naturally arranged in an ordered list, but for data that have no natural order, APE can be estimated as the mean over all orders (in theory), or over a few randomly selected orders (in practice). Figure 11.2 illustrates how CV and APE are estimated.

Formally, CV and APE are defined as

$$CV = -\sum_{i=1}^{n} \ln f\left(y_i | w^*\left(y_{\neq i}\right)\right)$$
$$APE = -\sum_{i=k+2}^{n} \ln f\left(y_i | w^*\left(y_{1, 2, \ldots, i-1}\right)\right) \quad (11.2)$$

In the aforementioned equation for CV, $-\ln f(y_i | w^* (y_{\neq i}))$, is the minus log likelihood for the calibration sample y_i evaluated at the best-fitting parameter values $w^*(y_{\neq i})$, obtained from the validation sample $y_{\neq i}$. The subscript signifies "all observations except for the ith observation." APE is defined similarly. Both methods prescribe that the model with the smallest value of the given criterion should be preferred.

The attractions of CV and APE are the intuitive appeal of the procedures and the computational ease of their implementation. Further, unlike AIC and BIC, both methods consider, albeit implicitly, all three factors that affect model complexity: functional form, number of parameters, and sample size. Accordingly, CV and APE should perform better than AIC and BIC, in particular when comparing models with the same number of parameters. Interestingly, theoretical connections exit between AIC and CV, and BIC and APE. Stone (1977) showed that under certain regularity

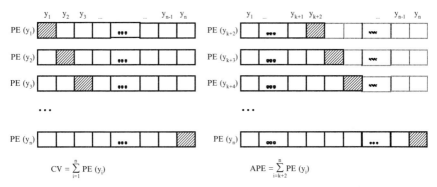

Figure 11.2 The difference between the two sampling-based methods of model comparison, cross-validation (CV) and accumulative prediction error (APE), is illustrated. Each chain of boxes represents a data set with each data point represented by a box. The slant-lined box is a validation sample, and plain boxes with the bold outline represent the calibration sample. Plain boxes with the dotted outline in the right panel are not being used as part of the calibration or validation sample. The symbol $PE(y_i)$, $i = 1, 2, \ldots n$, stands for the prediction error for the ith validation data point. k represents the number of parameters, and n the sample size.

conditions, model choice under CV is asymptotically equivalent to that under AIC. Likewise, Barron et al. (1998) showed that APE is asymptotically equivalent to BIC.

3.3. Bayesian Model Selection and Stochastic Complexity

Bayesian model selection (BMS; Kass and Raftery, 1995; Wasserman, 2000) and stochastic complexity (SC; Grunwald et al., 2005; Myung et al., 2006; Rissanen, 1996, 2001) are the current state-of-the-art methods of model comparison. Both methods are rooted on firm theoretical foundations; are nonasymptotic in that they can be used for data of all sample sizes, small or large; and, finally, are sensitive to all dimensions of complexity. The price to pay for this generality is computational cost. Implementation of the methods can be nontrivial because they usually involve evaluating high-dimensional integrals numerically.

Bayesian model selection and SC are defined as

$$BMS = -\ln \int f(y|w)\pi(w)dw \\ SC = -\ln f(y|w^*) + \ln \int f(z|w^*(z))dz. \quad (11.3)$$

Bayesian model selection is defined as the minus logarithm of the marginal likelihood, which is nothing but the mean likelihood of the data averaged across parameters and weighted by the parameter prior $\pi(w)$. The first term of SC is the minus log maximized likelihood of the observed data y. It is a lack of fit measure, as in AIC. The second term is a complexity measure, with the symbol z denoting the *potential data* that could be observed in an experiment, not the actually observed data. Both methods prescribe that the model that minimizes the given criterion value is to be chosen.

Bayesian model selection is related to the Bayes factor, the gold standard of model comparison in Bayesian statistics, such that the Bayes factor is a ratio of two marginal likelihoods between a pair of models. BMS does not yield an explicit measure of complexity but complexity is taken into account implicitly through the integral and thus avoids overfitting. To see this, an asymptotic expansion of BMS under Jeffrey's prior for $\pi(w)$ yields the following large sample approximation (Balasubramanian, 1997)

$$BMS \approx -\ln f(y|w^*) + \frac{k}{2}\ln\left(\frac{n}{2\pi}\right) + \ln\int\sqrt{\det(I(w))}\,dw, \quad (11.4)$$

where $I(w)$ is the Fisher information matrix of sample size 1 (e.g., Schervish, 1995). The second and third terms on the right-hand side of the expression represent a complexity measure. It is through the Fisher information in the third term that BMS reflects the functional form dimension of model complexity. For instance, the two models mentioned earlier, $y = ax^b + e$ and $y = ax + b + e$, would have different values of the Fisher information, although they both have the same number of parameters. The Fisher information term is independent of sample size n, with its relative contribution to that of the second term becoming negligible for large n. Under this condition, the aforementioned expression reduces to another asymptotic expression, which is essentially one-half of BIC in Eq. (11.1).

Stochastic complexity is a formal implementation of the principle of minimum description length that is rooted in algorithmic coding theory in computer science. According to the principle, a model is viewed as a code with which data can be compressed, and the best model is the one that provides maximal compression of the data. The idea behind this principle is that regularities in data necessarily imply the presence of statistical redundancy, which a model is designed to capture, and therefore, the model can be used to compress the data. That is, the data are re-expressed, with the help of the model, in a coded format that provides a shorter description than when the data are expressed in an uncompressed format. The SC criterion value in Eq. (11.3) represents the overall description length in bits of the maximally compressed data and the model itself, derived for parametric model classes under certain statistical regularity conditions (Rissanen, 2001).

The second (complexity) term of SC deserves special attention because it provides a unique conceptualization of model complexity. In this formulation, complexity is defined as the logarithm of the sum of maximized likelihoods that the model yields collectively for all *potential* data sets that could be observed in an experiment. This formalization captures nicely our intuitive notion of complexity. A model that fits a wide range of data patterns well, actual or hypothetical, should be more complex than a model that fits only a few data patterns well, but does poorly otherwise. A serious drawback of this complexity measure is that it can be highly nontrivial to compute the quantity because it entails numerically integrating the maximized likelihood over the entire data space. This integration in SC is even more difficult than in BMS because the data space is generally of much higher dimension than the parameter space.

Interestingly, a large-sample approximation of SC yields Eq. (11.4) (Rissanen, 1996), which itself is an approximation of BMS. More specifically, under Jeffrey's prior, SC and BMS become asymptotically equivalent. Obviously, this equivalence does not extend to other priors and does not hold if the sample size is not large enough to justify the asymptotic expression.

4. MODEL COMPARISON AT WORK: CHOOSING BETWEEN PROTEIN FOLDING MODELS

This section applies five model comparison methods to discriminating two protein-folding models.

In the modern theory of protein folding, the biochemical processes responsible for the unfolding of helical peptides are of interest to researchers. The Zimm–Bragg theory provides a general framework under which one can quantify the helix–coil transition behavior of polymer chains (Zimm and Bragg, 1959). Scholtz and colleagues (1995) applied the theory "to examine how the α-helix to random coil transition depends on urea molarity for a homologous series of peptides" (p. 185). The theory predicts that the observed mean residue ellipticity q as a function of the length of a peptide chain and the urea molarity is given by

$$q = f_H \cdot (g_H - g_C) + g_C. \quad (11.5)$$

In Eq. (1), f_H is the fractional helicity and g_H and g_C are the mean residue ellipticities for helix and coil, respectively, defined as

$$f_H = \frac{rs}{(s-1)^3}\left(\frac{n \cdot s^{n+2} - (n+2)s^{n+1} + (n+2)s - n}{n\left(1 + [rs/(s-1)^2][s^{n+1} + n - (n+1)s]\right)}\right)$$

$$g_H = H_0\left(1 - \frac{2.5}{n}\right) + H_U \cdot [urea] \quad (11.6)$$

$$g_C = C_0 + C_U \cdot [urea]$$

where r is the helix nucleation parameter, s is the propagation parameter, n is the number of amide groups in the peptide, H_0 and C_0 are the ellipticities of the helix and coil, respectively, at $0°$ in the absence of urea, and finally, H_U and C_U are the coefficients that represent the urea dependency of the ellipticities of the helix and coil (Greenfield, 2004; Scholtz et al., 1995).

We consider two statistical models for urea-induced protein denaturation that determine the urea dependency of the propagation parameter s. One is the linear extrapolation method model (LEM; Pace and Vanderburg, 1979) and the other is called the binding-site model (BIND; Pace, 1986). Each expresses the propagation parameter s in the following form

$$\text{LEM}: \ln s = \ln s_0 - \frac{m \cdot [urea]}{R \cdot T}$$

$$\text{BIND}: \ln s = \ln s_0 - d \cdot \ln\left(1 + k \cdot (0.9815 \cdot [urea] - 0.02978 \cdot [urea]^2 + 0.00308 \cdot [urea]^3)\right) \quad (11.7)$$

where s_0 is the s value for the homopolymer in the absence of urea, m is the change in the Gibbs energy of helix propagation per residue, $R = 1.987$ cal mol^{-1} K^{-1}, T is the absolute temperature, d is the parameter characterizing the difference in the number of binding sites between the coil and helix forms of a residue, and k is the binding constant for urea.

Both models share four parameters: H_0, C_0, H_U, and C_U. LEM has two parameters of its own (s_0, m), yielding a total of six parameters to be estimated from the data. BIND has three unique parameters (s_0, d, and k). Both models are designed to predict the mean residue ellipticity denoted q in terms of the chain length n and the urea molarity [urea]. The helix nucleation parameter r is assumed to be fixed to the previously determined value of 0.0030 (Scholtz et al., 1991).

Figure 11.3 shows simulated data (symbols) and best-fit curves for the two models (LEM in solid lines and BIND in dotted lines). Data were generated from LEM for a set of parameter values with normal random

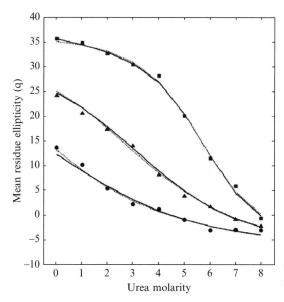

Figure 11.3 Best fits of LEM (solid lines) and BIND (dotted lines) models to data generated from LEM using the nine points of urea molarity (0, 1, 2, ..., 8) for three different chain lengths of $n = 13$ (●), 20 (▲), and 50 (■). Data fitting was done first by deriving model predictions using Eqs. (11.5)–(11.7) based on the parameter values of $H_0 = -44{,}000$, $C_0 = 4400$, $H_U = 320$, $C_U = 340$, $s_0 = 1.34$, and $m = 23.0$ reported in Scholtz et al. (1995). See text for further details.

noise of zero mean and 1 standard deviation added to the ellipticity prediction in Eq. (11.5) (see the figure legend for details). Note how closely both models fit the data. By visual inspection, one cannot tell which of the two models generated the data. As a matter of fact, BIND, with one extra parameter than LEM, provides a better fit to the data than LEM (SSE = 12.59 vs 14.83), even though LEM generated the data. This outcome is an example of the overfitting that can emerge with complex models, as depicted in Fig. 11.1. To filter out the noise-capturing effect of overly complex models appropriately, thereby putting both models on an equal footing, we need the help of statistical model comparison methods that neutralize complexity differences.

We conducted a model recovery simulation to demonstrate the relative performance of five model comparison methods (AIC, AICc, BIC, CV, and APE) in choosing between the two models. BMS and SC were not included because of the difficulty in computing them for these models. A thousand data sets of 27 observations each were generated from each of the two models, using the same nine points of urea molarity (0, 1, 2, ..., 8) for three different chain lengths of $n = 13$, 20, and 50. The parameter values used to

Table 11.1 Model recovery performance of five model comparison methods

		Data generated from	
Model comparison method	Model fitted	LEM	BIND
ML	LEM	47	4
	BIND	53	96
AIC	LEM	81	16
	BIND	19	84
AICc	LEM	93	32
	BIND	7	68
BIC	LEM	91	28
	BIND	9	72
CV	LEM	77	26
	BIND	23	74
APE	LEM	75	45
	BIND	25	55

Note: The two models, LEM and BIND, are defined in Eq. (11.7). APE was estimated after randomly ordering the 27 data points of each data set.

generate the simulated data were taken from Scholtz et al. (1995) and were as follows: $H_0 = -44{,}000$, $C_0 = 4400$, $H_U = 320$, $C_U = 340$, $s_0 = 1.34$, $m = 23.0$ and temperature $T = 273.15$ for LEM and $H_0 = -42{,}500$, $C_0 = 5090$, $H_U = -620$, $C_U = 280$, $s_0 = 1.39$, $d = 0.52$, $k = 0.14$ for BIND. Normal random errors of zero mean and standard deviation of 1 were added to the ellipticity prediction in Eq. (11.5).

The five model comparison methods were compared on their ability to recover the model that generated the data. A good method should be able to identify the true model (i.e., the one that generated the data) 100% of the time. Deviations from perfect recovery reveal a bias in the selection method. (The MatLab code that implements the simulations can be obtained from the first author.)

The simulation results are reported in Table 11.1. Values in the cells represent the percentage of samples in which a particular model (e.g., LEM) fitted best data sets generated by one of the models (LEM or BIND). A perfect selection method would yield values of 100% along the diagonal. The top 2 × 2 matrix shows model recovery performance under ML, a purely goodness-of-fit measure. It is included as a reference against which to compare performance when measures of model complexity are included in the selection method. How much does model recovery improve when the number of parameters, sample size, and functional form are taken into account?

With ML, there is a strong bias toward BIND. The result in the first column of the matrix shows that BIND was chosen more often than the true data-generating model, LEM (53% vs 47%). This bias is not surprising given that BIND, with one more parameter than LEM, can capture random noise better than LEM. Consequently, BIND tends to be selected more often than LEM under a goodness-of-fit selection method such as ML, which ignores complexity differences. Results from using AIC show that when the difference in complexity due to the number of parameters is taken into account, the bias is largely corrected (19% vs 81%), and even more so under AICc and BIC, both of which consider sample size as well (7% vs 93% and 9% vs 91%, respectively). When CV and APE were used, which are supposed to be sensitive to all dimensions of complexity, the results show that the bias was also corrected, although the recovery rates under these criteria were about equal to or slightly lower than that under AIC. When the data were generated from BIND (right column of values), the data-generating model was selected more often than the competing model under all selection methods, including ML.

To summarize, the aforementioned simulation results demonstrate the importance of considering model complexity in model comparison. All five model selection methods performed reasonably well by compensating for differences in complexity between models, thus identifying the data-generating model. It is interesting to note that Scholtz and colleagues (1995) evaluated the viability of the same two models plus a third, seven-parameter model, using goodness of fit, and found that all three models provided nearly identical fits to their empirical data. Had they compared the models using one of the selection methods discussed in this chapter, it might have been possible to obtain a more definitive answer.

We conclude this section with the following cautionary note regarding the performance of the five selection methods in Table 11.1: The better model recovery performance of AIC, AICc, and BIC over CV and APE should not be taken as indicative of how the methods will generally perform in other settings (Myung and Pitt, 2004). There are very likely other situations in which the relative performance of the selection methods reverses.

5. Conclusions

This chapter began by discussing several issues a modeler should be aware of when evaluating computational models. They include the notion of model complexity, the triangular relationship among goodness of fit, complexity and generalizability, and generalizability as the ultimate yardstick of model comparison. It then introduced several model comparison methods that can be used to determine the "best-generalizing" model among a set of competing models, discussing the advantages and disadvantages of each method. Finally, the chapter demonstrated the application of some of the comparison methods using simulated data for the problem of choosing between biochemical models of protein folding.

Measures of generalizability are not without their own drawbacks, however. One is that they can be applied only to statistical models defined as a parametric family of probability distributions. This restriction leaves one with few options when wanting to compare nonstatistical models, such as verbal models and computer simulation models. Often times, researchers are interested in testing qualitative (e.g., ordinal) relations in data (e.g., condition $A <$ condition B) and comparing models on their ability to predict qualitative patterns of data, but not quantitative ones.

Another limitation of measures of generalizability is that they summarize the potentially intricate relationships between model and data into a single real number. After applying CV or BMS, the results can sometimes raise more questions than answers. For example, what aspects of a model's formulation make it superior to its competitors? How representative is a particular data pattern of a model's performance? If it is typical, the model provides a much more satisfying account of the process than if the pattern is generated by the model using a small range of unusual parameter settings. Answers to these questions also contribute to the evaluation of model quality.

We have begun developing methods to address questions such as these. The most well-developed method thus far is a global qualitative model analysis technique dubbed *parameter space partitioning* (PSP; Pitt et al., 2006, 2007). In PSP, a model's parameter space is partitioned into disjoint regions, each of which corresponds to a qualitatively different data pattern. Among other things, one can use PSP to identify all data patterns a model can generate by varying its parameter values. With information such as this in hand, one can learn a great deal about the relationship between the model and its behavior, including understanding the reason for the ability or inability of the model to account for empirical data.

In closing, statistical techniques, when applied with discretion, can be useful for identifying sensible models for further consideration, thereby aiding the scientific inference process (Myung and Pitt, 1997). We cannot overemphasize the importance of using nonstatistical criteria such as explanatory adequacy, interpretability, and plausibility of the models under consideration, although they have yet to be formalized in quantitative terms and subsequently incorporated into the model evaluation and comparison methods. Blind reliance on statistical means is a mistake. On this point we agree with Browne and Cudeck (1992), who said "Fit indices [statistical model evaluation criteria] should not be regarded as a measure of usefulness of a model...they should not be used in a mechanical decision process for selecting a model. Model selection has to be a subjective process involving the use of judgement" (p. 253).

ACKNOWLEDGMENTS

This work was supported by Research Grant R01-MH57472 from the National Institute of Health to JIM and MAP. This chapter is an updated version of Myung and Pitt (2004). There is some overlap in content.

REFERENCES

Akaike, H. (1973). Information theory and an extension of the maximum likelihood principle. In "Second International Symposium on Information Theory" (B. N. Petrox and F. Caski, eds.), pp. 267–281. Akademia Kiado, Budapest.

Balasubramanian, V. (1997). Statistical inference, Occam's razor and statistical mechanics on the space of probability distributions. *Neural Comput.* **9,** 349–368.

Barron, A., Rissanen, J., and Yu, B. (1998). The minimum description length principle in coding and modeling. *IEEE Trans. Inform. Theory* **44,** 2743–2760.

Bozdogan, H. (2000). Akaike information criterion and recent developments in information complexity. *J. Math. Psychol.* **44,** 62–91.

Browne, M. W. (2000). Cross-validation methods. *J. Math. Psychol.* **44,** 108–132.

Browne, M. W., and Cudeck, R. C. (1992). Alternative ways of assessing model fit. *Sociol. Methods Res.* **21,** 230–258.

Burnham, L. S., and Anderson, D. R. (2002). In "Model Selection and Inference: A Practical Information-Theoretic Approach," 2nd Ed. Springer-Verlag, New York.

Dawid, A. P. (1984). Statistical theory: The prequential approach. *J. Roy. Stat. Soc. Ser. A* **147,** 278–292.

Greenfield, N. J. (2004). Analysis of circular dichroism data. *Methods Enzymol.* **383,** 282–317.

Grunwald, P., Myung, I. J., and Pitt, M. A. (2005). In "Advances in Minimum Description Length: Theory and Application." MIT Press, Cambridge, MA.

Jacobs, A. M., and Grainger, J. (1994). Models of visual word recognition: Sampling the state of the art. *J. Exp. Psychol. Hum. Perception Perform* **29,** 1311–1334.

Kass, R. E., and Raftery, A. E. (1995). Bayes factors. *J. Am. Stat. Assoc.* **90,** 773–795.

Myung, I. J. (2003). Tutorial on maximum likelihood estimation. *J. Math. Psychol.* **44,** 190–204.

Myung, I. J., Forster, M., and Browne, M. W., (eds.) (2000). Special issue on model selection. *J. Math. Psychol.* **44,** 1–2.

Myung, I. J., Navarro, D. J., and Pitt, M. A. (2006). Model selection by normalized maximum likelihood. *J. Math. Psychol.* **50,** 167–179.

Myung, I. J., and Pitt, M. A. (1997). Applying Occam's razor in modeling cognition: A Bayesian approach. *Psychon. Bull. Rev.* **4,** 79–95.

Myung, I. J., and Pitt, M. A. (2004). Model comparison methods. In "Numerical Computer Methods, Part D" (L. Brand and M. L. Johnson, eds.), Vol. 383, pp. 351–366.

Pace, C. N. (1986). Determination and analysis of urea and guanidine hydrochloride denatiration curves. *Methods Enzymol.* **131,** 266–280.

Pace, C. N., and Vanderburg, K. E. (1979). Determining globular protein stability: Guanidine hydrochloride denaturation of myoglobin. *Biochemistry* **18,** 288–292.

Pitt, M. A., Kim, W., Navarro, D. J., and Myung, J. I. (2006). Global model analysis by parameter space partitioning. *Psychol. Rev.* **113,** 57–83.

Pitt, M. A., and Myung, I. J. (2002). When a good fit can be bad. *Trends Cogn. Sci.* **6,** 421–425.

Pitt, M. A., Myung, I. J., and Altieri, N. (2007). Modeling the word recognition data of Vitevitch and Luce (1998): Is it ARTful? *Psychon. Bull. Rev.* **14,** 442–448.

Rissanen, J. (1996). Fisher information and stochastic complexity. *IEEE Trans. Inform. Theory* **42,** 40–47.

Rissanen, J. (2001). Strong optimality of the normalized ML models as universal codes and information in data. *IEEE Trans. Inform. Theory* **47,** 1712–1717.

Schervish, M. J. (1995). "The Theory of Statistics." Springer-Verlag, New York.

Scholtz, J. M., Barrick, D., York, E. J., Stewart, J. M., and Balding, R. L. (1995). Urea unfolding of peptide helices as a model for interpreting protein unfolding. *Proc. Natl. Acad. Sci. USA* **92,** 185–189.

Scholtz, J. M., Qian, H., York, E. J., Stewart, J. M., and Balding, R. L. (1991). Parameters of helix-coil transition theory for alanine-based peptides of varying chain lengths in water. *Biopolymers* **31,** 1463–1470.

Schwarz, G. (1978). Estimating the dimension of a model. *Ann. Stat.* **6,** 461–464.

Stone, M. (1974). Cross-validatory choice and assessment of statistical predictions. *J. Roy. Stat. Soc. Ser. B* **36,** 111–147.

Stone, M. (1977). An asymptotic equivalence of choice of model by cross-validation and Akaike's criterion. *J. Roy. Stat. Soc. Ser. B* **39,** 44–47.

Sugiura, N. (1978). Further analysis of the data by Akaike's information criterion and the finite corrections. *Commun. Stat. Theory Methods* **A7,** 13–26.

Wagenmakers, E.-J., Grunwald, P., and Steyvers, M. (2006). Accumulative prediction error and the selection of time series models. *J. Math. Psychol.* **50,** 149–166.

Wagenmakers, E.-J., and Waldorp, L. (2006). Editors' introduction. *J. Math. Psychol.* **50,** 99–100.

Wasserman, L. (2000). Bayesian model selection and model averaging. *J. Math. Psychol.* **44,** 92–107.

Zimm, B. H., and Bragg, J. K. (1959). Theory of the phase transition between helix and random coil. *J. Chem. Phys.* **34,** 1963–1974.

CHAPTER TWELVE

Desegregating Undergraduate Mathematics and Biology—Interdisciplinary Instruction with Emphasis on Ongoing Biomedical Research

Raina Robeva

Contents

1. Introduction	306
2. Course Description	309
2.1. Institutional context	309
2.2. Target audience and pedagogy principles	310
2.3. Features of the course	310
2.4. Course structure	311
2.5. Course topics	312
3. Discussion	317
Acknowledgments	320
References	320

Abstract

The remarkable advances in the field of biology in the last decade, specifically in the areas of biochemistry, genetics, genomics, proteomics, and systems biology, have demonstrated how critically important mathematical models and methods are in addressing questions of vital importance for these disciplines. There is little doubt that the need for utilizing and developing mathematical methods for biology research will only grow in the future. The rapidly increasing demand for scientists with appropriate interdisciplinary skills and knowledge, however, is not being reflected in the way undergraduate mathematics and biology courses are structured and taught in most colleges and universities nationwide. While a number of institutions have stepped forward and addressed this need by creating and offering interdisciplinary courses at the juncture of mathematics and biology, there are still many others at which

Department of Mathematical Sciences, Sweet Briar College, Sweet Briar, Virginia there is little, if any, interdisciplinary interaction between the curricula. This chapter describes an interdisciplinary course and a textbook in mathematical biology developed collaboratively by faculty from Sweet Briar College and the University of Virginia School of Medicine. The course and textbook are designed to provide a bridge between the mathematical and biological sciences at the lower undergraduate level. The course is developed for and is being taught in a liberal arts setting at Sweet Briar College, Virginia, but some of the advanced modules are used in a course at the University of Virginia for advanced undergraduate and beginning graduate students. The individual modules are relatively independent and can be used as stand-alone projects in conventional mathematics and biology courses. Except for the introductory material, the course and textbook topics are based on current biomedical research.

1. INTRODUCTION

In the last decade, the field of life sciences has undergone revolutionary changes spanning remarkable discoveries at all levels of biological organization—from molecules, to cells, to tissues, organs, organisms, and populations. A salient trait of these advances is the increased need for statistical, computational, and mathematical modeling methods. Scientific instruments are now, by orders of magnitude, more sensitive, more specific, and more powerful. The amounts of data collected and processed by these new-generation instruments have increased dramatically, rendering insufficient the traditional methods of statistical data analysis. To organize this information and arrive at a better fundamental understanding of life processes, it is imperative that powerful conceptual tools from mathematics be applied to the frontier problems in biology (BISTI Report, 2000).

For instance, automated DNA sequencing has given rise to data explosion, and the challenge now is to extract meaningful information from these data. The quest to better understand temporal and spatial trends in gene expression has led to a search for DNA sequences that have been conserved over time in a large number of different species. This is a tremendous task, as the human genome alone is approximately 3 billion base pairs. Comparing across species then requires comparisons of billions of sequences, over thousands of species. The sheer size of the data sets suggests that appropriate use of mathematical models coupled with statistical methods for data analysis and inference will play an exceptional role in modern biology and demonstrates that future advances in molecular biology will not be possible without the help of mathematics.

The field of molecular systems biology has emerged as equally mathematically driven. Broadly defined, this is a field that examines how "... large number of functionally diverse, and frequently multifunctional, sets of elements interact selectively and non-linearly to produce coherent behavior

(Kitano, 2002)." In other words, organismal function and behavior are determined by a tremendously complex set of interactions (protein–protein, protein–DNA, protein–RNA, and so on), and the complexity of the interactions requires the assistance of mathematics if we are to understand how living organisms function. The challenge is to combine the rich but disparate insights of molecular biology into a conceptual framework that would better allows us to see the overall structure of molecular mechanisms. Mathematical models have proved to be indispensable in this regard. Understanding these interactions has already led to significant progress in the treatment of various conditions such as heart disease, cancer, endocrine disorders, and diabetes mellitus.

These trends in contemporary biology have resulted in several high-profile reports containing recommendations regarding the education of future mathematicians and biologists. Thanks to such reports, the assessments that "...the main push in biology during the coming decades will be toward an increasingly quantitative understanding of biological functions..." (Committee on Mathematical Sciences Research, 2005) and that "...the traditional segregation in higher education of biology from mathematics and physics presents challenges and requires an integration of these subjects..." (Rising Above the Gathering Storm, 2007) are now widely accepted. Similar priorities were specified in the report *Bio2010* by the National Research Council (Bio2010, 2003).

Responding to those recommendations, the National Science Foundation (NSF), the National Institutes of Health (NIH), and the Howard Hughes Medical Institute have led the way by providing substantial funding for innovative ways to establish interdisciplinary ties and facilitate educational advances at the juncture of mathematics and biology in higher education. The Mathematical Association of America (MAA) and the American Mathematical Society now have numerous sessions and panels at their meetings devoted specifically to questions related to education and pedagogy in mathematical biology. In the last five years, the MAA (with NSF funding) has also sponsored faculty development workshops for mathematicians and biologists who wish to create and teach courses in mathematical biology. A new special interest group of the MAA with focus on mathematical biology (BIOSIGMAA) was launched early in 2007, and in June 2007 the Mathematical Biosciences Institute at Ohio State University offered its first ever educational workshop in mathematical biology *Over the Fence: Mathematics and Biology Talk about Bridging the Curriculum Divide.*

In response to these efforts, a rapidly increasing number of undergraduate and/or graduate programs in mathematical and computational biology are being offered by colleges and universities nationwide. Other institutions are adding new courses in mathematical biology and/or changing their existing mathematical and biology courses to accommodate the growing need for interdisciplinary crossover. By far the most popular change has been the addition of designated calculus sequences for biology majors where the traditional applications from physics and engineering are replaced with equivalents coming from biology and the life sciences.

Regardless of the progress made so far, however, the huge majority of undergraduate institutions have not fully embraced the need for major curriculum changes. Biology majors are generally not expected to take mathematics beyond the traditional requirements for a semester of introductory statistics and a semester of calculus, and even this minimal mathematical background is not being universally enforced, with many institutions only listing calculus and statistics in the "recommended" column for their biology majors. At the same time, as traditional mathematics courses do not place enough emphasis on applications, mathematics students often fail to recognize the potential utility of their skills and the variety of mathematics career opportunities in the fields of biology, ecology, and other life sciences.

Consequently, while the number of bachelor's degrees awarded in biology has about doubled in the last 35 years from 35,743 in 1971 to 69,178 in 2006, the total number of bachelor's degrees awarded in mathematics has dropped nationwide by about a half, from 24,937 in 1971 to 12,844 in 2006 (U.S. Department of Education, 1994; U.S. Department of Education, 2006). This growing curricular gap indicates that despite the ongoing clarion calls for integrated approaches, the strategic demand for educating undergraduates capable of providing integration of experimental, computational, and mathematical models in biology is not yet fully met.

The roots of this problem can be identified easily. On the one hand, traditional mathematics courses too often use simplistic material to exemplify applications to the life sciences. Studied out of context, as they normally are, these brief examples present little appeal to the students and are often downgraded to a couple of class sessions on "word problems" and applications. The average mathematics major completes her/his degree with a rich repertoire of mathematical skills but almost no experience in learning how to apply these skills to problems as they arise in actual biomedical research projects or other applied settings. With the emphasis in the mathematics courses being placed primarily on techniques and on theory, many undergraduates are left without an answer to one of the most important questions asked about primarily theoretical mathematical courses: *How can I apply this knowledge?*

On the other hand, many of the courses in the biology curriculum generally make no significant use of mathematical models and theory, regardless of their increasing importance. The students who take these courses have completed the required mathematical prerequisites (one or two semesters of calculus and a semester of statistics at best), have memorized a list of algorithms, and have mastered the technical use of some software packages but have not gained enough conceptual understanding to be able to either apply the algorithms in novel situations or interpret the computer output adequately. As a result, neither the students getting degrees in mathematics nor those getting degrees in biology are well equipped with the skills required for modern biological research that require the ability to look at an unfamiliar interdisciplinary problem, analyze it, discuss it with experts from the disciplines it relates to, and identify the underlying mathematical structure that will put it into a familiar framework. For the purposes of this chapter, we refer to those skills as *quantification skills*.

The pressing national need to educate undergraduates equipped with quantification skills requires a comprehensive solution that goes far beyond incorporating mathematical snippets into the relevant biology courses and populating mathematics courses with isolated examples and applications from biology. Such a comprehensive solution should go far beyond the mere addition of more traditional mathematics, statistics, and computer science courses to the lists of biology major requirements and should adhere to the following principles: (1) acquiring quantification skills is best achieved in the context of truly interdisciplinary courses and programs aimed at the gradual desegregation of undergraduate mathematics and biology; (2) as such skills are critically linked to the ability for effective interdisciplinary communication, they are best learned (very much like a foreign language) early in one's education, and (3) teaching quantification skills in the context of high-impact contemporary research increases student interest and involvement and every effort should be made to employ such applications after the necessary introductory concepts and material are introduced.

This chapter shares our own experiences in creating and teaching an interdisciplinary course in mathematical biology that has been designed around these pedagogical principles. The course has been created jointly by faculty from Sweet Briar College (SBC) and the University of Virginia (UVA) School of Medicine and offered for the first time at Sweet Briar College in the spring of 2002. Subsequently, the materials and projects for the course have been expanded into a textbook "An Invitation to Biomathematics" (Robeva et al., 2008a) and accompanying laboratory manual "Laboratory Manual in Biomathematics" (Robeva et al., 2008b) published by Academic Press. Selected chapters and projects have also been used in the UVA course "Computation in Endocrinology" designed for advanced undergraduate and beginning graduate students. The following presentation attempts to provide enough detail about the organizational and curricular structure of the SBC course for those who would consider adopting it or adapting parts of it for their own courses.

2. Course Description

2.1. Institutional context

Sweet Briar College, located in central Virginia, is a selective 4-year liberal arts and sciences college for women and awards Bachelor of Arts and Bachelor of Science degrees. SBC has approximately 650 students in residence and is consistently cited by the Princeton Review Guide to the Best 366 Colleges and U.S. News & World Report as one of the nation's leading liberal arts institutions. Since 2000, an average of 4% of the total number of degrees at SBC has been awarded in the mathematical sciences, a figure substantially higher than the national (combined male and female) average of less than 1%. Biology at SBC has also seen an increase in majors, comparable to the disciplinary national trends presented earlier, and an increased number of graduates who go on to graduate and professional school.

2.2. Target audience and pedagogy principles

Our course "Topics in Biomathematics" is designed for sophomores through seniors and requires the following prerequisites: one semesters of calculus, one semester of statistics, and one semester of biology. The pedagogy is to focus on the fundamental concepts, ideas, and related biological applications of existing mathematical tools rather than on a rigorous analysis of the underlying mathematical theory. We aim at emphasizing the hypothesis-driven nature of contemporary scientific inquiry and at developing students' ability to formulate, test, and validate hypotheses and mathematical models by exploring and analyzing data. We do not attempt systematic presentation of mathematical material, although there are important threads that run through several of the course topics. Instead, by using specialized software (e.g., Berkeley-Madonna, MATLAB, MINITAB, PULSE_XP) and avoiding most of the tedious mathematical details, our goal is to enable students to experience the "usefulness" of complex mathematical techniques as applied to a wide variety of biological problems.[1] For students with a taste for mathematical applications, the goal is to intensify the student interest in the field through demonstrating the meaning and practical use of the concepts. For students with a taste for mathematical theory, our goal is to provide motivation for further more detailed examination of the concepts through a more rigorous mathematical perspective.

2.3. Features of the course

Our course differs from other conventional mathematics and biology courses at SBC in the following ways.

---

[1] In this pursuit, we have been motivated by the recommendations given in the classical memorandum *On the Mathematics Curriculum in High School* (1962) that "...the introduction of new terms and concepts should be preceded by sufficient concrete preparation and followed by genuine, challenging application..." and that "...one must motivate and apply a new concept if one wishes to convince... that the concept warrants attention." The memorandum was prepared in the early sixties and signed by 55 mathematicians, including Garrett Birkhoff, Richard Courant, H. S. M. Coxeter, and George Polya.

1. The course is an interface course taught by a biologist and a mathematician to a mixed audience of students that have different mathematics and biology backgrounds. Usually, about 50% of the students are biology majors and about 50% are mathematics majors. Each group of students brings its own expertise into the brainstorming discussions about possible solution strategies.
2. The course uses projects based on ongoing biomedical research. In our choice of topics we have sought high-impact contemporary breakthroughs in the biomedical field brought about by clever uses of mathematical approaches. Such choice of topics serves as a highly motivating factor for students, as it boosts the level of interest and improves the quality of their involvement. This is particularly true for the mathematics majors who are unlikely to see much contemporary mathematical content covered in their conventional mathematics courses.
3. The course targets both mathematics students wishing to learn applications of mathematics in the life sciences and biology students wanting to improve their quantification skills. As many significant research projects now require real cross-disciplinary collaboration, our attempt is to replicate these experiences in the classroom to the extent possible.
4. The course reinforces students' mathematical background by exposing them to current ideas and presenting familiar mathematical topics from novel points of view. It is often the case that after taking a course in mathematics, students never have to use the learned material again, except in subsequent mathematics courses. This may, at times, convey the false impression that mathematics is present in the undergraduate curriculum and requirements solely as an abstract logical and algebraic exercise. In our course the focus is on substantive applications.
5. The course enables students to think in terms of mathematical models and motivates them to develop and further apply their quantification skills in different fields of the life sciences and in other courses. We also encourage the best students in the course to continue their work in biomathematics as part of SBC's summer research program and/or apply to related research experiences for undergraduate programs in other institutions.

2.4. Course structure

The format comprises of three 50-min lectures and one 3-h computer-simulation and data-analysis laboratory session per week. It is divided into six relatively independent biology and physiology topics: population studies, genetics, epidemiology, ligand binding, endocrinology, and circadian rhythms and biological clocks. Each of these topics consists of smaller projects and takes from 1 to 4 weeks of class time to complete. Students are divided into groups ("research teams"), the makeup of which varies between the topics of (ideally) two biology and two mathematics students each. Group work is required as a mechanism to develop collaborative analytical and evaluative skills in conjunction with specific knowledge and its applications. No preliminary solution strategies are provided for the projects and students are asked to discuss possible approaches and think of feasible mathematical techniques as potential solving tools.

In addition to introducing or reviewing appropriate information in biology and mathematics, during the lecture time we place emphasis on developing solid quantification skills: how to ask the right questions, how to answer these questions based on available information, and how to reevaluate the situation based on the newly drawn conclusions. The lectures are supplemented by regular class discussions targeting questions of relevance. At those sessions, through collaborative efforts and group discussions, students formulate hypotheses and propose solution strategies. During the laboratory periods the "research teams" examine the hypotheses generated at the discussions and formulate, compare, and contrast the outcomes. The results and final conclusions are presented in the teams' written laboratory report.

As our students gain proficiency throughout the semester, the course projects become increasingly more detailed, more realistic, and closer in nature to actual research projects. At the same time, the mathematics describing these models does not by design become more difficult, but becomes more diverse as the students discover what is needed to describe the biological processes being considered. Student grades are based on homework, laboratory assignments, two midterm projects, and one final project.

2.5. Course topics

The specific content of the broad biological topics listed previously includes the following.

2.5.1. Population studies

We use this classical topic to examine processes that change with time, introduce dynamical systems, and examine different ways of describing and modeling such processes. This is followed by some classical examples of complex behaviors that emerge from interactions between dynamical quantities.

We begin by exploring U.S. census data for the population of the United States from 1800 to 1860. These data are used to develop the concept of and justify the need for a mathematical model capable of describing and explaining these data. We adopt the extensive approach and guide students to the realization that a mathematical model should be hypothesis based and that the model parameters should have a clear biological meaning. We discuss the difference between time-discrete and time-continuous models and highlight examples for which the use of one of the types versus the other would be more appropriate. We develop the unlimited growth population model, which is followed by identifying its weaknesses and modifying it to arrive at the logistic growth model. In anticipation of a later discussion of biological oscillators, we examine some classical experiments, including populations of the water flea *Daphnia* that exhibit oscillating behavior (Pratt, 1943), and develop some models with delay.

During the laboratory sessions, students learn how to use the Berkeley-Madonna software for obtaining numerical solutions of difference and differential equations and work with census data from other countries and with yeast growth data to increase their skills in model development and their understanding of the role and importance of a mathematical model. We also begin to explore the idea of "best fit" for the model parameters to data, but we limit the initial discussion to visual perception. We return to the mathematical meaning of "best fit" later.

The students are then assigned a project that requires them to explore on their own the dynamics of drug elimination from the bloodstream. They are expected to formulate a single exponential decay model and then apply the model repeatedly in designing a drug intake schedule that maximizes the drug's therapeutic effect under certain requirements.

We next examine more complex dynamics emerging from interacting dynamical systems using predator–prey and competitive interaction examples. In this context, we examine phase profiles as an alternative to the conventional time plots and study the long-term behavior of the systems from both mathematical and biological perspectives. The concepts of equilibrium states, types of equilibria, points of bifurcation, and limit cycles are discussed in connection with the models and selected data sets.

2.5.2. Genetics

Here we consider the dynamics of genetic variability. Two apparently contradictory characteristics are observable: a tendency for populations to preserve variability by maintaining the genetic status quo versus the presence of continuing genetic adaptation and change. These characteristics represent the fundamental ideas of Mendelian genetics and Darwinian evolution that coexist in synthesis in contemporary biology. We develop the Hardy–Weinberg model describing the maintenance of genetic variability in a population, followed by a model that analyzes the change in the genetic constitution of a population subjected to natural selection. Special attention is paid to the models' underlying assumptions.

During the laboratory period, students work with cystic fibrosis[2] incidence data in the Old Order Amish of Holmes City, Ohio (who have the

---

[2] Cystic fibrosis is a genetic disease caused by a mutation in the gene for a chloride ion-transporting protein, the cystic fibrosis transmembrane regulator (CFTR). People who are homozygous for the recessive mutant CFTR allele will have CF, while those who carry only one CFTR allele will not.

tendency to marry within their community), to determine how many generations will be needed for the frequency of the harmful cystic fibrosis transmembrane conductance regulator allele to fall below certain thresholds. They also develop an understanding of how the models could be modified and used in the case of autosomal dominant diseases such as long QT syndrome and Marfan syndrome.

We further examine quantitative traits and the polygenic hypothesis, the mathematical concept of probability distributions, and specifically the binomial distribution and its good approximation by the Gaussian (Normal) distribution. This leads to a discussion involving the central limit theorem and the need to quantify the fractions of genetic and environmental contributions to the variation observed within a population. This, in turn, serves as motivation to a review of standard statistical methods for hypothesis testing, including the Student's t test, the χ^2 test, and the F test. Students then work on a project that includes determining the heritability ratio for human stature, together with its statistical significance, from a data set containing child stature data together with the average stature of each child's parents. We also examine some recent results reflecting heritability, including the heritability value of $h^2 = 0.31$ for insulin resistance, a major factor in the development of type 2 diabetes (Bergman *et al.*, 2003).

2.5.3. Epidemiology

This is a relatively short module focused on testing how the skills acquired by students up to this point in the course carry over to the new context of epidemiology. We begin with some classical epidemic models in a closed system: the SIS model, the SIR model, and the SIR models with intermediate groups. We then examine epidemic data showing periodic patterns such as the weakly case notifications of measles in England and Wales before general vaccination was initiated, followed by developing and discussing epidemic models with delay. Students use computer software to generate numerical solutions for the models, examine the long-term behaviors of the solutions for different values of the model parameters, and determine the threshold of an epidemic, both "experimentally" and analytically. They are subsequently assigned a project to select an infectious disease from the Centers for Disease Control and Prevention's list of reportable diseases, research the mechanism through which the infection is transmitted, and present an outline of some key features that a model aiming to describe the spread of this disease should contain. Students give oral presentations during which the proposed "concept" models are discussed by the entire class.

2.5.4. Endocrinology

This is the largest module in our course, including the following topics: diabetes; peaks in hormone time series; applications to treatment of infertility; modeling of hormone feedback networks; and endocrine oscillators.

We begin with an introduction on the pathogenesis and complications of diabetes, followed by the question of quantifying the excessive variability of blood glucose (BG) levels. Data collected by self-monitoring BG devices from Kovatchev *et al.* (2002) are used to examine quantitative differences in those patterns of variability between patients with type 1 and type 2 diabetes mellitus. Similar data are then used to formulate the questions of predicting upcoming episodes of severe hypoglycemia and build two mathematical functions, the low BG index and the high BG index, designed to quantify the risk for high and low BG. Following Kovatchev *et al.* (1998, 2000), those measures are subsequently validated as valuable quantitative characteristics for assessment and maintenance of glycemic control.

After these discussions, students are assigned a project that requires them to develop and validate similar risk measures using electrocardiogram (EKG) data in assessing the risk of sepsis in prematurely born babies.[3] Sepsis is a life-threatening illness caused by an overwhelming infection of the bloodstream by toxin-producing bacteria. Even in special neonatal care intensive units, infection is a major cause of fatality during the first month of life, contributing to 13–15% of all neonatal deaths. As a noninvasive approach based on EKG data, risk measures have been used successfully to quantify differences in the heart-rate patterns in illness and in health and to predict upcoming episodes of sepsis (Griffin and Moorman, 2001; Griffin *et al.*, 2003; Kovatchev *et al.*, 2003; Nelson *et al.*, 1998).

We next move to hormone secretion patterns and the medical importance of developing methods for recognizing both normal and pathological patterns of hormone production. A critical question from a medical point of view is to know the number of pulses (secretion events) in any given time interval. For example, in a healthy female, the number of luteinizing hormone (LH) pulses ranges from 14 to 24 in the early follicular phase, 17 to 29 in the late follicular phase, and 4 to 16 in the midluteal phases (Sollenberger *et al.*, 1990). In general, hormone secretion cannot be measured directly and information about it can only be inferred from data representing the hormone concentration in the blood. The latter measurements, however, do not present an accurate picture of the hormone secretion because once the hormone is secreted and has entered the bloodstream its physiological elimination from the blood begins immediately.

In the course, we consider both statistical and modeling approaches. For the statistical approach, we use two sets of actual hormone concentration data sets: one for luteinizing hormone and one for growth hormone (GH).[4] These are then analyzed by a number of approaches, including classical methods (e.g., Fourier method and signal-averaging methods), statistically

---

[3] Data courtesy of Dr. Randall Moorman, University of Virginia.
[4] Data courtesy of Dr. Michael Johnson, University of Virginia.

based pulse analysis algorithms [e.g., CLUSTER (Veldhus and Johnson, 1986)], and deconvolution algorithms that are based on a mathematical model for separating hormone secretion from hormone elimination in the bloodstream (Johnson and Veldhus, 1995; Johnson et al., 2004). The software PULSE_XP implementing many of these algorithms can be downloaded from http://mljohnson.pharm.virginia.edu/home.html. During the laboratory segment, students use this software to analyze and compare LH data from healthy women and women referred to the Reproductive Endocrinology Clinic at the UVA for infertility treatment.[5]

To illustrate the modeling approach, we study mathematical models of hormone networks. We explore some of the endocrine mechanisms that control the secretory glands and cell groups to assure precise hormone release with regard to amount, secretion times, and long-term secretion patterns. By these feedback mechanisms the body can sense that the concentration of a certain hormone has decreased and communicate to the secretion gland the amount of the additional hormone needed. The secretion rate will then be increased. A specific example we use here is the GH network. Students learn about models with control, where the control term is represented by a Hill function. They then work on a laboratory project to develop a model describing the GH concentration in the blood under the control of two pituitary hormones—growth hormone releasing hormone and somatotropin release-inhibiting factor—and use the model for obtaining GH concentrations similar to *in vivo* data in the male rat (Lanzi and Tannenbaum, 1992).

2.5.5. Ligand binding

Binding reactions are the most common type of molecular interactions taking place within living organisms, and accurate mathematical models that describe the mechanisms of ligand binding are important for nearly all fields of chemistry, biochemistry, physiology, medicine, and physics. In our biomathematics course, we study the history of mathematical models of hemoglobin–oxygen binding as well as models of drug–receptor binding. We then use these models and appropriate experimental data to formalize the mathematical concepts of data fitting (Johnson and Frasier, 1985).

2.5.6. Circadian rhythms

Circadian rhythms are by far the most studied internal timekeeping mechanisms, often called biological clocks, which give living organisms the ability to anticipate the cycles of day and night. In addition to their expression in whole organism behaviors, circadian rhythms have been demonstrated in organs and organ systems. They have been found to exist

---

[5] Data courtesy of Dr. William Evans, University of Virginia.

in the endocrine system, in the liver, pancreas and digestive system, and in muscle and adipose tissue, as well as in the circulatory and respiratory systems. Experimentation *in vitro* has shown that circadian rhythms may also be detected at the tissue, cellular, and molecular levels. In the course, we introduce some of the basic molecular bases of biological clocks and present some of the inherent challenges in analyzing circadian data due to confounds such as noise, period and/or phase instability, and mean and variance nonstationarity.

As a particularly engaging class example, we use a premise similar to a series of experiments by Ralph *et al.* (1990). We use synthetic data sets of the locomotor activity of a wild-type hamster and a homozygous super-short mutant hamster (tau^{ss}) over a period of time measured by the number of wheel rotations for every 6-min interval recorded for both types. Next, the wild-type hamster has had its suprachiasmatic nuclei (SCN) destroyed and SCN tissue from the tau^{ss} hamster transplanted into the wild-type hamster. After a recovery period, data are collected of the locomotor activity of the recipient wild-type hamster, providing a third data set. Using the three data sets from this simulated experiment, we discuss ways to quantitatively describe the effect induced on the circadian clock of the recipient animal. Following this example, students are asked to apply similar approaches to quantify the effect of a translational inhibitor on *in vitro* (therefore, free-running) *per-luc* activity of a sample isolated from an animal that had been entrained to a 12:12 light/dark cycle. We end this module and the course material with describing some quantitative methods for analyzing microarray experiments and especially the challenges for circadian analysis of microarray-derived time series.

3. Discussion

It is generally agreed that biomathematics, also called mathematical biology, has established itself as a discipline whose methods largely exceed a simple mix of conventional mathematics and biology tools. The specific needs of this discipline require a truly cross-disciplinary approach to educating the future generation of mathematical biologists, a task that is yet to be fully addressed in the undergraduate curricula nationwide.

In our opinion, the biggest challenge to overcome is the need to desegregate the undergraduate mathematics and biology curricula. Instead of teaching conventional mathematics and biology courses in the traditional way, even when a genuine attempt is made to "cross-pollinate" the courses by using methods and examples from the other discipline, we feel that a truly interdisciplinary education at this early stage can provide considerable benefits. The primary benefit is that by teaching biomathematics instead of mathematics and/or biology, the emphasis can be placed on inspiring meaningful questions and on searching for answers without concerns for the risk that in the process we may go beyond the curricular specifics of any conventional course in the curriculum. We believe that an approach that effectively removes the interdisciplinary boundaries has a better chance to be effective in preparing our undergraduate students for graduate programs or employment related to mathematical biology.

With the course outlined in this chapter we have attempted to respond to the need for a truly interdisciplinary curriculum that does not contain just mathematical topics that highlight biological applications or biology topics that present mathematical methods. We considered the first edition of the course to be an experiment and, in an attempt to attract only strong students, the course was offered at the honors level. The results were encouraging and since then the course has been open to all students and can be taken for either mathematics or biology credit.

Initially, the lack of a textbook for the course required students to work from notes, handouts, various mathematics and biology texts, research papers, and software documentation manuals. This presented students with yet another challenge in addition to that of learning how to navigate the interdisciplinary content of the lectures and projects. Having a textbook and laboratory manual for the fall 2007 edition of the course (Robeva et al., 2008a,b) alleviated most of these additional challenges. It is also hoped that the existence of a textbook would make our course more portable and motivate faculty to consider adopting this course or use parts of it in their own interdisciplinary courses.

The major goal for our course and textbook is to provide exposure to some classical concepts as well as to new and ongoing research. The course and textbook are not meant to be encyclopedic. Instead, they are designed to provide a glimpse into the diverse worlds of biomathematics and to invite students and readers to experience, through a selection of topics and projects that we consider highly motivating, the fascinating advancements made possible by the merging of mathematics, biology, and computer science. The laboratory component provides venues for hands-on exploration of model development, model validation, and model refinement that is inherent in contemporary biomedical research. The computer software used for the projects varies and each laboratory project in the "Manual" contains a section outlining the key features of the relevant software.

A critically important feature of the course is that it is taught by a team of a mathematician and a biologist, which underscores its truly interdisciplinary nature. We found that interdisciplinary team teaching is qualitatively different from teaching traditional courses and that teamwork creates synergy between students with backgrounds in different fields. A major finding was that when exposed to truly interdisciplinary problems, students with a minimal mathematics background were able to grasp ideas and concepts of mathematics and biology much more efficiently compared to students taught these concepts in the traditional environment of disciplinary separation. As anticipated, linking the coursework with contemporary biomedical research was a major motivating factor, helping the students feel more like contributors than pupils.

In creating and teaching the course, the collaboration between the SBC faculty teaching it and the UVA research faculty plays a critical role. Not only does the UVA faculty provide many of the projects and relevant data for the research-based projects but they are actively involved in the teaching and assessment processes as well. For instance, most of the faculty whose research projects have been incorporated into the course come to SBC during the semester to meet with the students and to present the opening lecture on the topic reflecting their research. We have strived to capture the spirit and essential features of these opening lectures in the introductory material to each of the research-based chapters of the textbook "An Invitation to Biomathematics."

With its minimal prerequisites, the course is generally accessible to students in their second though fourth year of study. We have determined that having students take the course at different stages of their college education enhances the learning environment as "student mentoring" emerges with upper classmen acting as liaisons between the lower classmen and the teaching faculty. Initially, this was an unexpected benefit that we have now fully embraced in our pedagogy. In our opinion, it reflects the peer-mentoring practices common in actual interdisciplinary research.

Judging from the student course evaluations, the course structure, topic selection, and teaching pedagogy were well received. All students stressed their increased understanding of the importance of interdisciplinary collaboration. One student described the course as "the learning environment of the future" and another one said the course was "the most intellectual class offered at SBC." We attribute our success to the small class size, the reduced formalism of the seminar-like format, and the use of specialized software that allows us to focus on the "big picture" while leaving technical details aside.

In closing, we would like to stress that the methods developed and discussed in our course were primarily calculus based. Alternative methods, including discrete mathematics, graph theory, Boolean algebra, abstract algebra, and polynomial algebra, have been used to successfully study genetic networks, DNA sequence evolution, protein folding, and phylogenetic trees, among others. Exposing undergraduate students to these alternatives is also very important for adequately equipping students to meet the challenges of modern biology. Our group has an ongoing project to design and implement appropriate curricular revisions addressing this need.

ACKNOWLEDGMENTS

The author thanks Robin Davies from the Department of Biology and James Kirkwood from the Department of Mathematical Sciences at SBC with whom she has developed the course materials, team taught the course "Topics in Biomathematics" outlined in this chapter, and coauthored the textbook and laboratory manual for the course. We thank Michael Johnson, Leon Farhy, and Boris Kovatchev from the UVA School of Medicine and Martin Straume from Customized Online Biomedical Research Applications (COBRA), Charlottesville, Virginia, for their generous help with identifying appropriate biomedical projects for the course, their visits and lectures at SBC, and their partnership in developing and coauthoring the textbook and laboratory manual. We also acknowledge the following current or former UVA faculty who took the time to visit SBC, give lectures on aspects of their research featured in the course, participate in discussions with our students, or provide data for the course projects: Stacey Anderson, Gene Block, William Evans, Daniel Keenan, William Knaus, Pamela Griffin, and Randal Moorman. Finally, we acknowledge the support of NSF under the Department of Education awards 0126740 and 0340930 and of NIH under award R25 DK064122.

REFERENCES

Bergman, R. N., Zaccaro, D. J., Watanabe, R. M., Haffner, S. M., Saad, M. F., Norris, J. M., Wagenknecht, L. E., Hokanson, J. E., Rotter, J. I., and Rich, S. S. (2003). Minimal model-based insulin sensitivity has greater heritability and a different genetic basis than homeostasis model assessment or fasting insulin. *Diabetes* **52,** 2168–2174.

Bio2010: Transforming Undergraduate Education for Future Research Biologists (2010). Committee on Undergraduate Biology Education to Prepare Research Scientists for the 21st Century, Board on Life Sciences, National Research Council, The National Academies Press, Washington, DC.

BISTI: Recommendations of the Biomedical Information Science and Technology Initiative Implementation Group (2000). National Institutes of Health, Bethesda, Maryland. http://www.bisti.nih.gov/bisti_recommendations.cfm.

Committee on Mathematical Sciences Research for Department of Education's Computational Biology (2005). "Mathematics and 21st Century Biology." The National Academies Press, Washington, DC.

Griffin, M. P., and Moorman, J. R. (2001). Toward the early diagnosis of neonatal sepsis and sepsis-like illness using novel heart rate analysis. *Pediatrics* **107,** 97–104.

Griffin, M. P., O'Shea, T. M., Bissonette, E. A., Harrell, F. E. Jr., Lake, D. E., and Moorman, J. R. (2003). Abnormal heart rate characteristics preceding neonatal sepsis and sepsis-like illness. *Pediatr. Res.* **53,** 920–926.

Johnson, M. L., and Frasier, S. G. (1985). Nonlinear least-squares analysis. *Methods Enzymol.* **117,** 301–342.

Johnson, M. L., and Veldhuis, J. D. (1995). Evolution of deconvolution analysis as a hormone pulse detection method. *Methods Neurosci.* **28,** 1–24.

Johnson, M. L., Virostko, A., Veldhuis, J. D., and Evans, W. S. (2004). Deconvolution analysis as a hormone pulse-detection algorithm. *Methods Enzymol.* **384,** 40–53.

Kitano, H. (2002). Computational systems biology. *Nature* **420,** 206–210.

Kovatchev, B. P., Cox, D. J., Gonder-Frederick, L. A., and Clarke, W. L. (2002). Methods for quantifying self-monitoring blood glucose profiles exemplified by an examination of blood glucose patterns in patients with Type 1 and 2 diabetes. *Diabetes Technol. Ther.* **4,** 295–303.

Kovatchev, B. P., Cox, D. J., Gonder-Frederick, L. A., Young-Hyman, D., Schlundt, D., and Clarke, W. L. (1998). Assessment of risk for severe hypoglycemia among adults with IDDM: Validation of the low blood glucose index. *Diabetes Care* **21,** 1870–1875.

Kovatchev, B. P., Cox, D. J., Straume, M., and Farhy, L. S. (2000). Association of self-monitoring blood glucose profiles with glycosylated hemoglobin. *Methods Enzymol.* **321,** Part C, 410–417.

Kovatchev, B. P., Farhy, L. S., Hanging, C., Griffin, M. P., Lake, D. E., and Moorman, J. R. (2003). Sample asymmetry analysis of heart rate characteristics with application to neonatal sepsis and systemic inflammatory response syndrome. *Pediatr. Res.* **54,** 892–898.

Nelson, J. C., Rizwan-Uddin, M. P., Griffin, M. P., and Moorman, J. R. (1998). Probing the order of neonatal heart rate variability. *Pediatr. Res.* **43,** 823–831.

On the Mathematics Curriculum of the High School (1962). *Am. Math. Month.* **69,** No. 3, 189–193.

Pratt, D. M. (1943). Analysis of population development in *Daphnia* at different temperatures. *Biol. Bull.* **85,** 116–140.

Ralph, M. R., Foster, R. G., Davis, F. D., and Menaker, M. (1990). Transplanted suprachiasmatic nucleus determines circadian period. *Science* **247,** 975–978.

Rising Above the Gathering Storm: Energizing and Employing America for a Brighter Economic Future (2007). The National Academies Press, Washington, DC.

Robeva, R. S., Kirkwood, J. R., Davies, R. L., Johnson, M. L., Farhy, L. S., Kovatchev, B. P., and Straume, M. (2008a). "An Invitation to Biomathematics." Academic Press, New York.

Robeva, R. S., Kirkwood, J. R., Davies, R. L., Johnson, M. L., Farhy, L. S., Kovatchev, B. P., and Straume, M. (2008b). "Laboratory Manual of Biomathematics." Academic Press, New York.

Sollenberger, M. J., Carlsen, E. C., Johnson, M. J., Veldhuis, J. D., and Evans, W. S. (1990). Specific physiological regulation of luteinizing hormone secretory events throughout the human menstrual cycle: New insights into the pulsatile mode of gonadotropin release. *J. Neuroendocrinol* **2,** 845–852.

Veldhuis, J. D., and Johnson, M. L. (1986). Cluster analysis: A simple, versatile, and robust algorithm for endocrine pulse detection. *Am. J. Physiol.* **250,** E486–E493.

U.S. Department of Education, National Center for Education Statistics (1994). "Degrees and Other Formal Awards Conferred-Bachelor's degrees conferred by institutions of higher education, by discipline division: 1970–71 to 1992–93." Washington, DC. http://nces.ed.gov/programs/digest/d95/dtab243.asp.

U.S. Department of Education, National Center for Education Statistics (2006). "Bachelor's, master's, and doctor's degrees conferred by degree-granting institutions, by sex of student and field of study 2005–06." Washington, DC. http://nces.ed.gov/programs/digest/d07/tables/dt07_265.asp.

CHAPTER THIRTEEN

MATHEMATICAL ALGORITHMS FOR HIGH-RESOLUTION DNA MELTING ANALYSIS

Robert Palais[*,†] and Carl T. Wittwer[†]

Contents

1. Introduction 324
2. Extracting Melting Curves from Raw Fluorescence 325
3. Methods Used for Clustering and Classifying Melting Curves by Genotype 332
4. Methods Used for Modeling Melting Curves 335
References 342

Abstract

This chapter discusses mathematical and computational methods that enhance the modeling, optimization, and analysis of high-resolution DNA melting assays. In conjunction with recent improvements in reagents and hardware, these algorithms have enabled new closed-tube techniques for genotyping, mutation scanning, confirming or ruling out genotypic identity among living related organ donors, and quantifying constituents in samples containing different DNA sequences. These methods are rapid, involving only 1 to 10 min of automatic fluorescence acquisition after a polymerase chain reaction. They are economical because inexpensive fluorescent dyes are used rather than fluorescently labeled probes. They are contamination-free and nondestructive. Specific topics include methods for extracting accurate melting curve information from raw signal, for clustering and classifying the results, for predicting complete melting curves and not just melting temperatures, and for modeling and analyzing the behavior of mixtures of multiple duplexes.

[*] Department of Mathematics, University of Utah, Salt Lake City, Utah
[†] Department of Pathology, University of Utah, Salt Lake City, Utah

1. Introduction

Fluorescent DNA melting analysis has been used for some time to validate the results of polymerase chain reaction (PCR, Mullis and Faloona, 1987). PCR amplification curves are obtained by acquiring fluorescence at a point in each temperature cycle at which product strands hybridize (Ririe et al., 1997; Wittwer et al., 2004). In their presence, intercalating dyes such as SYBR Green I become fluorescent and the strength of this signal can be used to quantify products as a function of cycle number. After the reaction reaches its plateau, a melting curve of fluorescence vs temperature can be obtained by acquiring fluorescence continuously as the temperature is raised, products denature, and fluorescence diminishes. The negative derivative of this curve exhibits peaks corresponding to reaction products, and the presence or absence of certain peaks indicates the success or failure of a clean reaction.

SYBR Green I fluorescence does not correlate well with the quantity of double-stranded DNA (dsDNA) present after PCR unless high concentrations that inhibit PCR are used. High-resolution melting analysis was made possible by the discovery and synthesis of "saturation" dyes that combine the desirable properties of accurate quantification of dsDNA and PCR compatibility. It is often useful to add short oligonucleotides called unlabeled probes to a reaction. These are similar to allele-specific primers in that they are synthesized to be complementary to one of the strands of a particular genotype but not the corresponding strands of other genotypes, though they are blocked to prevent extension. By binding with different stability to strands of different genotypes, changes in the associated fluorescence make different variants easier to distinguish. As a result, subtle differences in DNA sequence that are significant for molecular diagnostics are revealed.

The clinical and research applications of high-resolution melting analysis are numerous. We mention specifically genotyping (Erali et al., 2005; Graham et al., 2005; Liew et al., 2004; Palais et al., 2005; Wittwer et al., 2003), mutation scanning (Gundry et al., 2003; Willmore et al., 2004), and simultaneous genotyping and mutation scanning, (Dobrowolski et al., 2005; Montgomery et al., 2007; Zhou et al., 2005). High-resolution melting analysis has also been used as a rapid, economical means of screening close relatives for transplant compatibility by determining whether the highly variable HLA regions have identical genotypes, without actually genotyping either sample (Zhou et al., 2004).

The reagents, thermal control, and optical hardware that enable high-resolution melting analysis require algorithms of commensurate accuracy to extract clinically significant information from data. This chapter focuses on some of the most useful methods for analyzing, modeling and designing high-resolution melting assays. These include a method for filtering the melting signal from raw data based on a model of fluorescent background especially suited to the response of high-resolution dyes; methods for clustering, classifying, and quantifying the resulting melting curves; methods for using thermodynamic parameters to predict the melting curves of pure duplexes and their mixtures; and, conversely, methods for using experimental melting data to obtain better estimates of those parameters. Improvements in one of these aspects often lead to further refinement of the others. As a result, high-resolution melting analysis is currently highly sensitive and specific, as shown in blinded studies of genotyping and mutation scanning (Reed and Wittwer, 2004).

2. Extracting Melting Curves from Raw Fluorescence

A key property of thermal denaturation (melting) and hybridization (annealing) of DNA between its double-stranded helical state (dsDNA) and its single-stranded random coil state (ssDNA) is that it occurs in a narrow temperature interval about a melting temperature, T_M. According to accurate models discussed later, the entire melting curve representing the proportion of DNA that is hybridized vs temperature is characterized by the location and slope of this transition, which can be predicted from the base sequences of the opposite strands and their concentrations by thermodynamic methods, as discussed later. Outside of this transition region, the concentration of double-stranded DNA, [dsDNA], is effectively constant. High-resolution DNA melting curves quantify this process by the fluorescence of dsDNA dyes in solution with DNA.

The fluorescence of these dyes increases in approximate proportion to [dsDNA], and if this was the only source of variation, we would expect the fluorescence vs temperature response to reflect the smoothed step-function behavior of a hybridization melting curve. In practice, multiple factors affect this relationship, including fluorescence due to nontarget DNA, thermal variation in the response of dye in solution, and nonlinear response to hybridization. Below T_M, the raw fluorescence continues to increase as temperature decreases (Fig. 13.1). Because of these other factors, differences between raw fluorescence vs temperature curves belonging to different genotypes are difficult to genotype directly.

A baseline method has been used in many studies to extract melting curves that quantify hybridization as a function of temperature from raw data. These melting curves can be used for the classification of individual

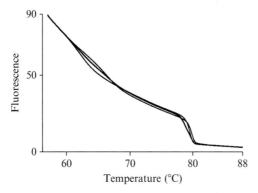

Figure 13.1 Raw high-resolution fluorescence vs temperature curves of three genotypes of a biallelic SNP. Control samples corresponding to each genotype, wild type (WT), homozygous mutant (MUT), and heterozygous (HET) of the Factor V Leiden mutation, along with a sample of unknown genotype, were amplified, and high-resolution melting was performed in the presence of saturating fluorescent DNA dyes and unlabeled oligonucleotide probes. Fluorescence data acquired for each sample are plotted as a function of temperature.

genotypes and quantification of their mixtures. This method has been used in conjunction with thermodynamic studies of denaturation of DNA and RNA using ultraviolet (UV) absorbance (Gray and Tinoco, 1970; SantaLucia, 1998; Uhlenbeck et al., 1973), as well as fluorescence from high-resolution dyes and labeled probes. The baseline method assumes that the raw signal may be well approximated by straight-line fits in temperature ranges above and below the melting transition, $L_0(T)$ and $L_1(T)$, respectively. It estimates the melting curve, $M(T)$, from the height of the raw signal, $F(T)$, above the lower baseline, $L_0(T)$, as a proportion of the difference between the upper and lower baselines (Fig. 13.2):

$$M(T) = \frac{F(T) - L_0(T)}{L_1(T) - L_0(T)} \quad (13.1)$$

In high-resolution fluorescent melting analysis, the baseline method often fails, most dramatically in experiments involving multiple small amplicons and unlabeled probes. In these cases, $L_0(T)$ and $L_1(T)$ intersect below the graph of $F(T)$ due to its concavity, and the denominator in Eq. (13.1) goes to zero, leading to divergence of $M(T)$. In these situations, an exponential background removal method produces superior results. In this method, the raw fluorescence $F(T)$ is modeled as the sum of the melting curve $M(T)$ plus an exponentially decaying background $B(T)$, that is,

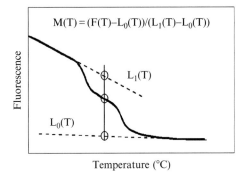

Figure 13.2 Baseline method for normalization of asymptotically linear raw fluorescence vs temperature curves. When raw fluorescence vs temperature data are linear outside of the melting transition, normalized melting curves can be obtained from the proportional vertical distance of the raw fluorescence between lower and upper fitting lines.

$$F(T) = M(T) + B(T) \qquad (13.2)$$

The exponential model has both physical and mathematical rationales in the familiar dependence of fluorescence on temperature, the thermodynamic behavior of statistical ensembles, and the universality of exponential dependence arising as the solution of the fundamental differential equation, $\frac{dy}{dt} = ay$. Background subtraction methods typically employ a model for the amplitude of the background that are fit using quantities that can be determined a priori, without knowing the decomposition of raw data into signal and background. What is unusual about the exponential background method is that it uses a fit to two slopes, and the amplitudes are then determined implicitly. In contrast, polynomial fits require at least one amplitude.

The exponential background removal method proceeds by the following steps.

1. Identify two temperatures, T_L and T_R, sufficiently below and above the oligonucleotide stability transition region, or melting temperature, T_M, so that the variation in fluorescence is solely due to background. This may be quantified in terms of a ratio of derivatives of $F(T)$ that is constant for a pure exponential plus a constant. In these temperature ranges, the melting curve is nearly flat,

$$\frac{dM}{dT}(T_L) = \frac{dM}{dT}(T_R) = 0 \qquad (13.3)$$

so the slope of the raw fluorescence is entirely attributable to the slope of the background. Differentiating Eq. (13.2), $\frac{dF}{dT} = \frac{dM}{dT} + \frac{dB}{dT}$ and combining with eq. (13.3) we find

$$\frac{dF}{dT}(T_L) = \frac{dB}{dT}(T_L) \text{ and } \frac{dF}{dT}(T_R) = \frac{dB}{dT}(T_R) \quad (13.4)$$

2. Use these two measured slopes to fit the background by an exponential model

$$B(T) = B_0 + Ce^{a(T-T_L)} \quad (13.5)$$

in which the argument of the exponential is shifted to T_L for numerical stability. To obtain the parameters of the model, we differentiate Eq. (13.5)

$$\frac{dB}{dT} = aCe^{a(T-T_L)} \quad (13.6)$$

and evaluate Eq. (13.6) at the lower and upper temperature values,

$$aC = \frac{dF}{dT}(T_L) \quad (13.7)$$

and

$$aCe^{a(T_R-T_L)} = \frac{dF}{dT}(T_R). \quad (13.8)$$

Since the two values on the right-hand sides have been measured, we divide Eq. (13.8) by Eq. (13.7) and take the logarithm to obtain

$$a = \frac{\ln \frac{dF}{dT}(T_R) - \ln \frac{dF}{dT}(T_L)}{T_R - T_L} \quad (13.9)$$

and substitute in Eq. (13.7) to get

$$C = \frac{\frac{dF}{dT}(T_L)}{a}. \quad (13.10)$$

3. Obtain the signal with the background removed by subtraction and normalization:

$$M(T) = F(T) - Ce^{a(T-T_L)}. \quad (13.11)$$

with the parameters a and C determined in Eqs. (13.9) and (13.10). The constant is determined implicitly by normalizing the extracted melting curve signal to the range 0–1 by applying the linear shift and rescaling

$$M_1(T) = \frac{M(T) - m}{M - m}, \qquad (13.12)$$

where $m = \min(M(T))$ and $M = \max(M(T))$ on the interval of interest, respectively. A common convention is to rescale by 100, thus reporting the melting curve as a percentage. Since exponentially decaying backgrounds and step function-like signals are also mathematically universal, for example, as the distribution derivative of the delta function, $\delta = \hat{1}$, where $\hat{1}$ is the Fourier transform, this framework is potentially useful in several other contexts.

Another convention in high-resolution melting analysis is the use of the negative derivative curve, $-\frac{dM}{dT}$, for feature analysis. In particular, multiple peaks of this curve indicate the presence of multiple species of DNA duplex with distinct thermodynamic properties, most commonly arising from amplification of heterozygous DNA. In that situation, four different types of duplex arise from nearly random annealing of strands with both their perfect complements, produced in the extension phase of PCR, and the nearly complementary strands belonging to the other allele. The modeling and analysis of the cumulative signal due to the simultaneous melting of these four duplexes are discussed in the final section. Since the derivative of an exponential is an exponential with the same decay rate, the same method may be applied to the derivative curve by simply subtracting an exponential fit of the values at the temperatures of interest. If one is not interested in the undifferentiated melting curve, it is possible to differentiate the raw signal and then subtract an exponential model of the background derivative,

$$\frac{dB}{dT} = De^{a(T-T_L)} \qquad (13.13)$$

where

$$D = \frac{dF}{dT}(T_L) \qquad (13.14)$$

and a is defined by Eq. (13.9) as before.

Since difference methods amplify grid-scale noise in data, the recommended method for differentiation of high-resolution melting data is to perform Savitzky–Golay filtering, in other words to differentiate a moving window least-squares polynomial fit (Press *et al.*, 1992). The resulting oscillations can dominate the intrinsic melting peaks in the negative derivative curves. Conversely, the smoothing aspect of the Savitzky–Golay method can make it difficult to resolve multiple nearby melting peaks that characterize derivative curves of heterozygous samples unless care is taken to keep the fitting window no larger than necessary.

Some examples of melting curves obtained using the exponential background subtraction method discussed earlier are shown in Figs. 13.3 and 13.4, corresponding to the raw fluorescence shown in Fig. 13.1. Four samples from an assay for the Factor V Leiden mutation, an SNP associated with hereditary hypercoagulation, were melted on a single-sample platform. The higher temperature melting of amplicons (Fig. 13.3) and lower temperature melting of unlabeled oligonucleotide probes (Fig. 13.4) of all three genotypes are displayed. The curves are in far closer agreement with the melting curves predicted by the thermodynamic models discussed later than those produced by the baseline method when it may be applied.

The robustness and effectiveness of the method are demonstrated by the fact that these curves were extracted from the raw fluorescence shown in Fig. 13.5, acquired from 48 samples of the same assay performed on a plate-based platform. Examples of Savitzky–Golay derivative curves with background removed using the exponential background removal are shown in

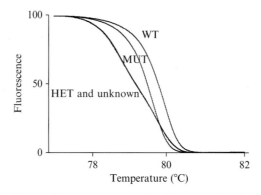

Figure 13.3 Amplicon melting curves normalized from raw data in Fig. 13.1 using the exponential method. The exponential background subtraction method was applied to raw data of Fig. 13.1 based on slope fitting at temperatures below and above the amplicon melting transition, and results are scaled to the range 0–100. The curves are substantially more distinguishable according to genotype, and the melting curve corresponding to unknown sample appears so similar to that of the heterozygous sample that they are difficult to resolve visually as nonidentical curves, suggesting that automatic classification would be successful. These melting curves are not located near the mean of the homozygous curves due to the comparable concentrations of two species of heteroduplex, formed by random reannealing late in PCR, that contribute to the observed melting curves.

Methods for High-Resolution Melting Analysis   331

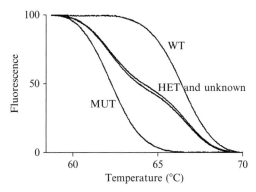

Figure 13.4 Unlabeled oligonucleotide probe melting curves normalized from raw data in Fig. 13.1 using the exponential method. The exponential background subtraction method was also applied to raw data of Fig. 13.1 based on slope fitting at temperatures below and above the unlabeled probe melting transition, and the results are scaled to the range 0–100. The curves are once again substantially more distinguishable according to genotype, although the melting curve corresponding to unknown sample appears similar but not identical to that of the heterozygous sample. Despite this, the heterozygous samples are relatively easier to classify using unlabeled probe melting curves since they are located near the mean of the two homozygous curves.

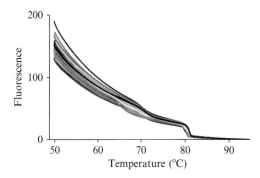

Figure 13.5 Raw high-resolution fluorescence vs temperature curves. These raw fluorescence vs temperature plots emphasize the increased difficulty of automatically clustering and classifying larger data sets without the benefit of exponential background removal. A plate containing 48 samples, including no-template controls and controls of each genotype at the Factor V Leiden locus as in Fig. 13.1, was amplified melted in the presence of unlabeled probes. Data from no-template control samples are not shown and can be excluded automatically by a signal-to-background removal method (not discussed here), as fluorescent background amplitude in these samples is still sufficiently large that it cannot be used as a sole criterion.

Fig. 13. 6. This assay is currently used for automated genotyping in clinical diagnostics, relying on exponential background removal in conjunction with clustering and classification methods discussed in the next section.

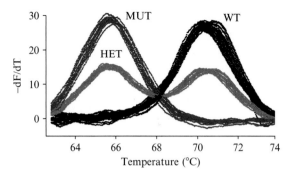

Figure 13.6 Unlabeled probe negative derivative curves normalized using the exponential method with the derivative taken by Savitzky–Golay polynomials. The exponential background subtraction method and Savitzky–Golay differentiation method were applied to raw data in Fig. 13.5 based on slope fitting at temperatures below and above the unlabeled probe melting transition. Curves were clustered by uniform-norm similarity, and the genotype cluster level was determined automatically using the method described in the text. Curves with a common genotype are shown with common grayscale values.

3. METHODS USED FOR CLUSTERING AND CLASSIFYING MELTING CURVES BY GENOTYPE

After applying the background removal method described in the previous section, melting curves are more easily distinguished by genotype. In certain relatively rare cases (Palais *et al.*, 2005), base sequences of wild-type and homozygous mutant samples exhibit a symmetry that makes the corresponding melting curves theoretically identical, according to the models in Section 4. Experimentally, their similarity is near the limits of resolution of current technology. However, aside from these exceptions, the variation in melting curves among samples of the same genotype is sufficiently small in comparison with differences between curves of different genotypes that the correspondence may be inverted reliably and uniquely.

Adaptations of two familiar methods in automatic classification to the particular features of melting curve analysis are described later and demonstrated by the figures of the previous section. The first is a form of agglomerative hierarchical clustering, the workhorse algorithm of unsupervised learning. In a common formulation known as centroid linkage, each of the initial melting curves is assigned a weight of one, and the distances between each pair are computed. This implies some particular choice of metric on the difference between melting curves that is discussed later. The closest pair is identified and replaced by a single melting curve consisting of the (equally) weighted average of the pair being replaced. It is assigned a weight of two, the sum of the original weights. This process reduces the total number of curves remaining by one, while the total weight remains unchanged. The same steps are now essentially repeated, although at each subsequent step, only the distances from the new melting curve to the other remaining melting curves need to be computed, and because the weights are no longer all equal to one, the weighted average of the pair being replaced is no longer necessarily a simple mean. After one less iteration than the original number of curves, all of the curves have been replaced by a single curve, the mean of the original set of curves. If this is the predetermined outcome of the algorithm, then what is the purpose? Somewhat atypically, this is an algorithm for which the intermediate results hold more interest than the ultimate one. At each stage, the original melting curves that were contributors to some remaining average curve are associated by greater similarity to each other than to the others. The problem that remains is to automatically identify the level of similarity of practical, in this case, clinical interest. However, there is also ambiguity in what this level should be, for example, do we need to know merely if a sample is wild type or not (two clusters) or its actual genotype (among three genotype clusters for a biallelic SNP). Therefore, a universal criterion for any canonically significant level is not possible. However, a natural criterion for selecting the level would be one that best separates the visually apparent scales of intraclass and interclass variation in melting curve data. A reasonable conjecture is that there should be a jump in the magnitude of the smallest distance between remaining melting curves after all curves of a common type have been joined. Still, it is observed that deriving the appropriate level from the largest ratio of minimum remaining distances under the implementation of clustering described earlier does not always choose a biologically significant level. What is more, the level this criterion chooses can often be unstable to small changes in the fitting intervals used in the background removal procedure. One way to understand this instability is to observe that by pure experimental chance, two random curves of the same genotype may be considerably more similar than the rest, and the ratio of distances joining their combination to the rest compared to joining them to each other will be a large outlier. Subclusters that divide a genotype level cluster into distinct "sides" may form, and the separation between the means of each subcluster mimics the separation of actual genotypes. Because of the formation of these subclusters, the ratio of successive minimum remaining distances grows gradually rather than remaining near unity except when jumping from subgenotype level to genotype level.

The remedy for this, in the method we describe, has the additional computational benefit that after the initial step, no weighted average curves and no new distances need be computed. Let $M_i, 1 \leq i \leq n$ be the melting curves to be clustered. We will do so using a method that at the kth stage, $k-1, \ldots, n-1$, defines $n-k$ clusters of melting curves $C_{k,l}, l = 1, \ldots, n-k$, where the $C_{k,l}$ form a partition of $\{1, \ldots, n\}$, that is, there are $n-k$ mutually disjoint subsets whose union is the full set $\{1,\ldots,n\}$.

1. Initialize the clusters for $k = 0$ by associating each curve to its own class: $C_{0,l} = \{l\}, l = 1, \ldots, n$.
2. Compute the half-matrix of distances between all distinct pairs of the original n melting curves,

$$D_{ij} = \| M_i - M_j \|, 1 \leq i < j \leq n. \tag{13.15}$$

Here $\| \bullet \|$ is the metric chosen to measure the distance between two curves. The discrete analogues of standard L^p-norms on the interval $[T_L, T_H]$ are all convenient to compute and capture the thermodynamically relevant features of high-resolution melting curves. The uniform, or max-norm, $\|\Delta M\|_{L^\infty} = \max_{T \in [T_L, T_H]} |\Delta M(T)|$, is more sensitive to deviations on the scale of data acquisition, whereas integral norms, such as the mean-absolute norm, $\|\Delta M\|_{L^1} = \frac{1}{T_H - T_L} \int |\Delta M(T)| dT$, and the root-mean-square norm, $\|\Delta M\|_{L^2} = \sqrt{\frac{1}{T_H - T_L} \int |\Delta M(T)|^2 dT}$, average variation with respect to the length of the interval on which they occur. Conceptually, the question that the user must ask is which is considered more or less significant. The L^∞-norm (max-norm) considers change by a smaller absolute magnitude over a wider interval less significant than a larger change over a narrower interval, whereas the L^1-norm behaves in exactly the opposite manner, and the L^2-norm (r.m.s.-norm) is a compromise. Depending on the resolution and parameters of a particular platform, the choice can be optimized empirically by comparing their performance on benchmark assays whose results have been validated independently.

3. For each stage $k = 1, \ldots, n - 1$, the clusters C_{k_2} are defined by merging the two clusters from the previous stage that contain the closest pair of curves belonging to distinct clusters at that stage. Symbolically, compute

$$d_k = \min D_{mn}, m \in C_{k-1,i}, n \in C_{k-1,j}, i < j, \tag{13.16}$$

and for the values of i and j corresponding to the minimizer, define

$$C_{k,i} = C_{k-1,i} \cup C_{k-1,j}, \quad C_{k,l} = C_{k-1,l}, 1 \leq l < j, l \neq i \text{ and}$$
$$C_{k,l} = C_{k-1,l+1}, j \leq l \leq n - k. \tag{13.17}$$

4. Identify the significant change of cluster scale by maximizing the ratio of successive d_k values. Set $d_0 = 1$ and let

$$r_k = \frac{d_{k+1}}{d_k}, k = 0, \ldots, n - 2. \tag{13.18}$$

The value of $K \geq 1$ such that $r_K = \max r_k, 1 \leq k \leq n - 2$ identifies the stage at which appropriate curve clusters are merged, and there are $n - K$ such clusters.

The method is vacuous when $n \leq 2$ and by definition it will always result in at least two clusters, even if all samples belong in the same cluster. Therefore, care should always be taken to provide control samples of at least two distinct types for satisfactory performance. The genotypic level depicted by the three grayscale values of the clusters in Fig. 13.5 was selected automatically using this method. After $n = 48$ samples were clustered, the algorithm identified $K = 45$. All prior distances $d_k, k \leq 45$ represent the distance between two original neighboring melting curves on the fine scale of a common genotype, while d_{46} and d_{47} are distances between two original curves with different genotypes, not distances between means of subclusters.

The caveat regarding inclusion of at least two genotypes has a natural extension that can be used to turn the clustering method given earlier into a classification or supervised learning method. By including a set of samples of known and thermodynamically distinct genotypes among the samples being assayed, any other sample whose genotypes is known to appear among the sample set will be automatically clustered with that standard sample at the level chosen by the method. The unknown sample in Figs. 13.3 and 13.4 was, in fact, automatically genotyped as heterozygous by this method, according to one of three standard samples with which it clustered. Many variations upon this method are possible, using direct nearest-neighbor classification, and training various feature-recognition algorithms, for example, a neural network, to recognize different genotypes through use of a training set of multiple standard samples.

4. METHODS USED FOR MODELING MELTING CURVES

The term DNA melting is used to refer to the sharp shift in the equilibrium

$$ssDNA + ssDNA' \leftrightarrow dsDNA \quad (13.19)$$

between the random coil state of two types of single-stranded DNA and their double helical double-stranded state, as temperature increases near T_M, although it differs mathematically from the discontinuous phase change of the same name. Predicting hybridization and denaturation as a function of temperature and the concentrations of various constituents of a reaction is essential for designing effective molecular diagnostic assays, in order that primers will anneal selectively to the desired segment of the genome under specified thermocycling conditions and yield amplification products whose melting curves may be accurately resolved according to genotype.

The fundamental thermodynamic relationship used to model this process is

$$T = \frac{\Delta H}{\Delta S + R \ln\left(\frac{[ssDNA][ssDNA']}{[dsDNA]}\right)}, \tag{13.20}$$

where T is the absolute temperature, R is Boltzmann's constant, and the two negative constants, ΔH and ΔS, are the standard enthalpy and entropy changes for the process, respectively. Equation (13.20) is derived from the van't Hoff equation, $\frac{d\ln K}{dT} = \frac{\Delta H}{RT^2}$, where $K = \frac{[dsDNA]}{[ssDNA][ssDNA']}$ is the equilibrium constant for Eq. (13.19), using two equivalent forms of the standard Gibbs energy change, $\Delta G = -RT \ln K = \Delta H - T\Delta S$.

For simplicity of exposition, we assume that the two types of strands are distinct, that is, the strands are not self-complementary (in the $5' \to 3'$ reversed sense) and that there are no excess strands, that is, $[ssDNA] = [ssDNA']$. Modifications for complete generality are lengthy but not difficult. Conservation of strands gives

$$[ssDNA] + [ssDNA'] + 2[dsDNA] = C. \tag{13.21}$$

If we divide by C, the quantity $p = \frac{2[dsDNA]}{C}$ gives the proportion of DNA in the hybridized state at a particular temperature. Substituting $[dsDNA] = \frac{Cp}{2}$ and $[ssDNA] = [ssDNA'] = \frac{C(1-p)}{2}$ we reach the basic inverse melting curve model equation

$$T(p) = \frac{\Delta H}{\Delta S + R \ln\left(\frac{C(1-p)^2}{2p}\right)} \tag{13.22}$$

The qualitative properties of melting curves can be derived from mathematical properties of the logarithm and the form of Eq. (13.22). Since $\Delta H, \Delta S < 0$, and as $p \to 1$, $u = \frac{C(1-p)^2}{2p} \to 0^+$, so $\ln(u) \to -\infty$ and $T \to \frac{-\Delta H}{-\infty} = 0^+$, in other words the temperature corresponding to complete annealing of strands is absolute zero.

Surprisingly, the proportion of hybridized DNA, p, does not go to zero as temperature grows arbitrarily large. As p decreases so that the argument of the log grows sufficiently positive to cancel ΔS, there is a small but finite value of p for which $T \to +\infty$. (For p smaller than this value, we have the unphysical situation of $T < 0$.)

If we differentiate and simplify Eq. (13.22), we obtain an "inverse derivative curve" equation,

$$\frac{dT}{dp} = \frac{\Delta H}{\left(\Delta S + R \ln\left(\frac{C(1-p)^2}{2p}\right)\right)^2} \frac{R(p+1)}{p(1-p)} \qquad (13.23)$$

Substituting the equation for the inverse melting curve gives

$$\frac{dT}{dp} = \frac{T(p)^2}{\Delta H} \frac{R(p+1)}{p(1-p)}. \qquad (13.24)$$

Equation (13.24) indicates the existence of large temperature regions where $p \approx 0$ and $p \approx 1$ and $\frac{dT}{dp} \approx -\infty$, separated by a sharp transition in a relatively small temperature region. In particular, setting $p = \frac{1}{2}$ defines the melting temperature, T_M, and gives the well-known formula for which strands are equally distributed between single-stranded and double-stranded states:

$$T_M = \frac{\Delta H}{\Delta S + R \ln\left(\frac{C}{4}\right)}. \qquad (13.25)$$

There are various methods for solving the inverse melting curve equation to compute the melting curve $p(T)$ explicitly from the thermodynamic values ΔH, ΔS. Rearranging Eq. (13.22) in the form

$$\ln\left(\frac{C(1-p)^2}{2p}\right) = \frac{\Delta H}{RT} - \frac{\Delta S}{R} = \frac{\Delta G}{RT}$$

and exponentiating both sides leads to a quadratic equation for p whose coefficients depend on T,

$$p^2 - 2\left(1 + \frac{1}{C} e^{\frac{\Delta G}{RT}}\right) p + 1 = 0.$$

The quadratic formula for its roots gives an explicit representation of the melting curve:

$$p = 1 + \frac{1}{C} e^{\frac{\Delta G}{RT}} - \sqrt{\frac{2}{C} e^{\frac{\Delta G}{RT}} + \frac{1}{C^2} e^{\frac{2\Delta G}{RT}}}. \qquad (13.26)$$

The negative sign of the square root is dictated by the requirement that $0 \leq p \leq 1$ and the fact that

$$1 + \frac{1}{C} e^{\frac{\Delta G}{RT}} \geq 1. \tag{13.27}$$

Numerical analysis cautions us that above the melting temperature where $p \to 0$, the form above suffers from loss of significance error due to cancellation among two nearly equal terms. To avoid this, compute the other root of the quadratic using the positive square root so that there is no cancellation, and since the constant term shows that the product of the roots is 1, taking the reciprocal gives the root we are looking for more accurately.

An alternate method for computing $p(T)$ involves reformulating Eq. (13.24) as an ordinary differential equation. Differentiation of $p(T(p)) = p$ using the chain rule gives the relationship of the derivatives of inverse functions, $\frac{dp}{dT}\frac{dT}{dp} = 1$ or $\frac{dp}{dT} = 1/\frac{dT}{dp}$. Applying this relation to Eq. (13.24) leads to the initial value problem,

$$\frac{dp}{dT} = \frac{\Delta H p(1-p)}{R(p+1)T^2}, \quad p(T_M) = \frac{1}{2} \tag{13.28}$$

Approximate solutions may be found to high accuracy using a numerical ODE solver, for example, fourth-order Runge-Kutta (or an implicit method such as the trapezoidal method, for stability). The advantage of such methods is that they are easily generalized to multicomponent systems describing the melting of more complex DNA, and they can be validated in this situation where explicit solutions are available. Note that the vector field corresponding to the differential equation Eq. (13.28) satisfied by the melting curve is completely determined by ΔH. It is solely through the dependence of T_M on ΔS that ΔS affects the melting curve, by selecting a particular integral curve.

Equation (13.22) may also be used to generate a melting curve with equally spaced or arbitrarily prescribed values of p instead of T directly, simply by reversing (p, T) pairs to (T, p) pairs. Regardless of the method chosen, the central lesson of the mathematics is that one characteristic quantity, T_M, is not sufficient to understand the competitive hybridization behavior of DNA, but two are. What the second parameter should be is somewhat arbitrary, but the slope of the melting curve at T_M seems like a natural choice.

For successful assay design, the methods described above can be used to predict thermodynamic similarity in terms of distance between melting curves, a physically and mathematically natural metric by which different sequences of DNA can be experimentally compared.

Setting $p = \frac{1}{2}$ in Eq. (13.28) gives

$$\frac{dp}{dT}\left(\frac{1}{2}\right) = \frac{\Delta H}{6RT_M^2} \tag{13.29}$$

and leads to a method for determining the two thermodynamic parameters that completely determine the melting process according to the two-state model.

1. Perform exponential background removal on raw fluorescence data and scale to the range $[0, 1]$ to obtain the normalized experimental melting curve, $p(T)$. Perform linear least-squares fitting of data in the range of $p = \frac{1}{2}$ to estimate T_M and $\frac{dp}{dT}(T_M)$.

2. Use the formulas

$$\Delta H = 6RT_M^2 \frac{dp}{dT}(T_M) \tag{13.30}$$

and

$$\Delta S = \frac{\Delta H}{T_M} - R \ln\left(\frac{C}{4}\right) = 6RT_M \frac{dp}{dT}(T_M) - R \ln\left(\frac{C}{4}\right) \tag{13.31}$$

derived from Eq. (13.25) to obtain estimates of the values of ΔH and ΔS. Estimating T_M from the peak of the negative derivative curve will lead to inconsistent results, since T_M as defined earlier and does not necessarily agree with the max of $-\frac{dp}{dT}$. Differentiating the inverse function relation $\frac{dp}{dT}\frac{dT}{dp} = 1$ one step further,

$$\frac{d}{dp}\left(\frac{dp}{dT}\frac{dT}{dp}\right) = \frac{d^2p}{dT^2}\left(\frac{dT}{dp}\right)^2 + \frac{dp}{dT}\frac{d^2T}{dp^2} = 0 \tag{13.32}$$

The negative derivative curve, $-\frac{dp}{dT}$ has its maximum where $\frac{d^2p}{dT^2} = 0$, so by Eq. (13.32) this occurs where $\frac{d^2T}{dp^2} = 0$. Differentiating Eq. (13.24), setting $p = \frac{1}{2}$, $T = T_M$, and using Eq. (13.29) gives

$$\frac{d^2T}{dp^2} = \frac{4RT_M^2}{\Delta H}\left(\frac{18RT_M}{\Delta H} + 1\right) \tag{13.33}$$

so unless $T_M = -\frac{\Delta H}{18R}$, that is, only in the unusual case that

$$\Delta S = 18 - R \ln\left(\frac{C}{4}\right), \tag{13.34}$$

will the two definitions of T_M coincide. Their difference may be exaggerated by varying ΔS and C to have peak occur at any $p > \sqrt{2}-1$.

3. In practice, these locally defined values of ΔH and ΔS are quite accurate, but they may optionally be refined iteratively by fitting the melting curve globally using the Levenberg–Marquardt nonlinear least-squares fitting algorithm (Press et al., 1992) using the local estimates as initial values.

There is considerable literature regarding the quantities ΔH and ΔS that are required to model melting curves. Results (Breslauer *et al.*, 1986; DeVoe and Tinoco, 1962; Dimitrov and Zuker, 2004; Fixman and Freire, 1977; Freier *et al.*, 1986; Gray and Tinoco, 1970; Lerman and Silverstein, 1987; Peyret *et al.*, 1999; Poland, 1974; SantaLucia, 1998) suggest that for DNA homoduplexes, consisting of fully complementary strands, the values may be predicted accurately by models of the form

$$\Delta H = \vec{p}_H \bullet \vec{n}_O \text{ and } \Delta S = \vec{p}_S \bullet \vec{n}_O + s|\vec{n}_O \bullet \vec{1}|. \tag{13.35}$$

Here, $\vec{n}_O$ is a vector of 10 nonnegative integers counting the number of tetrads (consecutive pairs of complementary base pairs) in the oligonucleotide, O, belonging to each of 10 thermodynamically distinct classes, $\vec{p}_H$ and $\vec{p}_S$ are vectors of nearest-neighbor enthalpy and entropy parameters corresponding to each class, s is a factor depending on concentration of various salt ions, for example, $[Na^+], [Mg^{++}]$ in solution, and $v \cdot w = \sum_{j=1}^{n} v_j w_j$ represents the dot product of two vectors. Ten classes of tetrads arise since, of the four times four possible pairs of base pairs, twelve occur in six thermodynamically $5' \to 3'$ symmetric pairs, for example, $\frac{AC}{TG} \leftrightarrow \frac{GT}{CA}$, and the remaining four are self-symmetric, for example, $\frac{AT}{TA}$. This same symmetry makes the mapping from DNA sequences to tetrad-count vectors that determine thermodynamic behavior (by the nearest-neighbor model) nontrivial. (Mathematically, we would like to know what conditions on a count vector guarantee that there is a sequence to which it corresponds and, in that case, how many different such sequences there are, but a discussion of these questions would take us too far afield for our current purposes.) If the salt concentration is considered fixed, there is a degree of indeterminacy in the model, as any portion of the salt contribution may be absorbed by adding a constant to each of the entropy parameters. Similarly, initiation–termination parameters corresponding to individual base-pair classes at the ends of the oligonucleotide are not completely independent, as there are always two ends; furthermore, tetrad counts can completely determine both end-pair types. There are also numerous published estimates of the parameter vectors, derived using methods such as UV absorbance that require experimental conditions far removed from those of high-resolution melting. Because of these differences, multiple empirically derived extrapolation factors for salts, dyes, and so on are necessary to make them applicable to the context of high-resolution melting.

The simplicity and accuracy of high-resolution melting suggest an alternative. Recasting the model Eq. (13.35) for fixed standard experimental conditions, with all such dependencies incorporated in the parameters as justified earlier, we view

$$\vec{n}_O \bullet \vec{p}_H = \Delta H \text{ and } \vec{n}_O \bullet \vec{p}_S = \Delta S \qquad (13.36)$$

as linear equations satisfied by the parameters adapted to these conditions, with coefficients obtained from the known sequence (hence known tetrad count vector) of the oligonucleotide, and right-hand sides derived from its experimental normalized melting using the method described earlier. To determine the parameter vectors, the mathematics requires that sufficiently many linearly independent oligonucleotides be chosen; additional equations lead to an overdetermined system that can give a superior best least-squares solution. The other criterion affecting the success of this approach is whether the melting behavior of the selected oligonucleotides is well approximated by the two-state model. This can be tested intrinsically by the goodness of fit of the best Levenberg–Marquart fit described previously to the closest theoretical melting curve it approximates, and candidates outside a prescribed tolerance excluded. A more systematic approach is possible based on minimizing alternate self-interactions other than the primary two-state structure using *in silico* sequence comparison algorithms and theoretical characterizations of maximally random finite sequences (Knuth, 1969). The two theoretical homozygous melting curve samples shown in Fig. 13.7 were obtained using the methods described earlier. The details of implementing this program in greater detail are discussed elsewhere.

High-resolution melting experiments on heteroduplexes formed with unlabeled probes (e.g., Figs. 13.4 and 13.5, MUT samples) and by other synthesized oligonucleotides can be used to extend the set of parameters obtained by the methods described previously incrementally and to determine additional thermodynamic parameters for mismatched tetrads. When heterozygous DNA is melted in the presence of synthesized unlabeled oligonucleotide probes complementary to a portion of a strand of one of the two homozygous genotypes, the resulting melting curves are well approximated by a weighted average of the melting curves of the probe with DNA belonging to these two genotypes individually (e.g., Figs. 13.4 and 13.5, HET samples). This suggests that a convex superposition of individual duplex melting curves is an effective method for predicting melting curves of mixtures. A full investigation of this method, which includes results of melting synthesized heteroduplexes, analysis and optimization across a range of unequally weighted artificial heterozygotes produced by pre-PCR mixing, and comparison with TGCE heteroduplex analysis, is reported elsewhere (Palais, 2007; Palais *et al.*, 2005). The theoretical heterozygous melting curve shown in Fig. 13.7 was obtained using this method to model the melting of duplex mixtures as convex superposition of individual duplex melting curves, applied to the heterozygous

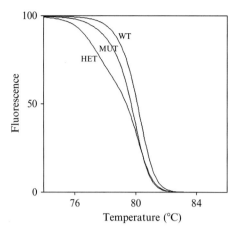

Figure 13.7 Theoretical amplicon melting curves corresponding to the normalized experimental curves in Fig. 13.3 predicted using the two-state nearest-neighbor method and the duplex mixing method. The theoretical melting curves of the two homozygous genotypes of the Factor V Leiden amplicons whose melting curves are shown in Fig. 13.3 were calculated using the method described in Eq. (13.26), with the quantities ΔH and ΔS computed using the standard experimental condition nearest-neighbor parameter method. The theoretical melting curve of the homozygous genotype was calculated using the method of weighted convex combination of the two homoduplexes curves described earlier and two heteroduplex curves.

sample whose melting curve is shown in Fig. 13.3. The coefficients for weighting heteroduplexes vs homoduplexes were obtained by least-squares fitting applied to a training set that did not include this sample. Agreement of the theoretical predictions and experimental results for all three genotypes using these methods encourages further study.

REFERENCES

Breslauer, K. J., Frank, R., Blocker, H., and Marky, L. A. (1986). Predicting DNA duplex stability from the base sequence. *Proc. Natl. Acad. Sci. USA* **83,** 3746–3750.
DeVoe, H., and Tinoco, I., Jr. (1962). The stability of helical polynucleotides: Base contributions. *J. Mol. Biol.* **4,** 500–517.
Dimitrov, R. A., and Zuker, M. (2004). Prediction of hybridization and melting for double-stranded nucleic acids. *Biophys. J.* **87,** 215–226.
Dobrowolski, S. F., McKinney, J. T., Amat di San Filippo, C., Giak Sim, K., Wilcken, B., and Longo, N. (2005). Validation of dye-binding/high-resolution thermal denaturation for the identification of mutations in the SLC22A5 gene. *Hum. Mutat.* **25**(3), 306–313.
Erali, M., Palais, B., and Wittwer, C. T. (2005). SNP genotyping by unlabeled probe melting analysis. *In* "Molecular Beacons" (O. Seitz and A. Marx, eds.). Humana Press, Totowa, NJ.
Fixman, M., and Freire, J. J. (1977). Theory of DNA melting curves. *Biopolymers* **16,** 2693–2704.

Freier, S. M., Kierzek, R., Jaeger, J. A., Sugimoto, N., Caruthers, M. H., Neilson, T., and Turner, D. H. (1986). Improved free-energy parameters for predictions of RNA duplex stability. *Proc. Natl. Acad. Sci. USA* **83,** 9373–9377.

Graham, R., Liew, M., Meadows, C., Lyon, E., and Wittwer, C. T. (2005). Distinguishing different DNA heterozygotes by high-resolution melting. *Clin. Chem.* **51**(7), 1295–1298.

Gray, D. M., and Tinoco, I., Jr. (1970). A new approach to the study of sequence-dependent properties of polynucleotides. *Biopolymers* **9,** 223–244.

Gundry, C. N., Vandersteen, J. G., Reed, G. H., Pryor, R. J., Chen, J., and Wittwer, C. T. (2003). Amplicon melting analysis with labeled primers: A closed-tube method for differentiating homozygotes and heterozygotes. *Clin. Chem.* **49**(3), 396–406.

Knuth, D. (1969). "The Art of Computer Programming" Vol. 2, pp. 147–150. Addison-Wesley, Reading, MA.

Lerman, L. S., and Silverstein, K. (1987). Computational simulation of DNA melting and its application to denaturing gradient gel electrophoresis. *Methods Enzymol.* **155,** 482–501.

Liew, M., Pryor, R., Palais, R., Meadows, C., Erali, M., Lyon, E., and Wittwer, C. T. (2004). Genotyping of single-nucleotide polymorphisms by high-resolution melting of small amplicons. *Clin. Chem.* **50,** 1156–1164.

Montgomery, J., Wittwer, C. T., Palais, R., and Zhou, L. (2007). Simultaneous mutation scanning and genotyping by high-resolution DNA melting analysis. *Nature Protocols* **2**(1), 59–66.

Mullis, K. B., and Faloona, F. A. (1987). Specific synthesis of DNA in vitro via a polymerase-catalyzed chain reaction. *Methods Enzymol.* **155,** 335–350.

Palais, R. (2007). Quantitative heteroduplex analysis. *Clin. Chem.* **53**(6), 1001–1003.

Palais, R., Liew, M., and Wittwer, C. T. (2005). Quantitative heteroduplex analysis for single nucleotide polymorphism genotyping. *Anal. Biochem.* **346,** 167–175.

Peyret, N., Seneviratne, P. A., Allawi, H. T., and SantaLucia, J. (1999). Nearest-neighbor thermodynamics and NMR of DNA sequences with internal A:A, C:C, G:G, and T:T mismatches. *Biochemistry* **38,** 3468–3477.

Poland, D. (1974). Recursion relation generation of probability profiles for specific-sequence macromolecules with long-range correlations. *Biopolymers* **13,** 1859–1871.

Press, W. H., Vetterling, W. T., Teukolsky, S. T., and Flannery, B. P. (1992). "Numerical Recipes in C: The Art of Scientific Computing" pp. 650–655. Cambridge Univ. Press, Cambridge.

Reed, G. H., and Wittwer, C. T. (2004). Sensitivity and specificity of single-nucleotide polymorphism scanning by high-resolution melting analysis. *Clin. Chem.* **50**(10), 1748–1754.

Ririe, K. M., Rasmussen, R. P., and Wittwer, C. T. (1997). Product differentiation by analysis of DNA melting curves during the polymerase chain reaction. *Anal. Biochem.* **245**(2), 154–160.

SantaLucia, J., Jr. (1998). A unified view of polymer, dumbbell, and oligonucleotide DNA nearest-neighbor thermodynamics. *Proc. Natl. Acad. Sci. USA* **95**(4), 1460–1465.

Uhlenbeck, O. C., Borer, P. N., Dengler, B., and Tinoco, I. (1973). Stability of RNA hairpin loops. *J. Mol. Biol.* **73,** 483–496.

Willmore, C., Holden, J. A., Zhou, L., Tripp, S., Wittwer, C. T., and Layfield, L. J. (2004). Detection of c-kit-activating mutations in gastrointestinal stromal tumors by high-resolution amplicon melting analysis. *Am. J. Clin. Pathol.* **122**(2), 206–216.

Wittwer, C. T., Hahn, M., and Kaul, K. (2004). "Rapid Cycle Real-time PCR: Methods and Applications." Springer, New York.

Wittwer, C. T., Reed, G., Gundry, C., Vandersteen, J., and Pryor, R. (2003). High-resolution genotyping by amplicon melting analysis using LCGreen. *Clin. Chem* **4,** 853–860.

Zhou, L., Vandersteen, J., Wang, L., Fuller, T., Taylor, M., Palais, B., and Wittwer, C. T. (2004). High-resolution DNA melting curve analysis to establish HLA genotypic identity. *Tissue Antigens* **64**(2), 156–164.

Zhou, L., Wang, L., Palais, R., Pryor, R., and Wittwer, C. T. (2005). High-resolution DNA melting analysis for simultaneous mutation scanning and genotyping in solution. *Clin. Chem.* **51**(10), 1770–1777.

CHAPTER FOURTEEN

BIOMATHEMATICAL MODELING OF PULSATILE HORMONE SECRETION: A HISTORICAL PERSPECTIVE

William S. Evans,[*] Leon S. Farhy,[†] and Michael L. Johnson[‡]

Contents

1. Introduction	346
2. Early Attempts to Identify and Characterize Pulsatile Hormone Release	348
3. Impact of Sampling Protocol and Pulse Detection Algorithm on Hormone Pulse Detection	349
4. Application of Deconvolution Procedures for the Identification and Characterization of Hormone Secretory Bursts	351
5. Limitations and Subsequent Improvements in Deconvolution Procedures	352
6. Evaluation of Pulsatile and Basal Hormone Secretion Using a Stochastic Differential Equations Model	353
7. Characterization of Regulation of Signal and Response Elements: Estimation of Approximate Entropy	355
8. Evaluation of Coupled Systems	356
9. Evaluation of Hormonal Networks with Feedback Interactions	360
10. Future Directions	362
Acknowledgments	362
References	363

Abstract

Shortly after the recognition of the profound physiological significance of the pulsatile nature of hormone secretion, computer-based modeling techniques were introduced for the identification and characterization of such pulses. Whereas these earlier approaches defined perturbations in hormone concentration-time series, deconvolution procedures were subsequently employed to

[*] Endocrinology and Metabolism Department of Medicine, and Department of Obstetrics and Gynecology, University of Virginia Health System, Charlottesville, Virginia
[†] Endocrinology and Metabolism Department of Medicine, University of Virginia Health System, Charlottesville, Virginia
[‡] Departments of Pharmacology and Medicine, University of Virginia Health System, Charlottesville, VA

Methods in Enzymology, Volume 454
ISSN 0076-6879, DOI: 10.1016/S0076-6879(08)03814-7

© 2009 Elsevier Inc.
All rights reserved.

345

separate such pulses into their secretion event and clearance components. Stochastic differential equation modeling was also used to define basal and pulsatile hormone secretion. To assess the regulation of individual components within a hormone network, a method that quantitated approximate entropy within hormone concentration-times series was described. To define relationships within coupled hormone systems, methods including cross-correlation and cross-approximate entropy were utilized. To address some of the inherent limitations of these methods, modeling techniques with which to appraise the strength of feedback signaling between and among hormone-secreting components of a network have been developed. Techniques such as dynamic modeling have been utilized to reconstruct dose–response interactions between hormones within coupled systems. A logical extension of these advances will require the development of mathematical methods with which to approximate endocrine networks exhibiting multiple feedback interactions and subsequently reconstruct their parameters based on experimental data for the purpose of testing regulatory hypotheses and estimating alterations in hormone release control mechanisms.

1. INTRODUCTION

The second half of the 20th century witnessed an extraordinary number of scientific advances that coalesced to provide a much improved understanding of how endocrine systems in general—and the reproductive axis in particular—function as "networks" in which individual components communicate with one another in a highly coordinated fashion. Central to the reproductive system are cells within the anterior pituitary gland, which secrete two so-called "gonadotropins"—luteinizing hormone (LH) and follicle-stimulating hormone (FSH)—which control both gametogenesis and hormone production in the testes and ovary. Although our recognition of the role played by LH and FSH in relation to gonadal function dates back to the 1920s (Fluhmann, 1929), it was midcentury before the possibility that the hypothalamus might serve to regulate pituitary function, including that of the gonadotropin-secreting cells, was put forward by Green and Harris (1947). As depicted in Fig. 14.1, these investigators suggested that neuronal systems within the hypothalamus could both synthesize and secrete chemical (hormonal) messengers, which could then travel to the pituitary gland via the hypothalamic–hypophyseal portal circulation where they would influence the secretion of pituitary hormones. However, another two decades would elapse before the first of these hypothalamic hormones, including gonadotropin-releasing hormone (GnRH), which stimulates the synthesis and release of LH and FSH, and thyrotropin-releasing hormone (TRH), which is responsible for thyrotropin stimulating hormone (TSH) production, were isolated, characterized, and synthesized by Guilleman (Burgus et al., 1972) and Schally (Schally et al., 1971; Matsuo et al., 1971).

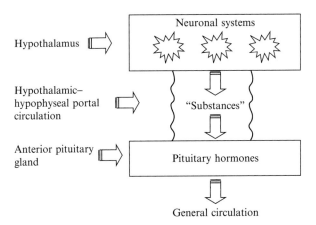

Figure 14.1 Control of pituitary hormone synthesis and secretion by the hypothalamus as hypothesized by Green and Harris (1947). Chemical messengers secreted by nerve cells within the hypothalamus were proposed to travel through the hypothalamic–hypophyseal portal circulation to the anterior pituitary gland where they would stimulate, inhibit, or otherwise modulate the synthesis and release of hormones.

The significance of this work was recognized by the Nobel Prize for Physiology or Medicine being awarded to these investigators in 1977.

Shortly after the discovery of GnRH, and capitalizing on the availability of a relatively new measurement technique known as radioimmunoassay, studies in the rat (Gay and Midgley, 1969), Rhesus monkey (Atkinson et al., 1970), and the human (Midgley and Jaffee, 1971; Santen and Bardin, 1973; Yen et al., 1972) suggested that LH and FSH were secreted in a pulsatile rather than a tonic fashion. The profound physiologic significance of this observation was demonstrated by Knobil and colleagues using the Rhesus monkey as a model (Knobil, 1980). Knobil's seminal studies demonstrated that whereas the administration of GnRH in a *pulsatile* fashion to monkeys in which the GnRH-secreting cells had been ablated did indeed effect the secretion of LH and FSH, the administration of the peptide as a *nonvarying infusion* failed to stimulate secretion of the gonadotropins (Fig. 14.2). Moreover, GnRH infused into monkeys with an intact hypothalamic–pituitary system actually *diminished* the secretion of the gonadotropins. These studies strongly suggested that GnRH itself might well be secreted in a pulsatile manner, thus stimulating the pulsatile secretion of LH and FSH. Approximately a decade later, studies in which GnRH and LH were measured simultaneously in the rat (Levine and Ramirez, 1982) and sheep (Clarke and Cummins, 1982; Levine et al., 1982) within the hypothalamic–hypophyseal portal circulation and in the vasculature draining the anterior pituitary gland confirmed this direct relationship.

Subsequent basic science and clinical studies have served to underscore the extraordinary significance of the pulsatile nature of hormone secretion

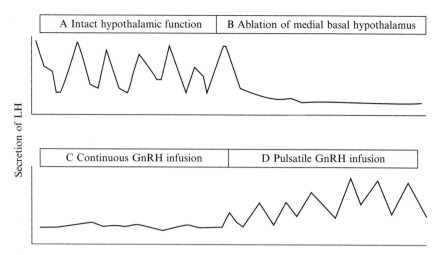

Figure 14.2 Luteinizing hormone (LH) secretion in Rhesus monkeys with (A) intact hypothalamic function, (B) after ablation of the medial basal hypothalamus, (C) in response to a continuous GnRH infusion after hypothalamic ablation, and (D) in response to pulsatile GnRH infusion after ablation of the hypothalamus. These studies (Knobil, 1980) were the first to demonstrate the profound importance of the pulsatile versus nonvarying nature of GnRH secretion with regard to LH and FSH secretion.

within the reproductive axis. Using the rat as a model, Marshall and colleagues have shown that the *gene expression* of LH and FSH is controlled by the frequency and amplitude characteristics of the pulsatile GnRH signal (Haisenleder *et al.*, 1994). Within the clinical setting, the significance of the characteristics of the GnRH signal has been exploited by clinicians who wish either to *restore* normal gonadotrope function by means of the exogenous pulsatile administration of GnRH (e.g., in cases of primary hypothalamic failure such as Kallmann's syndrome or hypothalamic failure-associated infertility) or to *inhibit* gonadotropin secretion by means of a long-acting GnRH agonist (e.g., in the setting of prostrate cancer, precocious puberty, or endometriosis), which provides a nonpulsatile signal to the pituatary and serves to desensitize the pituitary and essentially abolish LH and FSH secretion.

2. Early Attempts to Identify and Characterize Pulsatile Hormone Release

A logical extension of this improved understanding of both the fundamental physiological underpinnings of pulsatile hormone release noted earlier and the potential importance of pulsatile secretion in terms of clinical applications was the recognition of the need to define and subsequently characterize in an objective manner hormone pulses within concentration-time series. Examples of some initial efforts to achieve this goal included defining pulses as an increase followed by a decrease of an arbitrary number of units (e.g., 5 mIU in the case of LH; Yen *et al.*, 1972) or an increase in the serum concentration of a hormone of at least 20% over the proceeding nadir (the so-called "Santen and Bardin" algorithm; Santen and Bardin, 1973). For whatever the reason, however, nearly a decade passed before the power of computer-assisted algorithms for the identification and characterization of pulsatile hormone secretion—and importantly the impact of how blood sampling protocols may affect the ability to define such pulses—began to receive investigative attention.

3. IMPACT OF SAMPLING PROTOCOL AND PULSE DETECTION ALGORITHM ON HORMONE PULSE DETECTION

The early to mid-1980s saw a flurry of activity focused on the development of a range of sampling protocols (e.g., protocols in which blood samples were obtained anywhere from every 1 to every 60 min over intervals of 2–24 h) and of computerized pulse analysis techniques, each subserved by quite different mathematical constructs. The latter included *Ultra* developed by Van Cauter and colleagues (1981), *Pulsar* by Merriam and Wachter (1982), and *Cycle Detector* by Clifton and Steiner (1983). The use of sampling protocols of differing intensity and interval—and analysis of the resultant hormonal concentration-time series with computer-assisted methods based on nonuniform mathematical and statistical approaches—perhaps not surprisingly resulted in quite diverse estimates of pulsatile hormone secretion, in many cases under theoretically similar physiological conditions. Thus, application of these earlier pulse detection programs to LH concentration-time series obtained in normal women at several phases of the menstrual cycle resulted in the identification of anywhere from 14 to 24 LH pulses/24 h in early follicular phase women, from 17 to 29 in late follicular phase women, and from 4 to 16 in midluteal phase women (Backstrom *et al.*, 1982; Burger *et al.*, 1985; Crowley *et al.*, 1985; Filicori *et al.*, 1984, 1986; Murdoch *et al.*, 1989; Reame *et al.*, 1984; Rossmanith *et al.*, 1990; Schweiger *et al.*, 1989; Soules *et al.*, 1984).

To define the impact of sampling protocol and of pulse identification method on the ability to detect LH pulses in men and women, we applied several of the available pulse detection algorithms to serum LH concentration-time series obtained from normal men (Urban *et al.*, 1988) and normal women at different phases of the menstrual cycle (Evans *et al.*, 1992). Both the sampling protocol and the pulse detection algorithm were found to have a profound effect on the number of pulses detected. With regard to sampling intensity, up to an eightfold difference in the number of pulses detected within the same data set could be found when the complete data set (sampled at 4-min intervals) was progressively "deintensified" by omitting successive samples to simulate sampling at, for example, 12-, 20-, and 40-min intervals. Similar concerns were apparent when data from normal women during the midluteal phase of the menstrual cycle were analyzed with four of the independently developed computer-based techniques mentioned previously. As is shown in Fig. 14.3, when data obtained with a standard protocol (samples obtained at 10-min intervals over a 24-h interval) were analyzed, estimates for the number of pulses varied from 3 to 17 per 24 h.

Such divergent results prompted the development of yet more pulse detection algorithms. Nearly simultaneously, two novel programs became available to the scientific community: *Detect*, written by Oerter, Guardarraso, and Rodbard (Oerter *et al.*, 1986), and *Cluster*, provided by Veldhuis and Johnson (Veldhuis and Johnson, 1986). Although using distinctly different mathematical approaches, both programs shared the properties of being rigorously statistically based. Whereas *Cluster* identifies statistically significant increases and decreases in the concentration-time series with a sliding T test, *Detect* defines pulses as statistically significant increases and decreases in the numerical first derivatives of the concentration-time series.

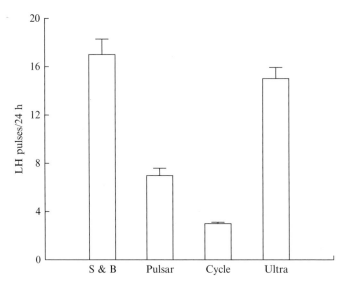

Figure 14.3 Number of LH pulses identified in concentration-time series obtained over a 24-h period of time during the midluteal phase of the menstrual cycle of normal women. The same data sets were analyzed with independent pulse detection algorithms, including *Santen & Bardin* (*S & B*), *Pulsar*, *Cycle Detector* (*Cycle*), and *Ultra*. Note the considerable difference in number of pulses detected as a function of the pulse analysis program used for the analysis (Evans *et al.*, 1992).

Of interest, when these programs were applied to the LH concentration-time series mentioned earlier, the number of LH pulses identified was statistically indistinguishable.

4. APPLICATION OF DECONVOLUTION PROCEDURES FOR THE IDENTIFICATION AND CHARACTERIZATION OF HORMONE SECRETORY BURSTS

Although computer-assisted pulse detection algorithms represented a significant advance in our ability to characterize hormone secretion, it was soon recognized that such methods identify *perturbations* in concentration-time series rather than describing actual underlying secretory events. The appreciation that a hormone "pulse" comprises both a secretory event and factors that affect clearance of the hormone, such as distribution, binding to proteins, metabolism, and excretion, stimulated interest as to whether mathematical methods might exist that would allow for the "separation" of a hormone pulse into its constituent secretory burst and clearance components. Indeed, Turner and colleagues (1972), Rebar and colleagues (1973), and MacIntosh and MacIntosh (1985) were among the first to employ *deconvolution* procedures to achieve this objective. The latter two groups determined that, for the case of LH, secretion did in fact occur in a burst-like fashion without the requirement for a tonic secretory component. The deconvolution procedures used in these earlier studies, however, had their own limitations, including the need for information about the half-life of the hormone under consideration. For example, in the case of one of the methods mentioned previously (Rebar *et al.*, 1973), the deconvolution procedure employed a sinusoidal function for which an estimated two-compartment half-life component was needed. Similarly, in a later report, Genazzani and colleagues (1990) applied "discrete" (sample-by-sample) deconvolution analysis, which again required a literature estimated LH half-life.

Recognizing the need for an approach that would not be constrained by the requirement for a priori information about the half-life of the hormone of interest, Veldhuis and colleagues (1987) developed a deconvolution procedure that allowed for simultaneous assessment of both the characteristics of the hormone secretory burst and provided an estimate of the hormone half-life. Known as *Deconv*, this multiparameter deconvolution method differed from earlier deconvolution procedures in that it describes secretion as a variable number of Gaussian-shaped secretion events convolved with a one- or two-component exponential elimination and then fit directly to the concentration-time series. Several of the earlier approaches as mentioned above evaluated secretion as a discrete amount at each data point but did not locate the distinct secretory events (McIntosh and McIntosh, 1985; Oerter, 1986; Rebar *et al.*, 1973). Algorithms such as *Cluster* or *Detect* were required to identify the position of the peaks in the secretion based on the evaluation at every data point produced by the earlier methods (Oerter, 1986; Veldhuis and Johnson, 1986). In contrast, *Deconv* allows for the identification of significant secretion events directly by the fitting process. The earlier approaches were also very susceptible to measurement errors within the concentration-time series.

Figure 14.4 shows a LH concentration-time series obtained from a normal woman during the early follicular phase of her menstrual cycle (top) and the LH secretory events as resolved by application of *Deconv* (bottom). In addition to identifying the secretory bursts, *Deconv* also provides information about secretory burst mass, half-duration, and an estimate of hormone half-life. Indeed, shortly after *Deconv* became available, this method was used to analyze data obtained from normal women at several phases of the menstrual cycle, given that our understanding of physiology predicted differences in (1) the *frequency* of secretory events, (2) the *characteristics* of these events underlying the secretory process itself, and (3) the *half-life* of the hormone (reflecting post-translation processing of the protein)(Sollenberger *et al.*, 1990). As can be seen in Fig. 14.5, although the total daily secretion of LH did not vary across the menstrual cycle, the deconvolution-resolved number of secretory bursts, their characteristics, and their half-life varied considerably as a function of the phase of the cycle in the younger women and in the postmenopausal group.

5. Limitations and Subsequent Improvements in Deconvolution Procedures

While deconvolution procedures represented a major advance in our ability to model hormonal data, it became readily apparent that the earlier versions had significant limitations, including the subjective nature of the choice of candidate secretory bursts, the lack of robust statistical verification of resolved secretory bursts, and the user-unfriendly interface of the programs. To address these concerns, a novel deconvolution procedure known as *AutoDecon* has been developed and validated for LH (Johnson *et al.*, 2008) and growth hormone. This program finds the optimal number of secretory bursts and initial parameter estimates while simultaneously performing the deconvolution. Moreover, the fact that *AutoDecon* is fully automated and statistically based addresses the issues noted earlier, including the user-unfriendly nature of prior deconvolution methods and the previous lack of statistical verification of resolved secretory events. Analysis with *AutoDecon* results in a substantially higher true-positive rate of identification of hormone secretory events than analysis with *Cluster*. Moreover, both

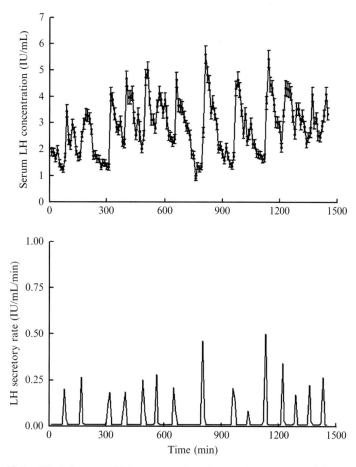

Figure 14.4 (Top) A serum LH concentration-time series constructed from samples obtained from a normal woman in the early follicular phase of her menstrual cycle. Samples were collected at 10-min intervals over a 24-h period of time. (Bottom) LH secretory bursts resolved from this concentration-time series using the deconvolution procedure *Deconv* (Veldhuis, Carlson and Johnson, 1987).

false-positive and false-negative rates are lower with *AutoDecon* compared to *Cluster* and sensitivity is substantially higher with *AutoDecon*.

6. EVALUATION OF PULSATILE AND BASAL HORMONE SECRETION USING A STOCHASTIC DIFFERENTIAL EQUATIONS MODEL

An alternative method proposed by Keenan and Veldhuis (1997) to appraise hormone secretion uses a combination of a model of random, but structured, variations in pulse amplitudes superimposed on basal hormone

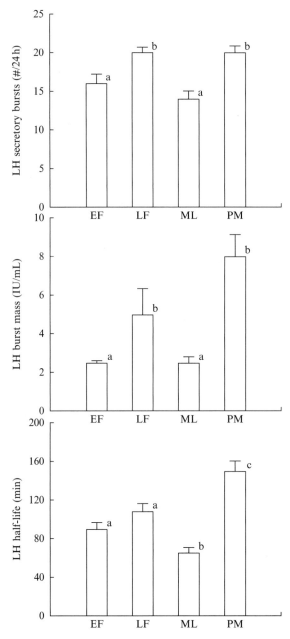

Figure 14.5 Deconvolution-resolved number of secretory bursts (/24 h; top), burst mass (IU/liter; middle), and half-life (minutes; bottom) in LH concentration-time series obtained from normal women in the early follicular (EF), late follicular (LF), and midluteal (ML) phases of the menstrual cycle and in postmenopausal (PM) women. For each group, values identified with different superscripts differ significantly ($p < 0.05$) (Booth et al., 1996; Sollenberger et al., 1990).

release with a nonstationary Poisson process generating the timing of the hormone pulses. This approach utilizes stochastic differential equations to evaluate certain properties of irregular hormone time series, including pulse onset times, basal secretion, and pulse shape, and has the advantage of using an automated procedure for initial estimate of the pulse positions. However, in this method the amplitude of each pulse is assumed to depend on the previous interpulse length, which is a restrictive condition that may not be valid in all hormone release control systems.

7. Characterization of Regulation of Signal and Response Elements: Estimation of Approximate Entropy

Following the development of methods for hormone pulse and secretory burst identification as described previously, attention focused on whether quantitative techniques could be utilized to appraise the regulation of individual "nodes" (i.e., hormone-secreting components) within overall hormone networks. Such components often respond to a primary input signal but are, at the same time, subject to modulation by a separate signal in the form of "feedback" from the target gland. Again using the reproductive network as an example, LH- and FSH-secreting gonadotropes within the anterior pituitary gland respond to GnRH from the hypothalamus (acting as the primary signal), but the pituitary response to the GnRH signal may be influenced significantly by ambient concentrations of hormones such as estrogen and progesterone secreted by the ovary (i.e., the target gland for LH and FSH). An early attempt to characterize regulation of a single hormone secretory component sought to determine, using quantitative methods, the degree of "orderliness" within a hormone concentration-time series, which was considered to reflect the degree to which the gland is regulated by both primary and feedback signals. To assess such orderliness, Pincus (1991) developed a mathematical approach with which to estimate "approximate entropy" within the data series. This approach (known as *ApEn*) is complementary to hormone pulse and secretory burst activity in that it evaluates both dominant and subordinate patterns in concentration-time series. Given its ability to detect changes in underlying episodic behavior, which will not be reflected in pulse occurrences or their characteristics, *ApEn* acts as an explicit "barometer" of feedback system change in many coupled systems. In addition to its application to components of the reproductive system (Pincus *et al.*, 1997), *ApEn* has been utilized to probe the regulation of other hormones, including growth hormone (Friend *et al.*, 1996; Hartman *et al.*, 1994), aldosterone (Siragy *et al.*, 1995), cortisol (van den Berg *et al.*, 1997), and insulin (Meneilly *et al.*, 1997).

8. Evaluation of Coupled Systems

In the same general time frame that attention was increasingly being focused on how the regulation of individual hormone-secreting components might be appraised in quantitative terms, methods with which to define the activity of *coupled hormonal systems* were being proposed. One of the earliest such approaches utilized cross-correlation, a standard and widely applied method for estimating the degree to which two data series are related. It was argued that cross-correlation could be used to support the existence of a relationship between two nodes in the hormone network and indeed this method has been applied successfully in various circumstances (Booth *et al.*, 1996; Veldhuis *et al.*, 1988). However, the straightforward use of cross-correlation suffers from several disadvantages, including spurious values due to autocorrelation and the linear nature of the assumed relationship. These limitations prompted the development of other methods, specifically adapted to endocrine applications.

Within this context, and shortly after the development of *ApEn* to appraise regulation of a single hormone-secreting node, Pincus and Singer (1996) suggested that the signal-response relationship between coupled nodes within a hormonal network could be described by estimating their cross-approximate entropy. Indeed, the program *Cross-ApEn* assesses the degree of asynchrony between two hormonal concentration-time series known to be coupled and determines whether the coupling is more "ordered" or more "random." Taking the LH–testosterone system as an example, application of *Cross-ApEn* demonstrated more irregularity or "asynchrony" in data obtained from older vs younger men, even though the mean serum concentrations of LH and testosterone were not found to differ as a function of age (Pincus *et al.*, 1996). Thus, *Cross-ApEn* reinforced the notion that information of potentially significant physiologic interest could be obtained by quantitating the interactions of coupled systems within the overall hormonal network.

Quite recently, approaches using distinctly different mathematical methodologies from those subserving cross-correlation or *Cross-ApEn* have been developed. Farhy and colleagues have proposed a method that seeks to define and quantitate relationships whereby one hormone modulates the release of a second hormone (Nass *et al.*, 2008). This method (referred to as *FeedStrength*) overcomes some of the problems associated with the application of the more traditional statistical methods (e.g., cross-correlation) for estimation of the relationships between two simultaneously sampled hormones. To test the hypothesis that hormone A modulates the secretion of hormone B, *FeedStrength* first determines the individual secretion bursts of hormone B using deconvolution procedures. The amplitude of each secretory burst of hormone B is compared to the average concentration of hormone A over a fixed time interval during and before each hormone B burst. Rather than comparing the entire hormone A and B concentration profiles (as would be the case if cross-correlation is used), the method relates the amplitude of an individual hormone B burst to the concentration of hormone A accompanying the development of this burst. This approach has the advantage that it disregards those parts of the release profile of hormone A during which there is no pulsatile secretory activity of hormone B. Consequently, the method is suitable for analyzing endocrine relationships in which one of the hormones is presumed to be an amplifier/modulator and not the sole regulator of the secretory activity of the second hormone.

The application of *FeedStrength* to estimate the amplifying effect of ghrelin on growth hormone secretion resulted in the first clear demonstration of a relationship between these two hormones in normal subjects (Nass et al., 2008). That this approach is also suitable for estimating the variability in feedback relationships was proposed when the method was applied to synthetic data (Farhy et al., 2006). As shown in Fig. 14.6, network feedback modeling was used to create a system mimicking the feedback effect of progesterone on LH secretion. Synthetic data were modeled to mimic serum LH (hormone B) stimulating progesterone (hormone A) secretion in normal midluteal phase women. The network also assumes a self-entrained GnRH pulse generator (G) stimulating the secretion of B, an ovarian-hypothalamic delayed feedback component, and a system-independent external regulation (F). Estimates of the level of feedback control

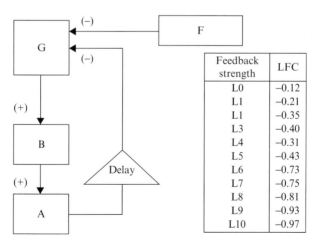

Figure 14.6 A synthetic endocrine network (left) and estimate of the level of feedback control (LFC) for the A-to-B negative coupling (table on right) corresponding to the gradual increase in the feedback strength and simultaneous decrease of the independent factor (Farhy et al., 2006).

(LFC) were computed using *FeedStrength* as described (Nass et al., 2008). Also shown in Fig. 14.6 are results of the application of this method to synthetic data sets in which the strength of the feedback signal (A-to-G) is progressively degraded and the external regulation (F-to-G) becomes dominant. As can be seen, the LFC for the A-to-B coupling identifies correctly gradual changes of the impact of A on the system relative to the effect of F across 11 different levels (L0–L10): (i) minimal LFC when only F, but not A, affects G (L0); (ii) a gradual LFC increase when the feedback becomes progressively dominant over the impact of F (L1–L9); and (iii) maximal LFC when only A, but not F, controls G (L10). Thereby, the LFC as calculated by *FeedStrength* provides a robust estimate of the feedback control exerted by one hormone on another when the main pulse-generating hormone is unknown and can estimate feedback alterations associated with such factors as age, gender, and pathophysiologic conditions.

An additional approach with which to assess the possibility that one hormone regulates the secretion of another hormone in a quantitative manner is to use dynamic modeling. To this end, Johnson and colleagues (2008) developed a computer-based approach for Signal-Response Quantification (*SRQuant*), which reconstructs the dose–response interactions between hormones by fitting parameters of a dynamic model to experimental data. Within this framework, several input signals can be individually delayed, spread in time, transformed by dose–response relationships, combined, and then convolved with a pharmacokinetic elimination function to predict the time course of the concentration of an output hormone. As an example, *SRQuant* has been used to estimate the effect of the endogenous opioid peptide antagonist naloxone on GnRH (and consequently LH) secretion using an animal model (the ovariectomized ewe) in which GnRH and LH were measured simultaneously in blood samples obtained from the hypothalamic–hypophyseal portal circulation (GnRH) and the jugular vein (LH)(Goodman et al., 1995). A minimal model was constructed based on the assumption that GnRH stimulates the secretion of LH, which is, in turn, eliminated after its release into the circulation. The model has several physiological parameters (rate of elimination and secretion of LH, ED_{50}, and slope of the GnRH–LH dose–response interaction)(Farhy, 2004), which were determined simultaneously by *SRQuant* via an iterative procedure to achieve the best possible fit of the model output to the LH data if GnRH was used as input. The result of this process is illustrated in Fig. 14.7 and depicts a very good coincidence between the model predicted and observed LH concentration: 80% of the variance was explained, confirming the existence of a strong feed-forward dose–response relationship between GnRH concentration and LH secretion in this animal model. Further analysis showed that before naloxone administration, the model provides a much better fit of data by explaining approximately 88% of the variation in LH, whereas during the treatment only 68% of the variation was

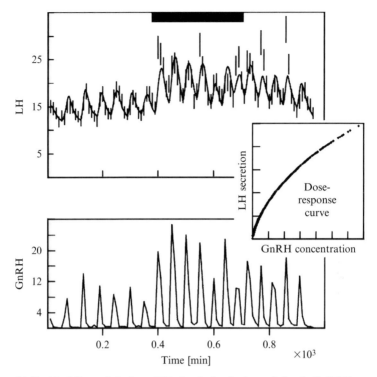

Figure 14.7 Modeling of pituitary LH stimulation by hypothalamic GnRH in a representative ewe treated with naloxone (solid bar). (Top) Model fit (solid line) and observed (error bars) LH (80% of the variance of LH is accounted for). (Bottom) GnRH concentration. (Inset) Reconstructed dose–response interaction between GnRH concentration and LH secretion over the range of observed GnRH.

explained. In addition, a clear change in the "shape" of the dose–response interaction was detected after the treatment. This outcome suggests that at least part of the effect of naloxone treatment on the GnRH–LH axis reflects a direct effect on GnRH–LH relationships. It is interesting to note that the analysis of such GnRH–LH relationships with other methods, including cross-correlation and *Cross-ApEn*, does not provide satisfactory results: In this particular case the cross-correlation was only 0.64, which would account for about 36% of the variance in LH by the linear correlation model. The (ratio) *Cross-ApEn* values were 0.94 ± 0.08 ($m = 1$) and 1.07 ± 0.2 ($m = 2$), which are typical for a random rather than a strong dose-dependent relationship between the two hormones. Thus, *SRQuant* provides a viable alternative and a powerful addition to the more traditional methods for the appraisal of coupled endocrine systems.

9. Evaluation of Hormonal Networks with Feedback Interactions

Whereas the methods described earlier have allowed for a detailed appraisal of systems in which hormone-secreting component A stimulates or inhibits hormone-secreting component B or vice versa, it is obvious that in many cases endocrine systems function in a *dynamic* fashion in which there are multiple feedback interactions among the system components. Given that such systems fulfill commonly accepted criteria to be considered as *networks*, an increasing amount of investigative interest has focused on defining such networks in quantitative terms.

A separate approach with which to address this issue and as adopted by Farhy and colleagues (Farhy, 2004; Farhy and Veldhuis, 2005; Farhy *et al.*, 2007, 2008; Kovatchev *et al.*, 1999) and others (Bergman *et al.*, 1979; Dalla Man *et al.*, 2005; Keenan and Veldhuis, 1997, 1998, 2004; Wagner *et al.*, 1998) suggests that endocrine networks can be appraised by deterministic differential equations that describe the rate of change of individual hormones and their interaction. Utilization of this method has the potential to test whether a physiological regulatory hypothesis is consistent with the experimental observations and typically results in formulation of physiological assumptions (additional to the main hypothesis) that would explain the observed data specifics (Farhy, 2004). In effect, this is *in silico* testing of physiological hypotheses, which addresses specific features of the *in vivo* behavior of endocrine systems. This approach permits estimation of the integrative action of individual inferred or postulated feedback and feedforward interactions within an endocrine axis and combines multiple steps (Farhy, 2004), including (i) preliminary analysis of available experimental data, (ii) design of a formal network describing the axis linkages and its approximation with coupled differential equations, and (iii) determination of the model parameters either functionally or by fitting the model output to available data. As an example, application of this approach may allow insights into the physiological parameters of a specific endocrine system, which may subserve its oscillatory behavior.

Evaluation of simple two- or three-hormone feedback networks, such as the one shown in Fig. 14.8, illustrates the capacity of delayed nonlinear feedbacks to generate hormone pulsatility (Farhy, 2004). In these examples (that may appear in different real endocrine axes), hormones (A, B, and C) are secreted continuously and are subject to elimination. However, in contrast to the previous examples, two or more (delayed; D_1, D_2) feedback control loops exist between the system components. It is these loops that generate the oscillatory secretory patterns. For example, the two delayed control loops in the network shown in the middle left panel of Fig. 14.8 can drive a particular hormone oscillatory pattern if certain physiological conditions are met. As is depicted in Fig. 14.9, this complex multiple-volley secretory pattern is typical for the growth hormone dynamics in the adult male rat, and results of the simulations show that the underlying construct (Fig. 14.8, middle left panel) wherein B may be thought to be growth hormone (GH), A GH-releasing hormone, and C somatostatin, represents

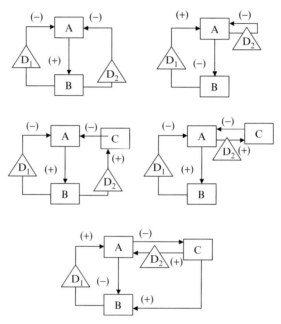

Figure 14.8 Examples of endocrine networks with multiple delayed feedback loops.

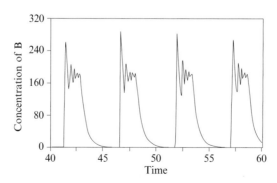

Figure 14.9 Computer-generated output (concentration of B vs time) of the core system shown in Fig. 14.8, middle left.

one minimal set of interactions capable of reproducing the typical *in vivo* GH pattern in the male rat. However, as evident from additional model-based analysis, this network is too simplistic and needs to be extended and modified to address other key experimental observations (Farhy and Veldhuis, 2005; Farhy *et al.*, 2007).

Another model-associated outcome could be a detailed reconstruction of the endocrine network dose–response interactions (similarly to the use of the aforementioned procedure *SRQuant*). This would be the case when suitable data exist to permit the precise iterative determination of the model parameters (Bergman *et al.*, 1979; Dalla Man *et al.*, 2005). In this case the physiological hypothesis that *the postulated network unifies the primary endocrine axis control mechanisms* would be strongly supported if the model can provide a good fit of data based on objective statistical criteria.

10. Future Directions

Given the complexity of endocrine axes, a logical extension of previous attempts to assess isolated signal–response endocrine interactions would include the development of automated methods with which to reconstruct the dose–response interactions within an entire hormone network. To this end, a dynamic modeling method, such as that described earlier, could be utilized within a computer-based model similar to *SRQuant*. This approach could make possible the approximation of endocrine networks exhibiting multiple feedback interactions and the subsequent reconstruction of their parameters based on existing experimental data. The result would be a powerful tool capable of testing regulatory hypotheses and estimating the alterations in hormone release control mechanisms related to various factors including, but not limited to, inter- or intraspecies differences, natural processes such as aging, pathophysiologic situations, and interventional procedures.

ACKNOWLEDGMENTS

We thank Martha Burner, David G. Boyd, and Paula P. Veldhuis for their excellent technical and editorial contributions to preparation of the manuscript. This work was supported in part by NIH Grant RR-00847 to the General Clinical Research Center at the University of Virginia; NIH Grant R01 RR019991 (to MLJ, WSE, and LSF); NIH Grant R25 DK064122 (to MLJ and WSE); and NIH Grants R21 DK072095, P30 DK063609, K25 HD001474, R01 DK076037 (to MLJ and LSF), and R01 DK51562 (to LSF).

REFERENCES

Atkinson, L. E., Bhattacharya, A. N., Monroe, S. E., Dierschke, D. J., and Knobil, E. (1970). Effects of gonadectomy on plasma LH concentration in the rhesus monkey. *Endocrinology* **87,** 847–849.

Backstrom, C. T., McNeilly, A. S., Leask, R. M., and Baird, D. T. (1982). Pulsatile secretion of LH, FSH, prolactin, oestradiol and progesterone during the human menstrual cycle. *Clin. Endocrinol.* **17,** 29–42.

Bergman, R. N., Ider, Y. Z., Bowden, C. R., and Cobelli, C. (1979). Quantitative estimation of insulin sensitivity. *Am. J. Physiol* **236,** E667–E677.

Booth, R. A. Jr., Weltman, J. Y., Yankov, V. I., Murray, J., Davison, T. S., Rogol, A. D., Asplin, C. M., Johnson, M. L., Veldhuis, J. D., and Evans, W. S. (1996). Mode of pulsatile follicle-stimulating hormone secretion in gonadal hormone-sufficient and -deficient women: A clinical research center study. *J. Clin. Endocrinol. Metab.* **81,** 3208–3214.

Burger, C. W., Korsen, T., van Kessel, H., van Dop, P. A., Caron, F. J., and Schoemaker, J. (1985). Pulsatile luteinizing hormone patterns in the follicular phase of the menstrual cycle, polycystic ovarian disease (PCOD) and non-PCOD secondary amenorrhea. *J. Clin. Endocrinol. Metab.* **61,** 1126–1132.

Burgus, R., Butcher, M., Amoss, M., Ling, N., Monahan, M., Rivier, J., Fellows, R., Blackwell, R., Vale, W., and Guillemin, R. (1972). Primary structure of the ovine hypothalamic luteinizing hormone-releasing factor (LRF). *Proc. Natl. Acad. Sci. USA* **69,** 278–282.

Clarke, I. J., and Cummins, J. T. (1982). The temporal relationship between gonadotropin releasing hormone (GnRH) and luteinizing hormone (LH) secretion in ovariectomized ewes. *Endocrinology* **111,** 1737–1739.

Clifton, D. K., and Steiner, R. A. (1983). Cycle detection: A technique for estimating the frequency and amplitude of episodic fluctuations in blood hormone and substrate concentrations. *Endocrinology* **112,** 1057–1064.

Crowley, W. F. Jr., Filicori, M., Spratt, D. I., and Santoro, N. F. (1985). The physiology of gonadotropin-releasing hormone (GnRH) secretion in men and women. *Rec. Prog. Horm. Res.* **41,** 473–531.

Dalla Man, C., Caumo, A., Basu, R., Rizza, R., Toffolo, G., and Cobelli, C. (2005). Measurement of selective effect of insulin on glucose disposal from labeled glucose oral test minimal model. *Am. J. Physiol.* **289,** E909–E914.

Evans, W. S., Sollenberger, M. J., Booth, R. A. Jr., Rogol, A. D., Urban, R. J., Carlsen, E. C., Johnson, M. L., and Veldhuis, J. D. (1992). Contemporary aspects of discrete peak-detection algorithms. II. The paradigm of the luteinizing hormone pulse signal in women. *Endocr. Rev.* **13,** 81–104.

Farhy, L. S. (2004). Modeling of oscillations in endocrine networks with feedback. *Methods Enzymol.* **384,** 54–81.

Farhy, L. S., Bowers, C. Y., and Veldhuis, J. D. (2007). Model-projected mechanistic bases for sex differences in growth hormone regulation in humans. *Am. J. Physiol.* **292,** R1577–R1593.

Farhy, L. S., Du, Z., Zeng, Q., Veldhuis, P. P., Johnson, M. L., Brayman, K. L., and McCall, A. L. (2008). Amplification of pulsatile glucagon counterregulation by switch-off of α-cell-suppressing signals in streptozotocin (STZ)-treated rats. *Am. J. physiol. Endocrinol. Metab.* **295,** E575–E585.

Farhy, L. S., Johnson, M. L., Veldhuis, P. P., Boyd, D. G., and Evans, W. S. (2006). *In* "Control Strength (CS) of Hormone Interactions in Endocrine Feedback Networks," pp. 2–21. 88th Annual Meeting of the Endocrine Society. [Abstract].

Farhy, L. S., and Veldhuis, J. D. (2005). Deterministic construct of amplifying actions of ghrelin on pulsatile growth hormone secretion. *Am. J. Physiol.* **288,** R1649–R1663.

Filicori, M., Butler, J. P., and Crowley, W. F. Jr. (1984). Neuroendocrine regulation of the corpus luteum in the human: Evidence for pulsatile progesterone secretion. *J. Clin. Invest.* **73,** 1638–1647.

Filicori, M., Santoro, N., Merriam, G. R., and Crowley, W. F. Jr. (1986). Characterization of the physiological pattern of episodic gonadotropin secretion throughout the human menstrual cycle. *J. Clin. Endocrinol. Metab.* **62,** 1136–1144.

Fluhmann, C. F. (1929). Anterior pituitary hormone in the blood of women with ovarian deficiency. *JAMA* **93,** 672–674.

Friend, K., Iranmanesh, A., and Veldhuis, J. D. (1996). The orderliness of the growth hormone (GH) release process and the mean mass of GH secreted per burst are highly conserved in individual men on successive days. *J. Clin. Endocrinol. Metab.* **81,** 3746–3753.

Gay, V. L., and Midgley, A. R. Jr. (1969). Response of the adult rat to orchidectomy and ovariectomy as determined by LH radioimmunoassay. *Endocrinology* **84,** 1359–1364.

Genazzani, A. D., Rodbard, D., Forti, G., Petraglia, F., Baraghini, G. F., and Genazzani, A. R. (1990). Estimation of instantaneous secretory rate of luteinizing hormone in women during the menstrual cycle and in men. *Clin. Endocrinol.* **32,** 573–581.

Goodman, R. L., Parfitt, D. B., Evans, N. P., Dahl, G. E., and Karsch, F. J. (1995). Endogenous opioid peptides control the amplitude and shape of gonadotropin-releasing hormone pulses in the ewe. *Endocrinology* **136,** 2412–2420.

Green, J. D., and Harris, G. W. (1947). The neurovascular link between the neurohypophysis and adenohypophysis. *J. Endocrinol.* 136–149.

Haisenleder, D. J., Dalkin, A. C., and Marshall, J. C. (1994). Regulation of gonadotropin gene expression. *In* "The Physiology of Reproduction" (E. Knobil and J. Neill, eds.), 2nd Ed pp. 1793–1832. Raven Press, New York.

Hartman, M. L., Pincus, S. M., Johnson, M. L., Matthews, D. H., Faunt, L. M., Vance, M. L., Thorner, M. O., and Veldhuis, J. D. (1994). Enhanced basal and disorderly growth hormone secretion distinguish acromegalic from normal pulsatile growth hormone release. *J. Clin. Invest.* **94,** 1277–1288.

Johnson, M. L., Pipes, L., Veldhuis, P. P., Farhy, L. S., Boyd, D. G., and Evans, W. S. (2008). AutoDecon, a deconvolution algorithm for identification and characterization of luteinizing hormone secretory bursts: Description and validation using synthetic data. *Anal. Biochem.* **381,** 8–17.

Keenan, D. M., and Veldhuis, J. D. (1997). Stochastic model of admixed basal and pulsatile hormone secretion as modulated by a deterministic oscillator. *Am. J. Physiol.* **273,** R1182–R1192.

Keenan, D. M., and Veldhuis, J. D. (1998). A biomathematical model of time-delayed feedback in the human male hypothalamic-pituitary-Leydig cell axis. *Am. J. Physiol.* **275,** E157–E176.

Keenan, D. M., and Veldhuis, J. D. (2004). Divergent gonadotropin-gonadal dose-responsive coupling in healthy young and aging men. *Am. J. Physiol.* **286,** R381–R389.

Knobil, E. (1980). The neuroendocrine control of the menstrual cycle. *Rec. Prog. Horm. Res.* **36,** 53–88.

Kovatchev, B. P., Farhy, L. S., Cox, D. J., Straume, M., Yankov, V. I., Gonder-Frederick, L. A., and Clarke, W. L. (1999). Modeling insulin-glucose dynamics during insulin induced hypoglycemia: Evaluation of glucose counterregulation. *Comput. Math Meth. Med.* **1,** 313–323.

Levine, J. E., Pau, K. Y., Ramirez, V. D., and Jackson, G. L. (1982). Simultaneous measurement of luteinizing hormone-releasing hormone and luteinizing hormone release in unanesthetized, ovariectomized sheep. *Endocrinology* **111,** 1449–1455.

Levine, J. E., and Ramirez, V. D. (1982). Luteinizing hormone-releasing hormone release during the rat estrous cycle and after ovariectomy, as estimated with push-pull cannulae. *Endocrinology* **111,** 1439–1448.

Matsuo, H., Baba, Y., Nair, R. M., Arimura, A., and Schally, A. V. (1971). Structure of the porcine LH- and FSH-releasing hormone. I. The proposed amino acid sequence. *Biochem. Biophys. Res. Commun.* **43,** 1334–1339.

McIntosh, R. P., and McIntosh, J. E. (1985). Amplitude of episodic release of LH as a measure of pituitary function analysed from the time-course of hormone levels in the blood: Comparison of four menstrual cycles in an individual. *J. Endocrinol.* **107,** 231–239.

Meneilly, G. S., Ryan, A. S., Veldhuis, J. D., and Elahi, D. (1997). Increased disorderliness of basal insulin release, attenuated insulin secretory burst mass, and reduced ultradian rhythmicity of insulin secretion in older individuals. *J. Clin. Endocrinol. Metab.* **82,** 4088–4093.

Merriam, G. R., and Wachter, K. W. (1982). Algorithms for the study of episodic hormone secretion. *Am. J. Physiol.* **243,** E310–E318.

Midgley, A. R. Jr., and Jaffe, R. B. (1971). Regulation of human gonadotropins: Episodic fluctuation of LH during the menstrual cycle. *J. Clin. Endocrinol. Metab.* **33,** 962–969.

Murdoch, A. P., Diggle, P. J., White, M. C., Kendall-Taylor, P., and Dunlop, W. (1989). LH in polycystic ovary syndrome: Reproducibility and pulsatile secretion. *J. Endocrinol.* **121,** 185–191.

Nass, R., Farhy, L. S., Liu, J., Prudom, C. E., Johnson, M. L., Veldhuis, P., Pezzoli, S. S., Oliveri, M. C., Gaylinn, B. D., and Geysen, H. M. (2008). Evidence for acyl-ghrelin modulation of growth hormone release in the fed state. *J. Clin. Endocrinol. Metab.* **93,** 1988–1994.

Oerter, K. E., Guardabasso, V., and Rodbard, D. (1986). Detection and characterization of peaks and estimation of instantaneous secretory rate for episodic pulsatile hormone secretion. *Comput. Biomed. Res.* **19,** 170–191.

Pincus, S. M. (1991). Approximate entropy as a measure of system-complexity. *Proc. Natl. Acad. Sci. USA* **88,** 2297–2301.

Pincus, S. M., Mulligan, T., Iranmanesh, A., Gheorghiu, S., Godshalk, M., and Veldhuis, J. D. (1996). Older males secrete luteinizing hormone and testosterone more irregularly, and jointly more asynchronously, than younger males. *Proc. Natl. Acad. Sci. USA* **93,** 14100–14105.

Pincus, S. M., and Singer, B. H. (1996). Randomness and degrees of irregularity. *Proc. Natl. Acad. Sci. USA* **93,** 2083–2088.

Pincus, S. M., Veldhuis, J. D., Mulligan, T., Iranmanesh, A., and Evans, W. S. (1997). Effects of age on the irregularity of LH and FSH serum concentrations in women and men. *Am. J. Physiol.* **273,** E989–E995.

Reame, N., Sauder, S. E., Kelch, R. P., and Marshall, J. C. (1984). Pulsatile gonadotropin secretion during the human menstrual cycle: Evidence for altered frequency of gonadotropin-releasing hormone secretion. *J. Clin. Endocrinol. Metab.* **59,** 328–337.

Rebar, R., Perlman, D., Naftolin, F., and Yen, S. S. (1973). The estimation of pituitary luteinizing hormone secretion. *J. Clin. Endocrinol. Metab.* **37,** 917–927.

Rossmanith, W. G., Liu, C. H., Laughlin, G. A., Mortola, J. F., Suh, B. Y., and Yen, S. S. C. (1990). Relative changes in LH pulsatility during the menstrual cycle; using data from hypogonadal women as a reference point. *Clin. Endocrinol.* **32,** 647–660.

Santen, R. J., and Bardin, C. W. (1973). Episodic luteinizing hormone secretion in man: Pulse analysis, clinical interpretation, physiologic mechanisms. *J. Clin. Invest.* **52,** 2617–2628.

Schally, A. V., Arimura, A., Baba, Y., Nair, R. M., Matsuo, H., Redding, T. W., and Debeljuk, L. (1971). Isolation and properties of the FSH and LH-releasing hormone. *Biochem. Biophys. Res. Commun.* **43,** 393–399.

Schweiger, U., Laessle, R. G., Tuschl, R. J., Broocks, A., Krusche, T., and Pirke, K. M. (1989). Decreased follicular phase gonadotropin secretion is associated with impaired estradiol and progesterone secretion during the follicular and luteal phases in normally menstruating women. *J. Clin. Endocrinol. Metab.* **68,** 888–892.

Siragy, H. M., Vieweg, W. V., Pincus, S., and Veldhuis, J. D. (1995). Increased disorderliness and amplified basal and pulsatile aldosterone secretion in patients with primary aldosteronism. *J. Clin. Endocrinol. Metab.* **80,** 28–33.

Sollenberger, M., Carlsen, E. C., Johnson, M., Veldhuis, J., and Evans, W. (1990). Specific physiological regulation of LH secretory events throughout the human menstrual cycle: New insights into the pulsatile mode of gonadotropin release. *J. Neuroendocrinol.* **2,** 845–852.

Soules, M. R., Steiner, R. A., Clifton, D. K., Cohen, N. L., Aksel, S., and Bremner, W. J. (1984). Progesterone modulation of pulsatile luteinizing hormone secretion in normal women. *J. Clin. Endocrinol. Metab.* **58,** 378–383.

Turner, R. C., Grayburn, J. A., Newman, G. B., and Nabarro, J. D. N. (1972). Measurement of the insulin delivery rate in man. *J. Clin. Endocrinol. Metab.* **33,** 279–286.

Urban, R. J., Evans, W. S., Rogol, A. D., Johnson, M. L., and Veldhuis, J. D. (1988). Contemporary aspects of discrete peak detection algorithms. 1. The paradigm of the luteinizing hormone pulse signal in man. *Endocr. Rev.* **9,** 3–37.

Van Cauter, E., L'Hermite, M., Copinschi, G., Refetoff, S., Desir, D., and Robyn, C. (1981). Quantitative analysis of spontaneous variations of plasma prolactin in normal man. *Am. J. Physiol.* **241,** E355–E363.

van den Berg, G., Pincus, S. M., Veldhuis, J. D., Frolich, M., and Roelfsema, F. (1997). Greater disorderliness of ACTH and cortisol release accompanies pituitary-dependent Cushing's disease. *Eur. J. Endocrinol.* **136,** 394–400.

Veldhuis, J. D., Carlson, M. L., and Johnson, M. L. (1987). The pituitary gland secretes in bursts: Appraising the nature of glandular secretory impulses by simultaneous multiple-parameter deconvolution of plasma hormone concentrations. *Proc. Natl. Acad. Sci. USA* **84,** 7686–7690.

Veldhuis, J. D., Christiansen, E., Evans, W. S., Kolp, L. A., Rogol, A. D., and Johnson, M. L. (1988). Physiological profiles of episodic progesterone release during the midluteal phase of the human menstrual cycle: Analysis of circadian and ultradian rhythms, discrete pulse properties, and correlations with simultaneous luteinizing hormone release. *J. Clin. Endocrinol. Metab.* **66,** 414–421.

Veldhuis, J. D., and Johnson, M. L. (1986). Cluster analysis: A simple, versatile, and robust algorithm for endocrine pulse detection. *Am. J. Physiol.* **250,** E486–E493.

Wagner, C., Caplan, S. R., and Tannenbaum, G. S. (1998). Genesis of the ultradian rhythm of GH secretion: A new model unifying experimental observations in rats. *Am. J. Physiol.* **275,** E1046–E1054.

Yen, S. S., Tsai, C. C., Naftolin, F., Vandenberg, G., and Ajabor, L. (1972). Pulsatile patterns of gonadotropin release in subjects with and without ovarian function. *J. Clin. Endocrinol. Metab.* **34,** 671–675.

CHAPTER FIFTEEN

AutoDecon: A Robust Numerical Method for the Quantification of Pulsatile Events

Michael L. Johnson,* Lenore Pipes,* Paula P. Veldhuis,* Leon S. Farhy,[†] Ralf Nass,[†] Michael O. Thorner,[†] *and* William S. Evans[‡]

Contents

1. Introduction	368
2. Methods	370
2.1. Overview of deconvolution method	370
2.2. *AutoDecon*-Fitting module	372
2.3. *AutoDecon* insertion module	372
2.4. *AutoDecon* triage module	373
2.5. *AutoDecon* combined modules	374
2.6. Variation on the theme	376
2.7. Outliers	378
2.8. Weighting factors	378
2.9. *AutoDecon* performance and comparison with earlier pulse detection algorithms	379
2.10. Experimental data	379
2.11. Growth hormone assay	380
2.12. Data simulation	380
2.13. Algorithm operating characteristics	382
2.14. Concordant secretion events	383
3. Results	384
3.1. Typical *AutoDecon* output	384
3.2. Experimental data analysis	385
3.3. Comparison between *AutoDecon* and other algorithms	390
3.4. Optimal sampling paradigms for *AutoDecon*	397

* Departments of Pharmacology and Medicine, University of Virginia Health System, Charlottesville, VA
[†] Endocrinology and Metabolism Department of Medicine, University of Virginia Health System, Charlottesville, Virginia
[‡] Endocrinology and Metabolism Department of Medicine, and Department of Obstetrics and Gynecology, University of Virginia Health System, Charlottesville, Virginia

4. Discussion	399
Acknowledgments	403
References	403

Abstract

This work presents a new approach to the analysis of aperiodic pulsatile heteroscedastic time-series data, specifically hormone pulsatility. We have utilized growth hormone (GH) concentration time-series data as an example for the utilization of this new algorithm. While many previously published approaches used for the analysis of GH pulsatility are both subjective and cumbersome to use, *AutoDecon* is a nonsubjective, standardized, and completely automated algorithm. We have employed computer simulations to evaluate the true-positive, the false-positive, the false-negative, and the sensitivity percentages of several of the routinely employed algorithms when applied to GH concentration time-series data. Based on these simulations, it was concluded that this new algorithm provides a substantial improvement over the previous methods. This novel method has many direct applications in addition to hormone pulsatility, for example, to time-domain fluorescence lifetime measurements, as the mathematical forms that describe these experimental systems are both convolution integrals.

1. INTRODUCTION

It has long been recognized that normal function within endocrine systems requires a highly coordinated interplay of events both within and among components of a given hormonal axis. Moreover, it has become increasingly evident that the secretory characteristics of certain endocrine signals are critical determinants of the target organ response. For example, it has been demonstrated that whereas the administration of gonadotropin-releasing hormone (GnRH) in a pulsatile fashion affected the release of luteinizing hormone (LH) and follicle-stimulating hormone (FSH), continuous administration not only failed to stimulate LH and FSH release, but actually inhibited secretion of these hormones (Nakai *et al.*, 1978). More recent studies have further demonstrated that the gene expression of LH and FSH is controlled by the frequency and amplitude characteristics of the pulsatile GnRH signal (Haisenleder *et al.*, 1994). The pulsatile nature of growth hormone (GH) secretion has also received considerable investigative attention and has provided insights into both stimulatory and inhibitory hypothalamic mechanisms that regulate GH secretion (Dimaraki *et al.*, 2001; Hartman, 1991; Maheshwari *et al.*, 2002; Thorner *et al.*, 1997; Van Cauter *et al.*, 1998; Vance *et al.*, 1985; Webb *et al.*, 1984). The secretion of numerous other hormones has also been shown to be pulsatile in nature, including, but not limited to, prolactin, thyroid-stimulating hormone, adrenocorticotropic hormone, parathyroid hormone, cortisol, and insulin.

Given the physiological significance of hormone pulsatility, the ability to separate a pulsatile signal from noise has become crucial with regard to understanding the mechanisms that control the dynamics of this signal. Although earlier attempts to apply computer-based methods to characterize such pulsatile signals provided enhanced objectivity, most methods were not statistically based (Merriam and Wachter, 1982; Santen and Bardin, 1973; Ultra and Cycle) and those that were (Oerter *et al.*, 1986; Veldhuis and Johnson, 1986) only identified perturbations in hormone concentration time series but could not separate pulses into their secretory and clearance components. The application of deconvolution procedures (Evans *et al.*, 1992; Johnson and Veldhuis, 1995; Johnson *et al.*, 2004; Urban *et al.*, 1988; Veldhuis *et al.*, 1987) has been employed effectively to address this limitation, that is, such procedures allow for the "unraveling" of hormone pulses into the secretory event itself and the mechanisms responsible for clearance.

The choice of the model-fitting approach for deconvolution analysis of experimental data is dictated by the properties of the measurement uncertainties that are inherent within such data. Measurement errors within hormone concentration time-series data are typically large, as compared to the size of the secretion events, and are heteroscedastic. In addition, missing values are fairly common and thus it cannot be assumed that the data values are equally spaced in time. Several deconvolution procedures [e.g., those based on the convolution theorem (Jansson, 1984)] may be utilized for this sort of analysis, but most other approaches are incompatible with the properties of data.

Another difficulty with previously published hormone pulsatility analysis approaches (Johnson and Veldhuis, 1995; Johnson *et al.*, 2004; Veldhuis *et al.*, 1987) is that they necessitate a prior knowledge or presumption of the number of secretion events, and initial estimates of the secretion event positions and amplitudes are required to be provided by the user. Finding the number of secretion events and initial parameter estimates is a *system identification* problem, which is addressed here via a fully automated statistically based approach, *AutoDecon*, which finds the optimal number of secretion events and the initial parameter estimates while simultaneously performing deconvolution. Moreover, a quintessential data analysis dilemma is in the identification and subsequent characterization of small pulsatile events within a data time series where the amplitudes of these events are comparable to the magnitude of the experimental measurement errors. Event identification is further complicated when the pulsatile events are aperiodic (i.e., they occur at apparently random intervals) and/or when the experimental measurement uncertainties are heteroscedastic (i.e., the measurement uncertainties are variable).

A Monte Carlo technique is utilized to provide information about the operating characteristics of the *AutoDecon* and other algorithms (e.g., true-positive, false-positive, and false-negative percentages for secretion event identification). This Monte Carlo procedure is an extension of procedures that we (Veldhuis and Johnson, 1988) and others (Guardabasso *et al.*, 1988; Van Cauter, 1988) have employed for this purpose.

The numerical methods outlined in this chapter are generally applicable to many biochemical systems, although this chapter describes this procedure and presents the analysis of GH concentration time-series data as an example of the use of *AutoDecon*. For example, time-domain fluorescence lifetime measurements (Lakowicz, 1983) are described by the mathematical forms (convolution integrals), which describe pulsatile hormone concentration time series.

2. Methods

AutoDecon implements a rigorous statistical test for the existence of secretion events (Johnson *et al.*, 2008). In addition, the subjective nature defining earlier deconvolution procedures is eliminated by the ability of the program to automatically insert and subsequently test the significance of presumed secretion events. No user intervention is required subsequent to the initialization of the algorithm. This automatic algorithm combines three modules: a parameter-*fitting* module (analogous to Johnson and Veldhuis, 1995; Veldhuis *et al.*, 1987), a new *insertion* module that automatically adds presumed secretion events, and a new *triage* module that automatically removes secretion events, which are deemed to be statistically nonsignificant.

2.1. Overview of deconvolution method

These deconvolution procedures function by developing a mathematical model for the time course of the hormone concentration and then fitting this mathematical model to experimentally observed time-series data with a weighted nonlinear least-squares algorithm. Specifically, this mathematical model (Johnson and Veldhuis, 1995; Veldhuis *et al.*, 1987) is

$$C(t) = \int_0^t S(\tau)E(t-\tau)d\tau + C(0)E(t) \quad (15.1)$$

where $C(t)$ is the hormone concentration as a function of time t, $S(t)$ is the secretion into the blood as a function of time, and $E(t-z)$ is elimination from the serum as a function of time. In Eq. (15.1) the limits of integration are from 0 to t and, as a consequence, a second term containing the concentration at time zero, $C(0)$, is included. For the time-domain fluorescence lifetime experimental system, the $S(t)$ function corresponds to the lamp function (i.e., the instrument response function) and the $E(t-z)$ corresponds to the fluorescence decay function.

The hormone concentration elimination function, $E(t-z)$ is assumed to follow a single compartment pharmacokinetic model [Eq. (15.2)]

$$E(t-z) = e^{-\frac{\ln 2}{HL}(t-z)} \qquad (15.2)$$

where HL is the one-component elimination half-life, or a two-compartment [Eq. (15.3)] model,

$$E(t-z) = (1-f_2)e^{-\frac{\ln 2}{HL_1}(t-z)} + f_2 e^{-\frac{\ln 2}{HL_2}(t-z)} \qquad (15.3)$$

where HL_1 and HL_2 are the elimination half-lives and f_2 is the amplitude fraction of the second component.

$C(0)$ in Eq. (15.1) is the concentration of the hormone immediately before the first data point, that is, at time equal to zero. This model is rigorously correct for a single-compartment pharmacokinetic elimination model, Eq.(15.2). However, for the two-compartment elimination model, Eq. (15.3), it is only correct under the assumption that all of the concentration at time $t = 0$ is the result of an instantaneous secretion event that occurs at this initial moment. This assumption is required as scant information pertaining to the secretion and elimination prior to the first data point is contained within data.

In the original formulation (Johnson and Veldhuis, 1995; Johnson et al., 2004; Veldhuis et al., 1987) the secretion rate is modeled by Eq. (15.4)

$$S(t) = S_0 + \sum_k e^{\left[\log H_k - \frac{1}{2}\left(\frac{t-PP_k}{SecretionSD}\right)^2\right]} \qquad (15.4)$$

where the secretion rate is assumed to be the sum of Gaussian-shaped events that occur at different irregularly spaced times, PP_k, have differing heights, H_k, and the same width, $SecretionSD$. Note that the size of the secretion events is expressed as the base 10 logarithm of the height, $\log H_k$, in order to constrain all heights to physiologically relevant positive values. The positive constant S_o is the basal secretion.

2.2. *AutoDecon*-Fitting module

The *fitting* module performs weighted nonlinear least-squares parameter estimations by the Nelder–Mead Simplex algorithm (Nelder and Mead, 1965; Straume et al., 1991). It fits Eq. (15.1) to experimental data by adjusting the parameters of the secretion function, Eq. (15.4), and the elimination function, either Eq. (15.2) or (15.3), so that the parameters have the highest probability of being correct. The module is based on the *Amoeba* routine (Press et al., 1986), which was modified such that convergence is assumed when both the variance-of-fit and the individual parameter values do not change by more than 2×10^{-5} or when 15,000 iterations have occurred. This is essentially the original *Deconv* algorithm (Johnson and Veldhuis, 1995; Veldhuis et al., 1987) with the exception that the Nelder–Mead Simplex parameter estimation algorithm (Nelder and Mead, 1965; Straume et al., 1991) is used instead of the damped Gauss–Newton algorithm, which was employed by *Deconv*. The Nelder–Mead algorithm simplifies the software, as it does not require derivatives.

2.3. *AutoDecon* insertion module

The *insertion* module inserts the next presumed secretion event at the location of the maximum of the probable position index, *PPI*:

$$PPI(t) = \begin{cases} -\dfrac{\partial[\text{Variance-of-Fit}]}{\partial H_z} & \text{if } \dfrac{\partial[\text{Variance-of-Fit}]}{\partial H_z} < 0 \\ 0 & \text{if } \dfrac{\partial[\text{Variance-of-Fit}]}{\partial H_z} \geq 0 \end{cases}$$

(15.5)

The parameter H_z in Eq. (15.5) is the amplitude of a presumed secretion event at time z. The index function $PPI(t)$ will have a maximum at the data point position in time where the insertion of a secretion peak will result in the largest negative derivative in the variance of fit versus secretion event size. It is important to note that the partial derivatives of the variance of fit with respect to a secretion event can be evaluated without any additional weighted nonlinear least-squares parameter estimations or even knowing the size of the presumed secretion event, H_z. Using the definition of the variance of fit given in Eq. (15.6), the partial derivative with respect to the addition of a secretion event at time z is shown in Eq. (15.7) where the summation is over all data points,

$$[\text{Variance - of - Fit}] = \sum_i \left(\frac{Y_i - C(t_i)}{W_i}\right)^2 = \sum_i R_i^2 \quad (15.6)$$

$$\frac{\partial[\text{Variance - of - Fit}]}{\partial H_z} = \sum_i \left[\frac{2}{W_i^2}(Y_i - C(t))\frac{\partial C(t)}{\partial H_z}\right]$$

$$\frac{\partial C(t)}{\partial H_z} = \left[e^{-\frac{1}{2}\left(\frac{t-z}{SecretionSD}\right)^2}\right] * E(t) \quad (15.7)$$

where W_i corresponds to the weighting factor for the ith data point and R_i corresponds to the ith residual (see Section 2.8). The inclusion of these weighting factors is the statistically valid method to compensate for the heteroscedastic properties of experimental data.

2.4. AutoDecon triage module

The *triage* module performs a statistical test to ascertain whether a presumed secretion event should be removed. This test requires two weighted non-linear least-squares parameter estimations: one with the presumed peak present and one with the presumed peak removed. The ratio of the variance of fit resulting from these two parameter estimations is related to the probability that the presumed secretion event does not exist, P, by an F statistic, as in Eq. (15.8). Typically, a probability level of 0.05 is used:

$$\frac{\text{Variance of Fit}_{removed}}{\text{Variance of Fit}_{present}} = 1 + \frac{2}{ndf}Fstatistic(2, ndf, 1 - P). \quad (15.8)$$

This is the F test for an additional term (Bevington, 1969) where the additional term is the presumed secretion event. The 2's in Eq. (15.8) are included, as each additional secretion event increases the number of parameters being estimated by 2, specifically the location and the amplitude of the secretion event. The number of degrees of freedom, ndf, is the number of data points minus the total number of parameters being estimated when the secretion event is present. Each cycle of the *triage* module performs this statistical test for every secretion event in an order determined by size, smallest to largest. If a secretion event is found to not be statistically significant, it is removed and the triage module is restarted from the beginning (i.e., a new cycle starts). Thus, the *triage* module continues until all nonsignificant secretion events have been removed. Each cycle of the *triage* module performs $m + 1$ weighted nonlinear least-squares parameter estimations where m is the current number of secretion events for the current cycle; one where all of the secretion events are present; and one where each of the secretion events has been removed and tested individually.

2.5. *AutoDecon* combined modules

The *AutoDecon* algorithm iteratively adds presumed secretion events, tests the significance of all events, and removes nonsignificant secretion events. The procedure is repeated until no additional secretion events are added. The specific details of how this is accomplished with the *insertion*, *fitting*, and *triage* modules are outlined here.

AutoDecon is generally initialized with the basal secretion, S_0, set equal to zero, the concentration at time zero, $C(0)$, set equal to zero, the elimination half-life (*HL*) set to any physiologically reasonable value, the standard deviation of the secretion events (*SecretionSD*) set to one-half of the data sampling interval, and zero secretion events. It is possible, but not required, that initial presumed secretion event locations and sizes be included in the initialization. Initializing the program with peak position and amplitude estimates might decrease the amount of computer time required. Note that for the examples presented in this work the initial number of secretion events is set to zero unless specifically noted otherwise.

The next step in the initialization of the *AutoDecon* algorithm is for the *fitting* module to estimate only the basal secretion, S_0, and the concentration at time zero, $C(0)$. The *fitting* module will then estimate all of the model parameters except for the elimination half-life and the standard deviation of the secretion events. If any secretion events have been included in the initialization, the second fit will also refine the locations and sizes of these secretion events. Next, the *triage* module is utilized to remove any nonsignificant secretion events. At this point the parameter estimations performed within the *triage* module will estimate all of the current model parameters, except for the elimination half-life and the standard deviation of the secretion events (*SecretionSD*).

The *AutoDecon* algorithm next proceeds with phase 1 by using the *insertion* module to add a presumed secretion event. This is followed by the *triage* module to remove any nonsignificant secretion events. Again, the parameter estimations performed within the *triage* module during phase 1 will estimate all of the current model parameters with exception of the elimination half-life, *HL*, and the standard deviation of the secretion events,

SecretionSD. If during this phase the *triage* module does not remove any secretion events, phase 1 is repeated to add an additional presumed secretion event. Phase 1 is repeated until no additional secretory events are added in the *insertion* followed by *triage* cycle.

Phase 2 repeats the *triage* module with the *fitting* module estimating all of the current model parameters, but this time including the elimination half-life, *HL*, and the standard deviation of the secretion events, *SecretionSD*.

Phase 3 will repeat phase 1 (i.e., *insertion* and *triage*) with the parameter estimations that are performed by the *fitting* module within the *triage* module estimating all of the current model parameters, again including the elimination half-life, *HL*, and the standard deviation of the secretion events, *SecretionSD*. Phase 3 is repeated until no additional secretion events have been added in the *insertion* followed by *triage* cycle.

Phase 4 examines the residuals (i.e., differences between the data points and the fitted curve) for trends that might indicate fitting problems. Specifically, if the first data point is a negative outlier (i.e., the data point is significantly lower than might be expected; see Section 2.7) or the last data point is a positive outlier (i.e., the data point is significantly higher than might be anticipated), then it is possible that the algorithm failed to identify a partial secretion event at either the start or the end of the time series, respectively. If only a portion of the secretion event is present within the data time series, then the *AutoDecon* algorithm might not resolve this partial event and thus an outlier may be present. When this is observed, then the offending data point is temporarily removed and the entire *AutoDecon* algorithm is repeated with the current values as the initialization.

An illustrative example of the results from this procedure is shown in Fig. 15.1. In this example, one secretion event has been located. The upper part of Fig. 15.1 shows the data points and the calculated concentration. The lower part of Fig. 15.1 depicts the observed secretion rate (lower curve) and the probable position index [upper curve calculated per Eq. (15.5)] for addition of the next presumptive secretion event. In this case, the next presumptive secretion event is not statistically significant at a level of 0.05. Thus, the *AutoDecon* algorithm added this presumptive event, tested it, and subsequently removed it (because it was not significant at a level of 0.05), and then stopped.

It is important to note that the peak of the secretion event occurs several minutes before the peak of the corresponding concentration peak. The shape of the secretion event, $S(t)$, dominates the rising portion of the concentration curve, whereas the falling portion of the concentration curve is dominated by the elimination function, $E(t)$. The location of the secretion event (asterisk in the top half of Fig. 15.1) corresponds to a position approximately halfway up the rising portion of the concentration curve.

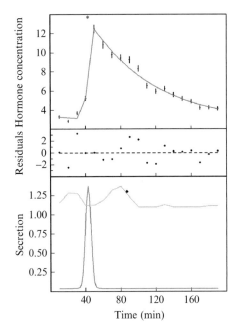

Figure 15.1 An illustrative example of the use of the *AutoDecon* algorithm. (Top) A set of data points with vertical error bars, with the solid curve corresponding to the predicted concentration, $C(t)$, from Eqs. (15.2), (15.3), and (15.5). The asterisk at the top of the panel marks the location of the single secretion event. (Middle) Weighted differences between the data points and the calculated concentration, with the weighting factors being the W_i in Eq. (15.10). (Bottom) Calculated secretion pattern, $S(t)$, from Eq. (15.5) as a lower curve and the probable position index, $PPI(t)$, as an upper curve. The diamond in this panel marks the location of the next presumed secretory event, which was discarded because it was not statistically significant. Note that the peak of the concentration event in the upper panel occurs significantly later in time than the actual peak of the secretion event. Basal secretion, S_0 in Eq. (15.5), is 0.037 per minute. Initial concentration, C_0 in Eq. (15.2), is 0.71 min. *SecretionSD* in Eq. (15.5) is 3.0 min. The elimination half-life, HL in Eq. (15.3), is 50.3 min. The secretion peak position, PP in Eq. (15.5), is 42.4 min.

2.6. Variation on the theme

The objective of phase 1 of the *AutoDecon* algorithm is to obtain a reasonable initial estimate of the level of basal secretion and find initial estimates of the secretion peak sizes and locations. As outlined, the algorithm works well when the data set being analyzed contains a sufficient amount of information. However, occasionally phase 1 will either overestimate the basal and underestimate the number of the small secretion events or underestimate the basal and overestimate the number of small secretion events. Effectively the algorithm has fallen into the wrong minima of the variance of fit. The later phases of *AutoDecon* will usually, but not always, correct this situation.

This is particularly a problem with limited amounts of data and when thousands of simulated data sets are being analyzed with the totally automated *AutoDecon* algorithm.

The *AutoDecon* algorithm is a *Systems Identification* approach involving multiple weighted nonlinear least-squares parameter estimations. When estimating the parameters of nonlinear equations, by any fitting procedure, the possibility of multiple minima in the variance of fit always exists. The trick to avoiding these incorrect minima is by initializing the fitting algorithm with values that are relatively close to the "correct answers." Two of the many potential solutions to this multiple minima problem are outlined here; both involve executing the *AutoDecon* algorithm twice for each data set.

If the *AutoDecon* algorithm appears to be overestimating the basal secretion and underestimating the number of the small secretion events, then it may be that the P value for the elimination of a secretion event is initially too high. Conversely, if the algorithm underestimates the basal and overestimates the number of small secretion events, then perhaps the initial P value is too small. One potential solution is to execute the *AutoDecon* algorithm twice. First, execute *AutoDecon* with a modified P value while keeping the initial estimates of the elimination half-life and *SecretionSD* constant. *AutoDecon* will stop after phase 1 since the elimination half-life and *SecretionSD* are being held constant. Second, execute the *AutoDecon* algorithm initialized with the results of the first execution, with the desired final P value this time estimating the basal secretion, elimination half-life, and *SecretionSD*.

For the present GH data it is obvious that the basal secretion is either very low or possibly even zero. In such a case, the *AutoDecon* algorithm can be executed twice to decrease the possibility of finding multiple minima in the variance of fit. First, execute the *AutoDecon* algorithm with a very low fixed value for the basal secretion and fixed initial estimates of the elimination half-life and *SecretionSD*. Again, the program will stop after phase 1 because the elimination half-life and *SecretionSD* are being held constant. Second, execute the *AutoDecon* algorithm initialized with the results of the first execution while now estimating the basal secretion, elimination half-life, and *SecretionSD*.

There are many possible perturbations for the use of *AutoDecon*. The optimal choice will be dependent on the specific hormone, data collection protocol, and assay characteristics. For the present work we simulated 500 data sets that mimic the results from our experimental results (see data simulation section later) and then tested several perturbations to decide which, if any, was optimal. This second approach was utilized routinely for all of the simulations presented in this chapter. In the current case the secretion rate was initially set to 0.0005 μg/liters/min.

2.7. Outliers

The detection of outliers is a complex statistical issue with a profuse number of divergent criteria depending on the specific application. The mathematical definition of an outlier for this chapter is when the absolute value of the Z score of the particular residual, Z_k, is greater than 4. These Z scores are calculated as the particular, kth, residual divided by the root mean squared average value of all of the other residuals, as shown in Eq. (15.9) where N is the total number of residuals:

$$Z_k = \frac{R_k}{\sqrt{\dfrac{\sum_i R_i}{N-1}}}, i \neq k. \tag{15.9}$$

There are a variety of reasons why an outlier may be present in the hormone concentration time series and, as a consequence, in the residuals. An outlier may be the result of a bad data point or may simply be a typographical error that occurred during creation of the data file. An occasional outlier is expected in data due simply to the random measurement uncertainties inherently present within experimental data. Outliers should never merely be discarded, as the apparent existence of such may be caused by some overlooked aspect of data. Thus, in *AutoDecon* phase 4, the presumed outliers provide useful information about a potential missing secretion event. It is imperative to investigate the cause of an outlier before arbitrarily removing it from the data series.

2.8. Weighting factors

The *AutoDecon* algorithm is based on a weighted nonlinear least-squares parameter estimation procedure. The required weighting factor corresponds to the inverse of the expected standard error of the mean (SEM) of the particular data point. If every data point were independently sampled and measured more than 15 or 20 times, then the precision of the measured hormone concentrations could be evaluated as the SEM of the replicate samples and measurements. However, with typically only 1 to 3 replicate samples, the SEM of the replicates is far too inaccurate to be used as a realistic estimate of the actual precision of the hormone concentrations. Thus, to estimate the experimental uncertainties in such cases, a variance model was utilized that accounts for the performance characteristics of the clinical laboratory assays. An empirical representation of the variance as a function of the hormone concentration is given in Eq. (15.10):

$$\text{Variance}([\text{Hormone}(t_i)]) = \frac{1}{w_i^2}$$
$$\approx \frac{1}{N}\left[\left(\frac{MDC}{2}\right)^2 + \left(CV \cdot \frac{[\text{Hormone}(t_i)]}{100}\right)^2\right], \quad (15.10)$$

where CV is the % *coefficient of variation* at the *optimal range* for the assay, MDC is the minimal detectable concentration, N is the number of replicates, and $[\text{Hormone}(t_i)]$ is the hormone concentration at time t_i. The MDC and CV are measured routinely as part of the quality control measurements performed by the clinical laboratories and thus their values are well known. The minimal detectable concentration is the lowest concentration that can be measured accurately and is measured experimentally as twice the standard deviation (SD) of approximately 15 or 20 samples that contain a hormone concentration of zero.

2.9. AutoDecon performance and comparison with earlier pulse detection algorithms

The present approach for validating pulse detection algorithms involves the creation of synthetic data sets where the number, position, and characteristics of the pulses are "known" a priori (similar to Guardabasso et al., 1988; Van Cauter, 1988; Veldhuis and Johnson, 1988). This chapter modeled these synthetic files upon "real" data obtained from normal control subjects in a clinical study. The pulsatile characteristics observed from these control subjects were incorporated into simulations. *AutoDecon* and other algorithms were applied to ascertain how well each pulsatile characteristic was recovered by each algorithm.

2.10. Experimental data

Clinical experimental data were obtained from a study approved by the institutional review board of the University of Virginia and the General Clinical Research Center (GCRC). All volunteers gave written informed consent before participating in the study. Sixteen healthy, obese individuals (eight male and eight female) between of the ages of 21 and 55 years of age were recruited by advertisement. The mean (±SD) age (years) of the subjects was 37.1 ± 9.3 [range: 24–55] and the mean (±SD) body mass index (kg/m^2) was 34.0 ± 3.0 [range: 30.5–38.4]. Study subjects had maintained a stable weight for the past 6 months (i.e., no change greater than ≈2 kg) and had no intention of losing or gaining weight during the entire study period. Subjects agreed to avoid unusual, unaccustomed, or exceptionally strenuous physical activity from 24 h prior to the start of the study until the conclusion of the study period. Female subjects of reproductive age demonstrated a serum β-human chorionic gonadotropin level consistent with the nongravid state at the prestudy visit. Hormone replacement therapy was allowed if the subject had been on a stable regimen for 3 months or more prior to the study start and the regimen remained stable during the entire study. All subjects had normal renal function and were judged to be in good health on the basis of medical history, physical examination, and routine laboratory tests.

All subjects agreed to avoid the use of any prescription and nonprescription medications and preparations, including any herbal, organic, or nutritional remedy other than a standard approved once-daily multivitamin for the entire study period. No alcoholic beverage consumption occurred for 48 h before or during days when blood samples were drawn. Subjects were permitted to consume two glasses of wine (4 ounces/glass) or two bottles (12 ounces/bottle) of beer or equivalent daily during other times during the study. Grapefruit juice consumption was prohibited for at least 2 weeks prior to the study start and throughout the entire study as grapefruit has been shown to inhibit CYP3A4 activity.

All volunteers were admitted to the GCRC for a 39-h stay. They were encouraged to sleep after 2100 h. Two forearm indwelling venous cannulae were placed in the morning at 0600 for blood sampling and were kept patent with saline and/or heparin flushes. Blood sampling was performed through an indwelling venous cannula every 10 min for 24 h, from 0800 h on day 1 until 088 h on day 2 of admission. Standardized meals were served at 0800, 1300, and 1800 h. Meals were consumed within 30 min with no snacks allowed.

2.11. Growth hormone assay

Serum GH concentrations were measured in duplicate by a fluoroimmunometric assay on an Immulite 2000 analyzer by Diagnostics Products Corporation (DPC). DPC claims an interassay CV of 3.8% at 2.5 μg/liter, 3.5% at 5 μg/liter, and 3.3% at 12.6 μg/liter and an intraassay CV of 3.4% at 2.4 μg/liter, 2.6% at 4.8 μg/liter, and 2.3% at 12 μg/liter and a sensitivity (MDC) of the assay stated to be 0.01 μg/liter. The variability from an analysis of our measured data sets indicates that DPC is conservatively overestimating the actual measurement errors. Our analysis indicates that our instrument has an intraassay CV of approximately 1.64% and a MDC of 0.0066 μg/liter.

2.12. Data simulation

The objective of data simulations is to provide a group of hormone concentration time series where the (1) correct answers (e.g., secretion peak areas and positions) are known a priori and (2) simulations closely mimic the actual experimental observations for a particular hormone within a specific experimental protocol. For this study we simulated 500 data sets for each test condition. These data sets were then analyzed by the *AutoDecon* algorithm together with several previously published algorithms. A comparison of the apparent results from the *AutoDecon* software and the correct answers yielded information about the operating characteristics of each of the analysis algorithms (e.g., *AutoDecon*, *Cluster*).

The most vital aspect of the data simulation approach is to model actual experimental data in the most exacting way possible. The first step is to analyze a group of actual experimental data with a specific analysis algorithm such as *AutoDecon*, *Cluster*, and *Pulse* in order to create a consensus analysis for each of the actual hormone concentration time series. For the present case, the group of actual experimental data consisted of sixteen 24-h GH concentration time series sampled at 10-min intervals and assayed in duplicate.

Apparent values, based on the analysis of each particular algorithm of the 16 data sets of the basal secretion [S_0 in Eq. (15.4)], the *SecretionSD* [Eq. (15.4)], and the half-life [HL in Eq. (15.2)], are assumed to follow lognormal distributions.

The base 10 logarithms of the individual interpulse intervals are potentially related to $\log_{10}$ of the previous interpulse interval, $\log_{10}$ of the secretion previous event area, $\log_{10}$ of the specific S_0, $\log_{10}$ of the specific *SecretionSD*, $\log_{10}$ of the specific HL, a 24-h cosine wave, and a 24-h sine wave. The relational coefficients were determined by multiple linear regression (MLR) analysis (as in Tables 15.3 and 15.4). The cross-correlations and covariances of the MLR parameters were calculated to verify that this use of the MLR was justified. Sine and cosine functions were included to compensate for the possibility of a 24-h rhythm induced by the sleep–wake cycles of the subjects. Note that the combination of a cosine and a sine wave is a linear model that is mathematically equivalent to a cosine wave with a variable phase. For these simulations, evaluation of the significance of the specific MLR terms is not required. If a term is not significant then it will have negligible amplitude and thus will not contribute to the simulations.

Similarly, the base 10 logarithms of the individual secretion event areas are potentially related to $\log_{10}$ of the previous secretion previous event area, $\log_{10}$ of the interpulse interval, $\log_{10}$ of the specific S_0, $\log_{10}$ of the specific *SecretionSD*, $\log_{10}$ of the specific HL, a 24-h cosine wave, and a 24-h sine wave.

For the actual simulations, the time series is generated to start 24 h prior to the first desired time point and continue for 24 h after the last desired time point. This will normally generate a partial secretion event at both the beginning and the end of the time series. All the secretion events with a position [PP_k in Eq. (15.4)] within the desired time are considered to be secretion events, which the various algorithms should be able to locate accurately, including the partial events at the beginning and end of the desired time series.

The simulation is initialized by evaluating unique values for the S_0, the *SecretionSD*, and the *HL* by generating pseudo-random values based on the corresponding lognormal distribution obtained from experimental data. Next, an interpulse interval and a secretion event area are simulated by generating pseudo-random values that follow the corresponding distributions and dependencies as determined by the multiple linear regressions of results from the analysis of actual experimental data. Initially the time is set to −24 h and the previous secretion event areas and previous interpulse intervals are assumed to be equal to the average values. This process of finding the next interpulse interval and secretion event area is repeated for 24 h past the last desired data point. Gaussian-distributed pseudo-random experimental measurement errors are then added according to Eq. (15.10). Data corresponding to the second of the three simulated days are subsequently utilized as the first simulated hormone concentration time series. This entire process is repeated 500 times to obtain the entire group of 500 simulated hormone concentration time series.

This simulation procedure is substantially different than previous hormone concentration time series simulation algorithms (Guardabasso *et al.*, 1988; Van Cauter, 1988; Veldhuis and Johnson, 1988). The previous algorithms assumed that the interpulse intervals and secretion event sizes each followed independent Gaussian.

2.13. Algorithm operating characteristics

The simulated data sets were analyzed by *AutoDecon* and several other algorithms for pulse detection. A subsequent comparison between apparent results returned by an algorithm and correct answers from the simulations translates into information about the operating characteristics of each algorithm (e.g., *AutoDecon*, *Cluster*). In particular, we used a concordant secretion events criterion (see next section) to evaluate the following.

True positive as the percentage of the apparent secretion events found by the algorithm that closely corresponds in time to actual secretion events

False positive (type I errors) as the percentage of the apparent secretion events found by the algorithm that do not closely correspond in time to actual secretion events

False negative (type II errors) as the percentage of simulated secretion events not located by the algorithm that occur between the first and the last data points

Sensitivity as the percentage of actual simulated secretion events located by the algorithm

Our definition of false negative (type II errors) includes a stipulation that the center of the secretion events must be between the first and the last data points to be considered as a false-negative incident. In approximately 10% of the simulated data sets a secretion event occurred slightly later than the last data point but close enough that the leading edge of the secretion event has a significant contribution to the hormone concentration before the last data point. This stipulation is included so that the failure to find secretion events outside the range of data is not considered as an error. These out-of-range secretion events are, however, included in the calculations of true positives. The analogous partial secretion events that occur before the time of the first data point are incorporated into the $C(0)$ term by the analysis algorithm and are thus not treated as a special case.

2.14. Concordant secretion events

Determining the operating characteristics of the algorithms requires a comparison of the apparent secretion event positions from an analysis of a simulated time series, with the actual known secretion event positions upon which the simulations were based. This process must consider whether the concordance of the peak positions is statistically significant or whether it is a consequence of a simply random position of the apparent secretion events. For example, when comparing two time series, each having approximately 20 randomly positioned events within a 24-h period, it is unlikely ($P < 0.05$) that 6 or more of these random events will appear to be coincident within a ± 5-min window, that 8 or more will appear to be coincident within a ± 10-min window, and that 11 or more will simultaneously occurring within a ± 20-min window.

Specifically, the following question could be posed: given two time-series with n and m distinct events, what is the probability that j coincidences (i.e., concordances) will occur based on a random positioning of the distinct events within each of the time series? The resulting probabilities are dependent on the size of the specific time window employed for the definition of coincidence. This question can be resolved easily by a Monte Carlo approach. One hundred thousand pairs of time series are generated with the n and m distinct randomly timed events, respectively. The distribution of the expected number of concordances can then be evaluated by scanning these pairs of random event sequences for coincident peaks where coincidence is defined by any desired time interval.

Obviously, as the coincidence interval increases so will the expected number of coincident events. Thus, the coincidence interval should be kept small. Coincidence windows of ± 5, ± 10, or maybe ± 15 min are reasonable for this study simply because the data sampling is at 10-min intervals. The expected number of coincident events will also increase with the numbers of distinct events, n and m.

3. Results

Growth hormone data chosen to test the *AutoDecon* algorithm were selected specifically to represent a particularly difficult case. The subjects are neither young nor lean. The size of the observed GH secretion events is expected to be decreased because of both of these factors. In addition, sampling at 10-min intervals makes the resolution of an approximately 14-min elimination half-life somewhat difficult. However, because the hormone concentrations were measured in duplicate with state-of-the-art instrumentation, data from this protocol are typical of the most pristine GH concentration time series data currently available.

3.1. Typical *AutoDecon* output

Figure 15.2 presents an example of the analysis of a typical GH time series by the *AutoDecon* algorithm. The upper part of Fig. 15.2 depicts the GH data points as vertical error bars and best-fit calculated GH concentrations. These are on a logarithmic scale to enhance visualization of the small secretion events. For this specific GH data set, the *AutoDecon* algorithm identified 24 secretion events with a median interpulse interval of 45.3 min, an elimination half-life of 13.1 min, and a secretion event standard deviation of 8.8 min. The median of the interpulse intervals is given because (as demonstrated later) the distribution of interpulse intervals is not Gaussian (skewness of 0.973 and kurtosis of -0.190), and thus the mean interpulse interval of 56.6 ± 7.9 min is not as informative. The median of 0.522 for the secretion event area (i.e., mass) is also preferable as its distribution is even more non-Gaussian (skewness of 1.594 and kurtosis of 1.281), and thus the mean secretion event area of 1.65 ± 0.46 is even less informative.

The probable position index (the upper line in the lower part of Fig. 15.2) indicates that the final presumptive secretion event was located at 666 min into the collection protocol. Visually it appears from the differences between the data points and the calculated curve (upper part of Fig. 15.2) that this might be an actual secretion event. However, this secretion event was not included because of its 0.0505 significance level. Having a significance level over 0.05 does not mean that the secretion event is not present, but simply means that it cannot be demonstrated at the 0.05 significance level. There is another possible secretion event at approximately 200 min into the collection protocol. If a less stringent significance level had been utilized, the number of secretion events would be increased and the apparent basal secretion level would, as a consequence, be lower.

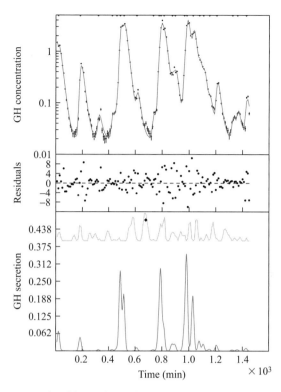

Figure 15.2 An example of the analysis of a typical growth hormone concentration time series by the *AutoDecon* algorithm. (Top) A set of data points with vertical error bars and a solid curve corresponding to the predicted concentration, $C(t)$, from Eqs. (15.2), (15.3), and (15.5). The asterisk at the top of the panel denotes the location of the single secretion event. (Middle) Weighted differences between the data points and the calculated concentration, with the weighting factors being the W_i in Eq. (15.10). (Bottom) The calculated secretion pattern, $S(t)$, from Eq. (15.5) as a lower curve and the probable position index, $PPI(t)$, as an upper curve. The diamond in this panel marks the location of the next presumed concentration event, which was discarded because it was not statistically significant. Note that the peak of the concentration event in the upper panel occurs significantly later in time than the actual peak of the secretion event.

3.2. Experimental data analysis

The combined results of the *AutoDecon* analysis of all 16 of the time series from this protocol are summarized in Table 15.1. In several cases the medians are not close to the means, which implies that the distributions are not Gaussian (i.e., distributed normally). Similarly, the standard deviations of these cases are also sufficiently large that a normal distribution will include a high percentage of values less than zero and thus not physically meaningful. This can clearly be seen in the distribution of interpulse

Table 15.1 Summary of secretion properties of the 16 hormone concentration time series analyzed with *AutoDecon*

	Median	Mean ± SEM	SD	N	Interquartile Range
S_0	0.00138	0.00187 ± 0.00024	0.00095	16	0.00138
SecretionSD	9.02	8.89 ± 0.58	2.31	16	2.72
HL	14.32	14.17 ± 0.43	1.72	16	2.08
# events	18	20.56 ± 0.93	3.71	16	7.0
# outliers	1	0.63 ± 0.16	0.62	16	1.0
Interpulse interval	50.49	66.82 ± 2.85	51.37	313	51.54
Event size (area)	0.51	2.14 ± 0.19	3.50	329	2.39
% pulsatile	93.87	93.13 ± 0.68	2.74	16	3.51

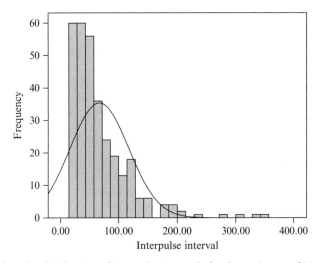

Figure 15.3 The distribution of interpulse intervals for the entire set of 32 concentration time series found by the *AutoDecon* algorithm. The solid line corresponds to the best normal (i.e., Gaussian) distribution.

intervals (Fig. 15.3) and in the logarithmic normal distribution of interpulse intervals (Fig. 15.4). The interpulse intervals appear to be distributed approximately lognormally.

The distributions of the secretion event sizes also cannot be analyzed as a normal distribution (Figs. 15.5 and 15.6). Here the secretion event sizes appear to follow a bimodal (or multimodal) lognormal distribution.

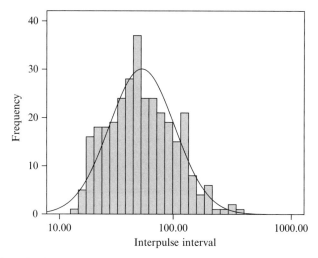

Figure 15.4 The distribution of interpulse intervals for the entire set of 32 concentration time series found by the *AutoDecon* algorithm but plotted on a logarithmic X axis. The solid line corresponds to the best lognormal distribution.

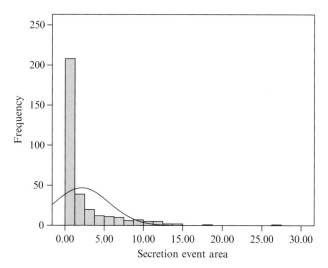

Figure 15.5 The distribution of secretion event sizes for the entire set of 32 concentration time series found by the *AutoDecon* algorithm. The solid line corresponds to the best normal (i.e., Gaussian) distribution.

The first column of Table 15.2 presents the apparent number of secretion events found by several of the previously published hormone pulse detection algorithms. The *Pulse* and *Pulse2* algorithms have two different

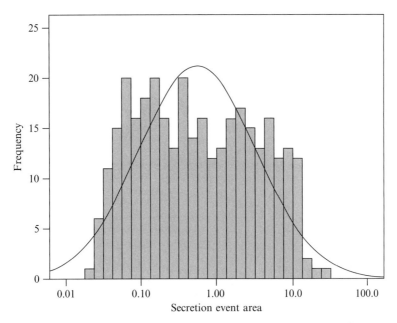

Figure 15.6 The distribution of secretion event sizes for the entire set of 32 concentration time series found by the *AutoDecon* algorithm but plotted on a logarithmic X axis. The solid line corresponds to the best lognormal distribution.

Table 15.2 Mean number of secretion events found for the 16 time series by several different algorithms

Algoritham	Number of secretory events		Reference
	Original data	Shuffled data[a]	
AutoDecon	20.6	0.0	
Cluster[b]	12.3	22.1	Veldhuis and Johnson (1986)
Pulse (zero basal)	13.8	18.2	Johnson and Straume (1999)
Pulse (variable basal)	7.2	15.6	Johnson and Straume (1999)
Pulse2 (zero basal)	11.2	1.4	Johnson and Straume (1999)
Pulse2 (variable basal)	12.9	0.3	Johnson and Straume (1999)
Pulse4	1.0	1.0	Johnson *et al.* (2004)

[a] Assuming $HL = 14.25$ and *Secretion SD* $= 8.95$.
[b] *Cluster* algorithm nadir size and peak size of 2, and both T statistics set to 2.0.

entries where the first assumes that the basal secretion is zero and the second assumes that the basal secretion is not zero. Obviously, the operating characteristics of these algorithms need to be determined when applied to GH data so that an appropriate choice of algorithms, and thus results, can be made. The choice of algorithm will be dependent on the specific hormone being investigated, the details of the specific experimental protocol, and the laboratory instrumentation utilized for the hormone concentration measurements.

One simple test of the algorithms is to see how they perform when data are shuffled (i.e., randomizing the time sequence of data). This is similar to a previously published procedure for the evaluation of the percentage of false positives (Van Cauter, 1988). The hormone concentration pulsatile events follow a distinct shape, that is, a fast rise and a somewhat slower fall (Johnson and Veldhuis, 1995; Veldhuis et al., 1987). Any operation, such as shuffling the data points, that disrupts the time sequence of the data time series should also disrupt the ability of the hormone detection algorithms to find distinct secretion events within the data sequence. Thus, it is surprising that the *Cluster* and the *Pulse* algorithms actually found more apparent events within shuffled (i.e. randomized) data than in original data (Table 15.2, column 2). This is sufficient reason not to use either the *Cluster* or the *Pulse* algorithm for GH data.

The *Pulse4* algorithm found the fewest peaks of all of the algorithms (Table 15.2), reflecting the conditions of this particular GH protocol, which are well outside the bounds of applicability for the *Pulse4* algorithm. Specifically, the *Pulse4* algorithm requires more than four data points per elimination half-life while the present protocol samples the data points at a 10-min interval even though GH is known to exhibit an elimination half-life of approximately 14 min, yielding only 1.4 data points per half-life. Thus, *Pulse4* clearly is not applicable for this particular GH protocol.

Multiple linear regressions were applied to the interpulse interval and secretion event areas, which yielded some interesting results. The apparent secretion event area is controlled significantly ($P < 0.05$) by the previous interpulse interval, the previous secretion event area, the basal secretion, and the *SecretionSD* (see Table 15.3). The interpulse interval is also controlled significantly ($P < 0.05$) by the previous interpulse interval, the previous secretion event area, and the *SecretionSD*, but with a much smaller magnitude. Both the intervals and the secretion event areas also appear to be controlled by a 24-h rhythm. These apparent controlling factors may be a reflection of the underlying molecular mechanisms of secretion or they may simply be a limitation of the numerical analysis.

Table 15.3 Significantly nonzero ($p < 0.05$) coefficients from multiple linear regressions in Table 15.4[a]

	Secretion event area	Interpulse interval
Log_{10} previous interval	-1.069 ± 0.123	-0.245 ± 0.061
Log_{10} previous event area	0.264 ± 0.050	-0.049 ± 0.025
$Log_{10} S_0$	1.088 ± 0.228	
$Log_{10} SecretionSD$	1.329 ± 0.343	0.665 ± 0.158
Cosine (2π TSS/1440)	-0.210 ± 0.051	0.056 ± 0.024
Sine (2π TSS/1440)	0.017 ± 0.049	0.117 ± 0.021
R^2	0.454	0.186
Largest cross-correlation	-0.351	0.441

[a] Backward removal was used to remove nonsignificant coefficients. The sine coefficient for the secretion event area is included because the cosine coefficient is significant.

Table 15.4 Simulation parameters from a MLR examination of *AutoDecon* analysis results[a]

$Log_{10} Sec.Area_n$	$= N(-0.250, 0.326) - 1.073 * [Log_{10} Interval_{n-1} - 1.729] + 0.267 * [Log_{10} Sec.Area_{n-1} + 0.253] + 1.082 * [Log_{10} S_0 + 2.78] + 1.324 * [Log_{10} SecretionSD - 0.925] + 0.688 * [Log_{10} HL - 1.151 - 0.203 * Cos [2\pi TSS/1440] + 0.026 * Sin [2\pi TSS/1440]$
$Log_{10} Interval_n$	$= N(1.754, 0.063) - 0.054 * [Log_{10} Sec.Area_{n-1} + 0.214] - 0.244 * [Log_{10} Interval_{n-1} - 1.723] + 0.091 * [Log_{10} S_0 + 2.78] + 0.674 * [Log_{10} SecretionSD - 0.925] - 0.033 * [Log_{10} HL - 1.151] + 0.043 * Cos [2\pi TSS/1440] + 0.108 * Sin [2\pi TSS/1440]$
$Log_{10} SecretionSD$	$= N(0.935, 0.00133)$
$Log_{10} S_0$	$= N(-2.770, 0.0.0356)$
$Log_{10} HL$	$= N(1.148, 0.00305)$
MDC	$= 0.00661$
CV	$= 1.644$

[a] The n subscript refers to the current interpulse interval and secretion event area, the $n-1$ subscript refers to the previous interpulse interval and secretion event area, and TSS is time from 8:00 AM, the start of the hormone sampling period. N (mean, variance) is a pseudo-random variable that follows a normal distribution with the specific mean and variance. Constants were determined by a multiple linear regression based on the experimental measurements. Duplicate samples were simulated unless otherwise noted.

3.3. Comparison between *AutoDecon* and other algorithms

Table 15.4 presents the simulation parameters utilized to create the first set of 500 simulated hormone concentration time series. This first set was generated to mimic results from the analysis of experimental data by the

AutoDecon algorithm. Table 15.4 was created by MLR analysis of the *AutoDecon* estimated results. In this, and all subsequent simulations, it is assumed that the individual interpulse intervals and secretion event areas could be a function of the previous individual interpulse intervals, secretion event areas, elimination half-life, basal secretion, secretion event standard deviation, and sum of 24-h sine and cosine waves.

Figure 15.7 presents a comparison of the operating characteristics of the *AutoDecon* and *Cluster* (Veldhuis and Johnson, 1986) algorithms when applied to these 500 simulated hormone concentration time series. In Fig. 15.7, true-positive events were determined for a series of different concordance windows. For example, if the algorithm locates a secretion event within a simulated hormone concentration time series to be within ±5 min of the exact location of the secretion event upon which the simulations were based, then it is a true positive with a concordance window of ±5 min or less. *AutoDecon* performs better by all measures with a ±2.5-min, or longer, concordance window than *Cluster* with a ±25-min concordance window. For example, with a ±10-min concordance window the sensitivity of *AutoDecon* is 97.5% and *Cluster* is 42.7%.

It is interesting to note that 25% of the false positives identified by *AutoDecon* corresponded to the first apparent secretion event and that 24% of the false negatives corresponded to the last simulated secretion event in

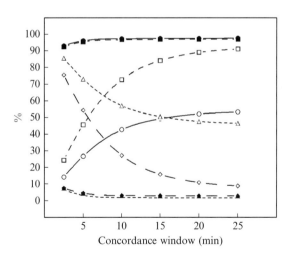

Figure 15.7 Performance characteristics of the *AutoDecon* and *Cluster* algorithms at different concordance windows. These were evaluated by the analysis of 500 simulated data sets generated as described in Table 15.4. Solid symbols are for *AutoDecon* and open symbols are for *Cluster*. Solid lines and circles describe the sensitivity %; medium-dashed lines and squares describe the true positive %; long-dashed lines and diamonds describe the false positive %; and short-dashed lines and triangles describe the false negative %.

the sequence. These are substantially higher that the ≈5% expected from random chance. It is, however, not unexpected, as the first and last simulated secretion events are contained only partially within the data sequence.

Figure 15.8 presents the corresponding mean true positive %, false positive %, false negative %, and sensitivity of the *AutoDecon* algorithm when different P values are used by the triage module. These were evaluated with the same 500 simulated hormone concentration time series used in Fig. 15.7. Clearly, the *AutoDecon* algorithm outperforms the *Cluster* algorithm for all triage P values between 0.02 and 0.08. The false positive % and false negative % are both approximately 2.1% when the triage P value is set to 0.0325. When the triage P value is set to 0.039, both the true positive % and the sensitivity are approximately 97.4%. Please note that these values only apply to GH data collected with this specific protocol. The actual choice triage P value is obviously dependent on which of these operating characteristics is deemed the most important. For the remainder of this chapter, the triage P value will be set to 0.0325, which represents a reasonable value to jointly optimize the false positive % and false negative %.

Table 15.5 presents a comparison of the operating characteristics of five of the analysis algorithms when applied to these 500 simulated hormone concentration time series. It is clear from Table 15.5 that the *AutoDecon* algorithm performs far better than the other algorithms when simulated data

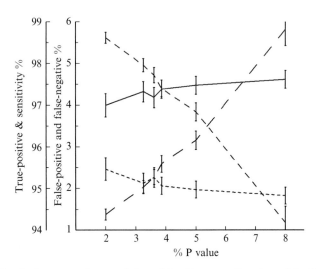

Figure 15.8 Performance characteristics of *AutoDecon* as a function of the P value used by the algorithm's triage module for the removal of nonsignificant secretion events. Solid lines describe the sensitivity %; medium-dashed lines describe the true positive %; long-dashed lines describe the false positive %; and short-dashed lines describe the false negative %.

Table 15.5 Comparison of performance of analysis algorithms evaluated from simulated data based on *AutoDecon* results

	Median	Mean ± SEM	Interquartile range
AutoDecon (P value = 0.0325)[a]			
%true positive	100.00	97.96 ± 0.16	4.76
%false positive	0.00	2.04 ± 0.16	4.76
%false negative	0.00	2.13 ± 0.24	0.00
%sensitivity	100.00	97.32 ± 0.25	5.26
Cluster[b]			
%true positive	90.00	89.13 ± 0.50	18.18
%false positive	10.00	10.87 ± 0.50	18.18
%false negative	47.37	47.64 ± 0.49	13.82
%sensitivity	52.00	52.08 ± 0.49	13.61
Pulse with no basal secretion and half-life = 14.25 min[a]			
%true positive	75.00	74.91 ± 0.54	16.67
%false positive	25.00	25.09 ± 0.54	16.67
%false negative	37.50	37.41 ± 0.47	13.11
%sensitivity	61.91	62.25 ± 0.47	13.19
Pulse[a] with basal secretion and half-life = 14.25 min[a]			
%true positive	100.00	99.00 ± 0.19	0.00
%false positive	0.00	1.00 ± 0.19	0.00
%false negative	55.28	53.97 ± 0.60	18.18
%sensitivity	44.44	45.78 ± 0.60	18.13
Pulse2 with no basal secretion[a]			
%true positive	76.92	76.73 ± 0.59	19.85
%false positive	23.08	23.27 ± 0.59	19.84
%false negative	23.91	25.51 ± 0.63	16.23
%sensitivity	75.00	74.09 ± 0.64	16.67
Pulse2 with basal secretion[a]			
%true positive	90.00	86.78 ± 0.57	21.43
%false positive	10.00	13.22 ± 0.57	21.43
%false negative	25.00	25.98 ± 0.58	17.19
%sensitivity	75.00	74.61 ± 0.58	21.43
Pulse4 with variable half-life[a]			
%true positive	0.00	41.15 ± 2.19	100.00
%false positive	100.00	58.85 ± 2.19	100.00
%false negative	100.00	94.14 ± 0.64	5.89
%sensitivity	0.00	5.84 ± 0.64	5.89
Pulse4 with half-life = 14.25 min[a]			
%true positive	100.00	66.33 ± 2.10	100.00
%false positive	0.00	33.67 ± 2.10	100.00
%false negative	94.59	94.23 ± 0.44	6.67
%sensitivity	5.26	5.74 ± 0.44	6.25

[a] Requiring alignment of secretion event positions to be within ±10 min.
[b] Requiring alignment of secretion event positions to be within ±20 min, evaluated with *Cluster* parameters of 2 × 2 and *T* statistics = 2.0.

are based on results obtained by the *AutoDecon* algorithm. It is important to note that the operating characteristics shown in Table 15.5. apply only to the analysis of GH sampled every 10 min with the experimental measurement errors shown in Table 15.4. These operating characteristics will undoubtedly vary with different hormones and possibly even with different experimental protocols for the measurement of GH.

Table 15.6 presents an analogous analysis based on 500 simulated hormone concentration time series generated to mimic the results of the analysis of experimental data by the *Cluster* algorithm. Clearly, the *AutoDecon* algorithm significantly outperforms the *Cluster* algorithm, even when data are simulated to mimic the *Cluster* algorithm results.

In both Tables 15.5 and 15.6 the window to accept a match between the simulated secretion event positions and the apparent event locations found was taken to be 20 min for the *Cluster* algorithm and 10 min for the *AutoDecon* algorithm. Thus, the *AutoDecon* algorithm outperformed the *Cluster* algorithm even with a significant test bias that favors the *Cluster* algorithm.

A similar analysis was performed for the *Pulse* and *Pulse2* algorithms with and without basal secretion and the results presented in Tables 15.7–15.10. In every case the *AutoDecon* algorithm performed significantly better than either the *Pulse* or the *Pulse2* algorithm.

The optimal *P* value for the removal of nonsignificant secretion events was determined (see Fig. 15.8) to be approximately 0.035 for simulated data based on the *AutoDecon* analysis of experimental data. However, for simulated data based on the *Pulse2* analysis of experimental data

Table 15.6 Comparison of performance of *AutoDecon* and *Cluster* algorithms evaluated from simulated data based on *Cluster* results

	Median	Mean ± SEM	Interquartile range
AutoDecon (*P* value = 0.0325)[a]			
%true positive	100.00	96.20 ± 0.25	8.33
%false positive	0.00	3.81 ± 0.25	8.33
%false negative	6.91	6.29 ± 0.31	9.09
%sensitivity	92.86	93.33 ± 0.31	10.00
Cluster[b]			
%true positive	87.50	85.15 ± 0.51	13.89
%false positiv	12.50	14.85 ± 0.51	13.89
%false negative	28.57	29.15 ± 0.51	14.54
%sensitivity	71.43	70.56 ± 0.51	14.14

[a] Peak alignment within 10 min.
[b] Peak alignment within 20 min.

Table 15.7 Comparison of performance of *AutoDecon* and *Pulse* algorithms evaluated from simulated data based on *Pulse* with no basal secretion results

	Median	Mean ± SEM	Interquartile range
AutoDecon (P value = 0.0325)[a]			
%true positive	100.00	97.85 ± 0.17	5.26
%false positive	0.00	2.16 ± 0.17	5.26
%false negative	0.00	0.67 ± 0.09	0.00
%sensitivity	100.00	98.96 ± 0.11	0.00
Pulse with no basal secretion and half-life = 14.25 min[a]			
%true positive	93.54	93.19 ± 0.35	10.00
%false positive	6.46	6.81 ± 0.35	10.00
%false negative	23.08	21.36 ± 0.63	22.03
%sensitivity	76.47	78.35 ± 0.63	10.00

[a] Peak alignment within 10 min.

Table 15.8 Comparison of performance of *AutoDecon* and *Pulse* algorithms evaluated from simulated data based on *Pulse* with basal secretion results

	Median	Mean ± SEM	Interquartile range
AutoDecon (P value = 0.0325)[a]			
%true positive	100.00	94.51 ± 0.36	10.00
%false positive	0.00	5.49 ± 0.36	10.00
%false negative	0.00	1.46 ± 0.17	0.00
%sensitivity	100.00	98.13 ± 0.19	0.00
Pulse with basal secretion and half-life = 14.25 min[a]			
%true positive	80.00	76.33 ± 0.96	41.89
%false positive	20.00	23.67 ± 0.96	41.89
%false negative	14.29	16.49 ± 0.52	15.00
%sensitivity	86.62	83.17 ± 0.53	15.00

[a] Peak alignment within 10 min.

(see Table 15.10), the optimal *P* value is substantially lower, on the order of 0.01 or less. The origin of this difference is that the *AutoDecon* algorithm is identifying more, and smaller, secretion events not identified by the *Pulse2* algorithm. Thus, *AutoDecon* will have fewer small secretion events to identify when it is applied to simulated data based on the *Pulse2* analysis. Hence, *AutoDecon* will have a lower false-negative percentage with these data. Consequently, the optimal *P* value is dependent on characteristics of the particular experimental data being analyzed.

Table 15.9 Comparison of performance of *AutoDecon* and *Pulse2* algorithms evaluated from simulated data based on *Pulse2* with no basal secretion results

	Median	Mean ± SEM	Interquartile range
AutoDecon (P value = 0.0325)[a]			
%true positive	100.00	97.40 ± 0.21	6.25
%false positive	0.00	2.60 ± 0.21	6.25
%false negative	0.00	0.08 ± 0.03	0.00
%sensitivity	00.00	99.56 ± 0.08	0.00
Pulse2 with no basal secretion[a]			
%true positive	100.00	96.54 ± 0.32	7.14
%false positive	0.00	3.46 ± 0.32	7.14
%false negative	5.41	6.65 ± 0.45	10.00
%sensitivity	93.33	93.02 ± 0.45	10.00

[a] Peak alignment within 10 min.

Table 15.10 Comparison of performance of *AutoDecon* and *Pulse2* algorithms evaluated from simulated data based on *Pulse2* with basal secretion results

	Median	Mean ± SEM	Interquartile range
AutoDecon (P value = 0.01)[a]			
%true positive	100.00	98.91 ± 0.12	0.00
%false positive	0.00	1.09 ± 0.12	0.00
%false negative	0.00	0.28 ± 0.15	0.00
%sensitivity	100.00	99.44 ± 0.16	0.00
AutoDecon (P value = 0.02)[a]			
%true positive	100.00	98.06 ± 0.17	0.00
%false positive	0.00	1.94 ± 0.17	0.00
%false negative	0.00	0.28 ± 0.15	0.00
%sensitivity	100.00	99.44 ± 0.16	0.00
AutoDecon (P value = 0.0325)[a]			
%true positive	100.00	96.98 ± 0.20	6.67
%false positive	0.00	3.02 ± 0.20	6.67
%false negative	0.00	0.24 ± 0.15	0.00
%sensitivity	100.00	99.47 ± 0.16	0.00
Pulse2 with basal secretion[a]			
%true positive	100.00	97.50 ± 0.21	5.56
%false positive	0.00	2.50 ± 0.21	5.56
%false negative	0.00	5.64 ± 0.35	8.71
%sensitivity	94.44	94.08 ± 0.35	9.09

[a] Peak alignments within 10 min.

3.4. Optimal sampling paradigms for *AutoDecon*

The general idea of examining a large number of simulated hormone concentration time series may also be employed to address questions such as: What if the samples were assayed in singlicate or triplicate instead of duplicate? Would 5-min sampled singlicates be better than 10-min sampled duplicates or 15-min sampled triplicates? What would happen if an older instrument with a larger measurement uncertainty (i.e., MDC and CV) was used for the assays? The easiest way to answer these questions is to simulate the appropriately corresponding hormone concentration time series and then determine how the particular analysis algorithm will perform with those series of data.

Table 15.11 addresses the question of singlicate vs duplicate vs triplicate assays. The data series samples were simulated as in Table 15.4 with the exception that three groups of 500 hormone concentration time series were simulated with singlicate, duplicate, and triplicate assays, respectively. It is, of course, expected that the results will improve as the number of replicates is increased. Based on the analysis of these simulated time series, it appears that the major effect is that the false-negative percentage improves as the number of replicates increases. Virtually every secretion event identified by

Table 15.11 Comparison of performance of the *AutoDecon*[a] algorithm when samples are simulated in singlicate, duplicate, and triplicate

	Median	Mean ± SEM	Interquartile range
Singlicate samples			
%true positive	100.00	97.76 ± 0.18	5.26
%false positive	0.00	2.24 ± 0.18	5.26
%false negative	0.00	3.27 ± 0.23	5.88
%sensitivity	00.00	96.06 ± 0.24	6.67
Duplicate samples			
%true positive	100.00	97.96 ± 0.16	4.76
%false positive	0.00	2.04 ± 0.16	4.76
%false negative	0.00	2.13 ± 0.24	0.00
%sensitivity	100.00	97.32 ± 0.25	5.26
Triplicate samples			
%true positive	100.00	97.56 ± 0.19	5.26
%false positive	0.00	2.44 ± 0.19	5.26
%false negative	0.00	2.06 ± 0.23	0.00
%sensitivity	100.00	97.47 ± 0.24	5.00

[a] Peak alignment within 10 min, *P* value = 0.0325.

Table 15.12 Performance comparison of the *AutoDecon* algorithm when samples are simulated in 5-min sampled singlicates, 10-min sampled duplicates, etc.

	Median	Mean ± SEM	Interquartile range
5-min singlicate samples[a]			
%true positive	100.00	97.44 ± 0.31	4.76
%false positive	0.00	2.36 ± 0.24	4.76
%false negative	0.00	2.84 ± 0.30	5.26
%sensitivity	100.00	96.89 ± 0.30	5.56
10-min duplicate samples[a]			
%true positive	100.00	97.96 ± 0.16	4.76
%false positive	0.00	2.04 ± 0.16	4.76
%false negative	0.00	2.13 ± 0.24	0.00
%sensitivity	100.00	97.32 ± 0.25	5.26
15-min triplicate samples[a]			
%true positive	100.00	97.26 ± 0.19	5.88
%false positive	0.00	2.74 ± 0.19	5.88
%false negative	0.00	5.36 ± 0.40	6.67
%sensitivity	97.92	93.94 ± 0.41	9.52
20-min quadruplicate samples[a]			
%true positive	94.12	94.24 ± 0.30	8.33
%false positive	5.88	5.76 ± 0.30	8.33
%false negative	7.70	12.12 ± 0.57	17.65
%sensitivity	90.69	87.20 ± 0.57	13.99

[a] Peak alignment within 10 min, *P* value = 0.0325.

the *AutoDecon* algorithm is an actual secretion event, and thus there is little room for improvement in the false-positive percentages.

Table 15.12 addresses the question of 5-min sampled singlicates vs 10-min sampled duplicates vs 15-min sampled triplicates. In this case, each of the groups incurs the same expense in terms of assay samples and thus addresses the question of the effects of the sampling interval at a constant cost for the assays. Again, the samples were simulated as in Table 15.4 with the exception that the groups of 500 hormone concentration time series were simulated with 5-min sampled singlicates, 10-min sampled duplicates, etc. The expectation was that the results would improve as the interval between the data points decreases, as this provides a better temporal resolution for the secretion events. Based on the analysis of these simulated time series, it is apparent that the false-negative and true-positive rates both improve as the sampling interval decreases.

Table 15.13 presents an analogous simulation except that the experimental measurement errors are simulated at three different levels: (1) the level observed in our experiments, (2) the level that the manufacturer claims for the Immulite 2000 instrumentation, and (3) the level that the manufacturer claims for the Immulite 1000 instrument. These measurement uncertainty levels represent the levels typical of the currently available instrumentation. Results from these simulations demonstrate that the true positive (%) and false positive (%) do not degrade rapidly as the measurement uncertainty increases. However, the percent false negative and percent sensitivity do diminish significantly with increasing measurement uncertainties.

4. Discussion

The process of validating *AutoDecon* or any other data analysis software involves analyzing a large number of data sets where the expected results are known a priori. Given data sets with known results, it is simple to calculate

Table 15.13 Comparison of performance of the *AutoDecon*[a] algorithm when samples are simulated with different measurement uncertainties

	Median	Mean ± SEM	Interquartile range
Our observed noise level MDC = 0.00661 and CV = 1.644%			
%true positive	100.00	97.96 ± 0.16	4.76
%false positive	0.00	2.04 ± 0.16	4.76
%false negative	0.00	2.13 ± 0.24	0.00
%sensitivity	100.00	97.32 ± 0.25	5.26
Immulite 2000 reported noise level MDC = 0.01 and CV = 3%			
%true positive	100.00	97.80 ± 0.17	5.26
%false positive	0.00	2.20 ± 0.17	5.26
%false negative	0.00	4.37 ± 0.29	6.67
%sensitivity	100.00	95.24 ± 0.30	7.14
Immulite 1000 reported noise level MDC = 0.01 and CV = 6%			
%true positive	100.00	97.25 ± 0.21	5.56
%false positive	0.00	2.76 ± 0.21	5.56
%false negative	5.88	7.03 ± 0.24	11.11
%sensitivity	93.93	92.55 ± 0.39	11.88

[a] Peak alignment within 10 min, *P* value = 0.0325.

how well the software and algorithms perform with the specific data sets. There are basically two methods of obtaining hormone concentration time series data where the answers are known: an experimental approach and a mathematical approach. The *experimental approach* involves manipulation of the biological system such that the hormone pulses are secreted at specific known times and with specific known amounts. This manipulation can theoretically be achieved by administering drugs or hormones at specific times that will induce and/or inhibit the secretion or elimination of the hormone of interest. Obviously, this experimental approach is extremely difficult, if not impossible, to control and achieve while also being time-consuming and expensive. The alternative *mathematical approach* is to simulate a large number of synthetic data sets with assumed (i.e., known) values of the model parameters and realistic amounts of random experimental measurement errors (Guardabasso et al., 1988; Van Cauter, 1988; Veldhuis and Johnson, 1988). The more realistic the *simulations*, the more realistic the *conclusions* about the operating characteristics of the software and algorithms can be.

Within this context, a particularly difficult set of GH data was chosen to test the *AutoDecon* algorithm. The size of the observed GH secretion events is small because the subjects are neither young nor lean. In addition, resolution of an approximately 14-min elimination half-life from data sampled at 10-min intervals is challenging. However, because the hormone concentrations were measured in duplicate with state-of-the-art instrumentation, data from this protocol are typical of the best GH concentration time series data currently available.

The best method for validation of the software is to try all possibilities for all of the variables involved in the simulation. Thus, the variables required for these simulations include the number of replicate assays, the sampling interval, the *MDC*, the *CV*, the mean interpulse interval, the standard deviation of the mean interpulse interval, the mean secretion event amplitude and its standard deviation, the *SecretionSD*, and the *HL*. However, it is clearly impossible to test all possible combinations and correlations for all of these variables. Indeed, generalizing these operating characteristics to all possible data sets cannot be achieved except in the simplest of cases simply due to the large number of potential variables involved. The more realistic approach is to create synthetic data that mimic the actual hormone concentration time series obtained by a specific experimental protocol and group of subjects. The resulting operating characteristics can then be applied to that specific hormone, experimental protocol, and its subjects.

Key to the simulation approach is the modeling of actual experimental data in the most rigorously precise way possible. Figures 15.3–15.6 clearly demonstrate that neither the interpulse intervals nor the secretion event areas can be described by Gaussian distributions. However, the interpulse intervals and the secretion event sizes are commonly modeled and reported in the literature as a Gaussian distribution (i.e., simply as a mean and standard deviation). If modeled this way, the actual number of secretion events will be underestimated significantly (because the distribution of intervals between secretion events will be overestimated) and the distribution of sizes of secretion events will be overestimated. Simulations with too few excessively large secretion peaks will be much easier to analyze and will, in all likelihood, falsely inflate the operating characteristics of the algorithm. This will, as a consequence, lead to an overestimation of the significance of experimental results.

This chapter modeled the lognormal distributions of interpulse intervals and secretion event areas by multiple linear regressions in order to generate data sets with known answers that closely resemble the original data set. The multiple linear regression approach is excellent for this purpose because it emulates the distribution and sizes of the measurable secretion events and simultaneously allows for the coupling of these to each other and other variables such as basal secretion, half-life, and a 24-h rhythm. This multiple linear regression model is arbitrary and thus provides only limited information about the underlying molecular mechanisms of GH secretion. It does, however, provide an excellent basis to generate simulated hormone concentration time series data (with known answers) that closely mimic actual experimental data.

Clearly, the *AutoDecon* algorithm outlined in this work performs significantly better than several of the commonly used algorithms when applied to the analysis of simulated GH concentration time series data (see Tables 15.2 and 15.5–15.10). Use of the *AutoDecon* algorithm is not subjective, as it is a totally automated algorithm. The user need only specify approximate values for the *SecretionSD* and the one- or two-component half-live(s). The algorithm then adjusts these parameters automatically and also simultaneously finds the locations and amplitudes of the secretion events that have the highest probability of being correct. Moreover, *AutoDecon* is not overly sensitive to the magnitude of the experimental measurement errors (see Table 15.13), and, as a consequence, to the required number of replicate measurements (see Table 15.12). Thus, within reasonable bounds, the choice of the frequency of data collection and required number of replicate assays is a cost-value judgment for the investigator.

This analysis reveals some interesting aspects about the mechanisms controlling the secretion of GH. The analysis presented here suggests that many more secretion events are present within the data series than previously thought and that these secretion events have a wide variation in size. An argument can be made that the numerous small secretion events are not clinically relevant and thus should be ignored. However, if the goal is to understand the molecular and physiological mechanisms that lead to the secretion and elimination of GH, then the accurate quantification of these small secretion events is critically important.

These results indicate that approximately 6% of the total GH secreted occurs in a basal, or constant, nonpulsatile mode. However, Fig. 15.2 shows small secretion events that are not considered as actual secretion events because they are not statistically significant at a 0.05 level. In all likelihood there are many of these small events that are not identified as distinct secretion events by the *AutoDecon,* or any other algorithm, which may in fact be secretion events that are simply too small to rise to the level of statistical significance. Under such circumstances, these small events are lumped into the apparent basal secretion. Thus, the 6% basal secretion should be considered as an upper limit for the basal secretion, which might in fact be substantially lower. Accompanying this, the estimate of approximately 18 secretion events per 24-h period should be considered as a lower bound for the actual number of secretion events.

It is tempting to conclude that the asymmetry of the lognormal distribution of interpulse intervals, as depicted in Fig. 15.4, is a consequence of the inability of the algorithm to detect small secretion events. These missed events would yield overestimates of the interval and, when averaged over the entire data set, may be the cause of the skewed distribution. This could change the shape of the distribution to an unknown extent. It is clear that with data sampled at a 10-min interval it will become increasingly difficult to resolve interpulse intervals as the interpulse intervals decrease to this 10-min limit. Thus, it is expected that the observed distribution of interpulse intervals will be underestimated for short interpulse intervals and similarly artificially high for the longer interpulse intervals. The bimodal (or multimodal) nature of the secretion event areas (Fig. 15.6) will also be affected by the probable existence of multiple small secretion events that are below the limit of detectability.

In summary, *AutoDecon* is a novel, user-friendly deconvolution method that provides both an objective approach to initial secretory burst selection and a statistically based verification of candidate secretory bursts. When applied to synthetic GH concentration time series, *AutoDecon* performs substantially better than a number of alternative pulse detection algorithms in terms of optimizing the detection of true-positive secretory events while at the same time minimizing the detection of false-positive and false-negative events. Although *AutoDecon* will require validation with regard to application to other hormonal systems, it would seem to hold substantial promise as a biomathematical tool with which to identify and characterize a variety of pulsatile hormonal signals.

The *Concordance* and *AutoDecon* algorithms are part of our hormone pulsatility analysis suite. They can be downloaded from http://www.mljohnson.pharm,virginia.edu/pulse_xp/.

ACKNOWLEDGMENTS

This work was supported in part by NIH Grants RR-00847 to the General Clinical Research Center at the University of Virginia, U54 HD28934, R01 RR019991, R25 DK064122, R21 DK072095, P30 DK063609, R01 DK076037, and R01 DK51562.

REFERENCES

Bevington, P. R. (1969). "Data Reduction and Error Analysis for the Physical Sciences." McGraw Hill, New York.

Dimaraki, E. V., Jaffe, C. A., Demott-Friberg, R., Russell-Aulet, M., Bowers, C. Y., Marbach, P., and Barkan, A. L. (2001). Generation of growth hormone pulsatility in women: Evidence against somatostatin withdrawal as pulse initiator. *Am. J. Physiol. Endocrinol. Metab.* **280**, E489–E495.

Evans, W. S., Sollenberger, M. J., Booth, R. A., Rogol, A. D., Urban, R. J., Carlsen, E. C., Johnson, M. L., and Veldhuis, J. D., (1992). Contemporary aspects of discrete peak-detection algorithms: II. The paradigm of the luteinizing hormone pulse signal in women. *Endocrine Rev.* **13**, 81–104.

Guardabasso, V., De Nicolao, G., Rocchetti, M., and Rodbard, D. (1988). Evaluation of pulse-detection algorithms by computer simulation of hormone secretion. *Am. J. Physiol.* **255**, E775–E783.

Haisenleder, D., Dalkin, A., and Marshall, J. (1994). Regulation of gonadotropin gene expression. *Phys. Reprod. Edition.* **2**, 1793.

Hartman, M. L., Faria, A. C., Vance, M. L., Johnson, M. L., Thorner, M. O., and Veldhuis, J. D. (1991). Temporal structure of *in vivo* growth hormone secretory events in humans. *Am. J. Physiol.* **260**, E101–E110.

Jansson, P. A. (1984). "Deconvolution with Applications in Spectroscopy." p. 22. Academic Press, New York.

Johnson, M. L., and Straume, M. (1999). Innovative quantitative neuroendocrine techniques, in (sex-steroid interactions with growth hormone). Serona Symposia. pp. 318–326.

Johnson, M. L., and Veldhuis, J. D. (1995). Evolution of deconvolution analysis as a hormone pulse detection algorithm. *Methods Neurosci.* **28**, 1–24.

Johnson, M. L., Virostko, A., Veldhuis, J. D., and Evans, W. S. (2004). Deconvolution analysis as a hormone pulse-detection algorithm. *Methods Enzymol.* **384**, 40–54.

Johnson, M. L., Pipes, L., Veldhuis, P. P., Farhi, L. S., Boyd, D., and Evans, W. S., (2008). Validation of *AutoDecon*, a deconvolution algorithm for identification and characterization of luteinizing hormone. *Anal. Biochem.* **381**, 8–17.

Lakowicz, J. R. (1983). "Principles of Fluorescence Spectroscopy." Plenum Press, New York.

Maheshwari, H. G., Pezzoli, S. S., Rahim, A., Shalet, S. M., Thorner, M. O., and Baumann, G. (2002). Pulsatile growth hormone secretion persists in genetic growth hormone-releasing hormone resistance. *Am. J. Physiol. Endocrinol. Metab.* **282**, E943–E951.

Merriam, G. R., and Wachter, K. W. (1982). Algorithms for the study of episodic hormone secretion. *Am. J. Physiol.* **243**, E310–E318.

Nakai, Y., Plant, T. M., Hess, D. L., Keogh, E. J., and Knobil, E. (1978). On the sites of negative and positive feedback actions of estradiol in the control of gonadotropin secretion in the rhesus monkey (*Macaca mulatto*). *Endocrinology.* **102**, 1015–1018.

Nelder, J. A., and Mead, R. (1965). A Simplex method for function minimization. *Comput. J.* **7**, 308–313.

Press, W. H., Flannery, B. P., Teukolsky, S. A., and Vettering, W. T. (1986). "Numerical Recipes: The Art of Scientific Computing." Cambridge Univ. Press, Cambridge.

Oerter, K. E., Guardabasso, V., and Rodbard, D. (1986). Detection and characterization of peaks and estimation of instantaneous secretory rate for episodic pulsatile hormone secretion. *Comp. Biomed. Res.* **19,** 170–191.

Santen, R. J., and Bardin, C. W. (1973). Episodic luteinizing hormone secretion in man: Pulse analysis, clinical interpretation, physiologic mechanisms. *J. Clin. Invest.* **52,** 2617–2628.

Straume, M., Frasier-Cadoret, S. G., and Johnson, M. L. (1991). Least-squares analysis of fluorescence data. *Top. Fluorescence Spectrosc.* **2,** 171–240.

Thorner, M. O., Chapman, I. M., Gaylinn, B. D., Pezzoli, S. S., and Hartman, M. L. (1997). Growth hormone-releasing hormone and growth hormone-releasing peptide as therapeutic agents to enhance growth hormone secretion in disease and aging. *Recent Prog. Horm. Res.* **52,** 215–244.

Urban, R. J., Evans W. S., Rogol A. D., Kaiser, D. L., Johnson M. L., and Veldhuis J. D., (1988). Contemporary aspects of discrete peak detection algorithms: I. The paradigm of the luteinizing hormone pulse signal in men. *Endocrine Rev.* **9,** 3–37.

Van Cauter, E. (1988). Estimating false-positive and false-negative errors in analysis of hormonal pulsatility. *Am. J. Physiol.* **254,** E786–E794.

Van Cauter, E., Plat, L., and Copinschi, G. (1998). Interrelations between sleep and the somatotropic axis. *Sleep.* **21,** 553–566.

Vance, M. L., Kaiser, D. L., Evans, W. S., Furlanetto, R., Vale, W., Rivier, J., and Thorner, M. O. (1985). Pulsatile growth hormone secretion in normal man during a continuous 24-hour infusion of human growth hormone releasing factor (1-40): Evidence for intermittent somatostatin secretion. *J. Clin. Invest.* **75,** 1584–1590.

Veldhuis, J. D., Carlson, M. L., and Johnson, M. L. (1987). The pituitary gland secretes in bursts: Appraising the nature of glandular secretory impulses by simultaneous multiple-parameter deconvolution of plasma hormone concentrations. *Proc. Natl. Acad. Sci. USA* **84,** 7686–7690.

Veldhuis, J. D., and Johnson, M. L. (1986). *Cluster* analysis: A simple, versatile, and robust algorithm for endocrine pulse detection. *Am. J. Physiol.* **250,** E486–E493.

Veldhuis, J., and Johnson, M. L. (1988). A novel general biophysical model for simulating episodic endocrine gland signaling. *Am. J. Physiol.* **255,** E749–E759.

Webb, C. B., Vance, M. L., Thorner, M. O., Perisutti, G., Thominet, J., Rivier, J., Vale, W., and Frohman, L. A. (1984). Plasma growth hormone responses to constant infusions of human pancreatic growth hormone releasing factor: Intermittent secretion or response attenuation. *J. Clin. Invest.* **74,** 96–103.

CHAPTER SIXTEEN

Modeling Fatigue over Sleep Deprivation, Circadian Rhythm, and Caffeine with a Minimal Performance Inhibitor Model

Patrick L. Benitez,* Gary H. Kamimori,[†] Thomas J. Balkin,[†] Alexander Greene,[‡] and Michael L. Johnson*

Contents

1. Introduction	406
2. Methods	407
2.1. The Walter Reed Army Institute of Research Stay Alert caffeine gum study	407
2.2. Data processing	408
2.3. Caffeine modeling	409
2.4. Performance inhibitor model	410
2.5. The minimal biomathematical model	412
2.6. Least-squares parameter estimation	412
3. Results	412
4. Discussion	414
Acknowledgments	419
References	419

Abstract

Sleep loss, as well as concomitant fatigue and risk, is ubiquitous in today's fast-paced society. A biomathematical model that succeeds in describing performance during extended wakefulness would have practical utility in operational environments and could help elucidate the physiological basis of sleep loss effects. Eighteen subjects (14 males, 4 females; age 25.8 ± 4.3 years) with low levels of habitual caffeine consumption (<300 mg/day) participated. On night 1, subjects slept for 8 h (2300–0700 h), followed by 77 h of continuous

---

* Departments of Pharmacology and Medicine, University of Virginia Health System, Charlottesville, VA
[†] Department of Behavioral Biology, Walter Reed Army Institute of Research, Division of Neuroscience, Silver Spring, Maryland
[‡] School of Medicine, University of Florida, Gainesville, Florida

Methods in Enzymology, Volume 454   © 2009 Elsevier Inc.
ISSN 0076-6879, DOI: 10.1016/S0076-6879(08)03816-0   All rights reserved.

405

wakefulness. They were assigned randomly to receive placebo or caffeine (200 mg, i.e., two sticks of Stay Alert gum) at 0100, 0300, 0500, and 0700 during nights 2, 3, and 4. The psychomotor vigilance test (PVT) was administered periodically over the 77-h period of continuous wakefulness. Statistical analysis reveals lognormality in each PVT, allowing for closed-form median calculation. An iterative parameter estimation algorithm, which takes advantage of MatLab's (R2007a) least-squares nonlinear regression, is used to estimate model parameters from subjects' PVT medians over time awake. In the model, daily periodicity is accounted for with a four-component Fourier series, and a simplified binding function describes asymptotic fatigue. The model highlights patterns in data that suggest (1) the presence of a performance inhibitor that increases and saturates over the period of continuous wakefulness, (2) competitive inhibition of this inhibitor by caffeine, (3) the persistence of an internally driven circadian rhythm of alertness, and (4) a multiplicative relationship between circadian rhythm and performance inhibition. The present inhibitor-based minimal model describes performance data in a manner consistent with known biochemical processes.

1. Introduction

In industrialized society, sleep loss has become a fact of life (Rajaratnam and Arendt, 2001). Ever more common for a wider swath of individuals, schedules that preclude adequate nightly sleep deleteriously impact alertness, performance, and quality of life. In certain operational (e.g., military and transportation) environments, the consequences of inadequate sleep can be especially dire. In such scenarios, the calculus involves weighing the need for pilots, firefighters, soldiers, and so on to operate during nighttime for sustained periods or under other conditions that prevent adequate sleep versus the increased risk of sleep loss–induced lapses in performance (Johnson *et al.*, 2007; Kryger *et al.*, 2000; Petrie and Dawson, 1997). Such lapses pose a threat to operator health and safety, odds of mission success, and so on.

Traditional biomathematical homeostatic models of sleep deprivation predict performance in terms of simple forms such as linear or exponential decrements and daily sinusoids (Achermann, 2004). Such homeostatic models assume a "performance capacity reservoir" that is depleted during wakefulness and replenished, at a rate proportional to sleep need, during subsequent sleep (Achermann, 2004; Akerstedt and Folkard, 1997). However, such models fail to accommodate data from acute sleep deprivation studies, which often reveal asymptotic and/or nonsinusoidal behavior (Kamimori *et al.*, 2006). Furthermore, such models lack a clear physiological basis, with the "performance capacity reservoir" remaining a hypothetical construct.

This chapter presents a biochemically relevant model that adequately describes the variation in performance data over a 77-h period of wakefulness. In this model, sleep deprivation-related fatigue (i.e., increased reaction time relative to baseline) occurs as a function of two processes: saturation of a performance inhibitor and a driven 24-h (circadian) rhythm. This model can be extended to account for the effects of caffeine, which decreases PVT reaction time for sleep-deprived subjects (Rupp et al., 2007), by assuming that the drug competitively inhibits the performance inhibitor (Boutrel, 2004). In this vein, the present "performance inhibitor"-based model constitutes the first attempt to describe, in physiological terms, the effects of pharmacological fatigue countermeasures in behavioral studies.

2. Methods

2.1. The Walter Reed Army Institute of Research Stay Alert caffeine gum study

Data were obtained from the Walter Reed Army Institute of Research (WRAIR) study on the efficacy of Stay Alert caffeine gum over a period of continuous (77 h) wakefulness (Kamimori et al., 2006). The 18 subjects, all mild or noncaffeine users (<300 mg/day), included 14 ($N = 14$) males and 4 ($N = 4$) females, none of whom used hormonal contraceptives. Subjects gave informed consent, and officials confirmed their health status with a review of medical history, routine laboratory tests, and a physical examination. The Human Subjects Research Review Board of the Office of the Surgeon General of the Army approved the protocol for the study. Study administrators provided standardized meals and sleep was encouraged from 2300 to 0700 h on the night prior to sleep deprivation. Water was provided *ad libitum*.

For the 77-h period of wakefulness, the protocol called for assigning the subjects randomly into two groups: caffeine and placebo. In a double-blind fashion, the caffeine group received the Stay Alert caffeine gum (two sticks = 200 mg caffeine) four times per night (at 0100, 0300, 0500, and 0700 h), while the placebo subjects received Amurol confectioners gum at those times. Over the 3 days, subjects undertook a 5-min psychomotor vigilance test (PVT) three times between 2400 and 0100 h, nine times every 2 h between 0100 and 0900 h, and once every 2 h between 0900 and 2400 h. For reference, initial time awake took place at 700 h, and the initial PVT battery took place at 0940 h, both on the first day.

As an assessment of performance, the PVT has several advantages over other tests (Kaida et al., 2007). The test itself measures the time it takes for a subject to push a button after presentation of a visual stimulus. Thus, learning curve and cognitive bias effects are minimized, increasing the signal-to-noise ratio (SNR), reducing systemic error, and making aggregate analysis (for the purposes of population parameter estimation) possible.

2.2. Data processing

Psychomotor vigilance test T data that fall into three categories thought to describe behavior other than canonical PVT reactions are excluded from analysis. First, PVT error events (such as anticipation, wrong key, or false start) are omitted. Second, tests with reaction times under 180 ms are omitted (i.e., considered too fast to represent actual reactions to the stimuli). Third, tests with reaction times over 2000 ms are omitted because independent studies have suggested that these PVT include periods of microsleep (Moller et al., 2006).

After filtering, reaction time data from each PVT battery are summarized on a distribution basis (Adam et al., 2006). As in Fig. 16.1, the distribution of reaction times for each battery is found to be threshold lognormal. This distribution can be described as Gaussian with mean μ and standard deviation σ under the shift θ (the threshold) and then a natural logarithm transform as in Eq. (16.1),

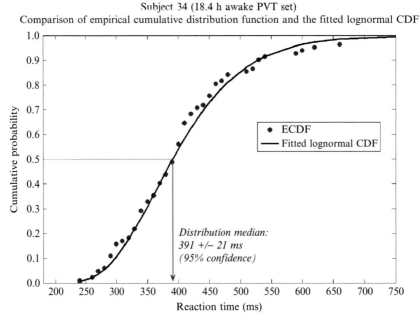

Figure 16.1 A graphical example showing a typical PVT set's empirical cumulative distribution function (ECDF), fitted lognormal CDF, and distribution median calculation. PVT data are shown to be lognormal (with a threshold) under several statistical tests. Using a nonlinear least-squares method, a lognormal CDF is fitted to data, which gives rise to a distribution median estimate. Confidence on the median is calculated with a Monte Carlo simulation.

$$\ln(\mathrm{RT} - \theta) \equiv \mathrm{Gaussian}(\mu, \sigma). \qquad (16.1)$$

Three statistical tests support this finding: (1) cumulative distribution function (CDF) comparison, (2) linearity in the threshold lognormal plot, and (3) Kolmogorov–Smirnov test. Given the distribution as described in Eq. (16.1), parameters θ, μ, and σ are calculated using a CDF-fitting method that took advantage of MatLab's (R2007a) nonlinear regression algorithm. With parameter estimates, a distribution median can be calculated,

$$\mathrm{median} = \theta + \exp(\mu), \qquad (16.2)$$

as well as its confidence interval (using a Monte Carlo simulation). This statistic is a robust measure of the center of the skewed distribution and is much more stable than the open-form calculation of the median, as it is resistant to the individual values in the center of the distribution. Figure 16.1 depicts a graphical representation of the distribution fitting and median estimation.

Finally, the distribution medians are shifted on a subject-by-subject basis so that the initial distribution's median (from the battery that took place at 0940 h on the first day) is 0 ms. Subtracting the baseline from each subject shifts data in a way that, to an extent, removes subject-to-subject variation and allows the medians from all subjects to be treated as one data set. Performing a regression on the aggregate data increases the SNR and yields reliable parameters for the population of subjects.

2.3. Caffeine modeling

In addition to time awake, a normalized caffeine concentration is included as an independent variable. (Placebo subjects always have a caffeine norm of zero.) Caffeine concentration is normalized so that one 200-mg dose resulted in 1 unit of normalized concentration. In Fig. 16.5, the solid line shows the caffeine norm over the period of wakefulness for caffeine subjects. In this method, it is assumed that body volumes and mastication/extraction rates are approximately equal across subjects.

The concentration of caffeine in the body is modeled using a simplified convolution integral of an input and an elimination process (Johnson and Straume, 1999; Johnson and Veldhuis, 1995; Syed et al., 2005 Veldhuis et al., 1987). The input process is approximated as an instantaneous dose_i at time T_i (the time of mastication), which allows the unbounded integral to be written as a sum of doses, one to $N = 12$. In this summation form, a Heaviside function H "turns on" dose_i at its mastication time T_i:

$$[C] = \sum_{d=1}^{N} H\{t - T_d\} \cdot 2^{-(t-T_d)/\lambda}. \qquad (16.3)$$

After time T_i, the elimination process is projected as a single-compartment exponential decay with half-life $\lambda = 8.25$ h and an offset so that the concentration due to the dose is unity at time T_i.

2.4. Performance inhibitor model

Whereas a performance reservoir that decrements over time awake is assumed in the classic homeostatic model, the assumption of a performance inhibitor that increases over time awake is central to the model described here. The proposed model is a more useful description of data since it is derived from physiologically relevant and theoretical concepts. The model includes two components: a binding equation and a circadian modulator.

The binding equation at the core of the model is derived from the equilibrium statement of free performance inhibitor X, unbound receptor R, and receptor–inhibitor complex RX, as well as a conservation statement on the receptor:

$$k_{eq} = \frac{[RX]}{[R][X]} \qquad (16.4)$$

$$[R_0] = [R] + [RX]. \qquad (16.5)$$

The total receptor concentration $[R_0]$, equilibrium quotient k_{eq}, and body volumes are constant. From here a binding equation can be derived, one that calculates the concentration of the inhibitor–receptor complex as a function of constants and the concentration of the inhibitor itself:

$$[R_X] = [R_0] \frac{k_{eq}[X]}{1 + k_{eq}[X]}. \qquad (16.6)$$

To extend the binding equation to include caffeine effects, it was assumed that caffeine acts as a competitive inhibitor of the performance inhibitor. Under this minimalist assumption, the binding equation includes a new term in the denominator—the normalized caffeine concentration $[C]$ and its coefficient k_c. The estimated coefficient reflects a combination of dose amount, body volume, and relative affinity between caffeine and the performance inhibitor:

$$[RX] = [R_0] \frac{k_{eq}[X]}{1 + k_{eq}[X] + k_c[C]}. \qquad (16.7)$$

To make the binding function relevant to data, it is necessary to express the concentrations' relation to reaction time and time awake. For the purpose of simplicity, it is assumed that the concentration of the performance inhibitor is proportional to time awake, with constant α, and that relative reaction time is proportional to the concentration of bound inhibitor, with constant β:

$$t_{reaction} = \alpha[RX] \qquad (16.8)$$

$$[X] = \beta t_{wake}. \qquad (16.9)$$

While the actual relationships between concentration and the physical analog may be more complex, the assumption of direct proportionality is the minimum required for the present model.

The model also includes a factor to allow for circadian modulation in reaction time (Dijk and Czeisler, 1994). This factor is a Fourier series to the fourth daily harmonic plus one, which allows for a nonperiodic, saturating time trend.

$$CircadianFactor = 1 + \sum_{i=1}^{4} \left(A_i \sin\left[\frac{2\pi \cdot i \cdot t_{wake}}{24}\right] + B_i \cos\left[\frac{2\pi \cdot i \cdot t_{wake}}{24}\right] \right)$$

$$(16.10)$$

Within this factor, the magnitude of the periodic component is normalized to the magnitude of the time trend. The Fourier series, while more complicated than a one-component sinusoidal oscillation, allows for the model to account for a greater variety of periodic patterns in data. Generalizing the form of the circadian oscillation allows the model to sift through the periodicity of data to better describe any potential evidence of a performance inhibitor's action. The Fourier series, due to the mutual orthogonality of the components, also allows the circadian oscillation's form to arise from data rather than to be imposed upon them.

The model multiplies the circadian factor by the binding equation and, for purposes of parameter estimation, combines several multiplicative constants. A baseline modifier is added to the model.

2.5. The minimal biomathematical model

$$t_{reaction} = K \frac{k_t t_{awake}}{1 + k_t t_{awake} + k_c [C]} \cdot CircadianFactor + RT_0 \quad (16.11)$$

Table 16.1 contains estimates for the aforementioned equation's parameters. Regarding the *CircadianFactor*, parameters A_i and B_i, i range [1, 4] are also estimated simultaneously.

2.6. Least-squares parameter estimation

Model parameter estimations were performed using MatLab's nonlinear least-squares regression method, a Gauss–Newton algorithm. Medians were weighted to account for the fact that some represented more individual PVT than others, but no weighting function is used in the regression. The reported parameter errors were estimated by taking the square root of the diagonal of the covariance matrix, that is, the asymptotic standard errors. Table 16.1 presents the parameters of the performance inhibitor model.

To enhance the stability of the nonlinear regression process (which can depend on the initial guess values of the parameters) the individual parameters were added one at a time to the model with the following algorithm:

1. Holding previously estimated coefficients constant, the new parameter was estimated.
2. Utilizing this estimate and the previous estimates as initial guesses, a simultaneous parameter estimation was performed.

Estimates were made in the order indicated in Table 16.1, with the initial guesses for K and RT_0 being derived from the coefficients of the linear regression with respect to time awake.

3. Results

Data plotted in Figs. 16.2 and 16.3 are reaction time distribution medians, normalized so that the first (in terms of time awake) median for each subject is zero (Adam *et al.*, 2006). The scatter plots in Fig. 16.2 and 16.3

Table 16.1 Parameter estimates with asymptotic standard error for the simultaneous fit of placebo and caffeine data

Parameter	Estimate	Error
K	230.9	1.4
RT_0	−51.7	1.0
K_t	0.0264	0.0006
K_c	0.356	0.004
A_1	−0.0474	0.0034
B_1	0.384	0.005
A_2	−0.0990	0.0034
B_2	0.0811	0.0031
A_3	−2.94E-03	2.89E-03
B_3	−1.43E-03	3.22E-03
A_4	0.0277	0.0026
B_4	0.0287	0.0026

are placebo and caffeine data, respectively; the fitted model is also shown. This analysis includes both caffeine and placebo data in a simultaneous, nonlinear least-squares parameter estimation.

The performance inhibitor model provides a sound description of caffeine and placebo data for the study. The model adequately characterizes the asymptotic (rectangular hyperbola) behavior (see Fig. 16.2) in the data: as the inhibitor saturates, it can only increase the concentration of bound receptor to the point where (almost) all receptors are bound. For comparison, a model assuming linear increments in reaction time would predict an interminable increase in reaction time, a behavior not supported by data. The inhibitor-based model, along with the multiplicative circadian modulation, provides an improved description of the data over the period of acute sleep deprivation.

Apart from the saturation component, the generalized circadian modulation estimate provides some insight into how the rhythm relates to the time trend. Data support a multiplicative interaction—the fatigue response to sleep deprivation depends on the time of day, as well as concentration of the inhibitor–receptor complex. In Fig. 16.2 the asymptotic increase in *amplitude* of the periodic function results from this multiplier effect. Moreover, data reject the relationship in which the concentration of performance inhibitor itself fluctuates in a 24-h cycle. This would predict attenuation in the circadian rhythm with regards to reaction time, that is, the rhythm would eventually disappear as the concentration of inhibitor approached saturation. As Fig. 16.2 demonstrates, the rhythm increases before stabilizing.

Using this minimal performance inhibitor model for sleep deprivation allows a natural segue into modeling the effect of caffeine. Figure 16.3 shows that assuming caffeine to be a competitive inhibitor of the performance inhibitor sufficiently describes data. In fact, the model does a superior

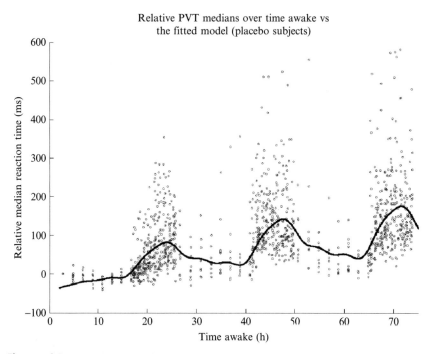

Figure 16.2 Plot of the placebo subjects' relative medians over time awake, along with the projected relative median reaction times according to the fitted model with a caffeine concentration of zero. Zero "relative reaction time" is defined as the subject's median reaction time for the first PVT set in terms of hours awake. In other words, relative reaction time measures the change in reaction time since the first PVT set. Overall, the model, which assumes a multiplicative relationship between circadian rhythm, succeeds in describing data. Figures 16.2 and 16.3 are less accurate for times awake under 5 h, perhaps due to sleep lag effects (Akerstedt and Folkard, 1997).

job describing caffeine data, which could possibly be attributed to the fact that the caffeine set contains fewer lapses. Furthermore, caffeine data confirm features found in placebo data. The amplitude of the circadian component on the first evening is much lower than its placebo counterpart. This finding makes sense in light of the multiplicative relationship and competitive inhibition. At this point in the study, the caffeine concentration sufficiently blocks binding of the performance inhibitor to the extent that circadian rhythm is suppressed.

4. Discussion

Results indicate a performance inhibitor that increases over time awake may actually underlie sleep loss mediated performance decrements. Viewed in this manner, caffeine can be seen to act as a competitive inhibitor

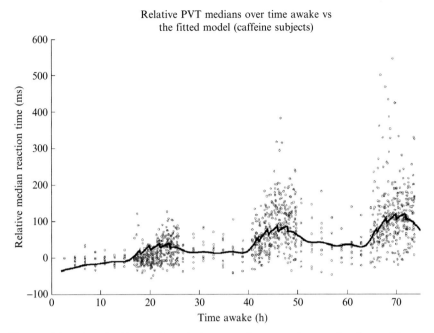

Figure 16.3 Plot of caffeine subjects' relative medians over time awake, along with projected relative median reaction times according to the fitted model with the projected normalized caffeine concentration. The plot is of the same form as the placebo plot. Again, the model succeeds in describing data, although multiplicity between saturation and circadian effects is even more apparent.

of this performance inhibitor. This model adequately describes data from this WRAIR sleep deprivation study.

The minimal inhibitor model can account for nonlinear features in data in a theoretically valid, physiologically based way. Saturation (i.e., beyond a certain point, an increased concentration of the inhibitor fails to increase the number of bound receptors) explains how reaction time (and amplitude of circadian variation in reaction time) approaches an asymptote. Moreover, the notion of a performance inhibitor as opposed to a reservoir fits into a modern understanding of the molecular biology of sleep deprivation (Boonstra and Stins, 2007). This model also unites the study of acute sleep deprivation with that of the effect of caffeine as a stimulant. Unlike previous models, it succeeds in simultaneously describing data under caffeine and placebo conditions, eliminating the artificial barrier between the two conditions of sleep deprivation. Using the estimated parameters, effective time awake and effective caffeine concentration can be estimated under the conditions of the study, if we ignore circadian effects. In Fig. 16.4, the

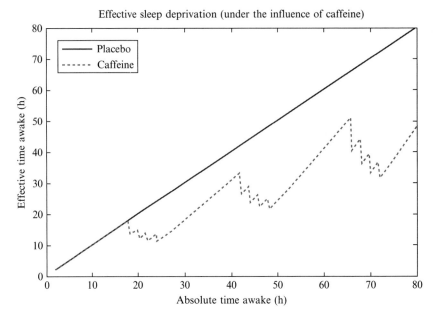

Figure 16.4 A plot of effective sleep deprivation under the placebo and caffeine regimens in the study. This figure is an implication of the model. Although caffeine seems persistently effective in this graph, it is important to remember that the difference between 20 and 30 h awake is more substantial than the difference between 40 and 70 h awake. This disparity can be explained in terms of saturation of the performance inhibitor.

dashed line plots effective time, which is, according to the assumptions in the model,

$$t_{\mathit{eff}} = t\left(\frac{1}{1 + k_c[C]}\right) \approx t\left(\frac{1}{1 + .356 \cdot [C]}\right) \quad (16.12)$$

and, in Fig. 16.5, the dashed line plots the effective normalized concentration:

$$[C]_{\mathit{eff}} = [C]\left(\frac{1}{1 + k_t t}\right) \approx [C]\left(\frac{1}{1 + .0264 \cdot t}\right) \quad (16.13)$$

Effective time awake is the time that would give rise to the same reaction time given a caffeine concentration of zero.

The model also confirms findings about the nature of the effect of caffeine. According to Fig. 16.5, the model-based analysis predicts that the half-life of the effect of the caffeine is 7.08 h on day 2 and 7.33 h on day 3. This diverges from the established half-life of caffeine in the body of 8.25 h (Syed *et al.*, 2005). This prediction, however, is in line with other studies that show that the half-life of the effect of caffeine is shorter than the half-life of its concentration in the blood (Ammar *et al.*, 2001; Kynastgales and Massey, 1994; Myers, 1988; Robertson *et al.*, 1978; Whitsett *et al.*, 1984). This common finding of the abbreviated effect of caffeine is consistent with the possibility of competitive inhibition.

This inhibitor-based minimal model predicts that the positive effect of caffeine will not be followed by a negative rebound effect as the caffeine concentration decreases. This is consistent with the observed effects of caffeine (Hewlett and Smith, 2007; Killgore, 2008).

While this minimal model can explain performance decrements over time awake, it has several limitations. The model takes the simplest possible route when dealing with concentrations. The relationship between time awake and the concentration of a performance inhibitor and between reaction time and the concentration of bound inhibitor may be nonlinear.

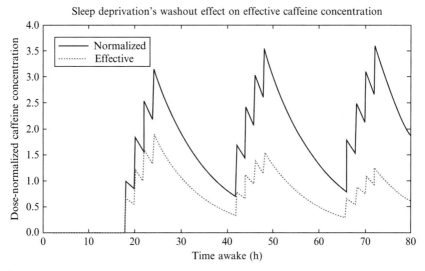

Figure 16.5 A plot of the effective strength of the caffeine concentration relative to the normalized concentration at zero hours awake. This figure is an implication of the inhibitor-based model. The normalized concentration is such that 1 unit equals the concentration that results from one 200-mg dose. As time awake increases, the performance inhibitor competitively swamps the caffeine so the relative strength decays. The established half-life of caffeine is 8.25 h, whereas the approximate half-life of the effect on the second day is 7.08 h and on the third day is 7.33 h. This is in agreement with other studies that have shown the half-life of the effect to be shorter than that of the concentration.

The equilibrium statement, moreover, could contain nonunity stoichiometry, necessitating Hill coefficients in the binding equation of values other than 1 (Farhy, 2004).

These limitations result from the desire to start from the simplest assumptions. The circadian component, estimated under a Fourier series to allow maximal generality, appears to be in the form of a driven oscillation. This oscillating factor persists despite abrogation of the sleep cycle and the light cycle, suggesting that the circadian rhythm originates internally (Dardente and Cermakian, 2007). According to Figs. 16.2 and 16.3, the driven oscillation is similar to the pattern found in daily modulation of the concentration of growth hormone in the blood (Straume et al., 1995), and perhaps similar mathematical methods can be applied to the sleep/alertness cycle.

More studies and analyses of other PVT data sets are needed in order to verify the performance inhibitor and the effect of caffeine. For the purposes of this analysis, data collected with a fixed frequency over the period of sleep deprivation are preferable as this would allow for time-series and variance analysis (Mulligan et al., 1994). Also, from a numerical analysis prospective, a better design for the PVT batteries would be to fix the number of valid response times instead of the time duration of the battery. This change, in addition to making data more comparable across time awake, would create an incentive for subjects to give maximal effort (so as to escape the tedium of a PVT battery), and thus help remove the effort bias (Kecklund et al., 2006). With regard to the WRAIR study, there was a dip in reaction times in some of the placebo subjects at around 12 h of continuous wakefulness (Kamimori et al., 2006). Experimental measurement uncertainties may explain why this dip occurs in some of the placebo subjects and in none of the caffeine subjects. This aberration occurs before caffeine administration, meaning that both groups should be identical at this point.

In addition to verification via replication, more distribution-based analysis, such as is depicted in Fig. 16.1, is needed to establish an accepted method of summarizing reaction time data (Adam, 2006). While summary statistics help account for variation and the inherent inadequacy of reaction time as a measure of performance over a period of sleep deprivation, there may be more appropriate ways by which to describe the center, spread, and shape of reaction time data.

Further areas of exploration include individualization, recovery from sleep deprivation, and chronic sleep restriction. Subjects have been shown to have interindividual variability in the following areas: chronotype (Mongrain and Dumont, 2007; Taillard et al., 1999), sensitivity to caffeine (Rétey et al., 2007), and sensitivity to sleep deprivation (Frey et al., 2004; Murray et al., 2006; Rétey et al., 2006; Tucker et al., 2007).

Extensions of this minimal model to these areas will make inhibitor-based modeling more applicable to real-world problems. For example, adding a second component to the recovery process (Johnson et al., 2004) of the homeostatic model allows the model to better reflect the cumulative effects of sleep deprivation. This second component has a theoretical interpretation that is consistent with the inhibitor hypothesis, specifically slow increments in the number of receptors over time awake and slow decay over time asleep.

Adenosine, which increments in concentration during acute sleep loss (Rétey et al., 2007) is a strong potential candidate for the performance inhibitor. Caffeine is a known adenosine antagonist (Fredholm et al., 1999). It is anticipated that further study along theses lines will lead to pharmacological therapies to regulate sleep pathology (Mignot et al., 2002).

ACKNOWLEDGMENTS

The authors acknowledge the support of the National Institutes of Health (NIH; RR019991, DK064122, RR00847, HD28934) and the Leadership Alliance (NIH T36 GM063480).

REFERENCES

Achermann, P. (2004). The two-process model of sleep regulation revisited. *Aviat. Space Environ. Med. S* **75,** A37–A43.

Adam, F., Rajib, S., Ganesan, G., and Kamimori, G. H. (2006). Life after lapses: Using distribution plots to characterize performance change across sleep deprivation. *Sleep S* **29,** A117–A343.

Akerstedt, T., and Folkard, S. (1997). The three-process model of alertness and its extension to performance, sleep latency, and sleep length. *Chronobiol. Int.* **14,** 115–123.

Ammar, R., Song, J. C., Kluger, J., and White, C. M. (2001). Evaluation of electrocardiographic and hemodynamic effects of caffeine with acute dosing in healthy volunteers. *Pharmacotherapy* **21,** 437–442.

Boonstra, T. W., Stins, J. F., Daffertshofer, A., and Beek, P. J. (2007). Effects of sleep deprivation on neural functioning: An integrative review. *Cell. Mol. Life Sci.* **64,** 934–946.

Boutrel, B., and Koob, G. F. (2004). What keeps us awake: The neuropharmacology of stimulants and wakefulness promoting medications. *Sleep* **27,** 1181–1194.

Dardente, H., and Cermakian, N. (2007). Molecular circadian rhythms in central and peripheral clocks in mammals. *Chronobiol. Int.* **24,** 195–213.

Dijk, D. J., and Czeisler, C. A. (1994). Paradoxical timing of the circadian rhythm of sleep propensity serves to consolidate sleep and wakefulness in humans. *Neurosci. Lett.* **166,** 63–68.

Farhy, L. S. (2004). Modeling of oscillations in endocrine networks with feedback. *Methods Enzymol.* **384,** 54–81.

Fredholm, B. B., Battig, K., Homen, J., Nehlig, A., and Zvarlaw, E. E. (1999). Actions of caffeine in the brain with special reference to factors that contribute to its widespread use. *Pharmacol. Rev.* **51,** 83–133.

Frey, D. J., Badia, P., and Wright, K. P. (2004). Inter- and intra-individual variability in performance near the circadian nadir during sleep deprivation. *J. Sleep Res.* **13,** 305–315.

Hewlett, P., and Smith, A. (2007). Effects of repeated doses of caffeine on performance and alertness: New data and secondary analyses. *Hum. Psychopharmocol. Clin. Exp.* **22,** 339–350.

Johnson, A., Umlauf, M., Brown, K., and Weaver, M. (2007). The influence of sleep deprivation on psychomotor performance in nurses who work the night shift. *Sleep S.* **30,** A133–A133.

Johnson, M. L., Belenky, G., Redmond, D. P., Thorne, D. R., Williams, J. D., Hursh, S. R., and Balkin, T. J. (2004). Modulating the homeostatic process to predict performance during chronic sleep restriction. *Aviat. Space Environ. Med. Suppl.* **75,** A141–A146.

Johnson, M. L., and Straume, M. (1999). Innovative quantitative neuroendocrine techniques, in (sex-steroid interactions with growth hormone). *Serona Symp.* 318–326.

Johnson, M. L., and Veldhuis, J. D. (1995). Evolution of deconvolution analysis as a hormone pulse detection algorithm. *Methods Neuroscie.* **28,** 1–24.

Kaida, K., Akerstedt, T., Kecklund, G., Nilsson, J., and Axelsson, J. (2007). The effects of asking for verbal ratings of sleepiness on sleepiness and its masking effects on performance. *Clin. Neurophysiol.* **118,** 1324–1331.

Kamimori, G., Fletcher, A., Johnson, D., Barton, N., and Thompson, T. (2006). Ability of caffeine to maintain night-time performance during three consecutive nights without sleep. *Sleep S.* **29,** A117–A118.

Kecklund, G., Axelsson, J., and Akerstendt, T. (2006). Five days with partial sleep deprivation and the effects on performance, sleepiness and effort. *Sleep S.* **29,** A134–A135.

Killgore, W. D., Rupp, T. L., Grugle, N. L., Reichardt, R. M., Lipizzi, E. L., and Balkin, T. J. (2008). Effects of dextroamphetamine, caffeine and modafinil on psychomotor vigilance test performance after 44 h of continuous wakefulness. *J. Sleep Res.* **17,** 309–321.

Kryger, M., Roth, T., and Dement, W. C. (2000). "Principles and Practice of Sleep Medicine,"3rd Ed. Saunders, New York.

Kynastgales, S. A., and Massey, L. K. (1994). Effect of caffeine on circadian excretion of urinary calcium and magnesium. *J. Am. College Nutr.* **13,** 467–472.

Mignot, E., Taheri, S., and Nishino, S. (2002). Sleeping with the hypothalamus: Emerging therapeutic targets for sleep disorders. *Nat. Neurosci. S.* **5,** 1071–1075.

Moller, H. J., Kayumov, L., Bulmash, E. L., Nhan, J., and Shapiro, C. (2006). Simulator performance, microsleep episodes, and subjective sleepiness: Normative data using convergent methodologies to assess driver drowsiness. *J. Psychosomat. Res.* **61,** 335–342.

Mongrain, V., and Dumont, M. (2007). Increased homeostatic response to behavioral sleep fragmentation in morning types compared to evening types. *Sleep* **30,** 773–780.

Mulligan, T., Veldhuis, J. D., and Johnson, M. L. (1994). Methods for validating deconvolution analysis of pulsatile hormone as a paradigm. *Methods Enzymol.* **240,** 109–129.

Murray, C. J., Killgore, D. B., Kamimori, G. H., and Killgore, W. D. (2006). Individual differences in stress management capacity predict responsiveness to caffeine during sleep deprivation. *Sleep S.* **29,** A43–A43.

Myers, M. G. (1988). Effects of caffeine on blood pressure. *Arch. Intern. Med.* **148,** 1189–1193.

Petrie, K. J., and Dawson, A. G. (1997). Symptoms of fatigue and coping strategies in international pilots. *Int. J. Aviat. Psychol.* **7,** 251–258.

Rajaratnam, S. M., and Arendt, J. (2001). Health in a 24-h society. *Lancet* **358,** 999–1005.

Rétey, J. V., Adam, M., Gottselig, J. M., Khatami, R., and Durr, R. (2006). Adenosinergic mechanisms contribute to individual differences in sleep deprivation-induced changes in neurobehavioral function and brain rhythmic activity. *J. Neurosci.* **26,** 10472–10479.

Rétey, J. V., Adam, M., Khatami, R., Luhmann, U. F., Jung, H. H., Berger, W., and Landholt, H. P. (2007). A genetic variation in the adenosine A2A receptor gene (adora2a) contributes to individual sensitivity to caffeine effects on sleep. *Clin. Pharmacol. Ther.* **81,** 692–698.

Robertson, D., Frolich, J. C., Carr, R. K., Watson, J. T., Hollifield, J. W., Shrand, D. G., and Oates, J. A. (1978). Effects of caffeine on plasma renin activity, catecholamines, and blood pressure. *N. Engl. J. Med.* **298,** 181–186.

Rupp, T., Grugle, N., Krugler, A., Balkin, T., and Killgore, W. (2007). Caffeine, dextroamphetamine, and modafinil improve PVT performance after sleep deprivation and recovery sleep. *Sleep S.* **30,** A44–A44.

Straume, M., Veldhuis, J. D., and Johnson, M. L. (1995). Realistic emulation of highly irregular temporal patterns of hormone release: A computer-based pulse simulator. *Methods Enzymol.* **28,** 220–243.

Syed, S. A., Kamimori, G. H., Kelly, W., and Eddington, N. D. (2005). Multiple dose pharmacokinetics of caffeine administered in chewing gum to normal healthy volunteers. *Biopharm. Drug Dispos.* **26,** 403–409.

Taillard, J., Philip, P., and Bioulac, B. (1999). Morningness/eveningness and the need for sleep. *J. Sleep Res.* **8,** 291–295.

Tucker, A. M., Dinges, D. F., and Van Dongen, H. P. A. (2007). Trait interindividual differences in the sleep physiology of healthy young adults. *J. Sleep Res.* **16,** 170–180.

Veldhuis, J. D., Carlson, M. L., and Johnson, M. L. (1987). The pituitary gland secretes in bursts: Appraising the nature of glandular secretory impulses by simultaneous multiple-parameter deconvolution of plasma hormone concentrations. *Proc. Natl. Acad. Sci. USA* **84,** 7686–7690.

Whitsett, T. L., Manion, C. V., and Christensen, H. D. (1984). Cardiovascular effects of coffee and caffeine. *Am. J. Cardiol.* **53,** 918–922.

Author Index

A

Ables, J. G., 155, 158
Achermann, P., 221, 406
Acker, C. D., 13, 17
Acton, P., 108
ADA, 74
Adam, F., 408, 412, 418
Adam, M., 418, 419
Agrawal, V., 181
Aita, T., 181
Ajabor, L., 347, 349
Akaike, H., 292
Akerstedt, T., 406, 407, 414, 418
Aksel, S., 349
Akwi, J. A., 71
Alberty, R. A., 34, 35, 36, 37, 39, 40, 41, 55, 61, 65
Albrecht, U., 2, 14, 15
Albritton, D. L., 263
Alexander, R., 4
Alexov, E. G., 203, 244
Allawi, H. T., 340
Almeida, A. M., 261
Alston, R. W., 181, 182, 185, 188, 192, 195
Alt, S., 163
Amat di San Filippo, C., 325
American Diabetes Association, 74
Ammar, R., 417
Amoss, M., 346
Amrein, H., 146
Andersen, N., 158
Anderson, D. E., 181, 192
Anderson, D. R., 292, 293
Anderson, S. M., 71, 73, 74
Anil, B., 185, 195
Antosiewicz, J., 203, 240
Antzelevitch, C., 11, 18
Antzutkin, O. N., 109
Arcus, V. L., 185, 195, 203
Arendt, J., 406
Arimura, A., 346
Arkin, A., 116, 117
Arnold, F. H., 181, 202
Aschermann, K., 104
Ashburner, M., 146
Ashcroft, F., 168
Asplin, C. M., 354, 356
Atkinson, L. E., 347

Aukhil, I., 198
Aurora, R., 193, 203
Ausborn, J., 14, 15, 16, 17
Aviles, F. X., 181, 198, 203
Avron, B., 181, 192
Axelsson, J., 407, 418

B

Baase, W. A., 176, 179, 180, 202
Baba, Y., 346
Bache, R. J., 35, 59, 61, 62, 63, 64, 65
Backstrom, C. T., 350
Badia, P., 418
Bagheri, B., 240
Bahr, M., 104
Baird, D. T., 350
Baker, D., 180, 182
Baker, N. A., 239, 241
Balaban, R. S., 63
Balasubramanian, V., 296
Balbach, J. J., 109, 201
Balding, R. L., 297, 298, 299, 300, 301
Baldwin, E. P., 180, 203
Baldwin, R. L., 193, 234
Balkin, T. J., 180, 407, 417, 418
Bang, D., 193
Banks, M. I., 13, 17
Baptista, A. M., 241
Baraghini, G. F., 351
Bardin, C. W., 347, 349, 350, 369
Barghorn, S., 104
Barkan, A. L., 368
Barlow, E., 104
Barrick, D., 297, 298, 299, 300, 301
Barron, A., 295
Barton, N., 406, 407, 418
Bassingthwaighte, J. B., 60
Basu, R., 360, 362
Battig, K., 419
Baumann, G., 368
Baumann, K., 261, 282
Baynard, M. D., 221
Beadle, B. M., 176, 179
Beard, D. A., 30, 31, 33, 35, 36, 56, 61, 62, 63, 64, 65
Beauchamp, K., 151, 153, 173
Beck, R. W., 71
Becktel, W. J., 181

423

Beek, P. J., 415
Beersma, D. G. M., 14, 16
Belenky, G., 213, 418
Belkas, J., 13, 19
Benitez, P. L., 180
Benzer, S., 149
Berezovsky, I. N., 181
Berge, P., 5
Berger, W., 418, 419
Bergman, R. N., 314, 360, 362
Berliner, L. J., 186, 188, 192, 193
Bernard, S., 4, 9, 23
Beroza, P., 237
Berry, D., 7
Bevington, P. R., 263, 277, 281, 374
Beylkin, G., 161, 162, 163
Bhattacharya, A. N., 347
Bhattacharya, S., 196
Bialek, W., 18
Bienert, R., 201
Bio 2010, 307
Bioulac, B., 421
Bishop, T., 155, 157, 158
Bissonette, E. A., 315
BISTI, 306
Bitner, R. S., 104
Black, M. E., 180, 182
Blackwell, R., 346
Blasie, C. A., 192
Blocker, H., 340
Bodmer, R., 142, 146
Bolon, D. N., 182
Boonstra, T. W., 415
Booth, R. A., Jr., 349, 350, 354, 356, 369
Boppana, R. V., 93, 95, 97
Borbély, A. A., 221
Borer, P. N., 326
Both, M., 71
Boulanger, B., 270
Boutrel, B., 407
Bowden, C. R., 360, 362
Bowers, C. Y., 360, 362, 368
Box, G. E. P., 215
Boyd, D., 370
Boyd, D. G., 352, 357, 358
Boyne, M., 71
Bozdogan, H., 292
Bragg, J. K., 297
Brauer, V., 14, 15
Bray, N., 144, 145, 168
Bray, W., 163
Brayer, G. D., 248, 255
Brayman, K. L., 71, 74, 78, 360
Breese, R., 108
Bremner, W. J., 349
Brems, D. N., 203

Brenan, K. E., 116
Brennan, M., 81
Breslauer, K. J., 340
Breton, M., 69, 71, 74, 78
Breton, M. D., 71, 73, 74, 80
Briggs, J. M., 240
Briggs, R. W., 63
Brocks, A., 349
Brookes, E., 90, 91, 92, 93, 94, 95, 97, 102
Brooks, C. L., 186, 188, 192, 193, 196
Brooks, C. L. III, 196, 241
Brown, K., 406
Brown, P. H., 109
Brown, S. A., 2, 3
Browne, M. W., 292, 293, 302
Brownlee, M., 70
Bruck, J., 120, 121
Brunger, A. T., 181
Bucher, D., 14, 19
Buck, J., 2
Bugg, C. E., 198
Bulmash, E. L., 408
Burg, J. P., 156
Burger, C. W., 349
Burgoon, P., 166
Burgus, R., 346
Burk, D., 36
Burnet, B., 164
Burnham, L. S., 292, 293
Burns, M. J., 261
Burrage, K., 125, 126, 135, 136, 137
Butcher, M., 346
Butler, J. P., 349
Bycroft, M., 180, 181, 192

C

Cabrera, M. E., 59
Callender, D., 180, 182
Cameron, F. J., 71
Campbell, S. L., 116
Canavier, C. C., 12, 17
Cann, P., 179, 204
Canonico, V., 71
Cao, W., 89, 92, 101
Cao, Y., 115, 117, 120, 121, 125, 129, 132, 135
Caplan, S. R., 360
Carlsen, E. C., 315, 349, 350, 352, 354, 369
Carlson, M. L., 351, 353, 369, 370, 371, 372, 389, 409
Caron, F. J., 349
Carp, S. A., 181, 182, 188, 192, 196
Carr, R. K., 417
Carroll, R. J., 282
Carter, W. R., 71
Caruthers, M. H., 340

Case, D. A., 241
Castel-Branco, M. M., 261
Caumo, A., 360, 362
Cavan, D. A., 71
Cermakian, N., 418
Chaffin, D., 93, 97
Chakrabartty, A., 193
Chapman, I. M., 368
Chassin, L. J., 71
Chatfield, C., 142, 143, 146, 148, 149, 154, 155, 161, 165, 170
Chatterjee, A., 125, 126
Chen, J., 324
Chen, X., 65, 179
Cheyne, E. H., 71
Chicheportiche, R., 2
Cho, J. H., 185, 195, 203
Choma, M., 146
Chon, K., 143
Christensen, H. D., 417
Christiansen, E., 356
Christiansen, T., 97
Chu, Z. T., 203, 234
Chui, C. K., 161, 162, 163
Clarke, E. C. W., 67
Clarke, I. J., 347
Clarke, W. L., 69, 71, 73, 74, 77, 78, 83, 315, 360
Clifton, D. K., 349
Clifton, J. G., 245
Clore, G. M., 179
Cobelli, C., 83, 360, 362
Cohen, N. L., 349
Cohen, S., 116
Coifman, R., 161, 162, 163
Coll, M., 198
Colot, H. V., 149
Comas, M., 14, 16
Committee on Mathematical Sciences Research, 307
Condron, M. M., 107
Conejero-Lara, F., 181, 203
Connolly, K., 164
Cook, W. J., 198
Cooley, J. W., 154
Copinschi, G., 349, 368
Corrent, C., 180, 182
Cox, C. D., 120, 122
Cox, D. J., 71, 77, 78, 83, 315, 360
Cox, M. M., 61
Craig-Schapiro, R., 185, 195
Crews, L., 104
Crouch, S. R., 267, 278
Crowley, W. F., Jr., 349, 350
Cudeck, R. C., 302
Cummins, J. T., 347
Currie, L. A., 262
Curtis, N., 145

Cymes, G. D., 234
Czeisler, C. A., 411
Czodrowski, P., 241

D

Daan, S., 2, 10, 14, 15, 16
Daffertshofer, A., 415
Dahl, G. E., 358
Dahlquist, F. W., 177, 181
Dalkin, A. C., 348, 369
Dalla Man, C., 360, 362
Dankesreiter, A., 176, 179
Dantas, G., 180, 182
Danzer, K., 262
Dao-pin, S., 192
Dardente, H., 418
Darwin, C., 71
Dash, R. K., 65
Datta, D., 192
Daubechies, I., 161, 162, 163
Davidian, M., 282
Davidson, A. R., 182, 186, 190, 197, 200, 201
Davies, M. N., 245
Davies, R. L., 309, 318
Davis, F. D., 317
Davis, M. E., 240
Davison, T. S., 354, 356
Dawid, A. P., 293
Dawson, A. G., 406
Dayer, J. M., 2
DCCT, 70
Debeljuk, L., 346
DeCoursey, P., 149
DeGrado, W. F., 180
Demchuk, E., 245
Demeler, B., 89, 90, 91, 92, 93, 94, 95, 97, 101, 102, 109
Dement, W. C., 406
Demott-Friberg, R., 368
Dengler, B., 326
De Nicolao, G., 370, 379, 382, 400
de Ruyter van Steveninck, R., 18
Desir, D., 349
Desjarlais, J. R., 179, 180
Deutschman, W. A., 177
DeVoe, H., 340
Diabetes Control and Compliance Trial, 70
Dierschke, D. J., 347
Diggle, P. J., 349
Dijk, D. J., 411
Dill, K. A., 176, 179
Dimaraki, E. V., 368
Dimitrov, R. A., 340
Dinges, D. F., 221, 222, 223, 228, 229, 418

Dobrowolski, S. F., 325
Dobson, G. P., 49, 50, 51, 52, 53, 61
Dodson, E. C., 108
Doig, A. J., 193
Dolan, J. W., 261
Dolinsky, T. J., 241
Dominguez, L. G., 13, 19
Dominy, B. N., 196
Donath, S. M., 71
Dong, H., 262, 278
Dooley, R., 93, 97
Dorval, A. D., 13, 17
Dowse, H., 142, 144, 145, 146, 149, 163, 164, 166, 168, 169
Dowse, H. B., 141, 142, 144, 145, 146, 149, 154, 163, 167, 168, 169, 170
Draper, R. N., 262, 263, 267
Du, Z., 360
Dulcis, D, 146
Dumas, A., 2, 3
Dumont, M., 418
Dunlap, J., 143, 149
Dunlop, W., 349
Durr, R., 418

E

Eastwood, L., 164
Ebert, U., 104
Eckberg, D. L., 18
Eddington, N. D., 409, 417
Elahi, D., 355
Elcock, A. H., 185, 195, 245
Eletr, Z. M., 180, 182
Enright, J. T., 154
Erali, M., 324
Erickson, H. P., 198
Ermentrout, B., 17, 19, 23
Ermolenko, D. N., 176, 179, 180, 193, 200, 203
Erra, R. G., 13, 19
Evans, M., 220
Evans, N. P., 358
Evans, P. R., 198
Evans, W., 352, 354
Evans, W. S., 315, 316, 345, 349, 350, 352, 354, 355, 356, 357, 358, 364, 367, 369–371, 389, 404

F

Fairman, R., 181, 182, 188, 192, 196
Falcão, A. C., 261
Faloona, F. A., 324
Farhy, L. S., 77, 78, 309, 315, 318, 345, 352, 356, 357, 358, 359, 360, 362, 367, 418
Farhi, L. S., 370
Faria, A. C., 368

Faunt, L. M., 268, 355
Federici, M. O., 71
Fee, L. R., 181, 182, 188, 192
Feher, G., 237
Fellows, R., 346
Fernandez, A. M., 181, 203
Fersht, A., 179, 181, 192, 203, 204
Fersht, A. R., 180, 181, 185, 192, 195, 202, 203
Fetz, E. E., 2, 13
Feuerstein, S., 104
Fezoui, Y., 107
Filicori, M., 349, 350
Filimonov, V. V., 181, 203
Fink, A. L., 107
Fiser, A., 186
Fishman, G. S., 225, 228
Fitch, C. A., 196, 203
Fixman, M., 340
Flanagan, M. A., 183
Flannery, B. P., 229, 263, 265, 268, 330, 339, 372
Fletcher, A., 406, 407, 418
Fleury-Olela, F., 2
Flower, D. R., 245
Fluhmann, C. F., 346
Folkard, S., 406, 414
Folta, T., 245
Fontana, W., 116
Forster, M., 292
Forti, G., 351
Foster, I., 109
Foster, R. G., 317
Francois, N., 270
Frank, R., 340
Frankenberg, N., 192
Frasier, S. G., 268, 316
Frasier-Cadoret, S. G., 372, 388
Fredholm, B. B., 419
Fredkin, D. R., 237
Freiberger, D. M., 213
Freier, S. M., 340
Freire, J. J., 340
Freskgard, P. O., 202
Frey, D. J., 418
Frieden, C., 36
Friend, K., 355
Frohman, L. A., 368
Frolich, J. C., 417
Frolich, M., 355
From, A. H., 59
Fuller, T., 324
Funes, P., 146, 149, 164, 168, 169
Furlanetto, R., 404

G

Gajiwala, K., 179, 203
Galán, R. F., 13, 17, 19
Ganesan, G., 408, 412, 418

Garcia-Mira, M. M., 188
Garcia-Moreno, B., 183
Garcia-Moreno, E. B., 185, 195, 196, 203
Garcia-Saez, I., 198
Garwood, M., 59
Gay, V. L., 347
Gaylinn, B. D., 356, 357, 358, 368
Genazzani, A. D., 351
Genazzani, A. R., 351
Genz, A., 213
Georgescu, R. E., 203, 244
Gershenson, A., 202
Geysen, H. M., 356, 357, 358
Gheorghiu, S., 356
Giak Sim, K., 325
Gibbs, C., 61
Gibbs, C. L., 60
Gibson, M., 120, 121
Gilkes, N. R., 176, 179
Gillespie, D., 117, 119, 120, 122, 123, 125, 127, 129, 132, 135
Gilso, M. K., 240
Gilson, M. K., 203
Giver, L., 202
Glass, L., 4, 5, 7, 8, 10, 13, 15, 17, 18, 21, 143, 168, 169
Glew, D. N., 67
Godoy-Ruiz, R., 188
Godshalk, M., 356
Godt, R. E., 49
Golding, E. M., 49, 61
Gonder-Frederick, L. A., 71, 77, 78, 83, 315, 360
Gong, Y., 193
Gonze, D., 4, 9, 23
Goodman, R. L., 358
Goodwin, B. C., 4, 23
Gorczyca, M., 164
Gordon, J. C., 245
Gottselig, J. M., 418
Govaerts, B., 270
Graham, R., 324
Grainger, J., 288
Granada, A., 1
Grattinger, M., 176, 179
Gray, D. M., 326, 340
Grayburn, J. A., 351
Green, J. D., 346, 347
Greene, A., 180
Greenfield, M., 5, 20
Greenfield, N. J., 298
Gribenko, A. V., 176, 178, 181, 182, 186, 187, 188, 190, 191, 192, 193, 194, 195, 196, 197, 198, 199, 200, 201, 202, 234
Griffin, M. P., 315
Griko, Y. V., 177, 179

Grimm, K. M., 108
Grimsley, G. R., 181, 182, 188, 192
Gromiha, M. M., 181
Gronenborn, A. M., 179, 193, 203
Grosman, C., 234
Grothaus, G., 237
Grugle, N. L., 407, 417
Grunwald, P., 291, 293, 295
Gu, G.-G., 150
Guardabasso, V., 350, 351, 352, 369, 370, 379, 382, 400
Guarnieri, F., 203
Guerritore, A., 201
Guevara, M. R., 5, 10, 13, 17, 21
Guillemin, R., 346
Gundry, C. N., 324
Gunner, M. R., 203, 238, 239, 244
Guo, M., 193
Gupta, R. C., 261
Gurd, F. R., 183
Guyton, A. C., 45
Guzman-Casado, M., 179
Gvritishvili, A. G., 196

H

Haaland, P. D., 282
Haas, J. S., 13, 17
Haffner, S. M., 314
Hahn, M., 324
Haisenleder, D. J., 348, 369
Hall, J., 143, 146, 149, 164, 168, 169
Hall, J. C., 142, 144, 149, 163, 164
Hall, J. E., 45
Hamamatsu, N., 181
Hamblen-Coyle, M., 149
Hamilton, W. C., 263
Hamming, R. W., 142, 144, 153, 164, 166, 167, 168
Hand, G., 253, 255
Handel, T. M., 179, 180
Hanging, C., 315
Hansel, D., 19
Hanson, R. J., 92
Haranczyk, M., 234
Hardy, J. A., 104
Harini, F., 71
Harlan, J., 104
Harper, J. D., 107
Harrell, F. E., Jr., 315
Harris, G. W., 346, 347
Harris, N., 261
Hartbauer, M., 5, 10, 14, 15, 20, 21
Hartley, D. M., 107
Hartley, R. W., 179
Hartman, M. L., 355, 368
Hastings, M. H., 5
Hastings, N., 220
Hauser, C., 2

Havranek, J. J., 180, 182
Heath, L. S., 245
Hei Masumoto, K., 11
Heinemann, U., 181, 182, 192, 196, 201
Hendrich, K., 59
Hendrickson, W. A., 198
Hendsch, Z. S., 181, 182, 188, 192, 196
Hennig, R. M., 1
Hepler, R. W., 108
Herz, A. V., 5
Herzel, H., 1, 2, 3, 4, 5, 7, 9, 18, 23
Hess, D. L., 368
Hewlett, P., 417
Higgins, G. A., 104
Hilgers, M. E., 73
Hille, B., 151
Hindmarsh, A., 116
Hirsch, I. B., 70
Hodgkin, A., 19, 22
Hokanson, J. E., 314
Holden, J. A., 324
Hollien, J., 179
Hollifield, J. W., 417
Holm Nielsen, E., 104
Holst, M. J., 239
Homen, J., 419
Honig, B., 203, 238, 239
Honma, K., 5
Honma, S., 5
Hopfield, J. J., 5
Hoppensteadt, F., 18
Horovitz, A., 181, 192
Hovorka, R., 71
Huang, J.-K., 221, 222, 223, 228, 229
Huang, M., 93, 97
Huang, X., 181
Hunter, P., 18
Hurley, J. H., 180
Hursh, S. R., 418
Husimi, Y., 181
Hutzli, I., 71
Huygens, C., 8
Huyghues-Despointes, B. M., 181, 182, 188, 192
Hwang, L.-J., 261, 282

I

Ibarra-Molero, B., 177, 181, 184, 186, 188, 191, 192
Ider, Y. Z., 360, 362
Ilin, A., 240
Ingle, J. D., Jr., 267, 278
Ingwall, J. S., 59
Iqbal, Z., 104
Iranmanesh, A., 355, 356
Isejima, H., 5
Isern, N. G., 180, 182
Ito, N., 198

Izatt, J., 146
Izhikevich, E., 18

J

Jackson, G. L., 347
Jacobs, A. M., 288
Jacquez, J. A., 261
Jaeger, J. A., 340
Jaenicke, R., 176, 179, 181, 192, 196
Jaffe, C. A., 368
Jaffe, R. B., 347
Jalife, J., 11, 18
Janson, B., 104
Jansson, P. A., 369
Jeneson, J. A., 65
Jennings, T., 170
Jensen, J. H., 241, 245
Jinagonda, S., 71
Johnson, A., 406
Johnson, C. H., 13, 16
Johnson, D., 406, 407, 418
Johnson, E., 142, 144, 145, 146, 168
Johnson, E. L., 261
Johnson, M. L., 180, 268, 309, 315, 316, 318, 321, 345, 349, 350, 351, 352, 353, 354, 355, 356, 357, 358, 360, 364, 367, 368, 369, 370, 371, 372, 379, 382, 388, 389, 391, 400, 409, 418
Johnson, P. E., 253, 255
Jones, J., 145
Joseph, S., 239
Joshi, M. D., 248, 253, 255
Joyce, J. G., 108
Juge, C., 2
Jung, H. H., 418, 419

K

Kaida, K., 407
Kaiser, D. L., 369
Kallenbach, N. R., 193
Kamen, P., 81
Kamimori, G., 406, 407, 418
Kamimori, G. H., 180, 408, 409, 412, 417, 418
Kantor, H. L., 63
Kao, Y. H., 196
Kaplan, D., 8
Kaplan, J., 71
Karnes, H. T., 261
Karsch, F. J., 358
Kass, R. E., 295
Kato, M., 234
Katsoulakis, M., 125, 126
Katsufumi, S., 143
Katsuno, Y., 5
Katz, A. M., 59, 60
Katz, L. A., 63
Kaufmann, H., 71

Kaul, K., 324
Kauzmann, W., 179
Kay, S. M., 155, 158
Kayumov, L., 408
Kecklund, G., 407, 418
Keenan, D. M., 353, 360
Keiffer, T. R., 181, 182, 186, 187, 188, 190, 191, 192, 197, 198, 199, 200, 201, 234
Kelch, R. P., 349
Keller, P., 104
Keller, P. M., 108
Kellis, J. T., Jr., 179, 204
Kelly, W., 409, 417
Kemp, M. L., 34, 35, 57, 58, 59, 65
Kendall, M. G., 154
Kendall Taylor, P., 349
Kent, S. B., 193
Keogh, E. J., 368
Kerr, D., 71
Khammash, M., 119
Khandogin, J., 241
Khatami, R., 418, 419
Khurana, R., 107
Kierzek, A., 135, 136
Kierzek, R., 340
Kilka, M., 166
Killgore, D. B., 418
Killgore, W. D., 407, 417, 418
Kim, K. S., 179
Kim, T., 93, 97
Kim, W., 302
King, C. R., 71, 73, 74
Kinney, G., 108
Kinney, K., 142, 144, 145, 146, 168
Kirkwood, J. G., 183
Kirkwood, J. R., 309, 318
Kirsten, C., 104
Kiser, M. K., 261
Kitano, H., 307
Klebe, G., 241
Klein, W. L., 104
Klonoff, D. C., 71, 72
Kluger, J., 417
Knobil, E., 347, 348, 368
Knuth, D., 341
Kobayashi, M., 5
Kobayashi, T. J., 11
Kollman, C., 71
Kolp, L. A., 356
Kondo, T., 11
Kongas, O., 33
Konishi, M., 52
Konopka, R. J., 149
Konopka, R. R., 149
Koob, G. F., 407
Kopell, N., 13, 17
Korkegian, A., 180, 182
Körner, M., 253, 255

Korsen, T., 349
Korzeniewski, B., 33
Kossiakoff, A. A., 193
Kovatchev, B., 69, 77, 78, 83
Kovatchev, B. P., 71, 73, 74, 77, 78, 83, 309, 315, 318, 360
Kramer, A., 1, 2, 3, 4, 9, 23
Krantz, C., 104
Kratzer, S., 5, 10, 14, 15, 20, 21
Krugler, A., 407
Krusche, T., 349
Kryger, M., 406
Kuhlman, B., 180, 182, 185, 195, 203
Kulcu, E., 71
Kumagai, H., 143
Kumar, A., 71
Kumar, S., 181
Kunysz, A., 21
Kunz, D., 2, 3
Kuramoto, Y., 18
Kuroki, R., 202
Kurths, J., 7, 18
Kushmerick, M. J., 30, 34, 35, 49, 52, 57, 58, 59, 65
Kynastgales, S. A., 417
Kyriacou, C. P., 142, 163

L

Laessle, R. G., 349
Lake, D. E., 315
Lakowicz, J. R., 370
Lamm, O., 89
Lanczos, C., 153, 154
Landholt, H. P., 418, 419
LaNoue, K. F., 61
Lansbury, P. T., 104
Lansbury, P. T., Jr., 107
Lassila, K. S., 192
Laughlin, G. A., 349
Laurents, D. V., 185, 195
Lawson, C. L., 92
Lawson, J. W., 49, 52, 53
Layfield, L. J., 324
Lazar, G. A., 179, 180
Leahy, D. J., 198
Leapman, R. D., 109
Leask, R. M., 350
Lecomte, J. T., 196
Lee, C. F., 186, 188, 192, 193
Lee, K. K., 196
Lee, P. S., 203
Lemba, M., 33
Lerman, L. S., 340
Leshchenko, Y., 13, 19
Lesho, M. J., 85
Levine, J., 146, 149, 164, 168, 169
Levine, J. E., 347

Levine, R., 146
Levy, M. N., 16
L'Hermite, M., 349
Li, H., 120, 122, 135, 241, 245
Liew, M., 324, 342
Liguri, G., 198
Linderstrom-Lang, K., 257
Lindsley, G., 166
Lineweaver, H., 36
Ling, N., 346
Linne, U., 104
Lipizzi, E. L., 417
Liu, C., 285
Liu, C. H., 349
Liu, H., 202
Liu, J., 356, 357, 358
Liu, Y., 285
Lo, S., 13, 19
Logothetis, D. E., 2
Loladze, V. V., 176, 178, 179, 180, 181, 182, 184, 186, 187, 188, 190, 191, 192, 193, 194, 195, 197, 198, 199, 200, 201, 202, 203, 204
Lombardi, A., 180
Lombardi, F., 143, 169
Longo, N., 325
Lopaschuk, G. D., 59
Lopez, M. M., 176, 178, 181, 186, 192, 193, 194, 195, 200, 202
Lu, L., 285
Luhmann, U. F., 418, 419
Luisi, D. L., 185, 195, 203
Luo, P., 93, 97
Luty, B. A., 240
Lyon, E., 324

M

Mackey, M. C., 143, 168, 169
Mackey, M. M., 4, 5, 7, 15, 18, 21
Madura, J. D., 240
Maheshwari, H. G., 368
Maislin, G., 221
Makhatadze, G. I., 175, 176, 177, 178, 179, 180, 181, 182, 184, 186, 187, 188, 190, 191, 192, 193, 194, 195, 196, 197, 198, 199, 200, 201, 202, 203, 204
Mallat, S., 161, 162, 163
Manion, C. V., 417
Marahiel, M. A., 181, 196
Marbach, P., 368
March, C., 261
March, K. L., 183
Marcus, J. S., 182
Marder, E., 14, 19
Marky, L. A., 340
Marple, S. G., Jr., 155, 158
Marqusee, S., 179, 192

Marr, R., 104
Marshall, J. C., 348, 349, 369
Marshall, S. A., 182, 193
Martin, A., 181
Martinez, J. C., 181, 203
Marti-Renom, M. A., 186
Masliah, E., 104
Massey, L. K., 417
Massi-Benedetti, M., 71
Mateo, P. L., 181, 203
Mato, G., 19
Matouschek, A., 179, 204
Matsuo, H., 346
Matsuo, T., 5
Matthew, J. B., 183
Matthews, B. W., 179, 180, 192, 202, 203
Matthews, D. H., 355
Maughan, D. W., 49
Max, K. E., 201
Mayo, S. L., 180, 182, 192, 193
Maywood, E. S., 5
McAdams, H., 116
McCall, A. L., 71, 74, 78, 360
McCammon, J. A., 185, 195, 203, 239, 240, 241, 242
McCollum, J. M., 120, 122
McDonnell, C. M., 71
McIntosh, J. E., 351
McIntosh, L. P., 248, 253, 255
McIntosh, R. P., 351
McKenzie, F. D., 221, 222, 223, 228, 229
McKinney, J. T., 325
McNeilly, A. S., 350
McNutt, M., 179, 203
McSparron, H., 245
Mead, R., 372
Meadows, C., 324
Mehler, E. L., 203
Meier, C. A., 2
Meiering, E. M., 202
Meister, M., 2
Melo, F., 186
Menaker, M., 317
Meneilly, G. S., 355
Mergell, P., 5
Merkel, J. S., 192
Merkle, H., 59
Merriam, G. R., 349, 369
Merritt, E. A., 180, 182
Messer, B., 202
Meunier, C., 19
Meyer, Y., 161, 162, 163
Midgley, A. R., Jr., 347
Mignot, E., 419
Miller, J. N., 262
Miller, M., 71, 80
Mindlin, G. B., 7

Author Index 431

Minoux, H., 196
Moe, B., 18
Moe, G. K., 18
Moller, H. J., 408
Mollicone, D. J., 221, 222, 223, 228, 229
Molt, O., 104
Monahan, M., 346
Mongan, J., 241
Mongrain, V., 418
Monroe, S. E., 347
Montgomery, J., 325
Moore, T., 4
Moorman, J. R., 315
Morawski, B., 181
Morgan, C. S., 193
Mortola, J. F., 349
Motono, C., 181
Mott, C. G., 221, 222, 223, 228, 229
Moulins, M., 14, 19
Mrosovsky, N., 4
Muegge, I., 203, 243
Mueller, U., 181, 182, 192, 196
Mueller, W. J., 18
Mulligan, T., 355, 356, 418
Mullis, K. B., 324
Munsky, B., 119
Murdoch, A. P., 349
Murray, C. J., 418
Murray, J., 354, 356
Murray, M., 93, 97
Musacchio, A., 198
Myers, J., 237
Myers, J. B., 245
Myers, M. G., 417
Myung, I. J., 289, 290, 291, 292, 293, 295, 301, 302
Myung, J., 287
Myung, J. I., 302, 303

N

Nabarro, J. D. N., 351
Naftolin, F., 347, 349, 351
Nagai, K., 198
Nagano, M., 11
Nagaraja, N. V., 261
Nagel-Steger, L., 90, 104
Nagi, A. D., 181
Nagoshi, E., 2
Nahas, D. D., 108
Nair, R. M., 346
Nakai, Y., 368
Nakajima, M., 181
Namihira, M., 5
Narayanan, S., 237
Nass, R., 356, 357, 358, 367
Nastri, F., 180
Natiello, M. A., 7

Navarro, D. J., 295, 302
Nehlig, A., 419
Neilson, T., 340
Nelder, J. A., 372
Nelson, D. L., 61
Nelson, J. C., 315
Netoff, T. I., 13, 17
Newman, G. B., 351
Nhan, J., 408
Nicholls, A., 203, 239
Nicholson, H., 192
Nielsen, J. E., 233, 238, 240, 241, 242, 244, 245, 247, 248, 251, 253, 255
Nilsson, J., 407
Nimmrich, V., 104
Nishino, S., 419
Nocedal, J., 220
Nomiya, Y., 181
Nordlund, P., 198
Norris, J. M., 314
Norusis, M., 261
Nybo, M., 104

O

Oates, J. A., 417
Oerter, K. E., 350, 351, 352, 369
Okamura, H., 2, 5
Okamura, M. Y., 237
Okura, R., 5
Oliveberg, M., 185, 195, 203
Oliveri, M. C., 356, 357, 358
Onami, T., 143
Ondrechen, M. J., 245
On the Mathematics Curriculum of the High School, 310
Onufriev, A., 237
Oprisan, S. A., 12, 17
Orwant, J., 97
Osaka, M., 143
O'Shea, T. M., 315
Oubridge, C., 198
Ovufriev, A., 245
Owenius, R., 186, 188, 192, 193

P

Pace, C. N., 179, 181, 182, 185, 188, 191, 192, 195, 203, 298
Pachla, L. A., 261
Pain, R. H., 202
Palais, B., 324
Palais, R., 323, 324, 332, 341
Palaniswami, M., 81
Paliwal, J. K., 261
Palmer, J. D., 143, 149
Parfitt, D. B., 358
Parody-Morreale, A., 179
Parson, W. W., 234

Patel, M. M., 197, 199, 200, 201, 202
Pau, K. Y., 347
Pavone, V., 180
Peacock, B., 220
Pearson, K. G., 15, 16, 17
Perez-Jimenez, R., 188
Perisutti, G., 368
Perl, D., 181, 182, 192, 196
Perlman, D., 351
Permyakov, S. E., 186, 188, 192, 193
Peterson, G. D., 120, 122
Petraglia, F., 351
Petrie, K. J., 406
Petzold, L., 117, 120, 121, 122, 125, 127, 129, 132, 135
Petzold, L. R., 116
Peyret, N., 340
Pezzoli, S. S., 356, 357, 358, 368
Philip, P., 421
Pikovsky, A., 7, 18
Pincus, S. M., 355, 356
Pipes, L., 352, 358, 367, 370
Pirke, K. M., 349
Pisliakov, A. V., 234
Pitt, M. A., 287, 289, 290, 291, 292, 293, 295, 301, 302, 303
Pittendrigh, C., 2, 10
Plant, T. M., 368
Plat, L., 368
Plaza del Pino, I. M., 188
Plesniak, L. A., 253, 255
Pohl, S. L., 71
Pokala, N., 180
Poland, D., 340
Pollastri, G., 253, 255
Pomeau, Y., 5
Pot, I., 248, 255
Potts, R. O., 85
Power, J., 142, 144, 145, 146, 168
Pratt, D. M., 313
Predki, P. F., 181
Press, W. H., 229, 263, 265, 268, 330, 339, 372
Privalov, P. L., 176, 177, 178, 179, 180, 203
Prudom, C. E., 356, 357, 358
Pryor, R. J., 324

Q

Qi, F., 65
Qian, H., 31, 33, 35, 36, 61, 62, 65
Quan, S., 181

R

Racchini, J. R., 73
Raftery, A. E., 295
Rahim, A., 368
Rajaratnam, S. M., 406
Rajib, S., 408, 412, 418

Raleigh, D. P., 181, 182, 185, 188, 192, 195, 196, 203
Ralph, M. R., 317
Ramaswami, M., 170
Ramirez, V. D., 347
Ramponi, G., 198, 201
Rao, C., 117
Raphael, L., 161, 162, 163
Rasmussen, R. P., 324
Rathinam, M., 132, 135
Reach, G., 85
Reame, N., 349
Rebar, R., 351
Rebrin, K., 71
Recchia, F. A., 59
Redding, T. W., 346
Reddy, A. B., 5
Redmond, D. P., 418
Reed, G. H., 324, 325
Reedy, M., 146
Rees, D. C., 177
Refetoff, S., 349
Regan, L., 181, 192
Reichardt, R. M., 417
Reichow, S. L., 180, 182
Reihle, T., 181, 182, 186, 187, 188, 190, 191, 192, 197, 198, 199, 200, 201
Reppert, S. M., 2
Rétey, J. V., 418, 419
Reverter, D., 198
Reyes, A. D., 2, 13
Reynolds, D. L., 261
Rezaee, S., 261
Rezakhah, S., 261
Rich, S. S., 314
Richmond, T. J., 184
Rieke, F., 18
Riesner, D., 104
Ringe, D., 245
Ringo, J., 144, 145, 146, 168
Ringo, J. M., 142, 144, 145, 146, 149, 163, 168
Rinzel, J., 19
Ririe, K. M., 324
Rising Above the Gathering Storm, 307
Rissanen, J., 295, 296, 297
Rivier, J., 346, 368, 404
Rizki, T., 145
Rizwan-Uddin, M. P., 315
Rizza, R., 360, 362
Robblee, J., 181, 182, 188, 192, 196
Robertson, A. D., 177, 185, 195, 245
Robertson, D., 417
Robertson, R. M., 15, 16, 17
Robeva, R., 305
Robeva, R. S., 309, 318
Robic, S., 179
Robinson, G., 154
Robinson, H. P., 13, 14

Robyn, C., 349
Roca, M., 202
Rocchetti, M., 370, 379, 382, 400
Rochet, J. C., 104
Rockenstein, E., 104
Rockman, H., 146
Rodbard, D., 350, 351, 352, 364, 369, 370, 379, 382, 400
Roelfsema, F., 355
Rogol, A. D., 349, 350, 354, 356, 369
Roizen, I., 20
Römer, H., 5, 10, 14, 15, 20, 21
Ronacher, B., 1
Rosbash, M. R., 149
Rose, G. D., 193
Rosenblum, M., 7, 18
Ross, J., 116
Ross, S. A., 182
Rossmanith, W. G., 349
Roth, T., 406
Rotter, J. I., 314
Roxby, R., 238
Rudolph, M. G., 198
Rupp, T., 407
Rupp, T. L., 417
Ruppert, D., 282
Ruskai, M., 161, 162, 163
Russell-Aulet, M., 368
Ryan, A. S., 355
Rzepecki, P., 104

S

Saad, M. F., 71, 314
Saber, H., 109
Sadray, S., 261
Saidel, G. M., 59
Saks, V., 33
Saleh, M., 7
Sali, A., 186
Sali, D., 180, 192
Samatova, N. F., 120, 122
Sampogna, R., 203, 238, 239
Samuels, D. C., 115
Sanchez, R., 186
Sanchez-Ruiz, J. M., 177, 179, 181, 184, 186, 188, 191, 192, 193
SantaLucia, J., 340
SantaLucia, J., Jr., 326, 340
Santen, R. J., 347, 349, 350, 369
Santiago, J. V., 72
Santoro, N., 349
Santoro, N. F., 350
Sanyal, S., 170
Saraste, M., 198
Sarkisian, C. J., 196
Saudek, C., 71
Sauder, S. E., 349
Sauer, R. T., 192

Sauer, U., 180
Schally, A. V., 346
Schefler, W., 145
Schervish, M. J., 296
Schibler, U., 2
Schindler, T., 181, 196
Schlundt, D., 315
Schmeltekopf, A. L., 263
Schmid, F. X., 181, 182, 192, 196
Schmidt, M., 104
Schoemaker, J., 349
Scholtz, J. M., 181, 182, 188, 192, 203, 297, 298, 299, 300, 301
Schoolwerth, A. C., 61
Schrader, T., 104
Schroder, K., 181, 196
Schuck, P., 91, 93, 109
Schulz, D. J., 14, 19
Schurig, H., 176, 179
Schuster, A., 154
Schuster, H. G., 21
Schutz, C. N., 203
Schwarz, G., 304
Schweiger, U., 349
Schweiker, K. L., 175, 182, 186, 190, 197, 200, 201
Scopes, R. K., 57
Scott, L. R., 240
Seidel, H., 5, 18
Selkoe, D. J., 107
Selverston, A. I., 14, 19
Seneviratne, P. A., 340
Sept, D., 239
Serrano, L., 179, 181, 192, 193, 202, 203, 204
Shakhnovich, E. I., 181
Shalet, S. M., 368
Sham, Y. Y., 203, 243
Shapiro, C., 408
Sharma, P. K., 234
Sharp, K., 203, 238, 239
Shaw, K. L., 181, 182, 185, 188, 192, 195
Shibanaka, Y., 181
Shields, D., 80
Shigeyoshi, Y., 11
Shirakawa, T., 5
Shire, S. J., 183
Shirley, B. A., 179, 203
Shoichet, B. K., 176, 179, 202
Shoja, V., 245
Shorey, H. H., 163
Shrand, D. G., 417
Shrier, A., 5, 10, 13, 17, 21
Shughrue, P., 108
Shukla, G. K., 262, 268
Sidhu, G., 248, 255
Sill, A., 93, 97
Silver, D., 71
Silverstein, K., 340

Simpson, M. L., 120, 122
Singer, B. H., 355
Singh, S., 150
Siragy, H. M., 355
Sismondo, E., 2, 10, 14, 17, 20
Sisosdia, S. S., 104
Smith, A., 417
Smith, A. D., 213
Smith, H., 262, 263, 267
Snedden, W., 5, 20
Soares, C. M., 241
Soderlind, E., 180
Solari, H. G., 7
Sollenberger, M. J., 315, 349, 350, 352, 354, 369
Søndergaard, C. R., 253, 255
Song, J. C., 417
Soules, M. R., 349
Sparacino, G., 83
Spector, S., 181, 182, 188, 192, 196
Spencer, B. R., 104
Spieth, H., 163
Spoelstra, K., 14, 15, 16
Spratt, D. I., 350
Staab, C. A., 181
Stafford, W., 90, 109
Stanley, W. C., 59
Steil, G. M., 71
Stein, W., 14, 15, 16, 17
Steiner, K., 5, 10, 14, 15, 20, 21
Steiner, R. A., 349
Stephenson, L., 166
Stewart, J. M., 297, 298, 299, 300, 301
Steyvers, M., 293
Stine, W. B., Jr., 104
Stins, J. F., 415
Stoddard, B. L., 180, 182
Stone, M., 293, 295
Stout, P. J., 73
Strange, P., 71, 80
Straume, M., 77, 78, 309, 315, 318, 360, 372, 388, 389, 409, 418
Street, A. G., 180
Strickler, S. S., 181, 182, 186, 187, 188, 190, 191, 192, 197, 198, 199, 200, 201, 234
Striebinger, A., 104
Strop, P., 192
Stuart, A. C., 186
Sturtevant, J. M., 177, 192
Sugimoto, N., 340
Sugiura, N., 292
Suh, B. Y., 349
Sujino, M., 11
Summa, C. M., 180
Sun, D. P., 180
Sun, Y., 285
Svehag, S. E., 104
Swint-Kruse, L., 185, 195
Syed, S. A., 409, 417

T

Taddei, N., 198
Tadros, S., 71
Taheri, S., 419
Taillard, J., 421
Talyn, B., 163
Tamada, J. A., 85
Tamanini, F., 2
Tamborlane, W. V., 71
Tanford, C., 183, 238, 252
Tang, Y., 287
Tannenbaum, G. S., 360
Tateno, T., 13, 14
Taylor, A. L., 14, 19
Taylor, M., 324
Teague, W. E., Jr., 49, 50, 51, 52, 53, 61
Teixeira, V. H., 241
Tellinghuisen, J., 259, 261, 262, 263, 265, 266, 267, 268, 275, 277, 278, 279, 280, 281, 282
Teo, C. H., 198
Teplow, D. B., 104, 107
Tereshko, V., 193
Teukolsky, S. A., 229, 263, 265, 268, 372
Teukolsky, S. T., 330, 339
Thirumalai, V., 12, 17
Thomas, S. T., 179, 193, 203, 204
Thominet, J., 368
Thompson, T., 406, 407, 418
Thomson, J. A., 185, 195
Thorne, D. R., 418
Thorner, M. O., 355, 367, 368, 404
Thunnissen, M. M., 198
Thurlkill, R. L., 181, 182, 188, 192
Thyagaraja, K., 93, 97
Tian, T., 125, 126, 135, 136, 137
Tiao, G. C., 215
Tidor, B., 181, 182, 188, 192, 196
Tilmann-Wahnschaffe, A., 2, 3
Tinoco, I., 326
Tinoco, I., Jr., 326, 340
Tissot, A. C., 192
Titze, I. R., 5, 7
Toffolo, G., 360, 362
Tomlinson, J., 181, 182, 186, 187, 188, 190, 191, 192, 197, 198, 199, 200, 201, 234
Toseland, C. P., 245
Tourellot, M., 5, 20
Trefethen, J. M., 203
Treves, C., 201
Tripp, S., 324
Tsai, C. C., 347, 349
Tucker, A. M., 418
Turner, D. H., 340
Turner, R. C., 351
Tuschl, R. J., 349
Tycko, R., 109
Tynan-Connolly, B., 251

U

U. S. Department of Education, 308
U. S. Senate hearing, 70
Ueda, H. R., 11
Ugurbil, K., 59
Uhlenbeck, O. C., 326
Ukai, H., 11
UKPDS, 70
UK Prospective Diabetes Study Group, 70
Ulrych, T., 155, 157, 158
Umlauf, M., 406
Urban, N. N., 17
Urban, R. J., 349, 350, 369
Uversky, V. N., 186, 188, 192, 193

V

Vadapalli, R. K., 93, 97
Valdivia, H., 261
Vale, W., 346, 368, 404
Van Cauter, E., 349, 368, 370, 379, 382, 389, 400
Vance, M. L., 355, 368, 404
Vandenberg, G., 347, 349, 355
Vanderburg, K. E., 298
van der Horst, G. T., 2, 14, 15
Vandersteen, J. G., 324
Van Dongen, H. P. A., 213, 221, 222, 223, 228, 229, 418
van Dop, P. A., 349
van Holde, K. E., 90, 91, 101, 109
van Kessel, H., 349
Van Nuland, N. A., 181, 203
Vanselow, K., 2, 3
Varani, G., 180, 182
Varley, P. G., 202
Veech, R. L., 49, 52, 53
Velazquez, J. L. P., 13, 19
Veldhuis, J. D., 315, 316, 321, 349, 350, 351, 352, 353, 354, 355, 356, 357, 358, 360, 362, 364, 368, 369, 370, 371, 372, 379, 382, 388, 389, 391, 400, 409, 418
Veldhuis, P. P., 350, 352, 353, 357, 358, 360, 367, 370
Vendelin, M., 33
Vendrell, J., 198
Venyaminov, S., 177
Vering, T., 71
Vetter, I. R., 198
Vettering, W. T., 229, 263, 265, 268, 330, 339, 372
Vidal, C., 5
Vidmar, S. I., 71
Vieweg, W. V., 355
Viguera, A. R., 193
Vijay-Kumar, S., 198
Villegas, V., 181, 203

Vinnakota, K. C., 30, 33, 34, 35, 49, 56, 57, 58, 59, 60, 62, 63, 65
Virostko, A., 316, 364, 371, 389
Vlachos, D., 125, 126
von Holst, E., 5
von Schilcher, F., 163
Vriend, G., 240, 241, 244
Vuilleumier, S., 192

W

Wachter, K. W., 349, 369
Wade, R. C., 240, 245
Wagenknecht, L. E., 314
Wagenmakers, E.-J., 292, 293
Wagner, C., 360
Wakarchuk, W. W., 253, 255
Waldorp, L., 292
Wall, L., 97
Walsh, D. M., 107
Waltermann, C., 4, 9, 23
Wang, H., 108
Wang, L., 324
Wang, M., 181, 182, 188, 192, 196
Wantanabe, M., 143
Warland, D., 18
Warshel, A., 202, 203, 234, 243
Wasserman, L., 295
Wasserthal, L., 146
Watanabe, R. M., 314
Watson, J. T., 417
Watson, L., 119
Wätzig, H., 261, 282
Weaver, M., 406
Webb, C. B., 368
Wehner, M., 104
Wei, G., 285
Weinzimer, S., 71
Weischet, W. O., 109
Weiss, R. G., 59
Welch, P., 158
Welker, C., 181, 192, 196
Welsh, D. K., 2
Weltman, J. Y., 354, 356
Werther, G. A., 71
Wessels, R. J., 142, 146
Westermark, P. O., 2, 3
Whitaker, E., 154
White, C. M., 417
White, J. A., 13, 17
White, L., 142, 144, 145, 146, 168
White, M. C., 349
Whitsett, T. L., 417
Whitten, S. T., 185, 195
Wiener, N., 148
Wilcken, B., 325
Wilinska, M. E., 71
Williams, B. G., 149
Williams, J. D., 418

Willmore, C., 324
Wilmanns, M., 198
Wilson, D. B., 176, 179
Wilson, D. M., 71
Wilson, K. P., 202
Winblad, B., 104
Winfree, A., 2, 11, 12, 18, 21, 22, 23
Winter, G., 203
Wintrode, P. L., 180
Withers, S. G., 248, 253, 255
Wittinghofer, A., 198
Wittwer, C. T., 323, 324, 325, 342
Wolf, H., 14, 15, 16, 17
Wolf, M., 146
Wong, K. B., 186, 188, 192, 193
Wong, M., 180, 182
Woodward, C., 179
Wozniak, J. A., 180
Wright, D. S., 261
Wright, K. P., 418
Wright, S. J., 220
Wu, F., 30, 33, 35, 56, 61, 62, 63, 64, 65
Wunderlich, M., 181
Wysocki, T., 71

X

Xiao, L., 203
Xiao, S., 285

Y

Yagita, K., 2, 5, 11
Yamaguchi, S., 5
Yang, A. S., 203, 238, 239
Yang, F., 33, 35, 62, 63, 65
Yang, T., 16
Yankov, V. I., 354, 356, 360

Yeager, M., 108
Yen, S. S., 347, 349, 351
Yen, S. S. C., 349
York, E. J., 297, 298, 299, 300, 301
Young, P., 185, 195, 203
Young-Hyman, D., 315
Yu, B., 295
Yu, X., 59
Yu, Y., 179
Yuen, C. K., 151, 153, 173
Yutani, K., 200

Z

Zaccaro, D. J., 314
Zadmard, R., 104
Zanderigo, F., 83
Zare, R. N., 263
Zarrine-Afsar, A., 182, 186, 190, 197, 200, 201
Zaugg, C., 71
Zeeb, M., 201
Zeng, Q. C., 262, 278, 281, 360
Zhang, E., 262, 278
Zhang, E. Y., 35, 61, 63–65
Zhang, J., 35, 59, 61, 62, 63, 64, 65, 119
Zhang, X. J., 202
Zhong, D., 285
Zhou, H. X., 181, 185, 193, 195
Zhou, L., 324, 325
Zhou, M., 285
Zimm, B. H., 297
Ziser, L., 253, 255
Zollars, E. S., 182
Zuker, M., 340
Zvarlaw, E. E., 419
Zwiebel, L. J., 149

Subject Index

A

Accumulative prediction error, model
  comparison technique, 293–295, 300
Acylphosphatase
  computational design of stability, 199
  stabilization effects on activity, 201
Adenosine, caffeine antagonism, 419
AIC, *see* Akaike information criterion
Akaike information criterion, model
  comparison technique, 292–293, 300
Amyloid aggregation
  analytical ultracentrifugation, 105
  experimental results and simulations,
    105–109
  overview, 104
  sample preparation, 104–105
  transmission electron microscopy, 105, 111
Analytical ultracentrifugation
  applications, 88
  data analysis using parallel
    distributed computing
    amyloid aggregation
      analytical ultracentrifugation, 105
      experimental results, 105–109
      overview, 104
      sample preparation, 104–105
      transmission electron microscopy, 105
    five-component system
      simulation, 100–104
    job submission from Web portal, 97–99
    overview, 89–91
    performance, 109–111
    steps
      basis function evaluation and noise
        elimination, 92–94
      parameter refinement and
        regularization, 94–96
      parameter space election, 91
      statistical evaluation, 96–97
  principles, 88–89
APE, *see* Accumulative prediction error
Approximate entropy
  coupled hormone system analysis, 356, 359
  estimation in pulsatile hormone
    release analysis, 355
AutoDecon
  advantages, 401–402
  combined modules, 374–375
  deconvolution technique, 370–371
  fitting module, 372
  gonadotropin system, 368
  growth hormone secretion analysis
    concordant secretion events and
      performance, 382–383
    data collection, 379–380
    data simulation, 380–382, 400
    experimental data analysis, 385–390
    immunoassay, 380
    output, 384
    pulsatile secretion, 368
  insertion module, 372–373
  outlier detection, 378
  performance
    accuracy in pulse analysis, 352–353
    comparison with other programs,
      379, 382, 390–396
    Monte Carlo evaluation, 370
  phases, 376–377
  sampling paradigm optimization, 397–399
  system identification problem
    and approach, 369, 377
  triage module, 373–374
  weighting factors, 378–379

B

Barnase, computational design of stability, 192
Bayesian information criterion, model
  comparison technique, 292–293, 300
Bayesian model
  cognitive performance impairment
    prediction during sleep deprivation
    95% confidence interval
      calculation, 223–227
    prediction model, 221–223
  confidence interval computation for
    predictions based on
    multidimensional parameter space
    advantages of algorithm, 227–229
    minimum and maximum over
      confidence region, 220–221
    overview, 214–215
    probability density function
      height at boundary of smallest
        multidimensional confidence
        region, 215–216
      proof that confidence region boundary is
        a level contour, 229–230

437

Bayesian model (*cont.*)
    slice approximation with normal curve spline pieces, 217–219
    smallest multidimensional confidence region boundary locating, 219–220
  forecasting applications, 214
  probability density function, 214
Bayesian model selection, model comparison technique, 295–297
BIC, *see* Bayesian information criterion
Biochemical systems simulation, *see also* Discrete stochastic simulation
  benefits of chemical detail, 33–34
  databases of biochemical information, 65–66
  enzyme kinetic data challenges, 35–36
  fundamental data, 34–35
  multiple cation equilibria in solutions
    apparent equilibrium constant, 42–43
    biochemical reaction with reference species, 41–42
    citrate synthase reaction simulation, 53–56
    creatine kinase reaction simulation, 48–53
    differential equations for ion concentrations, 43–48
    Haldane constant, 42
    overview, 36–37
    proton binding fraction calculation for each metabolite, 37–41
    proton flux, 42
    reference species, 41
  physiological systems
    cardiac energetics
      Gibbs free energy of ATP hydrolysis, 59–62
      ion concentration effects on mitochondrial matrix free energy change, 62–63
      physiological significance of simulations, 63–64
    glycogenolysis and pH dynamics in muscle, 57–58
  symbols, 38
  thermodynamics
    integration in model building, 64–65
    principles, 31–32
Biomathematics, *see* Mathematical biology interdisciplinary instruction
Blood glucose monitoring, *see* Continuous glucose monitoring
BMS, *see* Bayesian model selection
BRENDA database, 65–66

C

Caffeine, *see* Sleep deprivation
Calibration
  least squares fitting, 260, 270–271
  univariate calibration, 260
Cardiac arrhythmias, *see* Heart rhythms

Cardiac energetics
  Gibbs free energy of ATP hydrolysis, 59–62
  ion concentration effects on mitochondrial matrix free energy change, 62–63
  magnetic resonance spectroscopy, 63
  physiological significance of simulations, 63–64
Cardiac rhythms, *see* Heart rhythms
Catalytically competent protonation state, pH activity profile prediction, 248–249
CCPS, *see* Catalytically competent protonation state
CDC42
  computational design of stability, 199
  stabilization effects on activity, 201–202
Central pattern generator, phase response curve
  classification of generators, 19
  rhythms, 14–17
CGM, *see* Continuous glucose monitoring
Chemical master equation, discrete stochastic simulation, 117–119
Circadian rhythm, *see* Phase response curve; Sleep deprivation
Citrate synthase, multiple cation equilibria and reaction simulation, 53–56
Cluster
  AutoDecon performance comparison, 390–394
  pulse analysis, 350
Cognitive performance impairment, *see* Sleep deprivation
Cold shock proteins, computational design of stability
  CspB-Bs, 182
  stabilization effects on activity, 200–201
  TK-SA algorithm, 192–197
Computational model comparison, *see* Model evaluation and comparison
Confidence interval, computation
  for Bayesian predictions based on multidimensional parameter space
  advantages of algorithm, 227–229
  cognitive performance impairment prediction during sleep deprivation, 221–227
  minimum and maximum over confidence region, 220–221
  overview, 214–215
  probability density function
    height at boundary of smallest multidimensional confidence region, 215–216
    proof that confidence region boundary is a level contour, 229–230
    slice approximation with normal curve spline pieces, 217–219
    smallest multidimensional confidence region boundary locating, 219–220

Subject Index

Continuous glucose monitoring
  average glycemia and deviation from target measurements, 74–77
  data challenges and requirements, 71–73
  diabetes overview, 70
  rationale, 70–71
  risk and variability assessment, 75–79
  self-monitoring limitations, 71
  sensor error handling, 73–74
  system stability measures and plots, 80–81
  time series-based prediction of blood glucose values, 81–84
CPG, *see* Central pattern generator
Creatine kinase, multiple cation equilibria and reaction simulation, 41–42, 48–53
Cross-ApEn, coupled hormone system analysis, 356, 359
Cross-validation, model comparison technique, 293–295, 300
Csp, *see* Cold shock proteins
CV, *see* Cross-validation
Cycle Detector, pulse analysis, 349
Cytosine deaminase, computational design of stability, 182

D

Deconv, pulse analysis, 351–352
Detect, pulse analysis, 350
Diabetes, *see* Continuous glucose monitoring
Direct method, stochastic stimulation algorithm implementation, 120–121
Discrete stochastic simulation
  chemical master equation, 117–119
  comparison of approaches, 137–139
  Gillespie stochastic stimulation algorithm implementation approaches, 119–122
  hybrid stochastic stimulation algorithm/tau-leaping method, 125–127, 131
  LacZ/LacY model, 136–139
  reaction rate ordinary differential equations, 116
  Schlögl model, 136–137
  simulation error measurement, 132–134
  StochKit software, 134–135
  tau-leaping method, 122–125
  tau selection formula, 127–131
DNA melting analysis
  applications, 324, 330
  fluorescence data, 324
  melting curves
    clustering and classification by genotype, 332–335
    extraction from raw fluorescence data, 325–331
    modeling techniques, 335–342
  melting temperature, 325, 337
  thermodynamic principles, 336–340

E

Education, *see* Mathematical biology interdisciplinary instruction
Electron microscopy, amyloid aggregation, 105, 111
Electrostatic energy, calculation from protein structures, 239–240
Engrailed, computational design of stability, 182
EPUT, *see* Events per unit time
Events per unit time, biological rhythm data, 145

F

Fatigue, *see* Sleep deprivation
FeedStrength, coupled hormone system analysis, 356–358
First reaction method, stochastic stimulation algorithm implementation, 120–122
Follicle-stimulating hormone, *see AutoDecon*; Pulsatile hormone release
Fourier analysis, *Drosophila* cardiac rhythms, 153–155
Fyn, computational design of stability, 182, 199–200

G

Generalizability, model evaluation and comparison, 290–291, 302
Genetic algorithm
  analytical ultracentrifugation data analysis, 90, 94–95, 97, 102–103, 106–107
  protein stability optimization of surface charges, 188–190, 197–200
Genotyping, *see* DNA melting analysis
Gibbs free energy
  cardiac ATP hydrolysis, 59–63
  change in chemical reactions, 31–32
  DNA melting analysis, 336
  equation, 31
  flux relationship, 32
  Haldane relationship, 42
  protein stability, 176–179
  simulated biochemical systems, 32
Gillespie stochastic stimulation algorithm, *see* Discrete stochastic simulation
Glucose monitoring, *see* Continuous glucose monitoring
Glycogenolysis, simulation of pH dynamics in muscle, 57–58
Gondotropin-releasing hormone, *see AutoDecon*; Pulsatile hormone release
Gonze model
  circadian rhythm, 4–6
  entrainment regions, 8–10
  phase response curve, 7–8
  simulations, 23–24
Goodness of fit, model evaluation and comparison, 289–291

Growth hormone, secretion analysis
 with *AutoDecon*
 concordant secretion events and
  performance, 382–383
 data collection, 379–380
 data simulation, 380–382, 400
 experimental data analysis, 385–390
 immunoassay, 380
 output, 384
 pulsatile secretion, 368

H

Haldane constant, biochemical systems
 simulation, 42
Heart rhythms
 Drosophila cardiac rhythm analysis
  frequency domain analysis, 151–161
  signal
   conditioning, 164–168
   strength and regularity, 168–170
  time domain analysis, 145–151
  time/frequency analysis and wavelet
   transform, 161–164
 phase response curve
  cardiac arrhythmias, 18
  rhythms, 13, 15
Henderson–Hasselbach equation, 236
Homoscedasticity, least squares
 fitting data, 260–261

I

Igor, 268
Insect communication, phase response curve
 entrainment, 19–20
 rhythms, 14–17

K

KaleidaGraph, 268–270

L

α-Lactalbumin, computational design
 of stability, 192–193
LacZ/LacY model, discrete stochastic
 simulation, 136–139
λ-repressor, computational design
 of stability, 192
Least squares
 a priori variance–covariance matrix
  assumptions, 264
  evaluation, 265
 experimental design
  data transformation for
   simplification, 276–277
  large χ^2, 281–282
  nonlinear response function
   characterization, 268–274

 uncertainty in the unknown, 274–276
 variance function estimation, 277–281
calibration equation fitting, 260, 270–271
linear least squares
 formal equations, 263–266
 illustrations, 266–268
 overview, 260–261
NNLS, *see* Nonnegatively constrained
 least-squares fitting algorithm
nonlinear least squares, 268
ordinary least squares, 260–261, 264
parameter estimation in fatigue
 modeling, 412–414
practical concerns of commercial
 programs, 268–269
response function
 linearity, 261–263, 268
 statistical error evaluation, 270, 272–273
weighted least squares, 260
Linear least squares, *see* Least squares
Logarithmic direct method, stochastic
 stimulation algorithm implementation,
 120, 122
Luteinizing hormone, *see AutoDecon*;
 Pulsatile hormone release
Lysozyme, pH activity profile prediction, 248

M

Magnetic resonance spectroscopy, cardiac
 energetics, 63
Mathematical biology interdisciplinary
 instruction
 bachelor's degrees in biology and
  mathematics, 308
 Biomathematics undergraduate course
  features, 310–311
  institutional context, 309–310, 319
  structure, 311–312
  student evaluation, 319
  target students, 310
  textbooks, 318
  topics
   circadian rhythms, 316–317
   endocrinology, 314–316
   epidemiology, 314
   genetics, 313–314
   ligand binding, 316
   population studies, 312–313
 initiatives and funding, 307
 needs, 306–307, 309
 quantification skills, 309
MD, *see* Molecular dynamics
Mean absolute deviation, sensor error in
 continuous glucose monitoring, 73
MESA software, 158–160, 169
Model evaluation and comparison
 accumulative prediction error,
  293–295, 300

Subject Index

Akaike information criterion, 292–293, 300
Bayesian information criterion,
    292–293, 300
Bayesian model selection, 295–297
complexity, 290
criteria for evaluation, 288
cross-validation, 293–295, 300
generalizability, 290–291, 302
goodness of fit measures, 289–291
overview, 288–291
parameter space partitioning, 302
protein folding model selection
    example, 297–301
stochastic complexity, 295–297
Molecular dynamics, simulation for pK_a
    calculation, 243
Molecular weight analysis, see
    Analytical ultracentrifugation
Monte Carlo analysis
    analytical ultracentrifugation
        data, 90–91, 96–97, 102–104,
        106–107
    AutoDecon performance evaluation, 370
MRS, see Magnetic resonance spectroscopy
Multiple cation equilibria, see Biochemical
    systems simulation

N

Neuron spike, phase response curve
    dynamics, 13–17
    neuron classification by spiking patterns,
        18–19, 22
Next reaction method, stochastic
    stimulation algorithm
    implementation, 120–121
NMR, see Nuclear magnetic resonance
NNLS, see Nonnegatively constrained
    least-squares fitting algorithm
Nonnegatively constrained least-squares
    fitting algorithm, analytical
    ultracentrifugation data
    analysis, 92–93
Nonlinear least squares, see Least squares
Nuclear magnetic resonance
    fitting of pH-dependent
        characteristics, 253–255
    protein structure
        selection and analysis for pK_a
            calculation, 240
        stability studies, 186

O

Optimized direct method, stochastic stimulation
    algorithm implementation, 120–122
ORBIT, protein stability design, 182
Ordinary least squares, see Least squares
Origin, 268
Oscillators, see Phase response curve

P

Parallel distributed computing, see
    Analytical ultracentrifugation
Parameter space partitioning, model
    comparison and evaluation, 302
Pathway simulation, see Biochemical
    systems simulation
PBE, see Poisson–Boltzmann equation
PDF, see Probability density function
PEF, see Prediction error filter
pH activity profile
    decomposition analysis, 249–251
    electrostatic energy calculation from
        protein structures, 239–240
    fitting titration curves, activity profiles,
        and stability profiles
            classic equations, 253–254
            overview, 252–253
            statistical mechanical fitting, 255
    flow chart for characterization, 235–236
    importance of pH effects on enzymes,
        234–235
    ionization equations, 236–237
    ligand-binding profile prediction, 252
    pK_a
        calculation
            accuracy, 245–246
            calculated titration curve analysis,
                244–245
            molecular dynamics simulation, 243
            protein conformational change
                modeling, 243–244
            protein structure selection and
                analysis, 240–242
            sensitivity analysis, 246–248
            theory, 237–239
        equation, 237
        redesigning values, 249–251
    prediction, 248–249
    protein stability profile
        prediction, 251–252
Phase response curve
    applications, 2
    cardiac arrhythmias, 18
    central pattern generator
        classification, 19
    circadian clock example, 2–4
    components, 11–12
    coupled and entrained oscillators, 5, 7
    entrainment regions, 8–10
    glossary of terms, 22–23
    Gonze model, 4–8, 23–24
    insect communication entrainment, 19–20
    limit cycles, 10–11
    minimal protocol
        examples
            central pattern generator, 14–17
            circadian clock, 13–16

Phase response curve (cont.)
  heart rhythms, 13, 15
  insect communication, 14–17
  neuron spike dynamics, 13–17
  steps, 12–13
neuron classification by spiking patterns, 18–19, 22
phase normalization, 12
phase transition curve, 7, 9, 21
self-sustained biological rhythms, 4–6
types, 22
universality, 21
pKa, see pH activity profile
Poisson random number approximation, tau-leaping method, 124, 129
Poisson–Boltzmann equation
  electrostatic energy calculation from protein structures, 239–240
  protein stability design, 182
PRC, see Phase response curve
Prediction error filter, rhythm analysis, 158
Probability density function, Bayesian forecasting
  confidence interval computation for predictions based on multidimensional parameter space
    height at boundary of smallest multidimensional confidence region, 215–216
    proof that confidence region boundary is a level contour, 229–230
    slice approximation with normal curve spline pieces, 217–219
  overview, 214
Procarboxypeptidase, computational design of stability, 182, 199–200
Protein conformational change, modeling for pKa calculation, 243–244
Protein folding, model evaluation and selection, 297–301
Protein G, computational design of stability, 192
Protein stability, computational design
  caveats, 202–204
  comparison of design approaches, 182
  core versus surface interaction optimization, 181
  experimental verification
    enzyme activity effects, 200–202
    genetic algorithm optimization of surface charges, 197–200
    single-site substitutions in TK-SA model, 191–197
  intramolecular interactions, 179–180
  pH-dependence prediction, 251–252
  surface charge–charge interactions
    genetic algorithm optimization of surface charges, 188–190
    pair-wise charge–charge interaction energy calculation, 183–188
  thermodynamic principles, 176–177

PSP, see Parameter space partitioning
Psychomotor vigilance test, see Sleep deprivation
Pulsar, pulse analysis, 349
Pulsatile hormone release, see also *AutoDecon*
  gonadotropin system, 346–348
  history of biomathematical modeling
    approximate entropy estimation, 355
    coupled hormone system analysis, 356–359
    deconvolution techniques
      AutoDecon, 352–353
      Deconv, 351–352
    early attempts, 348–349
    feedback interaction evaluation, 360–362
    stochastic differential equations model, 353, 355
  sampling protocol and pulse detection algorithm impact on pulse detection, 349–351
Pulse, AutoDecon performance comparison, 393–395
Pulse2, AutoDecon performance comparison, 393–396
Pulse4, AutoDecon performance comparison, 393

R

Response function
  linearity in least squares fitting, 261–263, 268
  nonlinear response function characterization, 268–274
  statistical error evaluation, 270, 272–273
RF, see Response function
Rhythm, see also Phase response curve; *specific systems*
  biological data types and acquisition, 142–145
  frequency domain analysis, 151–161
  signal
    conditioning, 164–168
    strength and regularity, 168–170
  time domain analysis, 145–151
  time/frequency analysis and wavelet transform, 161–164
Ribonucleases, computational design of stability, 182
Rosetta, protein stability design, 182
Rubredoxin, computational design of stability, 192

S

SABIO, see System for the Analysis of Biochemical Pathways
SC, see Stochastic complexity, 295–297
Schlögl model, discrete stochastic simulation, 136–137
Sensor error, decomposition in continuous glucose monitoring, 73–74

Subject Index

Sequence-based design, protein stability design, 182
SigmaPlot, 268
Signal conditioning, rhythm data, 164–168
Sleep deprivation
  Bayesian forecasting of cognitive performance impairment
    95% confidence interval calculation, 223–227
    prediction model, 221–223
  fatigue modeling
    caffeine
      adenosine antagonism, 419
      competitive inhibition of performance inhibitor, 414–415
      concentration modeling, 409–410, 415–416
      half-life of effects, 417
    circadian factor, 411–412
    least squares parameter estimation, 412–414
    minimal biomathematical model, 412, 415–417
    performance inhibitor model, 410–412
    prospects for study, 418–419
    psychomotor vigilance test data processing, 408–409, 418
    Walter Reed Army Institute of Research Stay Alert caffeine study, 407–408
  homeostatic modeling, 406
SRQuant, coupled hormone system analysis, 358–359, 362
Stochastic complexity, model comparison technique, 295–297
Stochastic simulation algorithm, see Discrete stochastic simulation
StochKit software, 134–135
System for the Analysis of Biochemical Pathways, 66

T

Tanford–Kirkwood model, see TK-SA
Tau-leaping method, see Discrete stochastic simulation
Tenascin, computational design of stability, 199–200
Texas Internet Grid for Research and Education, analytical ultracentrifugation data analysis, 93, 97–98
Thioredoxin, computational design of stability, 182
TIGRE, see Texas Internet Grid for Research and Education

TK-SA, protein stability design
  caveats, 202–204
  cold shock proteins, 192–197
  comparison with other approaches, 182
  experimental verification of single-site substitutions, 191–197
  Fyn, 182, 199–200
  L30e, 93
  α-lactalbumin, 192–193
  pair-wise charge–charge interaction energy calculation, 183–188
  procarboxypeptidase, 182, 199–200
  tenascin, , 199–200
  ubiquitin, 186–187, 191–192
Two-dimensional spectrum analysis, analytical ultracentrifugation data, 90, 93–94, 96–97, 101–102

U

U1A, computational design of stability, 199
Ubiquitin, protein stability design, 186–187, 191–192, 199
Ultra, pulse analysis, 349
Ultracentrifugation, see Analytical ultracentrifugation

V

Variance function, estimation in least squares fitting, 277–281
VF, see Variance function

W

Walter Reed Army Institute of Research Stay Alert caffeine study, see Sleep deprivation
Wavelet transform, time/frequency analysis, 161–164
Weighted least squares, see Least squares

X

X-ray crystallography
  protein stability studies, 186
  protein structure selection and analysis for pK_a calculation, 240–242
Xylanase, pH activity profile prediction, 248–249, 255

Z

Zimm–Bragg theory, 297

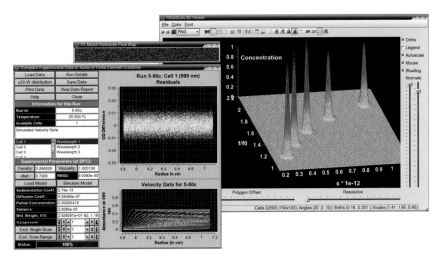

Borries Demeler et al., Figure 4.6 Visualization tools from the UltraScan GUI used to display models obtained in the supercomputer calculation, showing residual bitmap, residuals, experimental data and model overlay, and 3D solute distribution (f/f_0 vs s). The model shown here represents global GA analysis of Example 1.

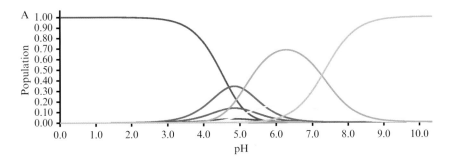

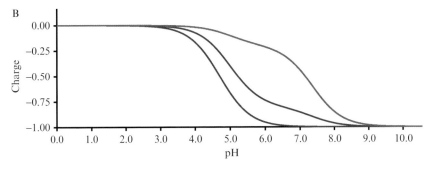

Jens Erik Nielsen, Figure 9.2 pH dependence of the relative populations of the eight possible protonation states for a system consisting of three titratable acids: I (black titration curve) intrinsic pK_a 4.7, II (blue titration curve) intrinsic pK_a value 5.1, and III (red titration curve) intrinsic pK_a value 5.7. II and III interact with an energy of 3.6 kT. All other pair wise interaction energies are zero. (Top) pH dependence of population of protonation states. The first set of bell-shaped curves corresponds to one deprotonated group, whereas the next set of bell-shaped curves corresponds to the protonation states with two groups deprotonated. The single curves at high and low pH correspond to protonation states with all groups protonated (low pH) and all groups deprotonated (high pH). (Bottom) titration curves.

Jens Erik Nielsen, Figure 9.3 Predicted pH dependence of the stability of hen egg white lysozyme (PDB ID 2lzt). The black curve shows the pH dependence of ΔG_{fold}, whereas colored areas show decomposition of the total charge difference. Figure produced with pKaTool (Nielsen, 2006a).

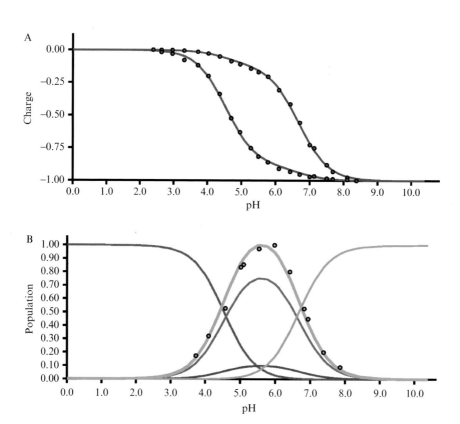

Jens Erik Nielsen, Figure 9.4 Fits of titration curves (top) and the pH activity profile (bottom) for *Bacillus circulans* xylanase using Eq. (9.6). Circles indicate experimental data points from McIntosh *et al.* (1996), whereas lines indicate functions fitted using GloFTE/pKaTool(Søndergaard *et al.*, 2008). (Top) Blue curve, Glu 172; black curve, Glu 78. (Bottom) black curve, Glu78H, Glu172H; red curve, Glu78-, Glu172H; blue curve, Glu78H, Glu172-; cyan curve, Glu78-, Glu172-; green curve is the red curve (the population catalytically competent protonation state) normalized.